Part of North America in the 18th Century

adapted from Bowen's map of the Treaty of 1763

NEW ENGLAND LIFE
IN THE 18TH CENTURY

Representative Biographies from
Sibley's Harvard Graduates

NEW ENGLAND LIFE IN THE 18TH CENTURY

Representative Biographies from
Sibley's Harvard Graduates

CLIFFORD K. SHIPTON

Custodian of the Archives of Harvard University
Director of the American Antiquarian Society
President of the Colonial Society of Massachusetts
Sibley Editor of the Massachusetts Historical Society

THE BELKNAP PRESS OF
HARVARD UNIVERSITY PRESS
Cambridge, Massachusetts

1963

The endpaper maps were contributed by
Colonel Edward P. Hamilton, Director of Fort Ticonderoga

Library of Congress Catalog Card Number 63–9562

Printed in the United States of America

CONTENTS

ILLUSTRATIONS

FOREWORD

By Samuel Eliot Morison

Sibley's Harvard Graduates is now over a century old. It really
began when John Langdon Sibley, a Maine boy who graduated
with the Class of 1825, became Assistant Librarian of Harvard
College in 1841. He was just in time to superintend the congenial
if exhausting labor of moving the biggest university library in the
country (41,000 volumes) from the upper west room of Harvard
Hall to Gore Hall, on the site of Widener. According to the rash
prediction of President Quincy this building was expected to contain the Harvard College Library for at least a century. And,
since he imagined that the Assistant Librarian would now have
plenty of leisure time, Josiah Quincy "prevailed upon" (the polite
term for a presidential order) Mr. Sibley to undertake the editing
of the Triennial Catalogue of Harvard graduates. The Triennials
of 1842 through 1869, and their successors the Quinquennials
through 1885, were Sibley's work. There was no end to this work.
No sooner was one Catalogue out than more young men graduated, old men died and had to have a *stella* and the year of decease added to their names, young men became "settled ministers"
or took holy orders and so had to be italicized; middle-aged alumni
who attained senatorial, gubernatorial, or college-presidential emi-
nence had to be capitalized; indignant graduates protested that
their honorary degrees and learned societies had not been mentioned; and suitable Latin abbreviations for the same had to be
thought up. For instance, under Jared Sparks of 1815 we find:

S. H. et S. Reg. Antiqq. Septentr. Hafn. Soc.

Which means:

Honorary Member and Fellow of the Royal Society
of Northern Antiquities of Copenhagen.

And, under *Georgius* Bancroft (for Christian names had to be
Latinized if possible, evidently Jared's couldn't) we find:

Reg. Scientt. Acad. Boruss. Berolin. et Institut.
Gall. Acad. Scientt. Moral. Politic. q.

Which means:

Socius der Königlich Preussischen Akademie der
Wissenschaften; Membre de l'Académie des Sciences
Morales et Politiques de l'Institut de France.

While he was engaged in this meticulous though hardly exhausting work, it occurred to certain alumni that Harvard, for
her future fame, needed an Anthony à Wood; and who could be
more suitable than Sibley? Dr. Danforth Phipps Wight (name
forever blessed) of the Class of 1815 first suggested this in a
letter of 1848. Sibley then declined to entertain any such idea.
His labors as librarian precluded it; his "advancing years" (he was
then forty-four years old) were an obstacle. But the tempting seed
had been planted. The more Sibley thought it over, the more he
felt the urge to fill this void in Harvard's *Athenae* (as university
biographical annals used to be called), and the more alumni pressed
him to fill it. Finally, "On Monday, 21 February 1859," the great
work began, and "after this, *nulla dies sine linea*" — these are Sibley's *ipsissima verba*, which may well be the motto of his successor.

Although Sibley lived to the ripe age of eighty-one, he succeeded in publishing only three volumes of the *Biographical
Sketches of Graduates of Harvard University* — I in 1873, II in
1881, and III in 1885. These covered Harvard's biographical annals only through the Class of 1689, an era when graduating classes
numbered only from one (1652) to fourteen (1685) members. At
Commencement 1885 Sibley for the first time in fifty years was unable to twang the official tuning fork and lead the alumni in singing Psalm 78, "Give ear, O my people." Before the end of that
year he had joined the great company of those who *e vivis cesserunt
stelligeri*, as the old Triennials designated the inevitable end that
Harvard graduates share with other less privileged mortals.

The three volumes were published by Charles William Sever
of the University Bookstore, Cambridge, and presumably at his
expense. Sibley received neither royalties nor honorarium; it was
all a labor of love to *alma mater*. Owing, it is said, to Sibley's dislike of President Eliot, who had never evinced the slightest en-

thusiasm for this devoted enterprise, Sibley made the Massachusetts Historical Society, membership in which gave him the distinction of S.H.S.* in the Triennial, his residuary legatee. Mrs.
Sibley added her own savings to his, the whole amounting to the
tidy sum of near $200,000. According to his last will and testament,
the Sibley Fund was to be devoted, first, to continuing the *Biographical Sketches*; second, to purchasing books by or about Harvard graduates; and third, to "the general purposes of the Society."

For over a quarter-century the Society completely ignored the
first object of Sibley's bequest. Julius H. Tuttle, the late librarian
of the Society, somewhat timidly called this dereliction of duty
to my attention around 1926, when work was being started on the
Tercentennial History of the University. And, as I also happened
to be chairman of the Society's publications committee, something
was done about it. A Sibley Editor was appointed — and what a
disappointment *he* was! About half of one biography was written
in a year. So that gentleman was discharged, and the search began
for a new editor. Fortunately the right man was available and willing to serve. He was Clifford Kenyon Shipton of the Class of 1926,
an even century younger than Sibley. He had taken his first degree
cum laude in history, and an A.M., and taught history at Brown
and Harvard Universities while studying for the doctorate (Ph.D.
1933).

Shipton in 1930 accepted the part-time job of Sibley Editor —
and what a happy choice that was! For Ted Shipton is not only
an historian in his own right (see his *Life of Roger Conant*), and
the Harvard Archivist, and successively librarian and director of
the great American Antiquarian Society at Worcester; he has a
delightful style, and he gets things done. Within three years, in
1933, the Society was able to publish Shipton's first offering: Sibley's *Harvard Graduates*, Volume IV, covering the Classes 1690–
1700. Since that time, eight more volumes have appeared, at in-
tervals of three to four years; Volume XII, covering the Classes
1746–1750, was published a few months ago, and the work moves
steadily and majestically forward.

Dr. Shipton continued, in the Sibley tradition, to write complete
biographies of each graduate, so far as the records and the impor-

---

* I.e., *Socius Historiae Societatis*, the M.H.S. being the only historical society
which the Triennial deemed worthy of mention.

tance of the subject permitted. He continued the bibliographies of each graduate's printed works and manuscripts (omitted in this selection). But he added to the Sibley stock by including, as far as possible, the non-graduates: students who died in college, or were withdrawn or expelled. And he has included recipients of honorary degrees as well; that is why George Washington (LL.D. 1776) appears in this particular gallery of notables. Shipton has also made a special point of publishing reproductions of portraits, whenever portraits could be found.

Covering, as these biographies do, the tag-end of the Puritan century, the century of Enlightenment and Revolution, and (in the case of long-lived graduates) a fair slice of the nineteenth century as well, they give a unique picture of New England (and to some extent New York and Pennsylvania) society when the colonists were growing up and full of growing pains. The Great Awakening, which rocked the Colonies spiritually and emotionally as the Great Earthquake rocked Lisbon materially, was scarcely over when the rumblings of the American Revolution began. Thus, some biographies are full of action; others simply tell the story of life in a country parsonage, the parson's readings and meditations, his children, cattle, and servants, and his desperate efforts to make both ends meet in an era of inflation. As Shipton describes it, this way of living, which seemed so thin to historians half a century ago, acquires a new depth and spiritual richness.

In one major matter Shipton refused to follow Sibley. In that kind old gentleman's estimate, no Harvard man could do wrong; or if he did, his misdeeds should be left decently to oblivion; he even managed to make a hero out of Sir George Downing, Bt., Class of 1642. Shipton, however, is made of sterner stuff. He believes that *Veritas* should be construed as telling all; and when he finds a Harvard graduate to be a villain, he says so. John Adams once blurted out that the lives of some Revolutionary leaders whom he knew very well would not bear looking into. Well, Shipton has looked sharp and deep into those who, as Harvard alumni, came within his ken; and some of his sketches of these gentry come fairly close to "debunking," the fashionable historical attitude of Shipton's salad days. *Per contra,* the Loyalists (about one-fifth of Harvardians living in 1775 were Tories) stand up remarkably well; one regrets that considerations of space have prevented the inclusion of Shipton's 68-page biography of the long-neglected Governor Thomas Hutchinson. But there are other representative Loyalist biographies, such as those of the Reverend Samuel Auchmuty, Attorney General Jonathan Sewall, and the pathetic but invincibly cheerful Marblehead merchant Benjamin Marston.

Time enough for the Master of Ceremonies. Go to it, reader! Dip in anywhere, and you will find enjoyment as well as instruction in what New England was really like between 1689 and the American Revolution.

INTRODUCTION

This volume is one of the brain children of my favorite poet, David McCord, who for years nursed the idea that the public would be interested in a selection of my biographical writings. It was Lyman Butterfield who organized the project and to assist him gathered a covin consisting of W. M. Whitehill, W. A. Jackson, S. T. Riley, W. J. Bell, T. J. Wilson, J. P. Boyd, David McCord, and William Bentinck-Smith. That brave crew stirred the pot of my writings and brought to the surface this selection of biographies. To shift similes, the process was something like using the broadside of the *Missouri* to shoot down butterflies.

The idea of writing collective American biography goes back to the *Magnalia* of Cotton Mather, who chose his subjects to prove his point rather than trying for an inclusive coverage to obtain a statistical base useful for historical purposes. The idea of collecting biographical data on all Harvard graduates probably originated with Nicholas Gilman, who in 1737 cut up a broadside alumni catalogue and pasted the lists of names into a little book in which he entered such facts as he could find. In the generations which followed, several other individuals made like collections of data. In 1841 President Quincy asked John Langdon Sibley, the assistant librarian, to edit the Triennial Catalogue of graduates. The mass of information accumulated in the course of correcting the catalogue suggested to the editor the desirability of using it as a basis for a biographical dictionary which would record the "intellectual and moral power, which, during more than two centuries, had been going out from the walls of Harvard." Sibley had worked his own way through college, so he was particularly interested in demonstrating that many other individuals had by this path made their way from poverty to places of great public usefulness, thus illustrating the working of the dream which the Americans had inherited from the Puritans and which Horatio Alger was to use for his inspirational fiction. On February 21, 1859, Sibley formally began work on the biographies, and in 1873 he brought out Volume I. When Volume III, carrying his work through the Class of

1689, came out in 1885, he laid down his pen, but in his will he provided that after the death of his widow his estate should go to the Massachusetts Historical Society for the purpose of financing the continuation of the series. For Sibley and his work the reader is referred to S. E. Morison's introduction to Volume IV of this series, to Rene Bryant's article in *American Heritage*, Volume IX, No. 4, June 1958, pp. 28 ff, and to my article in the *Harvard Library Bulletin*, Volume IX, No. 2, Spring 1955, pp. 236–261.

In the spring of 1930 I was teaching at Brown University when S. E. Morison invited me to come up to Boston to "spend the summer getting Volume IV of Sibley ready for the press." At the end of the summer I reported to the publications committee of the Massachusetts Historical Society that it would take three years to finish the volume, whereupon they told me to go ahead, and so I did. When in 1933 I reported that the volume was done, the committee replied, in effect, "Well, keep at it and stop bothering us with your reports"; I don't remember that I ever reported again. If the original committee could gather now — nine volumes, 1459 biographies, and 2,630,000 words later — it would probably feel like the traditional vacationist who returned at the end of the summer to find that he had forgotten to stop the milk delivery. I do not intend to imply, however, that the members of the committee were not individually of the greatest assistance in carrying on the work; my beloved friend Albert Matthews, in particular, poured out for me his vast knowledge.

In 1930 the Sibley work appealed to me because it afforded an opportunity to demonstrate that the intellectual history of New England as written by such influential historians as J. T. Adams, Beard, Jernegan, Parrington, and Wertenbaker was about as far from the facts of the case as it was possible to get. Surely there was no better way to answer them than to present the detailed biographies of hundreds of ministers, teachers, physicians, and lawyers. It just happened that the richest body of biographical material available for my purpose was that relating to Harvard men, and it was fortunate that most of them settled in New England, the richest area in the world in printed and manuscript historical materials, including the unrivaled archives of towns, churches, and courts.

When I got down to work I found in an index volume a note which Mr. Sibley had written to me thirty years before I was born:

I am aware that there will be great complaint about the obscurity in this volume, but I also know that the references can all be made out by any one who will try. Grumblers [who] fault indexes should be grateful for the immense amount of labor spent in preparing what there is without any cost or trouble to themselves. When any one has done as much and as well (imperfect and unsatisfactory as this is, and no body can judge of its unsatisfactoriness better than myself) there will be time for him to grumble, and when he has got to that he will cease to grumble and be heartily grateful for what has been done to his hand.

Thus forewarned, I never have grumbled, although when I have followed one of his references through an elephant folio newspaper of the 1840's to find, after an hour's searching, only a bare mention of the name I was seeking, I have sometimes looked coldly in his direction. His literary remains were a volume of references encountered in his tireless searching, two volumes of extracts and records of interviews, twenty-two volumes of correspondence about the graduates, and thirty-one volumes of clippings. The great bulk of this material relates to nineteenth century men who will never come under my pen, but with Volume XI, I reached the generation personally known by old men whom he interviewed; their memories will enrich the biographies to come.

For my volumes I have had access to masses of material not available in Sibley's day. With this in mind I checked the references in one of my early volumes and found that nine-tenths of the material I had used had not been within his reach. A busy generation of librarians has since his day built up vast collections of printed and manuscript source materials. Hundreds of historians have written useful books, great collections have been printed or indexed, and archivists have arranged and made usable the manuscripts in their care.

At the outset, I calculated that I might live to get from the Class of 1690 to that of 1800, so I prepared a single alphabetical list of their members. The research material I divided tentatively into three classes. The first, consisting of such general sources as newspapers and diaries, is of such bulk and low specific yield that it could be used only if searched for all hundred and ten classes at one time. I now have something like a quarter of a million eight-by-five slips of such material filed in my study. Class two material is that best worked for the biographies which are to appear in a single volume. It is mostly manuscript, and most of it requires some little travel, which is a serious problem for the editor who tries to do his work evenings and weekends. For my early volumes the court records were very useful because material was so sparse that every reference had to be worked into the fabric of the biographies; but for later volumes this kind of material has not been needed. The yield of manuscript town and church records seldom justifies the time spent in negotiation and travel, but when an important point does demand that such material be consulted for a single man, I make a day of giving the material class one or class two treatment. The third class of material is that which is attacked only in the preparation of a single biography. It consists of genealogies, town and county histories, the publications of historical societies, the printed writings of a subject's contemporaries, and his own manuscripts. When these are found in private hands or are unindexed I try to give them class one attention, and thus I spent a winter on the Robert Treat Paine manuscripts before they were given to the Massachusetts Historical Society. The genealogical periodicals were originally class three material, but the availability of research assistance enabled me to transfer them to class one, with a great saving in my time.

If progress is to be made on a biographical project of this magnitude, it is essential that work be kept going constantly at all three levels and a definite schedule of producing printer's copy maintained, even if it means abandoning the search of some low-yield sources. In my time similar work has been begun at several universities at the class one level, but nothing has been produced because the editors never did get to class three. It is quite true that maintaining a schedule in such a project means that quite a bit of material belonging in the early volumes will be missed. I could have wept when I read Lyman Butterfield's edition of the *Diary and Autobiography* of John Adams and saw the colorful material which he printed for the first time, but my consolation was that my volumes had been useful in his editing. My experience has been that by working three levels at once and publishing on schedule I have missed some amusing material, but little of factual importance. The errors in the earlier volumes which later research has exposed have been of a trivial nature.

Two or three European historians have sought me out saying that they wanted to shake the hand which produced such an incredible amount of biography. Indeed one could not turn out work in such volume if he had to use the relatively primitive bibliographical machinery and library arrangement usual in Europe. When in happier times I could devote entire days to research for Sibley, I found that I consulted about two hundred volumes a day; in few European libraries could they be located at such a rate. Research in Europe is hindered also by the caste separation between the scholars and the men skilled in library, archival, and editorial techniques. Frequently the libraries are managed by scholars who cannot produce the material wanted, or by technicians who, lacking personal experience in creative research, do not understand the needs of scholars. There are institutions from which I no longer attempt to obtain microfilm because, no matter how carefully worded the order, most of the film delivered has to go into the wastebasket as useless for any scholarly purpose. Fortunately America has bred in this generation a number of men who are at the same time learned scholars and skilled bibliographical technicians; without their help the Sibley series would indeed be poorer.

Some people who have looked at the long red line of Sibley volumes have remarked that I must have had the help of an army of assistants. In fact, any such large editorial project is rather like canoeing in a swift stream, where the utility of "assistance" is decidedly limited. No committee has a mind. At the outset of the task I had invaluable clerical help in indexing the Sibley correspondence and the Harvard Faculty Records, but such aid can be used only in class one material, and then primarily to locate, not to select and copy out. Turn assistants loose on twenty newspapers and they will make twenty copies of the same article. Most of their copy must be discarded as trivial or irrelevant, because they cannot possibly know what is in the editor's mind, or what he is likely to want thirty years hence. Many would-be helpers make a virtue of saying that they must be thorough, meaning that they refuse to use their judgment; such help only clutters the file. The elderly volunteers always insist on copying out what they consider important and ignore the editor's instructions. I remember one who copied out the list of a subject's grandchildren (of no possible use to me) and ignored his tale of being on a scalp-hunting party which chased its own tracks through the snow. My saddest tale is of a dear friend who saw how worn and battered was the typewritten list of names on which as a result of ten years of research I had entered hundreds of corrections of the errors which over two centuries had crept into the catalogue of graduates. When she brought me the nice clean copy she said, "What took me so long was correcting all the errors I found when I compared your copy with the catalogue." She had thrown away my list. It is too much to expect an untrained assistant to recognize a diamond when he sees one in the gravel bank. Once I turned a young Ph.D. loose with a class two list on the files of the Superior Court of Massachusetts, which had yielded so many jewels for the earlier volumes. After a week's work (I could have done the job in a day) he reported that he had not found a single item of interest.

I have been blessed above all other biographers in having a wife, Dorothy Boyd MacKillop Shipton, who, perhaps because she was plucked from the Deans' Office, took an immediate interest in the Sibley work and has for thirty years, on and off, before, between, and after children and grandchildren, made our class one material a hobby. Blessed with an accurate judgment of people, she has identified and demonstrated many oblique, implied slanders in the newspapers and manuscripts of the eighteenth century. A frown on her face at the breakfast table may mean the dawning solution of a problem of biographical identity which has bothered us for years. I have even known her, with those incredible powers of instinct which women share with animals, to identify portraits of men whom she has known only by their literary remains. Hardly a day goes by but I take out of the file and drop into place in a biography some note which she took three children and seven grandchildren ago.

I have also had the great good fortune to be librarian and director of the American Antiquarian Society, the library which has much the strongest collection of printed material for purposes of Sibley research, and the one which has its collections so arranged that, for my purposes, research can be done there in half the time it would take anywhere else. Its richness has meant that I could always have on my desk some volume of class one material to pick up whenever I had five minutes to wait for a visitor or a telephone call. This is the only way in which I could ever have turned up such juicy bits as the account of William Rand in the autobiography of his rival for the title of the champion scoundrel of New England.

Several institutions have asked me what kind of man they should select to do a Sibley of their graduates. He should be a man who finds the problem interesting enough either to devote full time to it, or to give up his bridge and golf, his evenings and week-ends, and make it his chief recreation and hobby. He must be consumed with curiosity to know what kind of man was each of those whose names appear in the long lists in the college catalogues. To me the most exciting moment is when I begin work on a new volume by searching the manuscript records of the College in order to find the names of non-graduates to add to the canonical list printed in the catalogue. At one such moment I came upon General John Ashe (Volume XII), of North Carolina Revolutionary War fame, who had never before been identified as a Harvard man. The artist, the musician, and the scientist will understand when I say that at such a moment of discovery I have been sick from excitement.

The editor of such a series must also be a man who can ruthlessly cut the thread of research which he can no longer afford to follow. About ten years after I began my task, I was one day standing comfortably at my study shelves patiently tracking an elusive subject when a realization struck me like a tomahawk blow: I did not have all eternity in which to finish this job; I must strike a pace and ignore the siren joys of research if the accumulation of odds and ends in the attic of my mind were ever to be recorded for the use of those who will follow. It is not enough to wander happily among the sources gathering posies with enthusiasm.

My covin, pardon me, my committee, insist that my readers will want to know the tricks by which I make the Sibley series more entertaining than the *Dictionary of American Biography*. The main thing, I think, is to have a thread of logic, or what seems to the reader to be logic. I lay out the slips for a single sketch in chronological order and then re-sort them following the thread of whatever narrative appears. So far as possible, I let the narrative develop by itself; this gives diversity. There are times, of course, when there is not enough material for a logical pattern of narrative, and then one has to help it by simple fraud. If one wants to make a connection between, for example, a paragraph about a man's writing and one about his pig-raising, it can be done by using the word "pen" in the last sentence of the first paragraph and the first sentence of the second paragraph. At times one has to present the material so that the reader's reaction will supply a logical connection which does not in fact exist. Thus I had nothing of interest to tell about an old bachelor schoolteacher but the fact that his inventory showed a drum. By saying that "he never married, but he had a drum," I implied things which so bemused a couple of reviewers that I have wondered about their wives.

Another trick which I have used to a greater extent than other biographers has been to load the sketches with as much direct quotation as they will stand, with the idea of making the eighteenth century tell its story in its own words. I soon found that this was giving my narrative a Congregationalist, liberal, hard-money, Old-Light bias, so I sought out and treated as class one sources the manuscripts of selected Yale men, New-Lights, Episcopalians, and inflationaries. These provided some caustic criticism of my subjects which I quoted without comment unless I knew the writers to be liars. Had I unnecessarily intruded my own comments I would have spoiled the atmosphere of these old brawls, which as much as anything make these volumes entertaining. One result of this technique has been that I have been accused of holding the opinions of the men whom I quoted. Shortly after the appearance of Volume IV two distinguished historians came to me, a day or so apart. One said, "This is good work except for your bias in favor of Cotton Mather." The other historian said, "This is a good job except for your prejudice against Cotton Mather." Of my account of James Otis in Volume XI another distinguished historian said that it was "one long, contemptuous sneer at the poor man," while others, almost as distinguished, have said it was gentle, understanding, and sympathetic. I have been lambasted for holding the views of men whom I quoted by reviewers who did not read the footnotes in which I did give my opinions. I suppose that I should take this as evidence that I have succeeded in exciting my readers.

Now that we have published the sketches of the members of sixty classes, the committee would like to know whether the evidence is that these men were really the moral and intellectual elite which they supposed themselves to be. The answer is a faint yes, faint because it involves an eliteness which does not have modern social and economic connotations. For various reasons, the social and financial status of the college man in the eighteenth century was far below that in the century preceding, but much above that of the nineteenth century. Did education assure material success for these men? By no means, but it did give them the advantage of being a part of an inner community more closely bound than any like group today. Every one of them could have gone from one end of New England to the other assured of a welcome, a meal, and a bed in the house of any college graduate. John Adams called this "classmating," but the bond was intellectual.

Why did these individuals go to college, and what distinguished them in later life from other Americans? Remember that I am talking about the whole sixty classes, not the few individuals whose sketches are reprinted in this volume. The graduates as a whole were not drawn from a socially and economically elite group. The median student in any class came from a family in average New England circumstances. In most classes the sons of poor men with common school educations exceeded in number the sons of wealthy or college-educated fathers. In general, the compulsion which sent them to college, whether they were rich or poor, was the old Puritan idea of "vocation": the pressing call to do the best one might with whatever talents one had, or in whatever position one might find oneself. The wealthy and educated took very seriously their duty to fill unpaid offices, and the whole "order of seniority" in the classes at Harvard in the eighteenth century was based on public service.

Did the members of these sixty classes have in common any intellectual quality different at least in degree from the rest of the community? They did, and it was this same tendency to seek and fulfill one's vocation. The "Veritas" which was Harvard's motto was no finished and packaged truth handed down from the Church and the fathers; it was an individual goal to be obtained only by an intellectual seeking like the progress of Bunyan's pilgrim. If Sir Galahad had first heard of the Holy Grail when he found it in his sock some morning, it would have had no validity for him. The significance of the Grail lay in the effect which the search for it had on the men who participated. The same was true of the education of the eighteenth-century New Englander. He believed that the Bible contained the whole Truth, the Word of God, but in essence and not in detail. He saw no conflict between science and the Bible because God had not used this means to reveal science; it was the spirit and not the letter which was the truth. He regarded creeds, dogma, and theology as worthy of respect as representing the journeys of pious men toward God, but they had no validity for him unless he made the long journey himself.

At Harvard neither science nor theology was taught as fixed dogma, as both were in most contemporary European universities. The professor of divinity taught by tracing the historical evolution of dogmas and rather hoped that his students would end up sharing his conclusions, but he did not regard it as particularly important if they did not. Heretical works were recommended for student reading with only the casual remark that they did contain heresies. The idea that students should be protected from heresy would have been taken as an insult to all concerned. What man could be absolutely sure that he knew the mind of God? Certainly not the professor of divinity. All that he could do was to help the student to find his own way to what for him was veritas, pointing out the more obvious pits into which others had fallen. In so far as the quarrels between ministers and their congregations, which fill these volumes, were due to theology, they had their origin in the fact that the clergy, being thus educated, could not accept the narrow and dogmatic truths of uneducated men.

The inevitable result of this kind of education was a highly individualistic product. Among the graduates were atheists, Deists, Congregationalists, Presbyterians, Episcopalians, obstinate Tories, conservative Loyalists, and Whigs. The one idea which they had in common was that it was their duty to follow the truth as they saw it, and obviously they could not deny to any other man his God-given right to be wrong. This was one reason why perhaps no civil war ever produced less hatred than did the Revolution in New England, why there was nothing here to compare with the savage fraticidal strife which soaked other colonies in blood. Here men died for their opinions on the field of battle only; there were no lynchings, no firing squads, no purges. The property of absentee Loyalists was seized over the bitter protests of some leading Whigs, and the State looked the other way while relatives rescued much of it. Most of those who hated, in a brief flash of rage, repented and apologized when the moment of anger had passed.

When Harvard-trained ministers went into other parts of the country they were frequently in trouble for their liberalism. In the early nineteenth century a president of Princeton who happened to attend a meeting of the Boston Association of Ministers was amazed at the damnable heresies which he heard around him, and still more amazed that men whom he regarded as orthodox would associate on terms of intimate friendship with such heretics. These men had been taught that it is only from diversity of opinion that knowledge is advanced. The heretic was a useful stimulant; surely God would not punish a man for an honest mistake which, properly understood by others, would lead them to a better knowledge of Him. Man was the noblest work of God, and although he was still relatively but a worm, he was capable of infinite improvement in his progress toward Him. It was this belief which gave the Founding Fathers, all of whom shared it in some degree, the faith in man without which they would not have dared to found the American republic. Only with such faith and respect for the beliefs of others can a democratic political system work.

This selection of biographies has not been made in order to demonstrate this moral, but to provide some interesting reading. For reasons of space we have had to leave out some of the longer sketches, such as those of Andrew Oliver (Volume VII), Thomas Hutchinson (Volume VIII), Sam Adams (Volume X), James Otis (Volume XI), and Robert Treat Paine (Volume XII). It has been found most convenient to manufacture this volume by offset lithography, which will explain why the changes in capitalization style and footnote form which have occurred over a period of thirty years are reproduced, and why the sketches contain some genealogical and similar statistical data which belong only in a work of reference. Let's take this as another illustration of the fact that the Harvard mind prefers reasonable diversity to rigid uniformity.

The reader of this volume will sometimes be baffled by the use of obscure terms with which the professionals who use the entire series can reasonably be expected to be familiar. Thus subscription for Prince's *Chronological History* (1736) is mentioned because it indicated a measurable degree of intellectual activity on the part of the subscriber. Subscription for Charles Chauncy's *Seasonable Thoughts* (1743) indicated that the subscriber was an Old-Light and opposed to the revivals. By and large, *Old-Lights* tended to be scholarly, dull, and liberal, while *New-Lights* tended to be hot and intolerant Calvinists. *The Testimony of July 7, 1743*, was a moderate stand on the revivals adopted by a majority of the clergy but not signed by the extremists. *Tate and Brady* was a modern translation of the Psalms with singable airs and a hymnal; liberal Old-Lights generally succeeded in having the use of it substituted for the practice of having the Psalms lined out from the painfully literal New England version. The *Election Sermon* was preached before the General Court on the day of the election of members of the Governor's Council; to be asked to give it was a considerable honor. The *Convention Sermon* was preached to the annual convention of the clergy a few days later; it was a lesser distinction. The *Artillery Election Sermon* was preached at the election of the officers of the Ancient and Honorable Artillery Company; that choice might fall on any minister.

In the colonial period the members of Harvard classes were placed in an *order of seniority* in which they recited, sat at table, marched in processions as long as they lived, and were embalmed in the catalogue of graduates. No doubt they still maintain it in Heaven; it was far more important than piety or sectarianism. When the members of the earlier classes in this volume were *placed*, a considerable weight was given to the intellectual promise of the individual. In the later period the order of seniority became a reflection of the order employed at any public function in those days. First came the sons of governors, then the sons of lieutenant governors, then the sons of the members of the upper branches of the legislatures in sequence of the fathers' elections, then the sons of Justices of the Peace in the order of the dates of the fathers' commissions, then the sons of other college graduates roughly in the sequence of the fathers' graduations, and finally the bulk of the class in an order based on neither wealth nor social standing in the modern sense, except that elderly charity students of much piety but little intellectual promise were at the foot. Pending the publication of a study of this system now being made under a Ford

Foundation grant, the fullest description of it is in my article in the *Harvard Alumni Bulletin* for December 11, 1954, pp. 258–263.

Students who had taken the B.A. and were in residence while waiting for the M.A. were addressed by the title of *Sir* prefixed to their family names: *Sir Smith*. Aside from occasional lectures by the professors, each class was carried through its four years by a single *Tutor*, usually a young man preparing for the ministry, medicine, or law. Originally the Tutors were members of the Corporation, but in the period of this volume the practice grew up of electing older men to that body, and the Tutors, now called *Members of the House*, with the professors and the president formed the *Immediate Government*, or Faculty. For further information on colonial college practices the reader is referred to Samuel Eliot Morison's fascinating volumes of Harvard history. Hamilton V. Bail's *Views of Harvard* (1949) reproduces all known pictures of the college in colonial times. Albert Matthews' two volumes of Harvard College Records, issued as XV and XVI of the *Publications* of the Colonial Society of Massachusetts, contain the Corporation Records through 1750 and several very useful appendices giving such material as a list of Commencement dates. Volume XXXI of the *Publications* of the same Society contains several other useful documents. For other bibliographical information the reader is referred to the footnotes in the sketches which follow.

Clifford K. Shipton

Harvard University Archives
August 16, 1962

BENJAMIN WADSWORTH

BENJAMIN WADSWORTH, eighth President of Harvard College, was born in a barn at Milton on February 28, 1669/70, shortly after the destruction of the family homestead by fire. His father was Captain Samuel Wadsworth and his mother was Abigail, daughter of James and Mary Lindall of Marshfield. When Benjamin was six years old, his father was killed by the Indians in the Sudbury fight.[1] He inherited no warlike traits from his father, for his undergraduate record was absolutely unspotted. Curiously enough, after he had taken his first degree he was charged 4s in one quarter for broken window glass, and 9d more after he had proceeded Master of Arts. In the exercises for that degree he took the affirmative to a scientific question — "An Atomi ob Solam Imporositatem sint Indivisibiles?" — which has only recently been answered in the negative. Sir Wadsworth served as butler in 1690 and 1692, while studying for his second degree, and remained in residence after receiving it.

On November 28, 1693, the First Church in Boston voted that Mr. Wadsworth be invited to assist the teacher by preaching once a month, or rendering any other needful help. His service there was interrupted in August, 1694, by a journey to Albany as chaplain to the Massachusetts commissioners sent to treat with the Five Nations. His journal shows that the rocky wilderness in western Massachusetts did not dampen his quaint humor: "This day we din'd in the woods. Pleasant descants were made upon the dining room: it was said that it was large, high, curiously hung with green." One night they built a lean-to and "The next morning, it was queried whether the house should be pulled down or sold." [2]

The First Church in Boston was highly pleased with young Wadsworth, and in the winter of 1695 invited him to move in town and live among them with a view to eventual settlement in office. His sermons had already met with marked success,

[1] In 1730 Benjamin Wadsworth erected a monument to the memory of those who had died in the battle.   [2] 4 *Coll. Mass. Hist. Soc.* I, 102, 103.

1

according to Judge Sewall. Already Wadsworth was favored by the friendship of the great, for once when Sewall dropped in at his chamber Governor Usher of New Hampshire and the province secretary were already there.[3]

On February 12, 1695/6, Mr. Wadsworth moved to Boston and went to live with Deacon Bridgham.[4] He was ordained as the "teaching officer" of the church on September 8.[5] On December 30 he married Ruth, daughter of Andrew Bordman of Cambridge, and sister of the college steward. A week later the young couple returned to the Bridgham house, where they lived until the parsonage became vacant somewhat over a year later.[6] The Wadsworths were not blessed with children, but their home was enlivened by a stream of boys and girls who boarded and studied there. The parson records the names of more than fifty young persons to whom he doled out spending money, making note of the purpose. Sometimes the girls paid in service, and sometimes the parents paid in butter and eggs and like commodities.[7] Besides teaching these children he went about catechizing the children of the parish, ceasing his visits with regret (on his part) only when winter closed in.[8] He urged their parents, not only to bring them up well, but so to dispose of them in trade and in marriage that they might be least liable to temptation.[9]

Benjamin Wadsworth had succeeded early to one of the most important pulpits in New England, and was eager to assume the historic leadership of his office. This he accomplished without antagonizing either religious faction. His simple, earnest sermons accomplished much good and did not provoke quarrels.[10] He was neither enthusiastic, nor quick to pass judgment upon his fellows. He preached against health-drinking, illuminations, and bonfires without antagonizing the liberals; yet appeared at lecture in a periwig without exciting more than grief among the conservatives, some of whom still believed that this particular vanity had been responsible for King William's War. Both factions liked to claim him for their own. Increase Mather hailed his orthodoxy,[11] but Cotton Mather

[3] Samuel Sewall, *Diary*, I, 421, 422.
[4] Wadsworth's Ms. Account Book, M. H. S., p. 2.
[5] *Id.* 10.   [6] *Id.* 9.   [7] *Id. passim.*   [8] *Id.* 60.
[9] *The Saint's Prayer to Escape Temptations* (Boston, 1715), p. 30.
[10] Joseph Sewall, *When the Godly Cease* (Boston, 1727), p. 29.
[11] Sewall, *Diary*, I, 432.

BENJAMIN WADSWORTH

came to believe that his lack of aggressiveness indicated want of ability, and, blaming him for the declining and languishing state of the First Church, set out to remedy it: "I saw, there was but *one Way* to do it; and that was, by commending to them, and procuring for them, a Minister of some *Age*, and great Ability, and Authority, and Experience, and of eminent Piety. . . ." With more than his usual officiousness Cotton Mather selected a candidate, and after consulting some brethren of the First Church, sent him a call, and then informed Mr. Wadsworth of his action. " . . . That young Man, was very angry with mee, and with them, for the Action; and stirr'd up a Storm of most unworthy *Reproaches* on mee, from a Party in the Town. This was the *Reward* of my sincere and zelous Labours, to save the *Old Church* from a dreadful *Convulsion*, that I see hastening on them. . . ." [12] The young minister's prompt defence of his rights against the famous arbiter of New England's religious affairs shows a spirit beneath his smooth exterior which helps to explain his success.

Mr. Wadsworth was always closely interested in college affairs. First named a Fellow of Harvard College in the Charter of 1697, he was continued in office by the Charter of 1700, and in ten years missed only two Corporation meetings. He must have been considered one of the Mather party who voted against the election of President Leverett, for he was one of the fellows who were dropped in 1708 when the Charter of 1650 was restored.[13] The fact that he was reëlected in 1712 suggests that the Leverett administration accounted him liberal; yet he joined Judge Sewall and other conservatives in rebuking President Leverett for his alleged neglect of religion at the College.[14] And three years later the Mather-Sewall party tried to expel him from the Corporation on the ground of non-residence, along with Benjamin Colman and Nathaniel Appleton. From all of which we may conclude that Mr. Wadsworth was something of a trimmer.

After the death of President Leverett in May, 1724, first Joseph Sewall and then Benjamin Colman were elected President, but both declined. On June 28, 1725, the Corporation

[12] Cotton Mather, *Diary*, I, 317.

[13] Albert Matthews, Introduction to Harvard College Records (*Publications Colonial Soc. Mass.* xv), pp. lxi, lxvi, cl, cli.

[14] Sewall, *Diary*, III, 203.

met at the house of Mr. Wadsworth, and chose him to the presidency. To accept was against his instincts and his interests, for his ministerial salary of 3*l* 5*s* a week was more than any Harvard president had received; but unlike Increase Mather, he considered the claims of the College superior to those of his Boston parish, which consented to release him.

As Cotton Mather and the orthodox party who controlled the lower house considered Mr. Wadsworth one of themselves, the General Court increased the presidential salary and voted 1000*l* for the building of a "handsome wooden dwelling house" for his residence.[15] The inauguration took place on Commencement day, July 7, 1725, with the approbation of both parties. The General Court continued its generosity, adding to his salary the rents of Massachusetts Hall, which brought the total to 400*l* for his first year and nearly as much for those that followed.

Late in July the Wadsworth family moved to Cambridge, and for over a year lived uncomfortably in various lodgings. The building of Wadsworth House, as it is still called, progressed slowly, despite the 2*s* which Judge Sewall gave a workman to drive a nail for him, for good luck. Late in October, 1726, the President "boought a Negro wench (thought to be under 20. years old) of mr Bulfinch of Boston, Sail-maker," on credit for 85*l*,[16] and two days later some of the Wadsworth family "lodged at the New-House." "Nov. 4. at night was the first time that my wife and I lodg'd there," wrote the President. "The House was not half finished within." [17] Uncomfortable housing was no new experience, for the Wadsworths some years before had complained that the parsonage was "in Despair."

The great increase in the number of students, and the new wine of the Eighteenth Century, made Wadsworth's administration one of disorder, and, to the religious conservatives, one of decline. These disorders were but symptoms of the social changes taking place in New England life. The really significant aspect of his administration was the fact that the College kept abreast, and even gained upon, contemporary social and intel-

[15] *House Journals* (M.H.S., 1925), VI, 417.
[16] Wadsworth's Ms. Book Relating to College Affairs, H.U. Archives, p. 37.
[17] *Id.* 37.

lectual progress. Isaac Greenwood was called from the private
school in which he taught science and practical mathematics,
to become the first Hollis Professor of Mathematics and Natural
Philosophy. It was a significant step in a time when students
were turning impatiently from the classical learning of the old
grammar schools to the 'practical education' offered by private
teachers. So great was the change that at the time of President
Wadsworth's death Nathaniel Appleton (A.B. 1712) spoke very
slightingly of the curriculum when Wadsworth was a student,[18]
as compared with that of his administration.

President Wadsworth died in office, March 16, 1736/7, leaving to the College a legacy of 110*l*, old tenor, for the benefit of
poor scholars, "tho to no dunce or rake." His contemporaries
found much to praise in his even temper, his care to avoid
quarrels, and his good deeds;[19] and the latest historian believes
that, while not to be numbered among the great presidents,
he worthily maintained the liberal tradition established by
Leverett.

HENRY FLYNT

*Why any man in his right mind wanted to be a college teacher in
the colonial period passes all understanding. The food was often
frightful, always monotonous. Tradition has it that at Yale every
supper served over a period of a hundred years consisted of deep-
dish apple pie. The students, much younger then, lived in a constant state of warfare with the Faculty. Cutting off the tail of the
president's horse was a neat trick. The situation at New Haven was
even worse than that at Cambridge, as the sketches of presidents
Clap (A.B. 1722), Daggett (Class of 1748), and Stiles (Class of
1746) will show.*

HENRY FLYNT, the famous 'Tutor' or 'Father' Flynt of Harvard annals, was born May 5, 1675, the son of Josiah and Esther
(Willet) Flynt. His father (A.B. 1664), the minister of Dorchester, died when Henry was five years old. His maternal
grandfather was that Thomas Willet of Plymouth who took
part in the English 'conquest' of New York in 1664, and be-

[18] Nathaniel Appleton, *Reviving Thoughts in a Dying Hour* (Boston, 1737), p. 30.
[19] Appleton, pp. 30–1; *Boston Weekly News-Letter*, March 17–24, 1737.

HENRY FLYNT

came first mayor of the city.[1] Sickly as a child, quiet and studious as an undergraduate, Henry from his sophomore year served as Scholar of the House with a stipend of 4*l* a year. The College Library owns his copy of Maccovius' Metaphysics, and the Massachusetts Historical Society has a curious rhymed *Isagoge* to the New Testament that he inherited from his father.

Sir Flynt remained in residence until July, 1695, and proceeded M.A. in course, responding to a question that had long troubled theologians, but had largely ceased to worry business men: 'An Aliqua Usura sit licita?' (Whether any interest on money be permitted). Henry believed that it might.

As a preparation to entering the ministry, Henry was admitted to full communion with the Dorchester church in April, 1695. The following year he was preaching on probation at Norwich, Connecticut. In April, 1697, he was invited to settle as permanent pastor on a salary of 52*l* while a bachelor, or 70*l* plus 60 loads of wood if he married.

He twice declined this offer, but promptly accepted an appointment on August 7, 1699, as a tutor of Harvard College at a salary of 35*l* which was made up largely by docking the two other members of the teaching staff, Ebenezer Pemberton and Jabez Fitch. In October, 1699, Henry Flynt's name was back in the steward's book, and there it remained for over half a century.

Hitherto, Harvard had followed the medieval tradition of appointing as tutors a succession of recent graduates, who were expected to move into some eligible pulpit in four or five years' time and make room for their juniors. But Henry Flynt had tried preaching once, and that was enough.[2] He was clearly born to be a teacher of youth, a college character, a 'don' in the eighteenth-century manner. President Mather did not particularly like Tutor Flynt — he was suspected of Episcopal leanings — and Judge Sewall once caught him saying 'Saint' Luke. But Tutor Flynt outstayed President Mather and out-

---

[1] In 1718 Henry Flynt appealed to the Privy Council for the reversal of a judgment of the General Court of Trials at Newport, R. I., respecting his rights to the Willet estate. *Acts of the Privy Council, Colonial*, II, 744.

[2] He was a candidate for the pulpit of the Old South in 1712, but was defeated by a former pupil, Joseph Sewall, by a vote of 44 to 4; a few years later he was supporting Sewall for the Harvard presidency.

lived Judge Sewall. Mr. John Leverett regarded Mr. Flynt's original appointment as an "unnecessary addition," [3] but Tutor Flynt served all through President Leverett's administration, and President Wadsworth's, and well into President Holyoke's. He was named one of the Fellows of the Corporation in the Charter of 1700, which soon went the way of all other attempts to supersede the Charter of 1650; but he remained a Fellow of Harvard College until 1760. In 1713, probably through the influence of his brother-in-law Edmund Quincy, he had a taste of diplomatic life in negotiating a treaty with the Eastern Indians at Portsmouth, and four years later, another of the same sort; but he was always back in College when the first bell rang to prayers at the end of summer vacation. In 1718 he was offered the Rectorship of Yale, but he preferred to remain a tutor at Harvard. From his college window he saw the last of the Puritan century, and the Yard filled with boys who like himself had been born under the rule of the Saints; he watched the eighteenth century roll on toward revolution, and one of his last pupils was John Hancock.

This 'century of enlightenment' was not many years old before Tutor Flynt had become a college character, and before long 'Father' Flynt, as the students called him, was a Harvard institution. When the College was being attacked by the 'New-Lights' for its godlessness, Mr. Flynt's former pupils rushed into print in order to testify that he had taught them to fear God, and obey His commandments.[4] When Governor Burnet visited the College in 1728, "he went first to Mr. Flynt's chamber," and after a Latin oration in the Hall, returned to that chamber again. When Governor Shirley came out from Boston thirteen years later, he saw three or four philosophical experiments, then "From the Apparatus he went to Mr. Flynt's Chamber, where he tarried till Dinner was ready." [5] Whenever two or three Harvard men gathered together, stories were told of 'Father' Flynt, of his hospitality and parsimony, his quick temper and warm heart, his ripe learning, antique manners, and phrases of the bygone century.

One of the popular stories was this. At one morning recita-

3 *Publications Colonial Soc. Mass.* xv, lii n.
4 *Boston Gazette*, Apr. 13–20, May 18–25, June 22–29, 1741.
5 *Publ. C.S.M.* xv, 565, 711.

tion in his chamber, while the students were standing around, he chanced to look in the glass and see one of them behind him lift a keg of wine from the table and take a satisfying drink from the bung-hole. "I thought," said Father Flynt, "I would not disturb him while drinking; but, as soon as he had done, I turned round and told him he ought to have had the manners to have drank to somebody." [6]

Sad tricks were played on him by the young hellions of the time: there was a case of "puting a living Snake into mr Flynt's chamber" and another of mutilating his mare; [7] but he was always the one to plead for offenders in faculty meeting, saying that "wild colts often made good horses." Particularly did his old pupils relish his salty retort to Whitefield. That eminent evangelist, when visiting Harvard, expressed the opinion that one of her favorite authors, Bishop Tillotson, for his liberal theology had gone to hell. "It is my opinion that you will not meet him there!" said Father Flynt.[8]

During college vacations, Father Flynt stayed at the home of his brother-in-law, Judge Edmund Quincy of Braintree. There he maintained a study and chamber which are still to be seen very much as he left them. But occasionally he defied the provincial roads and travelled about. David Sewall, A.B. 1755, has left a long and amusing account of a chaise journey at Portsmouth with the venerable Tutor, who was then eighty years old. But Father Flynt was still sharp; when asked to contribute something for a clergyman whose wife had presented him with twins, he replied, "Aye, that is no fault of mine." He declared that he could not comprehend the young couple whom he saw park their horse chair in a lonely road, but he could accost a comely lady with "Madam, I must buss you!" followed by a hearty kiss.[9]

As Tutor Flynt grew older, his eyesight, which had always troubled him, grew steadily worse. His diary records pitiful attempts to cure the ill with such remedies as powdered sugar. He was blind in one eye as early as 1719, probably temporarily. Because of this infirmity he was excused from being in commons at supper. On September 25, 1754, he resigned his tutorship,

[6] 1 *Proc. Mass. Hist. Soc.* xvi, 10. [7] Ms. Faculty Records, i, 39, 54.
[8] 2 *Coll. M.H.S.* iii, 211 n. [9] 1 *Proc. M.H.S.* xvi, 5-11.

although he continued to serve as secretary of the Board of Overseers until 1758, and as Fellow of the College until 1760. John Adams left an account of him in his last sickness: "Father Flynt has been very gay and sprightly this sickness. Colonel Quincy went to see him a Fast day, and was or appeared to be, as he was about taking leave of the old gentleman, very much affected; the tears flowed very fast. 'I hope sir,' says he in a voice of grief, 'you will excuse my passions.' 'Ay, prithee,' says the old man, 'I don't care much for you nor your passions neither.' Morris said to him, 'You are going, sir, to Abraham's bosom; but I don't know but I shall reach there first.' 'Ay, if you are going there I don't want to go.'" [10] Having achieved the distinction of being the oldest living graduate, Flynt died February 13, 1760, in the eighty-fifth year of his age.

There are numerous contemporary descriptions of his character. According to Charles Chauncy (A.B. 1721), "He was not contemptible for his learning; he might have excelled in it, considering his advantages, had he not been of an indolent temper to a great degree." [11] Paine Wingate of the Class of 1759 wrote: "I remember very distinctly, hearing him preach for Dr. Appleton, when I was a freshman. He was the slowest speaker that I ever heard preach, without exception. He hardly kept connected in his discourse so as to make progress. However he made some amends for this defect by the weight and pertinency of his ideas. He was thought to be a judicious and able preacher, but not very popular. . . . He was rather short and thick-set in corporal appearance. . . ." [12]

Father Flynt managed to accumulate a considerable estate upon his salary of from 50 to 80*l*. Unlike most of his contemporaries he owned little land, although his holdings on Lake Quinsigamond caused it to be called Flynt's Pond for some years. Instead of holding for speculation the lands which he acquired, he sold them at once, taking interest-bearing bonds from the purchasers. He left little personal property but books to the value of 340*l*, fourteen jackets, seven wigs, thirty-one pairs

---

[10] John Adams, *Works* (Boston, 1850–56), II, 62.

[11] 1 *Coll. M.H.S.* x, 165.

[12] Benjamin Peirce, *History of Harvard University* (Cambridge, 1833), p. 263. See also James Lovell, *Oratio in Funere Henrici Flyntii* (Boston, 1760); and Nathaniel Appleton, *Discourse on the Death of Henry Flynt* (Boston, 1760).

of gloves, and twenty gold rings — these last doubtless souvenirs of the funerals he had attended. He bequeathed to the College 700*l*, old tenor, to supplement the salaries of the tutors, and fifty Spanish dollars for a scholarship.

WILLIAM VESEY

The rectorship of William Vesey is particularly significant in that under it the Church of England was more rigidly established in New York than the Congregational church ever was in New England. Presbyterian ministers were jailed for daring to preach without permission, and clergymen of all denominations were among those taxed for Vesey's support. For this system in action, see the account of his successor, Samuel Auchmuty, Class of 1742, below. It was largely responsible for what would otherwise seem to be the pathological fear of Episcopacy in New England and Pennsylvania.

WILLIAM VESEY, first Rector of Trinity Church, New York, was born August 10, 1674,[1] to William and Mary Vesey, who lived in the part of Braintree now included in the town of Quincy. The Vesey family belonged to that little group which believed with the Church of England,[2] although they outwardly conformed to the established Congregational church, where William was baptized.[3] The elder Vesey was, according to Lord Bellomont, "the most impudent and avowed Jacobite . . . known in America," [4] and was once ordered to be placed

[1] *Records of Braintree* (Randolph, 1886), p. 652.
[2] Henry W. Foote, *Annals of King's Chapel* (Boston, 1882), I, 256 n.
[3] *N. E. Hist. Gen. Reg.* LIX, 156.
[4] *Documents Relating to the Colonial History of New York* (Albany, 1856–83), IV, 581.

WILLIAM VESEY

in the pillory in the Boston market place for defiantly plough-
ing on a day appointed for public thanksgiving for the escape
of King William from assassination.[5] As might have been expected, William was a somewhat rebellious undergraduate.

From shortly after graduation until the early part of 1696,
Sir Vesey preached on Long Island, "6 months at Sag and 2
years at Hempstead," both congregations being rather notorious for their dissenting views. The Reverend John Miller
mentions "Mr. Vesy, without any Orders," as preaching at
Hempstead in 1695.[6] It was as a Dissenter that on January 26,
1695, he was called to the church recently organized in New
York under an act of the provincial assembly. But at the beginning of 1696, under the influence of Governor Fletcher, the
Church of England sympathizers gained a majority of the
churchwardens and vestrymen.

This was no loss to Vesey. He went to Boston and studied
theology under the direction of the Reverend Samuel Myles
(A.B. 1684), Rector of King's Chapel. Judge Sewall records in
his diary for July 26, 1696: "Mr. Veisy preach'd at the Church
of England; had many Auditors. He was spoken to to preach
for Mr. Willard; but am told this will procure him a discharge."[7]
Vesey's former teacher, Increase Mather, undoubtedly had
him in mind when at the ordination of Benjamin Wadsworth
he "Spake notably of some young men who had apostatized
from New England principles, contrary to the Light of their
education."[8]

Vesey returned to New York, after three months as an assistant at King's Chapel, with the warm recommendations of Mr.
Myles and the wardens. On November 2, 1696, the church-
wardens and vestrymen of "the Citty of New Yorke" called
him to officiate over the new organization. Three days later
they lent him 95*l* on bond, to defray the expense of a voyage
to England to obtain episcopal ordination.

[5] *Historical Magazine*, 2nd ser., I, 331; *Publications Colonial Soc. Mass.* III, 65;
Suffolk County Court Files, 3443.

[6] John Miller, *New York Considered and Improved. 1695* (Cleveland, 1903), p. 54.
Friends of the Church of England have attempted to prove that Vesey was acting as
a lay reader and preacher of that organization at this early date, but have not assembled a convincing argument. See Morgan Dix, *History of the Parish of Trinity Church*
(New York, 1898), I, *passim*; Charles A. Briggs, *American Presbyterianism* (New York,
1885), pp. 144–53.

[7] Samuel Sewall, *Diary*, I, 430–1.   [8] *Id.* 432.

There was no hitch in the plans. Vesey arrived safely in England, was presented by Merton College for the degree of Master of Arts in the University of Oxford, and was ordained a deacon and priest on August 2, 1697. Next Christmas day he was inducted into the priestly office at New York, the ceremony taking place in the Dutch Church on Garden Street. Trinity Church was completed and opened for public worship on March 13, 1697/8, Mr. Vesey officiating. He was married about the same time, for on March 1 he took out a license to be united to Mary, daughter of Lawrence Reade.

Governor Fletcher had labored to bring about the establishment of the Church of England in the province, and had made to Trinity Church a grant for a term of years of a tract of land known as the King's Farm, which was reckoned as one of the perquisites of the governor's office. His successor, the Earl of Bellomont, considered this and other grants to be extravagant and unwarranted, and persuaded the legislature to declare them null and void. Mr. Vesey entered the controversy which followed by publicly praying for the Dutch minister, Dellius, who had enjoyed one of Fletcher's grants, instead of for the existing Governor; and in general so conducted himself that Lord Bellomont out of self-respect had to abstain from attending church while in New York. The Governor had no desire to injure Trinity Church, and allowed Mr. Vesey 26*l* per annum for house rent as a compensation for the loss of the income from the farm, which was only 7*l* per annum; but this arrangement was soon terminated by a state of open hostility between the two.[9] The minister plunged into politics on the side of the aristocratic anti-Leisler faction, whose practices were such as to give color to Lord Bellomont's oft-repeated opinion that Mr. Vesey was wicked, base, unchristian, and dishonest. The Governor wrote to the Board of Trade requesting that it send "a good moderate Divine of the Church of England to supply the cure at New Yorke in the room of Mr. Vesey," for as long as he and Dellius were there, there would be no business for any honest governor in New York.[10] The Whig writers paid considerable attention to Mr. Vesey's practices and to his

---

[9] *Docs. Col. Hist. N. Y.* IV, 510, and *passim*. Most of the papers relating to this controversy are printed in: *Ecclesiastical Records of the State of New York* (Albany, 1901–14), II–IV.

[10] *Docs. Col. Hist. N. Y.* IV, 535.

conversion to the Church of England: "As to *Mr. Vesey's Learning*, 'tis no secret, that he went raw from the *Boston* University of *Dissenters* to the *Church of England* and was among them bred up to a popular way of Preaching, in which his Schoolboys Memory, and heated Fancy, gave him a reputation for his Talent at Invectives, which then were against the *Church of England*." They suggested that he supported Bayard from the pulpit because Bayard supplied him with beer, and that he "had been concerned with the *Pyrates* of *Madagascar*." [11]

The death of the Earl of Bellomont suddenly changed the complexion of politics. The Bayard-Vesey faction had an uneasy feeling that Lieutenant-Governor Nanfan might not be as moderate as his predecessor; and Mr. Vesey, knowing that he had once told the Lieutenant-Governor that "this Dutch King won't live always," uneasily reflected on the fate of Leisler. [12] So they fled to meet Governor Cornbury, and returned in triumph with him. His Excellency restored the farm to Trinity Church, and thus laid the foundation for its present immense wealth, although the rental then was 35*l* a year.

Except that "Doctor Fiezee," "Pheasy," or "Phesy" occasionally appears as preaching before the Common Council, there is no mention of him in the political annals of New York under Governor Cornbury. There is evidence, however, that he was opposing the interests of Harvard College in England, for he had recently written to the Bishop of London: "There have been, and are great endeavours to Establish the Charter of New England College, on which I beseech your Excellency to have an intense eye, as it, which if granted will be of a very fatal consequense to that very glorious work . . . for the . . . good of his Church in all these American parts. . . . I hope in some time we shall be able to send some scholars to reap the benefit of it." [13] At a later period he was busy in an effort to obtain an annual allowance of 50*l* from the British government, showing in the transaction an eighteenth-century politician's appreciation of the importance of money well placed. [14]

[11] "The Case of William Atwood, Esq." (London, 1703), reprinted in: *Coll. N. Y. Hist. Soc.* 1880, pp. 240–319.   [12] *Id.* 276.

[13] Transcript from Fulham Palace Archives in Harvard University Archives, Chronological File, Dec. 2, 1699.

[14] *Docs. Col. Hist. N. Y.* v, 465–7.

Governor Robert Hunter reported that Mr. Vesey had "grossly and openly abused" him before his arrival.[15] However that may be, they were at odds soon after he landed. The Rector desired the Governor to interfere in the quarrel between the Episcopalians and the Presbyterians for the meeting-house at Jamaica, Long Island, and to dispossess the latter by use of the prerogative. Hunter insisted on leaving the question to the courts, which decided in favor of the dissenters. The angry Rector was soon deep in politics again, giving aid to the opposition party in the Assembly which was trying to starve out the Governor by withholding appropriations.[16] Hunter's friends complained to the Bishop of London of Mr. Vesey's attitude and of the overbearing manner in which he conducted the church, with the result that he was obliged to repair to London to defend himself in June, 1714. The following year he returned, triumphant, with an appointment as the Bishop's commissary and with episcopal instructions to the church to pay his arrears of salary. Curiously enough, he was soon reconciled with the Governor, who wrote to the Board of Trade: ". . . That gentleman has thought fit to humble himself of late, to acknowledge his errors and promiss very warmly a more commendable conduct for the future I hope he is sincere, he has owned that he was put upon going to England by Mr Nicholson who used him ill for declining it when he first proposed it. . . . I am labouring hard to get the City vestry to pay him his salary, which hitherto they absolutely refuse affirming his disertion. . . ." [17]

In the years which followed, Mr. Vesey abstained from politics and enjoyed a more peaceful life. His days were busy, his work useful, and his parishioners so satisfied that they paid his relatively large salary (160*l*) with unusual regularity. He was, however, troubled by the Reverend John Sharpe, chaplain of the military forces stationed in the province, and had to appeal the case to the Bishop of London early in 1722. A few years later he was involved in a private dispute,[18] and in 1732 took part in the Zenger case. So far as one can determine from

[15] *Id.* 311.
[16] *Id.* 311, *et passim; Documentary History of New York* (Albany, 1850–51), III, *passim.*
[17] *Docs. Col. Hist. N. Y.* v, 477.
[18] *New York Gazette*, Mar. 15–22, 1730/1.

the pamphlet [19] directed against him by the adherents of Zenger, he had carried secular matters into the pulpit again: "I am very sorry that when *Don Com Phiz* left his Pastoral Charge amongst the Dissenters, and conform'd to the *Church of England*, he did not leave his Malice, Avarice, and Disposition to Slander behind him; Vices too epidemical or common among the more ignorant and selfish Part of the Dissenters; but it seems they have been too strongly rivetted in him, to be ever eradicated, and therefore, tho' his Body and external Services are for the Church, his Soul cannot be disentangled from the predominant Vices of his first Associates. . . ."

George Whitefield, in his Journal, under date of 15 November, 1739, speaks of the inhospitable reception given him by Vesey.

Waited upon Mr. *Vessey*, but could wish, for his own Sake, he had behaved in a more Christian Manner. — He seem'd to be full of Anger and Resentment, and before I asked him for the Use of his Pulpit, denied it. — He desired to see my Letters of Orders, I told him they were left at *Philadelphia*. — He asked me for a License. — I answered, I never heard that the Bishop of *London* gave any License to any one that went to preach the Gospel in *Georgia*, but that I was presented to the Living of *Savannah* by the Trustees, and upon that Presentation had Letters Dimissory from my Lord of *London*, which I thought was Authority sufficient. — But this was by no Means satisfactory. — He charged me with breaking my Oath, for breaking the Canon, which enjoins Ministers and Church-Wardens not to admit Persons into their Pulpit without a License. Alas! How can I break that, when I am neither a Church-Warden, nor have any Church hereabouts to admit any one into? Upon this, hearing he was a Frequenter of Publick Houses, I reminded him of that Canon which forbids the Clergy to go to any such Places. — This, tho' spoke in the Spirit of Meekness, stirr'd up his Corruptions more and more. — He charged me with making a Disturbance in *Philadelphia*, and sowing and causing Divisions in other Places. — But you, says he, have a Necessity laid upon you to preach; I told him I had. For the Clergy and Laity of our Church seem'd to be settled on their Lees, but my End in Preaching was not to sow Divisions, but to propagate the pure Gospel of JESUS CHRIST. — He said they did not want my Assistance; I replied, if they did preach the Gospel, I wished them good Luck in the Name of the Lord. — But as he had denied me the Church without my asking for the Use of it, I would preach in the Fields, for all Places were alike to me. — Yes, says he, I find you have been used

[19] *To the Reverend Mr. Vesey and his two Subalterns, viz. Tom Pert the Beotian and Clumsy Ralph the Cimmerian* [New York, 1732].

to that. After this, he taxed me with censuring my Superiors. I told him I was no Respector of Persons; if a Bishop committed a Fault, I would tell him of it; if a common Clergyman did not act aright, I would be free with him also, as well as with a Layman. — Whilst we were talking, he called for some Wine, and I drank his Health; soon after, he rose up, said he had Business to do; and (as we were going out) full of Resentment, said to Mr. *Noble*, who accompanied me, with Brother *Seward*, — Mr. *Noble*, as you sent for this Gentleman, so I desire you will find him a Pulpit. — Alas! alas! what manner of Spirit are the Generality of the Clergy *possessed* with? . . . Their Bigotry, if it were nothing else, in Time would destroy them.[20]

Under Mr. Vesey's care as Commissary the Church of England so grew that by the time of his death on July 11, 1746,[21] there were twenty-two churches under him. He left 50*l* to the poor of New York, like sums to his relatives in Braintree, and the bulk of his estate to his widow, who two years later married Judge Daniel Horsmanden. Rector Street, New York, was so named because he lived upon it, and Vesey Street was named for him.

JEDEDIAH ANDREWS

JEDEDIAH ANDREWS, first Presbyterian minister at Philadelphia, was born in Hingham, July 3, 1674,[1] the son of Thomas and Ruth Andrews. The year before Jedediah entered college his father died of smallpox while on Sir William Phips's expedition to Quebec. According to the college quarter-bill book "Androsse" entered during the third quarter of his freshman year. Part of the time he lived out of college, which probably contributed to the fact that he was the most law-abiding individual of his college generation.

During 1696 and 1697 Sir Andrews taught school in Hingham. At his Master's Commencement in 1698 he had already proclaimed himself a member of the Mather party, by defending the affirmative of the proposition "An Politeia Ecclesiastica, sit quoad substantiam Immutabilis?" There was probably some connection between this performance and the fact that

---

[20] *A Continuation of . . . Mr. Whitefield's Journal from his Embarking after the Embargo to his Arrival at Savannah in Georgia* (London, 1740), pp. 36–7.

[21] *Boston News-Letter*, July 24, 1746.

[1] *History of the Town of Hingham* (Hingham, 1893), II, 12.

Increase Mather recommended Andrews and his classmate
Robinson when a call came from Pennsylvania for a New
England preacher.

Mr. Andrews was licensed to preach and went to Philadelphia
in the autumn of 1698. There he found a congregation of Baptists, English Nonconformists, New England Congregationalists, and New York Reformed Dutch, worshiping, when they
could get a minister, in a store which belonged to the Barbados
Company, on the northwest corner of Second and Chestnut
Streets. The Baptist element promptly withdrew to meet in a
brew-house.[2] About the same time there arrived at Philadelphia an Anglican missionary, Thomas Clayton, who battled
with Andrews to attach the religiously foot-loose to their respective congregations. Andrews in the spring of 1699 was considering giving up and returning to New England, when he
came to a non-proselyting agreement with the Episcopalian. In
the end he gained some influential parishioners from the rival
congregation because Clayton was not sufficiently deferential.

Andrews was apparently ordained by the neighboring ministers in 1701. In view of his early insistence on the immutability of the New England Way, it is interesting to find that
Andrews is generally considered by historians of Presbyterianism to be the first minister of that sect ordained in this country.
In doctrine he clung to the tenets of his New England fathers,
at least until 1729; but circumstances if not the tolerant atmosphere of Philadelphia forced him to accept Presbyterian polity.
The steady influx of Scotch-Irish gave his congregation a Presbyterian tinge, and contributed to its growth, until in 1704 it
left the Barbados Company store and built the Old Buttonwood
Church.

When the Presbytery of Philadelphia was formed, Andrews
became one of its most useful members. All the minutes, both
of the Presbytery and the Synod from 1708 to 1746, are in his
handwriting.[3] He was treasurer, corresponding secretary, and a
famous arbiter of disputes.

When the Presbytery of Philadelphia was formed, Andrews
was the only member who had not been educated or ordained in

---

[2] Their correspondence with the congregation is preserved in Morgan Edwards,
Materials Toward a History of the Baptists in Pennsylvania (Philadelphia, 1770).

[3] Published in the *Records of the Presbyterian Church* (Philadelphia, 1841).

Scotland or Ireland. Against the majority he opposed the requirement of subscription to the Westminster Confession. On April 7, 1729, he wrote to Benjamin Colman (A.B. 1692) of the Brattle Square Church: "Means were then used to stave it [the Confession] off, and I was in hopes, we should have heard no more of it. . . . The proposal is, that all ministers and Intrants shall sign it, or be disowned as members. Now what shall we do? . . . I am not so determined, as to be uncapable to receive advice, and I give you this account, that I may have your ju[dgment] as to what I had best do in the Matter." [4] The Synod of that year voted to require the Westminster Confession and Catechisms, but, as a concession to those of Andrews's views, permitted dissenters to state their cases before Presbytery or Synod that the seriousness of the difference might be judged.

Colonial Congregationalists and Presbyterians had not yet really separated; for their theology was identical. Andrews's correspondence constantly refers to the contributions of Boston Congregationalists to his Philadelphia church. He wrote Thomas Prince (A.B. 1707) in 1730 that "The help that was kindly afforded us, from Boston, was of singular use to us, in enlarging our house, which would not, I think, have been done, without it. — It is now in a manner, finished, and proves very favorable for enlarging our Congregation." [5]

The growing congregation led Andrews to ask the Synod in 1733 for an assistant in the ministry. The request was unanimously granted if "first sufficient provision be made for an honourable maintenance of Mr. Andrews, during his continuance among this people. . . ." This, after long discourse by the Synod and after conference with some gentlemen of the congregation, was modified so as to allow the congregation to call an assistant, provided Mr. Andrews's stipend were not diminished. [6] This led to what Andrews described as "the most trying Time . . . that ever I met with in all my Life."

There came from Ireland one Mr. Hemphill at that Time to sojourn in Town for the winter, as was pretended, till he could fall into Busi-

[4] Colman Mss., Mass. Hist. Soc.; Charles Hodge, *Constitutional History of the Presbyterian Church* (Philadelphia, 1851), I, 142.

[5] *Hazard's Register of Pennsylvania*, xv, 200–2.

[6] *Records of the Presbyterian Church* (Philadelphia, 1841), I, 105–6.

ness among some People in the Country (tho' some think he had other views at first, considering the infidel disposition of two many here) some desiring I should have assistance and some Leading men not disaffected to the way of Deism so much as they should be that Man was imposed upon me and the Congregation. Most of the best of the People, were soon so dissatisfied that they would not come to meeting. Freethinkers, Deists and Nothings getting a scent of him flocked to him. I attended all winter, but making Complaint brought the Ministers together. . . . Never was there such a Tryal known in the American world. I was obliged, tho' with great Regret . . . [to appear] against him, and as Providence ordered, all my charges came out fair. Tho' he had promised to produce his notes, yet he fell back, and put me upon the Proof of my articles.[7]

When haled before the Synod, Hemphill replied "I despise the Synod's claim of authority. . . . I shall think you will do me a deal of honour, if you entirely excommunicate me." [8] They did him the lesser honor of a suspension.

Benjamin Franklin, who was not in Andrews's estimation one of the "best of the People," has left an account that shows him to have been eminently fair to Hemphill. According to Franklin, Hemphill

delivered with a good voice, and apparently extempore, most excellent discourses, which drew together considerable numbers of different persuasions, who join'd in admiring them. Among the rest, I became one of his constant hearers, his sermons pleasing me, as they had little of the dogmatical kind, but inculcated strongly the practice of virtue, or what in the religious stile are called good works. Those, however, of our congregation, who considered themselves as orthodox Presbyterians, disapprov'd his doctrine, and were join'd by most of the old clergy, who arraign'd him of heterodoxy before the synod, in order to have him silenc'd. I became his zealous partisan, and contributed all I could to raise a party in his favour, and we combated for him a while with some hopes of success. There was much scribbling pro and con upon the occasion; and finding that, tho' an elegant preacher, he was but a poor writer, I . . . wrote for him two or three pamphlets, and one piece in the Gazette of April, 1735. . . .

During the contest an unlucky occurence hurt his cause exceedingly. One of our adversaries having heard him preach a sermon that was much admired, thought he had somewhere read the sermon before, or at least a part of it. On search, he found that part quoted at length, in one of the British Reviews, from a discourse of Dr. Foster's. This detection gave many of our party disgust, who accordingly aban-

[7] Curwin Mss., Am. Antiq. Soc., II, 107.
[8] *Records*, I, 117.

doned his cause, and occasion'd our more speedy discomfiture in the synod. I stuck by him, however, as I rather approv'd his giving us good sermons compos'd by others, than bad ones of his own manufacture, tho' the latter was the practice of our common teachers. He afterwards acknowledg'd to me that none of those he preach'd were his own; adding, that his memory was such as enabled him to retain and repeat any sermon after one reading only. On our defeat, he left us in search elsewhere of better fortune, and I quitted the congregation, never joining it after, tho' I continu'd many years my subscription for the support of its ministers.[9]

As might be expected, Franklin's opinion of Andrews was not of the highest:

He us'd to visit me sometimes as a friend, and admonish me to attend his administrations, and I was now and then prevail'd upon to do so, once for five Sundays successively. Had he been in my opinion a good preacher, perhaps I might have continued, notwithstanding the occasion I had for the Sunday's leisure in my course of study; but his discourses were chiefly either polemic arguments, or explications of the peculiar doctrines of our sect, and were all to me very dry, uninteresting, and unedifying, since not a single moral principle was inculcated or enforc'd, their aim seeming to be rather to make Presbyterians than good citizens.

At length he took for his text that verse of the fourth chapter of Philippians, 'Finally, brethren, whatsoever things are true, honest, just, pure, lovely, or of good report, if there be any virtue, or any praise, think on these things.' And I imagin'd, in a sermon on such a text, we could not miss of having some morality. But he confin'd himself to five points only. . . . These might be all good things; but, as they were not the kind of good things that I expected from the text, I despaired of ever meeting with them from any other, was disgusted, and attended his preaching no more.[10]

There was a less kindly criticism in the *Weekly Mercury* for June 12, 1729, representing one who had been distracted by the young ladies in Christ Church in the morning, trying to concentrate on the afternoon sermon in the Presbyterian church:

Now will I guard against my morning's fall;
Eyes, by your leave, now ears shall have it all.
This said, I closed them, and in posture sate
Like devotee, to hear and meditate:
But now 'twas worse and worse; the priest did creep
So dull and slowly that I fell asleep.[11]

[9] *Writings* (Smyth edition), I, 345-7.   [10] *Id.* I, 325-6.
[11] Richard Webster, *History of the Presbyterian Church* (Philadelphia, 1857), p. 317 n.

Like most plain and sober preachers, Andrews opposed the Great Awakening. Those who preferred activity of the White-field sort seceded from the congregation and formed the Second Presbyterian Church. The necessity of unity which the revival excitement made imperative led Andrews to accept finally the Presbyterian form of organization. As moderator of the Synod of 1741 he voted to exclude those ministers who denied the power of the Synod to bind dissenting members, those who licensed and ordained preachers in opposition to the acts of the Synod, those who persuaded persons to believe that a call of God to the ministry did not consist in being regularly ordained, those who denied the benefit of the preaching of 'unconverted' ministers, and those who preached so as to cause men to cry out or fall into convulsion. When these radicals had left the Synod, Andrews and the remainder passed an acknowledgment of the Westminster Confession and the Longer and Shorter Catechisms, and eliminated the provision of 1729 that dissenters might be heard and judged.

At the end of a long, honorable, active, and useful (if not inspiring) ministry, Andrews was accused of some disgraceful act the exact nature of which we do not know. He was put on trial, and with his own hands recorded his statement of the matter in the records — his denial of drunkenness, criminal intent or act, and his confession of imprudence and foolish tampering with evil. He deplores the shame brought on the ministry by a levity so unbecoming to his advanced age. He closed his labors as clerk of the Presbytery by recording that the sentence of suspension was passed on him. Probably the offence was trivial, for he was restored in a few months, but died not long after. His will, proved May 25, 1747, left his property to his widow during her life; and in case his only son should die without issue, all should go to John Andrews of Boston, son of his brother Benjamin. His library consisted of 363 volumes.

His wife's first name was Helena. They had at least two children; Mary, born September 21, 1707; and Ephraim, born January 28, 1708/9. Ephraim, who became a physician in Baltimore, was his only surviving child.[12]

[12] Colman Mss., Apr. 7, 1729.

RICHARD SALTONSTALL

The Saltonstall family, beginning with Henry of the Class of 1642, has established a record without rival for consecutive generations graduated from an institution of higher learning. The University used a genealogical chart of the family as a part of its exhibition at the World's Fair of 1892, and might well have added that perhaps no other family has such an unbroken record of solid, honest public servants filling minor offices faithfully and without thought of personal gain. From their fortified house in Haverhill they protected the northern flank of the colony against the savages as the medieval lords of the marches had protected England from the raids of the Scotch and Welsh. Royal governors sent them suggestions as to military policy, not orders. Yet the Saltonstalls were not rough Indian fighters, but gentlemen who in decades of peace worked their way as high as the chief justiceship. In spite of intermarriage with the Cookes, the leaders of the popular party, they were without political ambition. Leverett Saltonstall, the present Senator, was the first of the family to reach the governorship. When, early in his political career, I read to him this Richard's letter about the family home "thronged with Children and Lice" he roared with laughter, and then admonished me, "But don't tell anyone that I have an ancestor." This news will not harm his political career now.

RICHARD SALTONSTALL, soldier and magistrate, born April 25, 1672, was the second son of Colonel Nathaniel Saltonstall (A.B. 1659) of Haverhill and his wife Elizabeth, daughter of the Reverend John Ward. His elder brother Gurdon graduated in 1684. Sir Richard Saltonstall was his great-grandfather. After a destructive freshman year Richard settled down to a normal undergraduate career, of which the only surviving relic is a copy of

Pierre du Moulin's *Elementa Logica* in which Richard inscribed his name followed by this half-hearted attempt at Greek: "Biblios autou." A letter written shortly after his Commencement suggests that Richard was not homesick for the Saltonstall residence, then serving as a garrison house in King William's War: "Last week we heard from Haverhil, all Well and in health, but much thronged with Children and Lice; which discourages our taking a Journey thither. . . . Doctor Oliver . . . returned on Wensday, garded for fear of the Enemy; Varnums Garrison on Last Tuesday-night was besett with about fourscore Indians. . . ." [1] But after one more term at peaceful Cambridge he returned to prepare himself to take his father's place as a colonial counterpart of a lord of the Marches.

The subject of his Master's *quæstio* at Commencement, 1698, "An Mendici sint tolerandi in Republica? [whether beggars should be tolerated in the Commonwealth] Negat Respondens Richardus Saltonstall," suggests that the importunate demands of frontiersmen on the supposedly inexhaustible wealth of the Saltonstall family furnished material for humorous treatment. The following year he was elected to the General Court, but military life was more to his taste than a political career, and he declined reëlection.

On March 25, 1702, Richard was married to Mehitable, daughter of Captain Simon and Sarah Wainwright.[2] After serving as captain and major of militia, upon the death of his father in 1707 he succeeded so quietly to his father's military leadership of the district as well as his justiceship of the peace, that local historians have failed to detect the change. During Queen Anne's War he commanded the large company which was kept constantly armed and prepared for service, and was supplied with snowshoes by the General Court. He reported to the Governor and Council that on the night of September 25, 1708, some thirty savages were discovered in the town; he gave the alarm, gathered a number of his soldiers, and drove out the enemy without loss.

All the dangers of life in Haverhill did not come from Indians. One night the Saltonstall garrison house was blown up with gun-

---

[1] Mass. Hist. Soc. Misc. Mss., Aug. 18, 1695.
[2] Captain Wainwright's deed of the marriage portion is in Letters and Papers, Mss., M. H. S., 71, I, 14.

powder. The Colonel and his wife, tradition says, were thrown in their bed some distance from the house, and the soldiers scattered in all directions; but by a marvelous providence no one was injured. The exasperated Colonel leaped from his now *al fresco* bed and rushed to the servants' quarters in the farm house, where he found all hands up and alarmed but a certain negro wench, whom he had lately punished, and who feigned sleep.[3] The following day he reported the crime to the Court of General Sessions at Ipswich, and asked the "assistance of a Justice or Justices to examin suspected person with respect to the aforesaid high handed Crime." [4] The crime was never proven, and no pressure was placed upon the negress to confess.

After all these alarums and excursions, it seems a pity that Colonel Saltonstall could not have enjoyed the long peace on the frontier that followed Queen Anne's War. He died on April 22, 1714. Of his children,[5] Richard graduated in 1722 and Nathaniel in 1727. He left a considerable estate, including "Artillery and house Furniture. . . . 19*l* 18*s*." [6]

SAMUEL VASSALL

The Vassalls were another ancient Harvard family, but entirely unlike the Saltonstalls. They were wealthy merchants with plantations in the West Indies, and their beautiful mansions in Boston, and particularly in Cambridge, reminded General Washington of the great homes along the rivers of Virginia. Their houses were thronged with black slaves, not Indians, and their owners felt none of the Puritan compulsion to serve the community in relatively minor offices.

SAMUEL VASSALL, Speaker of the Jamaica Assembly, was a grandson of William Vassall, the member of the Massachusetts Bay Company, and a son of Colonel John Vassall, sometime of Scituate in the Plymouth Colony, of South Carolina, and finally

---

[3] George W. Chase, *History of Haverhill* (Haverhill, 1861), p. 233; 1 *Proc. M.H.S.* XIV, 149.

[4] *Essex Institute Hist. Coll.* XXVIII, 185.

[5] See Leverett Saltonstall, *Ancestry and Descendants of Sir Richard Saltonstall* (Cambridge, 1897), pp. 21–2.

[6] Essex Probate Records, cccxi, 146.

of the Island of Jamaica. His mother was Anna, a daughter of John Lewis, an English resident of Genoa. He was probably born about 1676 in Jamaica, where Colonel Vassall died ten years later, providing in his will that: "£20 sterling per annum is to be layed out at Port Royal or from thence shipped to New England yearly for the maintenance of my sonn, Samuel Vassall in College there." [1] In view of the fact that young Sam was the aristocrat of the Class, it is interesting to see how far his allowance went. In the course of his freshman year he spent $9l$ $13s$ $6d$ for commons and sizings, $1l$ for room rent, $2l$ for tuition, $1s$ $4d$ for monitor, $6d$ for wood and candles for the College Hall, $3s$ $4d$ for gallery money, $2d$ for a sweeper, and $11d$ for window glass. His expenses were somewhat heavier for the later years.[2] Some idea of the social distinctions prevalent in the English colonies may be gathered from this letter which his classmate Joseph Green wrote to him shortly after graduation:

Honoured Sir.

When I consider the many kindnesses I have received from you, I cannot but condemn myself for horrid ingratitude in that I never wrot above once vnto you; and that is so long since that I have forgotten Every word etc: but when I call to mind your excelling wit, your comprehensive mind, your great learning and knowledge, your argumentative skill, your solid and discerning judgment, your affable pleasant and amiable conversation and other your incomparable endowments, then I account myself unhappy in that I cannot so much as receive one line from you concerning your welfare.

But although heretofore I have been negligent in my duty to you, yet I considering your great ingenuity am notwithstanding encouraged to Send my non-sensical Scrawls unto you. . . .[3]

Sir Vassall returned to Jamaica upon taking his Bachelor's degree, and applied himself to the law.[4] He served as a member of the Assembly for Kingston from 1707 to 1711. In the latter year he was returned from Vere and served as Speaker *pro tem*. He seems to have been a judge as early as 1709,[5] and deep in the questionable politics of the Island.[6]

[1] Charles M. Calder, *John Vassall and His Descendants* (Hertford, England, 1920), p. 9.

[2] The handsome Hebrew Bible which he acquired in 1692 was used by successive Harvard students until 1810, and is now in the Andover-Harvard Library.

[3] Joseph Green, Ms. Commonplace Book, H. C. L., pp. 116–7.

[4] *Id.* 118.

[5] *Calendar of State Papers, Am. and W. I.*, 1708–1709, p. 274.

[6] *Id.*, 1711–1712, p. 160.

On March 15, 1710/11, Judge Vassall was married to Elizabeth Young of Vere, apparently twice a widow. They are not known to have had any children. He died at Canterbury, England, and was buried in St. Alphege's Church, where a tablet marks his grave. His will was proved on June 9, 1714.

PETER THACHER

The Thachers were a third common type of Harvard family, but in their group they exceeded all others in the number of ministers which they produced during the colonial period. As individuals they tended to be obscure and useful country parsons, disinclined to dogmatism and controversy. In days when few small towns had professional lawyers, physicians, surveyors, or schoolmasters capable of preparing boys for college, the ministers filled all of these functions. It was thanks to men like these that the Puritan towns were bits of the ancient civilization of England transplanted and flourishing in the wilderness, not rough frontier settlements with the violence and vices all too common therein.

PETER THACHER, minister, was the son of Thomas and Mary (Savage) Thacher of Boston, where he was baptized August 26, 1677. His father, a merchant, died in 1686 leaving the family in poor circumstances. Mrs. Thacher in 1691 took out a license to sell liquor "out of Dores," [1] but her brother-in-law, the Reverend Peter Thacher (A.B. 1671) of Milton, seems to have come to the family's aid with the gift of a house and land.[2] The Reverend Mr. Thacher, sometime tutor and Fellow of the College, probably watched over the education of his nephew and namesake, who, under the influence of Samuel Willard of the Old South, showed an inclination toward religion. When Peter entered college the steward placed him eighth in the Class, but the Faculty soon moved him up two places, and by the end of his freshman year he had been advanced to third place. Probably this was not due to family standing, but as a recognition of

[1] *Boston Town Records 1660–1701* (Record Commissioners, VII), p. 207.
[2] Suffolk Deeds, XV, 185.

piety, for according to a contemporary, Peter, "sometime after his Admission into *College* . . . had such an amazing Sight of the odious and accursed Nature of Sin, and such overwhelming Apprehensions of the tremendous Wrath of God, that is due for Sin, that he was thrown hereby into the Agonies of a dreadful Dispair . . . so insupportable, that . . . he almost lost the Possession of himself. . . ." [3]

Although this edifying experience may have been responsible for the conspicuous absence of Peter's name from the records of fines and misdemeanors, it was not sufficient to obtain him a scholarship, and he was forced by poverty to leave college after taking his first degree and take charge of the grammar school at Hatfield. He proceeded M.A. at Commencement, 1699, denying the Arminian proposition "An Gratia sufficiens ad salutem concedatur omnibus?" At Hatfield he studied divinity and "largely partook of the Blessing" of a revival at Northampton. ". . . Having a large and comprehensive *Genius*, he was soon taken Notice of, by the Learned and Judicious, as a very thorough and accomplish'd Divine . . . a great Master of most of the unhappy Controversies which perplex the great Doctrines of Christianity. . . ." [4] Thus he came to the notice of the General Court, which on April 11, 1705, "Advised. That Mr. Peter Thacher Minister be imploy'd as Chaplain on board the Province Galley." [5]

On July 21, 1707, the church of Weymouth elected Peter Thacher of Boston their minister, perhaps inclined toward him because his grandfather had preached there. The town approved the choice and offered a salary of 70*l*. Probably there was some dickering, for in October the offer was raised to 80*l* "and convenient firewood for his fuel when he has a family of his own, or as soon thereafter as he shall call for the same." [6] This was found satisfactory, for the ordination took place on November 26, 1707. Although his salary was small, it was half the entire town budget. On October 14, 1708, he was married to

[3] John Webb, *The Duty of Survivors* (Boston, 1739), pp. 23–4. Thacher's copy of Schrevelius' edition of Hesiod's Poetry, which was presented to him freshman year, is, with other books that he owned in college, in the Prince Collection at the Boston Public Library.

[4] Webb, pp. 25–6.

[5] *Acts and Resolves of Province of Massachusetts Bay*, VIII, 504.

[6] *History of Weymouth* (Weymouth Hist. Soc., 1923), II, 539.

Hannah, daughter of John Curwin and granddaughter of Governor John Winthrop of Connecticut.

Certain Bostonians seem to have kept an eye on the young parson. Judge Sewall wrote in his diary: "Hear Mr. Peter Thacher of Weymouth, who prays and preaches well; though he had been at Boston to see his new-married Sister, which might occasion his preparations to be less full. Sup at Mr. Thacher's."[7] In 1712 he was called from Weymouth to preach the sermon before the Artillery Company. At one time some friends tried to place him in a vacancy in the Old South of Boston, a proposition toward which he was probably not unfriendly, for he was afflicted with asthma and found the climate of Weymouth unpleasant, and the people uncongenial.

Mr. Thacher's uneasiness made the people of Weymouth angry, for it was held a disgrace to have a pastor leave his flock before death took him. Benjamin Wadsworth and Benjamin Colman both thought that Mr. Thacher did not have sufficient reason for leaving his parish; but he was dismissed on February 23, 1718/19. About this time, he was approached in regard to settlement at the New North of Boston. This increased the storm of criticism, and spread it to Boston; for it was generally regarded as unethical, as it was certainly then unprecedented, for a large and wealthy church to entice away the minister of a smaller parish.

At first Cotton Mather had taken an attitude of neutrality, writing that "Dear Mr *Thacher*, is not without some view of a good Reception anon, at least with the New North, in Boston."[8] The New North itself was so bitterly divided that the final choice of Mr. Thacher was ratified only by a majority of one, and that, apparently, by the vote of the minister, John Webb (A.B. 1708). The ministers of the town, other than Mr. Webb, were unanimously agreed in disapproving the call, and tried to prevent it by carrying the matter to a church council. But the Thacher faction was obdurate, and the quarrel developed into a question of congregational independence against quasi-Presbyterianism desired by the Mathers.[9] When that issue became clear, Increase and Cotton Mather "turned his utter enemys

[7] Samuel Sewall, *Diary*, II, 304.
[8] 4 *Coll. Mass. Hist. Soc.* VIII, 434 (Mather Papers).
[9] Henry M. Dexter, *Congregationalism of the Last Three Hundred Years*, p. 500.

and obstructed his settlement all that in them lay. However, after all their spite and melice, a certain day was sett apart and appointed for his installment. . . ." [10] The Boston clergy let it be understood that they did not wish to be invited, so on the day appointed, January 27, 1719/20, the installation council, which gathered at Mr. Webb's house, consisted of Thomas Cheever (A.B. 1677) of Rumney Marsh and Peter Thacher of Milton. The opposition assembled so as to intercept Mr. Thacher and the council on the way to the meeting-house, intending to use force if necessary in order to prevent the installation. But Mr. Webb led the council out of a back gate, through Love Lane, and into the pulpit, before the opposition awoke to the situation and came storming into the galleries — apparently the Thacher party was already in possession of the floor. The induction ceremony was carried through despite the din and confusion. "I dont believe," wrote Lechmere, "the like was ever heard of or seen before; no bear garden certainly was ever like it; such treatment and language had they that hardly ever was given to the vilest of men, and had they been such they could not have done worse." [11]

Both sides took to the printing press and gave their versions wide circulation.[12] The tone of the anti-Thacher argument may be gathered from this text on the title-page of the pamphlet entitled *An Account of the Reasons*: "Ministers shall not be Vagrants, nor have the Liberty to Intrude themselves of their own Authority, into any Place which best pleaseth them." [13] Thacher's defence is based entirely upon the principle of congregational independence. The unreconciled minority withdrew, formed a new organization, and considered calling it the Revenge Church; but it became known as the New Brick Church, on Hanover Street. In the New North, Mr. Webb generously stepped down and permitted the older man to become senior pastor.

The day after the induction riot, one of Mr. Thacher's parti-

---

[10] 6 *Coll. M.H.S.* v, 393 n. (Winthrop Papers).

[11] *Id.* 393 n. In later years a lady who was present at the ceremony used to tell her grandchildren with asperity that the mob "did sprinkle a liquor, which shall be nameless, from the galleries, upon the people below," and spoiled her new velvet hood. — Ephraim Eliot, *Historical Notices of the New North* (Boston, 1822), p. 45.

[12] There is a bibliography of these pamphlets in Samuel G. Drake, *History and Antiquities of Boston* (Boston, 1856), p. 546.

[13] The Harvard copy contains caustic Ms. comments by one of the participants.

sans on the Council propounded him for the honor of delivering the election sermon; but Judge Sewall "thought it not convenient at this time." [14] A year later when the good Judge wrote to invite Mr. Thacher to open the Council with prayer, he added the note: "I purposed to have waited on you my self, and have spoken my mind to you; That seeing so many were offended at your departure from Weymouth, if you could have done any thing for the Satisfaction of the Aged Ministers in the Town, it would have been both safe and honourable. For where the generality of Christians are offended, 'tis to be fear'd GOD is offended." [15] It was not long before Mr. Thacher was received into the good graces of the Boston clergy, although his own troubles were recalled in 1735, when, as moderator of the Boston convention of ministers, he bitterly opposed honoring a clergyman who had been practically discharged by his church. [16]

Mr. Thacher's contemporaries thought very highly of him. According to one,

He had a strong masterly Genius: His Apprehension was quick, his Judgment penetrating, his way of thinking extensive and close. . . . He had read much, and laid up a large Fund of useful Learning out of the best of Books. . . . His common Pulpit Stile was strong and manly; but his great Mind was above the Affectation of Language; tho' sometimes . . . he could be polite and elegant. . . . his Utterance was not the most clear, and his Method and Train of Reasoning were not so easily taken by common Hearers . . . But the attentive and judicious . . . were his greatest Admirers: And every one could see, by his Action and Voice, that his Heart was engag'd in what he spoke. . . . [17]

Benjamin Colman, his erstwhile opponent, praised his powers of prayer:

Then he breath'd his *native Air*! this his *Asthma* left him! he *soar'd* up as on *Eagles Wings*, and with its *Eye* he could look into the *Light* of the *Throne* of the *Majesty* in the *Heavens*! [18]

John Eliot (A.B. 1772) heard old men say that Mr. Thacher

mingled too much metaphysical speculation with his theology . . . in his practical discourses a ray of light was necessary to make certain

[14] *Diary*, III, 242. [15] *Letter-Books*, II, 132.
[16] *Boston Weekly News-Letter*, May 29–June 5, 1735.
[17] William Cooper, *Compendium Evangelicum* (Boston, 1739), pp. 29–30.
[18] Benjamin Colman, *Faithful Pastors Angels of the Churches* (Boston, 1739), p. 15.

things intelligible — That, like most preachers of this description, he was thought profound, because obscure.[19]

Mr. Thacher spent the last days of his life in his chair by the fireside, unable to sleep because of his asthma. He died on February 26, 1738/9,[20] and his wife followed him on April 9, 1749.[21]

JOHN READ

The Puritans had hoped to create in New England a society without lawyers, whom they regarded as social vampires and creators of discord in the hope of gain from it. In their place all college graduates were expected to acquire enough legal knowledge to serve as gentleman Justices who would dispense justice and not law, and families like the Saltonstalls did this well. The growing complexity of life after the first generation gave rise to a group of drawers of legal papers who gradually worked into practice before the courts. They had little education and no social position; one Connecticut law classed them with prostitutes. Read represents the rediscovery of English law by the college community, and he did much to prove that it was respectable and worthy of the intellectual powers of an educated man. It was another generation, however, before lawyers began to be accepted as Judges on the colonial Bench.

JOHN READ, 'father of American law,' was a son of William and Deborah (Baldwin) Read of Fairfield, Connecticut, where he was born January 29, 1679/80.[1] According to the college quarter-bill book he was a lively undergraduate, once paying a fine of

[19] *Sermon . . . Before . . . the New-North Religious Society* (Boston, 1804), p. 19.

[20] *Boston Weekly News-Letter*, Feb. 22–Mar. 1, 1739. The issue for March 8–15, 1739, contains a 'character' of him.

[21] *Id.* Apr. 6, 1749.

[1] George B. Reed, *Sketch of the Life of the Honorable John Read of Boston* (Boston, 1903), p. 5.

JOHN READ

7s 9d. He spent even more freely than his father's many acres would justify, one quarter-bill for commons and sizings being no less than *5l 2s 11d.* Peter Burr (A.B. 1690) was employed by the elder Read to deal out John's spending money, and in his account book John is charged with such articles as a pair of gloves, two penknives, six pewter spoons, three "inkorns," four handkerchiefs, and "1½ lb Liquorish Balls." It is not clear whether John or his father consumed the three dozen pipes and beer and cider to the value of *9s.*[2] With Collins and Southmayd he played a trick on their classmate, Hugh Adams, who with ready Celtic wit retorted in verse

> Blest is the man who hath not lent
> To wicked Reed his ear.[3]

Years later Read gave Thomas Prince a somewhat different impression of his undergraduate diversions: "When I was senior sophister at college in, 1696, there being a day of prayer, kept by the association at Newtown . . . I and several others went from college to attend the exercise. . . ." After a day of solid preaching and praying in relays, "Mr. Torrey stood up and prayed nearly two hours; but all his prayer so entirely new and various, without tautologies, so exceedingly pertinent, so regular, so natural, so free, lively, and affecting, that towards the end of his prayer, hinting at still new and agreeable scenes of thought, we could not help wishing him to enlarge upon them. But time obliged him to close, to our regret; and we could gladly have heard him an hour longer."[4] Read did not take his second degree in course, and is first credited with it in the Triennial of 1721. No copy of the *Quæstiones* of that year survives, but it is probable that he was not required to perform for the degree.

From Cambridge, Read went to preach at the little Connecticut town of Waterbury, then only fourteen years old and having a rateable estate of but *1700l.* "Febeuary: 8: 1698/9 the town hauing by a comity giuen Mr. John Reed a Call to the worck of the ministrey amongst us acsept what they haue done in it and do now renew our call to him in order to the worck of the ministrey a mongst us."[5] The town renewed the call on

[2] Peter Burr, Ms. Account Book, Mass. Hist. Soc., pp. 19, 39.
[3] See under Hugh Adams, *Sibley*, IV, 322.
[4] Joseph Anderson, *The Churches of Mattatuck* (New Haven, 1892), pp. 185–6.
[5] Henry Bronson, *History of Waterbury* (Waterbury, 1858), p. 212.

July 10, offered to build him a house and to give him a salary of 50*l* in provisions, but he declined. He left probably because of an invitation to preach at East Hartford, where a rate of three halfpence in the pound was laid "to satisfy the Rev. John Reed for his pains in the ministry among us, and to defray charges about providing for him." [6] To qualify him for the ordination, he was admitted into the First Church of Hartford on November 12, 1699,[7] but when he received a call a few months later, he refused it. He was in the pulpit there in 1702, and the following year was paid "ffor preathing Two Sabaths" at the First Church.[8] About the time of his coming to Hartford he had married Ruth, daughter of John and Mary (Cook) Talcott, and half-sister of Governor Joseph Talcott.

The town of Windsor on April 27, 1703, voted to ask Mr. Read to preach there for a quarter year. A few days later the town of Stratford chose an agent "to go to Hartford and endeavor, by all lawful means, the obtaining of Mr. Reed. . . ." [9] Perhaps the means of the Stratford agent did not immediately prevail, for in August the town voted to lay aside the other candidates for the pulpit and to try Mr. Read. He gave satisfaction, and the town voted him 40*l* in provision pay and 6*l* for firewood for a half year, besides undertaking to bring his family from Hartford and to find housing for them. The proposal to settle him provoked long quarrels, during which he was accused of having leanings toward the Church of England.[10] Apparently he denied this, for on November 20, 1706, the town voted: "Whereas, the Reverend Elders in their advice to the town of Stratford, recommended to take all suitable care to purge and vindicate Mr. Reed from such scurrilous and abusive reflections (if any be) that such sentiments may reasonably be supposed to being upon him; and Mr. Reed in order thereto, having laid before the town his request that the town would be pleased to call a Council of Elders to hear what shall be proper to lay

[6] J. Hammond Trumbull, ed. *Memorial History of Hartford County* (Boston, 1886), II, 91.

[7] George B. Reed, *The Honorable John Read* (Boston, 1879), p. 5. The date given in George L. Walker, *History of the First Church in Hartford* (Hartford, 1884), p. 253, is probably a printer's error.   [8] Walker, p. 253.

[9] Samuel Orcutt, *History of Stratford and Bridgeport* (Fairfield Co. Hist. Soc. 1886), I, 296.

[10] Francis L. Hawks and William S. Perry, *Doc. Hist. Prot. Epis. Church in U. S. A.*, Conn. (New York, 1863–64), I, 27.

before them in order for a clearing of his name from those abusive reflections that he is apprehensive have been put upon him." [11] But he must have given the Church of England party some encouragement, for Caleb Heathcote wrote back to the S. P. G.: "By reason of the good inclination he shews for the Church, he has undergone persecution by his people, who do all which is in their power to starve him, and being countenanced and encouraged therein by all the Ministers round them, they have very near effected it; so that if any proposal could be made to encourage his coming over for ordination, his family, which is pretty large, must be taken care of in his absence." [12] He was no meek martyr, for May, 1707 he haled certain members of his congregation before a church council for "contempt of him and his ministry." [13] Two suspected Episcopalians stood by him and in July, 1708, deeded him "for good will and affection" all of their undivided lands in Stratford for a space of ten thousand years. The local brand of Anglicanism was peculiarly infectious, for Read was succeeded by Timothy Cutler.

For a time Read seems to have preached in New Milford, where he built a house and settled. He expressed willingness to take holy orders, but despite the urging of Caleb Heathcote and Evan Evans that he was "well worth the gaining, being by much the most ingenious man" among the Dissenters, nothing came of it.[14] Possessing much spare time, a keen mind, and pressing legal disputes over land, he naturally turned to the law. In later life he used to say, "My knowledge of the law cost me seven years' hard study in that great chair." [15] In May, 1708, the Connecticut Assistants began the practice of admitting attorneys to the bar, and five months later Read became the sixth licensed lawyer in the colony. His first experiences in practice were not happy, for on May 11, 1709, the Assistants voted:

Mr John Read attorney at the Barr of this Court behaveing himself with a manifest contempt of the Court, by objecting in a rude and unbecoming manner against the Court as partial . . . the Court . . . do resolve that the said Read be admonished in this Court, for his misdemeanour and contempt, by the Chief Judge of the Court, and that he be not admitted to plead in this Court untill he shall have made an

[11] Orcutt, I, 297.   [12] Hawks and Perry, I, 20-1.
[13] Letter of B. L. Swan, Feb. 3, 1868.
[14] Hawks and Perry, I, 20, 38.
[15] John Adams, *Works* (Boston, 1850–56), IX, 572.

acknowledgment in the said Court to their acceptance of such his contempt. This resolve was read to said John Read, and he admonished
by the Honourable the Governour, and by him forbid to plead untill
such acknowledgment be made as above.[16]

The next day he appeared before the General Assembly and
apologized, but it was the following October before he was re-
admitted to practice.[17]

These years were discouraging to John Read. In 1710 he
petitioned the General Assembly:

Misfortunes in my adventures have undone me utterly, for as I
thought with a prudent foresight I purchased about twenty thousand
acres of land in Wiantinock [New Milford and vicinity] . . . had spent
much to settle and defend it; settled some inhabitants with me there
afterwards, tried the title and defended it against home pretenders.
Sixteen times have I been to Court about it, ever gaining till the last
Courts Assistants wherein I finally lost; and am utterly discouraged
and broken — finding two things, 1st that I am not able to maintain
suits forever, and that Indian titles are grown into utter contempt,
which things make me weary of the world. Wherefore I pray, seeing I
nor my father have had not one foot of land by division or grant of
town or country, tho' spending all our days in it, that I may have
liberty if I can find a place in the colony . . . that by your allowance I
may settle it with some others of my friends, where in obscurity we
may get a poor living, and pray for your health and prosperity with
great content.[18]

Perhaps it was in compensation that Mr. Read was chosen the
following year to act as Queen's Attorney in the prosecution of
John Rogers, leader of the Rogerenes, at New London. He was
formally commissioned in that office on May 22, 1712, and was
thereafter much in public practice. On occasion he was attorney
for Yale College, and in 1717 presented to the Governor and
Council a humble proposal "for Setling the disputes concerning
the place of the Collegiate School. . . ." [19] He was employed in
settling colony boundary disputes, and was one of the commissioners who met from the New England colonies in Boston in
1720 to discuss the currency problem. He reported:

The paper money now abroad daily sinks in its value and estimation, that already it don't serve as a just medium of trade, but the

[16] C. J. Hoadly, Letter of Dec. 12, 1865.
[17] *Public Records of Colony of Conn.*, 1706–1716, p. 104.
[18] George B. Reed (ed. 1903), p. 7.
[19] *Publications Colonial Soc. Mass.* VI, 183.

merchants raise their goods, I believe, to what they think it will sink to before they are paid, and so the husbandry (the stay of the land) always come off the worst by it. . . . doubtless it is best to give our fathers at home no occasion to reform any real evils among us, lest we be grieved at the measures taken with us.[20]

In 1714 Mr. Read secured title to some land south of New Milford by this whimsical deed:

Know all men by these crooked scrawls and seals that we *Chickens*, alias Sam Mohawk, and *Naseco* do solemnly declare that we are owners of the tract of land called Lonetown, fenced around between Danbury and Fairfield; and John Read, Governor and Commander-in-chief thereof and of the Dominions thereupon depending, desiring to please us have plied the foot and given us three pound in money, and promised us a house next autumn. In consideration thereof we do hereby give and grant to him and his heirs forever, the farm above mentioned and corn appurtaining and further of our free will, motion, and soverain pleasure make the land Manour; Indowing the land thereof, and creating said John Read, Lord Justice and Soverain Pontiff of the same to him and his heirs forever. Witness our crooked marks and borrowed seals this seventh day of May 1714.[21]

At Lonetown Manor Mr. Read lived for six years, the out-standing man of the little settlement which was later named Redding in his honor. He served on church committee and school committee, collected taxes, built and kept the pound, and represented the town before the General Assembly. Although there were Church of England people among the first settlers, he remained a Congregationalist. He was an active land specula-tor, his most notable purchase being 10,000 acres of the 'equiva-lent lands' at Ware, Massachusetts, which the colony of Con-necticut auctioned off in 1716. He paid but a little more than a farthing an acre, and the colony realized but 683*l* on the whole transaction, which was thought a great scandal.[22]

It was probably the visit to Boston on the currency commis-sion which determined Read to move from Lonetown to the provincial metropolis. He took a house on Hanover Street, where later the American House stood, and subsequently moved to a more pretentious mansion on the present Court Street.

[20] George B. Reed (ed. 1879), p. 7.
[21] *Id.* (ed. 1903), p. 8.
[22] W. A. Beers, "John Reed, the Colonial Lawyer," in Fairfield County Hist. Soc. *Fifth Anniversary Volume.*

Here he remained for the rest of his life, although he maintained the 'Manor of Peace' as a gentleman's estate on the equivalent purchase lands.

So great was Mr. Read's reputation when he moved to Boston that he was at once elected to the office of attorney general, and as promptly negatived by the Governor on constitutional grounds. For the next fifteen years with almost perfect regularity he was alternately negatived and permitted to serve. After his last term, in 1735, he was employed as special agent before the boundary commissions of 1737 and 1741. In 1738 he was chosen to represent Boston in the General Court, being, apparently, the first lawyer ever elected to that body. In 1741 and 1742 he was elected to the Council. Although the greatest lawyer of his day, he did not become a political leader, partly because of the lingering New England prejudice against men of the law, and partly because of his temperament. He had little personal ambition, and was too independent and outspoken to follow the devious paths of politics with success.[23]

Read's interest in legislation turned chiefly to the currency question. Although a hard money advocate, he saw clearly that the shock of suddenly recalling the outstanding paper would make matters worse. He proposed, therefore, that the outstanding paper be replaced by notes loaned at five per cent interest payable in silver. The specie, as it came in, was to be used to retire the notes. He several times pressed Connecticut to issue him a patent to coin copper as a means of avoiding inflation, and tried to get the Simsbury copper mining company headed by Adam Winthrop (A.B. 1694) to agree to give him a share of their stock in return for such a patent.[24]

No other lawyer had a private practice to compare with that of Leather-Jacket John, as the Bostonians called him to distinguish him from contemporaries of the same name. One fee from the colony was 60*l*; another was 200 acres of land. The fees which he charged the town increased gradually from 3*l* to 40*l*, so in 1741 he was given a permanent retaining fee of 10*l* per annum. Judge Sewall, who must have strongly disagreed with Read on many subjects besides religion, was one of those who retained

[23] Samuel L. Knapp, *Biographical Sketches of Eminent Lawyers.* . . . (Boston, 1821), p. 157.

[24] Trumbull Mss., M. H. S., 81, I, 43.

him permanently. Read, having associated himself with King's Chapel, where he was vestryman for many years, defended John Checkley, and in so doing seems to have anticipated a more liberal law of libel.[25] He was counsel for the Episcopal ministers who asserted their right to sit as Overseers of the College, arguing for them: ". . . we account the College a common interest, and beg leave, with the answerers, to call it *our College*. . . ." [26] For this his fee was only 10*s*. He held, moreover, "a retaining fee of three pounds for the Service of the Suffering members of the Church of England in this Province." [27]

John Winthrop (A.B. 1700), who seems to have engaged Read inadvertently in the famous suit of Lechmere *v*. Winthrop, had a low opinion of his methods:

My brother [Paul Dudley] at Roxbury writt a power of attorney to Mr Robinson and Read, and I . . . never read or considered it, but signed it, concluding it was only an ordinary power of attorney to answer at that Court. Now I am lately told that I am trap't and ensnared, and that Read and others have dropt some strange words about it which I never dream't of. . . .

A friend of mine here tells me, if you demand it point blank, he will promise you he will look for it and send it to you, but disappoint you; or elce will say it is mislaid, or anything to put you off. . . .[28]

Although Read was given charge of the case in the Connecticut courts and accepted a fee for fighting it, he failed to appear at the trial, with the result that the Winthrops roundly accused him of having sold out to their opponents.[29] There is no question that his knowledge of the fine points of the law made him one of the pioneers in the use of technicalities in winning cases.[30]

To Dudley and Read go the credit of bringing order out of the chaos of early New England legal practice and making it a fit medium for a rapidly developing society. Read found in use a confused mass of prolix and obscure English forms of procedure and conveyances, and organized them into the more simple and

[25] Clyde A. Duniway, *Freedom of the Press in Massachusetts* (New York, 1906), pp. 109–10.

[26] Josiah Quincy, *History of Harvard University* (Cambridge, 1840), I, 574.

[27] Henry W. Foote, *Annals of King's Chapel* (Boston, 1882–96), I, 462.

[28] 6 *Coll. M.H.S.* v, 406–7 (Winthrop Papers).

[29] *Id.* 418, 426–7.

[30] See the case in Knapp, pp. 155–7.

direct forms still used.[31] James Otis said that he was "perhaps justly esteemed the greatest common lawyer this continent ever saw," [32] while John Adams said that he "had as great a genius and became as eminent as any man." [33] To this Hutchinson added that Read was a man of great integrity and firmness of mind.[34]

Read had always been a man of whimsical temperament,[35] and was now in a position to indulge it. He made a journey through the colonies incognito and on foot, volunteering to act as attorney for poor folk, and amazing the courts with his knowledge.[36] Despite his remunerative law practice, his continued land speculations brought him into difficulties in his later years. He led the legal battle of Redding when the lands claimed by the settlers for commons were granted to other parties by the colonial authorities. In 1737 he paid the town of Boston 1020*l* for Township No. 1, comprising 23,040 acres in the Berkshires. For a part of this sum he gave a bond, which the town had to put in suit four years later. His Boston property was frequently mortgaged to obtain money for land dealings. He died at Boston, February 7, 1749, "esteemed one of the greatest Lawyers in this Country." [37] The most interesting item of his estate was a library of over fifty volumes of legal works.[38] The portrait reproduced here hangs in the Addison Gallery of American Art at Phillips Academy, Andover.[39] A tentative list of his children may be constructed from various records:

(1) Ruth, b. 1700; m. Sept. 14, 1737, Rev. Nathaniel Hunn (A.B. YALE 1731), minister of Redding. (2) John, b. in Hartford in 1701; m. Mary Hawley, and then Sarah Bradley; lived in Redding, but was probably the son who died in Boston Feb. 4, 1733, according to the *Journal* of Feb. 5. (3) Rachel, b. at

---

[31] William Willis, *The Law, the Courts, and the Lawyers of Maine* (Portland, 1863), p. 77; John Adams, *Works*, III, 533.

[32] William Tudor, *Life of James Otis* (Boston, 1823), p. 12.

[33] *Works*, IX, 572.

[34] *History of Massachusetts Bay* (Boston, 1795), II, 336 n.

[35] See anecdote in Jacob W. Reed, *The Reed Family* (Boston, 1861), p. 209.

[36] Knapp, pp. 159–61.

[37] *Boston Weekly Post-Boy*, Feb. 13, 1749, p. 2/1.

[38] A list of them is printed in Reed (ed. 1879), pp. 17–18.

[39] Bartlett H. Hayes, Jr., director of the gallery, is confident that Smibert's signature on the portrait is a forgery, and is doubtful whether the portrait is indeed of Read.

Stratford, Feb. 1704. (4) William, b. 1710; a bachelor of
Boston and a lawyer; d. 1780. (5) Mary, b. Apr. 14, 1716;
m. Capt. Charles Morris of Boston and later Chief Justice
of Halifax. (6) Abigail, m. Joseph Miller of Boston. (7)
Deborah, m. ——Wellstead, and then Henry Paget of Smith-
field, R. I.

JONATHAN BELCHER

*Nothing better illustrates the gulf of time between us and Jonathan
Belcher than the fact that Captain Frederick Marryat, the British
naval officer and caustic critic of America, gives no evidence, in his
volume of travels in this country, that he was aware that he was
one of Belcher's descendants. Since this sketch of Belcher was
written, the Massachusetts Historical Society has acquired his
journals of European travels. These and the unused mass of material in his letter-books would provide the means for a full-length
biography.*

THE paternal grandfather of Governor Jonathan Belcher was
Andrew, one of the settlers of Sudbury, whence he moved to
Cambridge and took out a license to "sell beer and bread, for
entertainment of strangers and the good of the Town." His son
Andrew, the father of Jonathan, was a merchant of Connecticut, Cambridge, and Boston, as well as soldier, member of the
Committee of Safety at the deposition of Governor Andros,
Representative, and Councillor. It is curiously suggestive of the
public lives of Captain Andrew Belcher and his son Jonathan,
that the Captain was not present at the Great Swamp Fight, but
arrived after it was over with the provisions which saved the
army. The Belchers always kept an eye on the bread and beer.
While on business at Hartford, Captain Belcher met and married Sarah, the daughter of Jonathan Gilbert. Their son, Jona-

JONATHAN BELCHER

than, the future governor, was born in Cambridge on January 8, 1681/2.

Captain Andrew Belcher had by this time become a wealthy merchant of Boston and was determined to give his only son all possible advantages of education and travel. Consequently Jonathan "was early instructed in the learned *Languages*, and liberal *Arts* and *Sciences*, in which he made good Proficiency." [1] From the Latin School he went to the College, where he was "chambermate and bedfellow" of Tutor William Brattle. He spent money very freely, and after the marriage of Mr. Brattle had removed a certain restraint, he provided much work for the college glazier. Sir Belcher resided at the College until October, 1699, and proceeded M.A. at Commencement, 1702, taking the affirmative to the metaphysical question "An Creaturæ Existentia sit Contingens?"

In 1704 Jonathan Belcher was sent to Europe, and characteristically gravitated to the court of Hanover. He wrote John White (A.B. 1685): "Since my last I have made a Tour of about 1000 miles in those Countries [Germany and Holland] and Its too tedious to Relate here, What Variety I saw in those places ... at Hannover I was Entertain'd by the Princess Sophia (who Is next heir to the Crown of England) as If she had been my mother, She has done our Countrey the honour of her picture, Which I shall bring with Me." [2] While in England Belcher took another step to his future advantage; he became a Mason, and was, so far as we know, the first native New Englander to join the craft.

Upon his return from Europe, Belcher participated in a social event which attracted considerable attention in the provincial capital:

Piscataqua, January 11th [1705/6]. On Fryday the 4th. Currant several Gentlemen went from hence as far as Hampton to meet Mr. *Jonathan Belcher* Merchant of Boston, where he was met being accompanyed by several Gentlemen, and arrived here the said night in order to his Marriage on Tuesday the 8th. Instant, being his Birthday, unto Mrs. *Mary Partridge* Daughter to William Partridge Esq. late Lieutenant Governour of this Province; But at the motion of the Gentlemen that accompanyed him, they were Marryed the same night as he came off his Journey in his Boots: The Wedding was Celebrated on the

---

[1] Aaron Burr, *A Servant to God dismissed from Labour to Rest* (N. Y. 1757), p. 12.
[2] *Publications Colonial Soc. Mass.* xx, 97.

Tuesday following, where there was a Noble and Splended Entertainment for the Guests, and honoured with a Discharge of the Great Guns of the Forts, &c.

In Boston likewise "There was several great Guns discharged at his Father Capt. Andrew Belcher Esqr's Wharffe, and aboard of several Ships." [3] On the 23rd the couple arrived at Boston with a noble procession of "About 20 Horsmen, Three Coaches and many Slays," and dined with the Lieutenant-Governor. [4] Shortly after, he was elected to his first public office, that of constable of Boston for 1707. But he declined the service and paid a fine instead.

Mr. Belcher returned to Europe in March, 1708, being careful to inform the Electress of Hanover: "having now made a Journey purposely to throw myself at your Royal Higness's feet . . . I herewith, offer you the thanks of Her Majesty's Governour in New England for your Respect to that Countrey, and have brought with me the Candles of which I formerly spoke, and an Indian Slave, A native of my Countrey, of which I humbly ask your Royal Highness's Acceptance. . . ." [5] He was in Europe again in 1715, spending there about six years in all. It was reported that in his travels "he preserved his Morals unsullied . . . and even maintained a sacred Regard to that *holy Religion* which he made an early profession of." [6]

Hereditary influences, education, and travel had by this time developed in Belcher the attributes of a polished, plausible, and clever politician; or, as contemporaries expressed it, his moral and mental worth were "set of, by a peculiar Beauty and Gracefulness of Person, in which he was excelled by no Man in his Day," joined to a dignified, frank, polite, and easy deportment, in which "the *Scholar*, the accomplished *Gentleman*, and the true *Christian*, were seldom ever more happily and thoroughly united. . . ." [7] He began his political career by serving as a tithingman of Boston in 1714, then as town accountant, and thence with a great leap into the Council itself where he served from 1718 to 1720, in 1722–23, and in 1726–27.

Meanwhile the mercantile house of Andrew and Jonathan Belcher had prospered greatly. They imported such articles as

[3] *Boston News-Letter*, Jan. 7–14, 1705/6.
[4] Samuel Sewall, *Diary*, ii, 153.
[5] *Publ. C.S.M.* xx, 100.   [6] Burr, p. 13.   [7] *Id.*

"Choice Good Madera Wines," furnished supplies for the colonial armies, participated in provincial loans, invested in the Simsbury copper mines, and speculated in Maine lands. Once in time of scarcity the house insisted on exporting grain to Curaçoa, and provoked a riot. The selectmen asked Captain Andrew not to send it, and "he told them, The hardest Fend off! If they stop'd his vessel, he would hinder the coming in of three times as much." [8] Even Jonathan was not always popular: "Capt. Paxton reviles Mr. Jonathan Belchar upon the Parade, calls him Rascal, many times, strikes him with his Cane: Mr. [Paul] Dudley upon his view fines him 5s. He carried it off insolently, and said, He would doe so again." [9]

An interesting picture of the rising young merchant and politician is given in the *Boston News-Letter* for February 1–8, 1720:

> On the last Day of January last in the Nighttime the Warehouse of Jonathan Belcher of Boston, Esq; was broken open, and from thence was Stolen and carried away a White Hair Camlet Cloak lined with Blue, a Suit of Cinnamon Coloured Broad Cloth lined with Silk, a Drab Riding Coat, and sundry other things, also a Book Entituled, Magnalia Christi Americana, A New-England Law Book, and several other Printed Books, which Goods and Books were of the Value of near Seventy or Eighty Pounds.

Jonathan Belcher climbed into high office by adroitly stepping from one issue to another. In 1712, there was pending in the General Court a bill by which the inflationists intended to make compulsory the acceptance of the depreciated bills of credit. Judge Sewall records: "Mr. Jonathan Belchar comes to me and speaks very freely for passing the Act about Bills of Credit; said I should do well to be out of the way rather than hinder so great a good." [10] Four years later it was largely because it was feared that Colonel Elizeus Burgess would favor the inflation party, that Belcher and Jeremiah Dummer bribed him with £1000 out of their own pockets not to accept a commission as governor of Massachusetts. The town of Boston was displeased by this action and instructed its Representatives "That if the Payment of the Thousand pounds Starling advanced by the Agent and mr Belcher be again moved for payment, they use their utmost Indeavor that the Same be not granted, because Such a Presedent will be Attended with many inconveniences." [11] Between

---

[8] *Diary*, II, 384–5.   [9] *Id.* 343.   [10] *Id.* 365.
[11] *Boston Records 1700–1728* (Record Commissioners, VIII), p. 155.

the merchants and the debtors of Boston, Belcher finally preferred the merchants.

Having opened the way for the appointment of Samuel Shute
as governor by bribing Burgess not to take the office, it was not
unnatural that Belcher should attach himself to the Shute administration. He became generally known as a 'prerogative
man' and as such was left out of the Council by the 'popular'
party in 1728. But suddenly, to the great surprise of the province, he avowed anti-prerogative views and became intimate
with Elisha Cooke (A.B. 1697) and the other popular leaders.
Consequently the House, on December 20, 1728, voted that:
"*Whereas* Jonathan Belcher *Esq; is intended on a Voyage shortly
to* Great Britain . . . he be desired and impowred upon his
Arrival there to be aiding and assisting *Francis Wilks*," the
regular agent, in resisting those instructions.[12] The Council refused to concur in the appropriation of funds for Belcher's
agency, and the radicals resorted to popular subscription, more
to give popular color to the action than because Belcher needed
the money. It was rightly protested by the conservatives that
he was not the agent of the province but only of the radical
majority in the House.[13]

Now enjoying the votes of the radicals, Belcher was elected to
the Council in the spring of 1729. Governor Burnet at once disallowed his election, designating him as the leader of the opposition. Then came a curious twist of fate. Burnet died on September 7, 1729. For a moment the future of the colony swung
in the balance. The home government hesitated between a
policy of moderation and one of reducing the Colony to "a more
absolute Dependancy on the Crown." Shute was at first considered for reappointment, but did not care for it. He remembered that his term of office had been made possible by 500*l*
which had never been paid back to Belcher, and he made a
motion in that gentleman's direction. Belcher was a promising
candidate; he was a provincial, yet known at Court; he was
known as a supporter of the prerogative, yet the chosen representative of the radical party. Moreover, Belcher gave the
Board of Trade to believe that the House had refused a permanent salary to Burnet only for personal reasons, and that no

12 *Journals of the House* (M. H. S., 1927), VIII, 391.
13 *New England Journal*, Mar. 10, 17, 1728/9.

such problem would arise were he governor. He wrote to Benjamin Colman (A.B. 1692) that when Shute refused the office, "... I fixt My Eye Upon the Matter ... When it Was first Mentioned to the King by a great Minister the King Askt if it Was the same Mr Belcher Who Was twice at Hannover, It Was Answered Yess. ..." [14] Belcher was appointed on November 27, 1729, and two days later kissed hands, "After which his Excellency and the Gentlemen trading to New-England, dined elegantly at Pontack's." [15] His commission was dated January 8, 1729/30.

A few months before, Belcher had written Colman: "I Am Sensible You have been all Along for fixing the Salary And I really believe from the Same principle That I have Opposed it — Vizt The preservation of our liberties. ..." [16] Matters now appeared in a different light to His Excellency the Governor than they had to Mr. Belcher, agent of the radical party. He proclaimed his love of the prerogative, promised the ministry that he would abandon the popular cause, abandoned his friend Cooke, and permitted William Dummer to be replaced as Lieutenant-Governor by William Tailer, thus abruptly ending the skeptical friendship of his classmate Jeremiah Dummer. Nevertheless the prospect of a pious administration seemed so bright that Dr. Isaac Watts the hymn writer broke out into an adulatory poem which was thought to be so extravagant as to border on impiety.[17]

The news of the appointment amazed and rejoiced the province, although it must have caused no little scurrying among politicians and merchants. About the middle of the afternoon of Saturday, August 8, 1730, it was announced by signal from Castle William, that the *Blandford* man-of-war with His Excellency on board was entering port, but could get no farther that night than the entrance of the Narrows. There His Excellency was waited on by a committee from the General Assembly and "other Gentlemen, who were all received and entertained with that Nobleness and Affability which" was "natural" to him.

The usual Services of the Sabbath were attended by His Excellency at the Castle. ... At the opening of the following Day, was the Town

14 Colman Mss., M. H. S., Feb. 7, 1729/30.
15 *Boston News-Letter*, Feb. 12–19, 1729/30.
16 John Davis Mss., M. H. S., July 7, 1729.
17 *Broadsides, Ballads, &c., Printed in Massachusetts* (M. H. S., 1922), No. 589.

of *Boston* in a voluntary Alarm. . . . The Troop and Militia were collected and ranged in the Street below the Town-House, in martial
Order. . . . The Turrets and Balconies were hung with Carpets, and
almost every Vessel was blazon'd with a rich variety of Colours. [Between ten and eleven o'clock His Excellency disembarked] with a
great Number of Boats and Pinnaces to attend Him, while His Majesties Cannon were playing, to inform the Town of his Approach. At
length the great Object of our Hopes, and reverent Affection, was received and congratulated at the End of the Long-Wharffe, by the
Honourable Lieut. Governour and Council, the Judges and Justices,
and an almost numberless Multitude of Gentlemen Spectators.
Several Standards and Ensigns were erected on the Top of Fort-Hill,
and at *Clark's* Wharffe at the North part of the Town, and a number
of Cannon planted; which were all handsomely Discharged at the
arrival of His Excellency, and followed with such Huzza's as inspired
the whole Town: The Bells all ringing on the joyful Occasion.

While the Pomp was making its orderly Procession, the Guns,
which were bursting in every part of the Town, were answered in mild
and rumbling Peals by the Artillery of Heaven, which introduced the
plentiful and refreshing Showers that succeeded a dry Season.

After His Majesties Commission was opened and exhibited in the
Council Chamber, and the ususal Ceremonies were concluded, the vast
Multitude of Spectators without, express'd, in their united Shouts, an
unusual Joy and Elevation of Soul. The Troops and Regiment discharged their Duty in triple Vollies; and again were the Cannon roaring at the Batteries in every part of this Town, and at *Charlestown*, by
which the Country was acquainted with the joyful Proclamation.

From the Court-House, His Excellency was conducted by his Civil
and Military Attendents to a splendid Entertainment at the *Bunch
of Grapes*, and after Dinner to his own pleasant Seat.

The Afternoon was spent in firing, and other Expressions of Joy;
and the Evening concluded with a Bonfire and Illuminations.[18]

On August 11, the Governor was addressed formally by
Timothy Cutler (A.B. 1701), minister of Christ Church, in behalf of himself, his churchwardens and vestry. On the next day
he was waited upon by the associated pastors of the town, who
had "saluted him near the town House" two days before. In
their behalf he was addressed by Benjamin Colman, who told
him that "The KING could not have chosen any One of it's
[the country's] *Sons*, more worthy to represent His Royal Person, nor more accepted of" his "Brethren." [19] To the clerical

---

[18] *Boston News-Letter*, Aug. 6–13, 1730. According to a Philadelphia fashion note
"Governour Belcher made his Entry in Crimson Velvet trim'd with Gold." — *American Weekly Mercury*, Aug. 13–20, 1730.

[19] Appendix to Colman, *Government the Pillar of the Earth* (Boston, 1730).

gentlemen he replied: "I do assure you it has not been from any Self interested Views, That I have sought His Majesty's Favour in the Station He has been pleas'd to place Me; but from a Hope of Advancing His Majesty's Service, and the Interest and Prosperity of this Country. . . ." [20] On the next day the selectmen in an address informed him: ". . . we Adore that Providence that has stirr'd up Your Excellency to appear so Couragiously and Early for us at the Court of *Great Britain*. . . ." [21] With the same misapprehension the delegation of merchants congratulated him upon his "inviolable Attachment to the Privileges of this People. . . ." [22] Likewise the poetic muse hailed him as the bearer of liberty. [23]

Both political parties greeted Belcher as their own, the conservatives believing that his popularity would enable them to break down the obstructive tactics of Cooke, and the radicals, that there would be no more question of a fixed salary and the prerogative. They learned the true situation when he opened the meeting of the General Court at Cambridge, September 9, by informing it that "The Honour of the Crown, and the Interest of *Great Britain*, are (doubtless) very compatible with the Privileges and Liberties of her Plantations," and that, despite his personal efforts, he was given stringent orders to obtain a fixed salary, and that — this threateningly — His Majesty was inclined to give "one more Opportunity of paying a due Regard and Deference to what in his Royal Wisdom he thinks so just and reasonable." [24] The General Court eventually voted him 1500*l* for his service as agent, but held out on the question of a permanent salary as governor. He threatened to return to England in obedience to his instructions, but reported to the home government that the struggle was hopeless. Torn between recalling the newly appointed Governor and surrendering on the issue, the Crown chose the latter. In effect Belcher had betrayed the British government into one of the most serious political defeats it ever sustained in the colonies. The radical party had little fear or dislike of him and would apparently have been willing to grant a regular salary had it not been for the precedent.

[20] *Boston News-Letter*, Aug. 6–13, 1730.
[21] *Id.* Aug. 13–20, 1730.   [22] *Id.*
[23] *New England Journal*, Aug. 11, Aug. 17, and Sept. 7, 1730.
[24] *House Journals*, ix, 239–40.

On the question of the prerogative, Belcher reverted to his earlier stand. He took business before the Council for its formal sanction, rather than for its advice, and gave it less freedom than any other governor had done. The radicals in the House, far from being able to make capital out of this situation, rapidly lost power, leaving the province in a relative state of peace. Merchants rallied to support the Governor, and the mob was not sufficient support for the opposition, as it was to become in the days of Samuel Adams. Elisha Cooke was driven into an unnatural alliance with Lieutenant-Governor David Dunbar of New Hampshire, which aroused the suspicions of his followers. By 1735 the power of the Cooke faction was in decline, and for the rest of Belcher's administration his difficulties were of an economic rather than a political nature.

Although Governor Belcher generally favored the interests of the merchants, he was careful to keep his skirts clear of the mire of financial influences. He had an estate valued at about 60,000*l* and wisely retired from all business activity other than the Simsbury mining project. He supported the popular views on the King's Woods, but there is no evidence that he profited from the spoliation of them. Personally, Belcher appears to have been honest; but he realized that bribery and the buying and selling of offices were as much a part of the British Constitution as the cabinet itself, and acted accordingly.

Governor Belcher was gifted with an amazing power of invective and a sharp temper which caused more trouble to his administration than political issues. On his first visit to New Hampshire he was entertained by Lieutenant-Governor John Wentworth, but learning that that gentleman had written letters to London while the governor's chair was empty, mildly favoring Shute for the office, he accused him of duplicity. He refused Wentworth's next invitation, reduced his income, and removed from office his son Benning (A.B. 1715) and his son-in-law Theodore Atkinson (A.B. 1718). Upon Wentworth's death, these two began a more bitter opposition to Belcher, and welcomed the appointment of Colonel David Dunbar as Lieutenant-Governor of New Hampshire. That official, as Surveyor-General of the King's Woods, had already been savagely attacked by Belcher and his friends, who desired to convert the royal mast trees into planks and boards. Within a few weeks of Dunbar's appointment there was sent to England a petition from New Hampshire alleging that Belcher's government was grievous, oppressive, and arbitrary, and praying the King for his removal.

As Lieutenant-Governor, Dunbar continued his efforts to conserve the potential masts, to have a new province set up between the Kennebec and Nova Scotia, and to obtain the complete separation of Massachusetts and New Hampshire; in all of which he was savagely opposed by Belcher and his friends. The quarrels were carried to the Board of Trade and fill the official publications with accusations and invective, Belcher being clearly superior in the latter, while Dunbar's friends told the greater number of untruths.

The really serious issue of Belcher's administration was inflation. He was willing to violate his instructions in regard to a fixed salary, but as a merchant he stood firmly upon his instructions against the issue of paper currency to run beyond 1741, and for the redemption of all outstanding at that date. The House, in order to force the Governor's hand, refused to vote taxes to meet current expenses until there was a great deficit that could be met, they calculated, only by an issue of paper currency. In the meantime the steady retirement of the outstanding paper in accordance with instructions straitened the merchants until some of them combined to issue private notes as a circulating medium. The Governor issued a proclamation against the plan, and condemned it in his next speech to the General Court. Another project for relief was a revival of the Land Bank scheme, which Belcher suppressed so harshly as to bring the colony to the verge of revolution.

With very little disturbance Massachusetts under Belcher took long steps toward the achievement of religious liberty. The Governor's family connection with Richard Partridge inclined him toward the Quaker interest, as did the obvious political influence of that sect with the Whigs at home. He was greeted with joy by the Quakers of the province and in the first year of his term secured them relief from compulsory church rates.

The situation in regard to the Church of England in the province was somewhat different. The Governor's personal leanings were puritanical, although he attempted to obtain a share of the royal bounty for Christ Church, Boston, and celebrated Christmas there. His need of Anglican support led him to seek for Episcopalians such relief as the Quakers enjoyed. Yet he indulged in an undignified quarrel with the Rector of King's Chapel, which led him to say of his daughter, who was being courted by a member of the Church of England, that "he would rather cut her legs of than see her go into the Church . . . another time, [and] when he was asked the reason of his going to Church in England and not here, his answer was that the Churchmen there were a sober and regular People, but here only a loose disaffected Company." [25]

Governor Belcher was ever a friend to the College. He called the attention of the General Court to the "Complaint of the Sons of the Prophets, that they are straitned for Room . . . that *Stoughton* College is gone much to Decay," and asked that "a Committee of this Court might be chosen to View it, and Report what may be proper to be done for the better Accomodation of the Students there." [26]

On another occasion he recommended "the bringing forward of a Law for the better preventing as well as punishing Excessive Drinking, which grows daily, and is so ruinous to the Families and Estates of this Province. . . ." [27] He even took up the case of those who had suffered from the witchcraft persecution of a half century before, urging the General Court to do something "for retrieving the Estates and Reputations of the Posterities of the unhappy Families that so suffered." [28]

The death of Mrs. Belcher, October 6, 1736, removed one of the few checks on the Governor's temper. "Her Education in every Female Accomplishment was the best Her Country could give. . . . Tho' she ever liv'd in the Splendor of a Gentlewoman, in her OEconomy it was her constant Maxim, *Let nothing be lost*. The Poor can witness to the Compassions of her Heart, and the largness of her Munificence." [29]

The Governor's honest if violent opposition to popular measures soon raised a storm about his ears.[30] His enemies drew up

---

[25] Henry W. Foote, *Annals of King's Chapel* (1882–96), I, 417.
[26] *Boston News-Letter*, Dec. 10–17, 1730.
[27] *Boston Evening-Post*, Feb. 6, 1738.
[28] 2 *Proc. M.H.S.* I, 97.
[29] *Boston News-Letter*, Sept. 30–Oct. 7, Oct. 7–14, 14–21, 1736.
[30] 2 *Coll. M.H.S.* x, 40.

petitions against his tyranny, signed with forged signatures of prominent colonials. The most steady and persistent opposition came from the agent for New Hampshire, who never ceased to urge the claims of that "poor, little, loyal, distressed province" over the "vast, opulent, overgrown" colony of Massachusetts. Thomas Hutchinson relates that Belcher's removal was the price of a petty political deal in England. Lord Euston, the son of the Duke of Grafton, stood for election from Coventry but seemed likely to lose when Belcher's enemies offered to secure his election if the Duke would oblige them by removing Belcher. A day or two after Euston was returned Belcher was removed.[31] His own version was that he fell because he would not bow to the popular demand for inflation: ". . . what was my great sin? why, that I would not indolently and wickedly fall into that vile maxim, *si populus vult decipi decipiatur*. No, by no means! It is the duty of governours and rulers to stand upon the watch towers and warn their poeple of their danger and to hide them from the evil. A tender parent wont let a foolish, mad child run into the fire. Well their idol was *stampt paper*, and they were so daring as to prophane the word *money* in calling that so." [32]

On August 14, 1741, Jonathan Belcher was succeeded by William Shirley as Governor of the Province. "In the Council Chamber, the Commission in form was read. . . . Upon which finished, Gov. Belcher rose up and taking Gov. Shirley by the hand, surrendered the chair to him with a very friendly, cheerful and courteous congratulation, wishing him all happiness and prosperity. . . ." [33]

Thanks to his refusal to make money from his office, Belcher was left at the age of sixty with a dependent family and a seriously depleted fortune. Yet Benjamin Colman could describe him as enjoying "himself far more in his retirement than he could in his Government, his Heart seems happily, steadily set upon Heavenly things, as a view of his approaching Dissolution." [34] This enjoyment of retirement did not prevent Belcher from going to England in March 1743/44. There he was told

---

[31] Thomas Hutchinson, *History of Massachusetts Bay* (London, 1768), II, 398–9.

[32] 2 *Proc. M.H.S.* IX, 12–13.

[33] *Diaries of Benjamin Lynde and of Benjamin Lynde Jr.* (Boston, 1880), p. 115.

[34] Archives of the Society for Promoting Christian Knowledge, Private Letters, VII, 73–4.

by the ministry: "Mr Belcher, no charge or imputation lyes against you, nor need you give yourself or us a trouble of the nature you mention. It has been the King's pleasure to remove you, and you must submit, as we all must in such cases, and when there may be a proper opportunity we shall not forget to serve you." [35] Dr. Avery wrote from England to Dr. Colman: "Governour Belcher is bright gay and high in Spirits, and the greatest Instance of application to Business that he ever knew at his years; and tho' he had indeed discouraged his coming over, the Expence being certain and gaining anything most precarious, yet it did not appear now unlikely but some Compensation for his Losses might be made him; but in what shape or Form he was not able to say. The King was pleased to remember Him, when he was admitted to kiss his Hand." [36] It was Mr. Belcher's object to become Governor of South Carolina, alleging that the present incumbent was a "person of very little merit or pretensions," who had held office long enough.[37] Belcher was suggested as governor of Louisbourg, and was finally offered New Jersey, which he accepted in preference to a more profitable post in a "horrid torrid zone."

On August 8, 1747, Governor Belcher arrived at New York, and two days later published his commission at Perth Amboy. According to his letter-book he found the inhabitants of New Jersey, "the greatest part of them in a Wretched State of Ignorance Unpolisht and of bad Manners." Although New England was graven upon the palms of his hands, he was determined not to bestir himself to become viceroy of America, but to die governor of New Jersey, His Majesty permitting. Of his new home in Burlington he wrote: "I have fixt this little city for the place of my residence, which is about 16 short miles from the charming city of Philadelphia, and as I keep 4 good trotters I can be there in two hours with ease, for the road is a bowling green, and in that city there are a sett of gentlemen of good virtue, sense, and learning . . . so that in the summer, when tired with my garden and books I may now and then take a turn thither." [38] His letters indicate that for the first time in his life he found peace

[35] 2 *Proc. M.H.S.* IX, 15.
[36] *Law Papers (Coll. Conn. Hist. Soc.* XI), I, 251.
[37] Leonard W. Labaree, *Royal Government in America* (New Haven, 1930), p. 47.
[38] 2 *Proc. M.H.S.* IX, 16.

and contentment in the quiet of this residence, which he called World's End.

While in England, Belcher became betrothed to Mary Louisa Emilia Teal, whom Josiah Cotton called a Quaker gentlewoman. From the Colony he wrote back impatiently urging "Her imperial royal Majesty the Empress-Queen of Cesarea" to join him as he was becoming "Something oldish."[39] Mrs. Teal, whom the Philadelphia papers called "a Lady of great Merit, and a handsome Fortune,"[40] answered the call and at Burlington on September 9, 1748, became Mrs. Governor Belcher.

New Jersey was not, as Belcher hoped, a safe and quiet haven for weary governors. His predecessor had died in the midst of a bitter salary quarrel, in which he promptly surrendered by accepting an annual salary bill without a suspending clause for review. Governor Belcher's chief difficulty was the land riots, which he permitted to get out of hand in a manner which the home government thought negligent.[41] He likewise became involved in the struggle between the two houses of the legislature, on one occasion informing the upper house: "When I want the Advice of the Councill, I shall ask it, to Which Mr. Moriss again reply'd 'I believe Sir the Councill will hardly wait for that', And then His Excellency Left the Councill Chamber."[42] The Council complained and intrigued in England against him, and he in turn kept writing home for the removal of the Councillors. As a New England man he was naturally interested in the French and Indian War, and attempted to whip New Jersey into doing her part in the common defence. He thought that a close union of the colonies was the only hope of defeating the French.

Jeremiah Dummer, who had sat next to Belcher for four years at Harvard, was the man who interested Elihu Yale in the struggling Collegiate School of Connecticut, and, next to Yale himself, was that college's greatest early benefactor. Curiously enough Jonathan Belcher had a leading part in the building of Princeton. He was actuated by a laudable desire to provide educational facilities for the sons of New Jersey. He secured a new and liberal charter in place of the questionable one under

[39] Ms. Letter-Books, M. H. S., June, 1748, *passim.*
[40] *Documents Relating to the Colonial History of the State of New Jersey*, XII, 488.
[41] *Id.* VII, 239.   [42] *Id.* 456.

which he found the College existing. He pulled all his English political strings in the interest of the College, and sought to procure sons of prominent Massachusetts families as students. He chose Princeton as its site and took a prominent part in the Commencements there. The first college building was to have been named Belcher Hall, but the Governor declined the honor and suggested the name of Nassau Hall. The College received from his estate 474 volumes, his portrait, and ten portraits of the kings and queens of England.

In 1750 Governor Belcher was seized with an attack of palsy while attending Commencement at Princeton; this may account for his weakness and indecision in political matters thereafter. In search of health he removed to Elizabethtown accompanied by his "numerous train of Bostonians, whineing, praying and canting continually. . . ." From Benjamin Franklin he received an electrical apparatus which it was hoped might help him. He remained active until his death on August 31, 1757. Throughout his life he had one abiding friendship, that for Judge Jonathan Remington of the Class of 1696, who had died thirteen years before. Belcher provided for the building of a tomb in the burial ground opposite the College in Cambridge in which his remains were placed beside those of Remington. His widow removed to Milton, where she died in 1778. Of his children, Andrew graduated from Harvard in 1724 and Jonathan in 1728, while Sarah married Byfield Lyde of the Class of 1723.[43]

Governor Belcher does not deserve the abuse which has been heaped upon him by historians. His fawning for favors and his petty political schemes were those of the time, employed by English courtiers and statesmen such as Bolingbroke, Marlborough, Harley, and Walpole. In our own day the successful politician must harp on worn-out political phrases which sound familiar to the voters and have the sanction of tradition; and for the same reason Belcher constantly employed the religious phraseology of the Puritan century. Hence he acquired a reputation for cant and hypocrisy among eighteenth-century intelligentsia. As administrator he was at once more able and more honest than the popular obstructionists who have gone down in history as the patriots of their era.

Jonathan Belcher seems to have been fond of having his por-

43 For genealogies see *N. E. Hist. Gen. Reg.* xxvii, 239–45.

trait painted. In addition to the Hudson portrait reproduced here, there is one by 'F. Liopoldt' (Franz Lippold?) at the Massachusetts Historical Society, one by Nathaniel Emmons in the collection of the American Arts Association in New York, and a Copley in the Massachusetts State House.

JEREMIAH DUMMER

JEREMIAH DUMMER, colonial agent, was born early in 1681, the son of Jeremiah and Ann, or Hannah, (Atwater) Dummer of Boston. His father was that silversmith and engraver who learned to paint such excellent portraits that he deserves to be called the first native Anglo-American painter of competence.[1] Richard Dummer, his grandfather, had been an Assistant under the old charter; Jeremiah, his father, was a member of the Council of Safety in 1689 and was appointed justice of the peace in 1692.

According to President Mather, Jeremy was by far the best scholar of his time in Harvard College.[2] He spent liberally, broke a few windows, was once fined the large sum of 6s 8d, and once fell into the river on horseback at midnight. He remained in residence until April, 1701; and in March, 1702, was being congratulated for his safe arrival in Holland, whither he had gone for study. He entered the University of Leyden in July, 1702.[3] The Dutch universities allowed students to come up for degrees as soon as they were properly prepared, and on February 3, 1703, Professor Hermann Witsius of the University of Leyden signed a certificate to the effect that JEREMIAS DUMMER, *Anglus Americanus*, had under him completed the philosophical and theological studies which he had so happily begun in his own country (*quæ in patria feliciter inchoaverat*), and had his professors' approval *morum comitate, ingenii elegantia, judicii acrimonia, multa eruditionis varietate, et quæ cuncta ista exornat, insigni modestia*. He had, moreover, performed all his scholastic acts, and in particular defended a lengthy disputation on the descent of Christ into hell, with the applause of all. Armed with this formidable recommendation Dummer emigrated to the

[1] F. W. Coburn, in *D.A.B.*

[2] Mather's Introduction to Dummer's *Discourse on the Holiness of the Sabbath-Day* (Boston, 1763), p. 7.

[3] *Album Studiosorum Lugd. Bat.* (1875), p. 770.

JEREMIAH DUMMER

neighboring University of Utrecht, where, after being examined in Universal Philosophy and defending theses on the transmigration of souls (de annimorum μεταγγισμῶ), he was granted the degree of A.M. and Ph.D. on February 13, 1703.[4]

About a year later Dr. Dummer returned to Boston with copies of his doctoral dissertation, a disquisition on the Jewish Sabbath, a dissertation on the integrity of the Scriptures, and the disputation on the descent into hell, which he had published at Utrecht and Leyden, with complimentary poems in his honor by English, Swiss, Hungarian, and Transylvanian students. These occasioned great amazement as to his erudition, no little envy, and some suspicion as to the orthodoxy of his views. Increase Mather at once determined to have a place made for Dr. Dummer on the Faculty, and wrote to Governor Dudley ". . . That for the said Mr *Jeremiah Dummer*, to serve the students of that society as professor of *philology* There, may be A thing that would have a good Aspect on the Interest of Learning; and that such a prelector doing his Duty well, may be A benifit and Ornament unto the Society." [5] Mather then strove to find a place for Dummer in the ministry: "Dr *Witsius* . . . Professor of *Theology* in the University of *Leyden* . . . has commended his Industry, and blameless Conversation during his abode in that Celebrious *University*, withal acknowledging his good accomplishments as to the Knowledge in *Divinity* as well as in *Philosophy*. . . . Nor will it be for the honour of *New-England* that one so well qualified should for want of due Encouragement, after his return to his Country, be constrained to leave it again." [6]

These efforts failed. Judge Sewall was most discouraging to the first Harvard man to return from the Continent with a Ph.D: "As to your Professorship, I am still of the same mind; considering the way in which you obtain it, it will be hurtfull to yourself, to the College and to the province. And as to your

---

[4] Certificate and diploma were copied into the Harvard records, and are printed in *Publications Colonial Soc. Mass.* xv, 305–8. Dummer is credited in the Triennial and Quinquennial Catalogues as having a Harvard A.M., but as his name does not appear on the Quæstiones with those who took the degree in course, and as there is no record of his being given such a degree out of course, it is likely that an earlier cataloguer mistakenly assumed that his Utrecht M.A. was taken at Harvard. In the continental universities the second degree in Arts was called indifferently M.A. or Ph.D., or both.

[5] Harvard College Papers, I, 76.

[6] Introduction to Dummer's *Sabbath*, pp. 7, 8.

Title of Dr. of Philosophy; seeing the very ancient and illustrious Universities of England, Scotland and Ireland know nothing of it;[7] I am of the Opinion that it would be best for you not to value your self upon it, as to take place any otherwise than as if you had only taken the Degree of Master."[8] In later years Dummer himself commented on this situation: "They have a saying here in England, that when a man is put up for Knight of the Shire or going to be marryed, all his faults are publish't with additions. . . . I had a tryall of this my selfe, when I was going to be a professour at Cambridge. . . ."[9] The truth seems to have been that Dummer had too much knowledge for the community. The saints were displeased with his display of continental theology, and the sinners were unmoved by his brilliant sermons. No church would give him a call. In 1708 the town of Boston chose him constable, but excused him "upon reasons by him given." Undoubtedly the reasons given were an intention to return to Europe. Sewall records under the date of August 23, 1708: "Go to Cousin Dummer's, where Mr. Wadsworth, Mr Cotton Mather pray'd excellently: then Mr. Bridge and Dr. Mather pray'd for Cousin Jeremiah Dummer going to England."[10]

Sometime in the fall of 1708 Dummer arrived in London. For about a year he wavered between an offer to minister to the English church at Amsterdam, his friends' advice to enter the legal profession, an opportunity to serve his Colony as agent, and certain openings to participate in the intrigues of Henry St. John and the Tories. In the meantime he applied himself to trade. A manuscript diary which Dummer kept from 1709 to 1711, now preserved in the Harvard College Library, shows clearly the breaking down of his Puritanical principles under the influences of London society. At first he records repentantly: "I have divers times of late been Engag'd at Cards (which I could not well avoid, it being customary in this City to use that diversion after dinner) . . ."; and he resolves to lay aside that pastime forever.[11] Such entries become fewer and less con-

---

[7] Sewall did not know that the Philosophy Faculty in the Scotch and most of the continental universities corresponded to the Faculty of Arts at Oxford and Cambridge.

[8] *Letter-Book*, I, 302.

[9] Dummer to Treasurer John White, Belknap Mss., Mass. Hist. Soc., VII, 9.

[10] *Diary*, II, 232.   [11] Diary, Sept. 18, 1709.

science-stricken, and mention of the court ladies who caught his eye become more frequent, until the record ceases. Apparently he determined to attach himself to St. John, in whose service moral scruples would have been inconvenient. But as late as January 13, 1709/10, he recorded in his diary: "Of Late the Service of my Countrey has oblig'd me to be much at Court . . . I've Seen so much of the Court and like it so ill, that I think my selfe happy in having no further to doe there."

Sewall had suggested to Dummer that he might be of use to Sir William Ashurst, the Massachusetts agent at London. In 1709 Dummer presented to the government a memorial setting forth the English claim to Canada and Nova Scotia.[12] He wrote to Sewall that further mistreatment of Quakers and Baptists by the magistrates of Massachusetts would be likely to bring trouble upon the province,[13] and he urged the English Dissenters not to resist the Tory lash.[14] At the same time he entered into the struggle to obtain the post of judge advocate and secretary on the Shannon expedition to Canada.[15] In May, 1710, Sir William Ashurst urged him upon the General Court for the post of agent.[16] A similar idea had occurred to John Barnard (A.B. 1700) in Massachusetts, and he told some great merchants "that Mr. Jeremiah Dummer, a courtier, and one intimate with the excellent ministry then at the helm, was as proper a person as we could get, being our own countryman, of admirable capacity, and diligent application." [17] He likewise wrote to the same effect to Governor Dudley and other persons of importance in the colony. But when the question came before the House "At the first bringing in the Vote to make Mr. Dummer Agent, the Governour grew very warm, and said he would be drawn asunder with wild Horses before he would be driven as last year." [18] But he signed Dummer's commission on November 11, 1710.

For a time Dummer served the province and the Tory party simultaneously, being employed by Bolingbroke in some secret negotiations, and having every prospect of rising in English politics, like George Downing of the Class of 1642. So

---

[12] In 3 Coll. Mass. Hist. Soc. I, 231–4.
[13] Susan M. Reed, Church and State in Massachusetts (University of Illinois, 1914), p. 111.
[14] A Letter from a Dissenter in the City. . . . (London, 1710).
[15] Calendar of State Papers, Am. and W. I., 1710–1711, p. 133.
[16] Id. 91.   [17] 3 Coll. M.H.S. v, 208.   [18] Sewall, Diary, II, 288.

completely were his interests identified with those of Boling-broke, that with the fall of that statesman at the death of Queen Anne, all Dummer's own prospects were blasted.

Dummer served the province none the less well for being active in English politics. He wrote home to Tutor Henry Flynt: "In the midst of the noise and hurry which attend my station at court, I often think of your happy collegiate life, where you have a sweet air, good company, time to study, and a calm retreat from the business and vexatious cares of life." [19] Affection as well as interest drove him to unusual efforts. He foresaw the struggles of the next generation and constantly preached the destruction of Canada as the American Carthage.[20] In 1712 he published a pamphlet [21] vindicating Massachusetts of the charges brought against her by the leaders of the expedition under Admiral Walker.

The increasing influence of the Board of Trade and Plantations in England made the period of Dummer's agency the most critical for the colonial liberties that was to be seen until the eve of the Revolution. In 1713 he warned the colonies "of a design to obtain a new modelling the plantations, and make alterations in their civil government." Massachusetts, Connecticut, and Rhode Island consulted together and sent him funds with which to combat the plan.[22] Connecticut as well as Massachusetts employed him as agent.

The appointment of Elizeus Burgess as Governor of Massachusetts Bay interrupted Dummer's labors for the province by private quarrels. Burgess, who never went out to his post, gave out that Dummer had been accustomed to go in disguise by night to the house of Lord Oxford where he obtained great quantities of money as "an instrument of his rogueries." [23] At home, Colonel Nathaniel Byfield led an attack upon Dummer for betraying him after giving him his word and hand in a political matter,[24] and efforts were made to cut off his compensation; [25] but in the following April Burgess sold his commission

---

[19] I *Coll. M.H.S.* VI, 79.

[20] W. T. Morgan, "Some Attempts at Imperial Co-operation During the Reign of Queen Anne," 4 *Transactions Royal Hist. Society* X, 171–94.

[21] *A Letter to a Noble Lord Concerning the Late Expedition to Canada* (London, 1712).

[22] *Public Records of Colony of Conn.*, 1706–1716, pp. 410, 414.

[23] Dummer to John White, June 25, 1715. Belknap Mss., M. H. S., VII, 13.

[24] Byfield was trying to get rid of Dummer and feather his own nest.

[25] Sewall, *Diary*, III, 66, 69, 70.

for £1000 which Dummer and his classmate Jonathan Belcher advanced out of their own pockets.[26] The new appointee, Samuel Shute, was favorable to Dummer; but at home the campaign against him continued in full swing. Sewall records: "The Lt. Gov. comes to my House in the morn, and shows me the Accusation of Sir Alexander Brand against Mr. Agent Dummer, as if he had made the Knight drunk, and pick'd his pocket of 26. Guineas and brought in two Lewd Women into the Cross-Keys Etc. I presently thought on the Soldiers set to guard our Saviour's Tomb, their Tale; and said, If Sir Alexander were drunk, how could he tell who pick'd his Pocket? And as to the Women, I said, My Kinsman might be seen going in, and vile Women might press in so close after him, as to make a semblance of his introducing them. Seem'd to ask my advice Whether he ought not to acquaint the Governour of Connecticut that they might discard him from being their Agent. In the Letter Shewed, Mr. Agent Dummer is call'd this Fellow, Rascal. . . ." [27] Although Dummer was vindicated by documents sent over under the seal of the Lord Mayor,[28] the 'popular' party led by Elisha Cooke continued to bay on his scent. It was Cotton Mather who championed Dummer:

> Three or four of the Representatives, (those particularly that act for the city of Boston,) have been extremely disaffected unto our Agent Mr. D——r. . . . These Gentlemen being sufficiently noisy and subtil and Master of all the Arts which were necessary on such an Occasion, caused much Distemper in the General Assembly at their first comming together. About nineteen or twenty principal Members of the House, together did me the Honour of a Visit. . . .
>
> At what time. . . . I sett before them, in as engaging a manner as I could, what Reasons there were for our publick Respects to be still continued unto Mr. D——r. How amply and fully he had been vindicated from Aspersions; and how copiously he had been recommended unto us, by our best Friends . . . for his Fidelity and Assiduity, in our Service.[29]

Two years later Dummer, in the course of his duties as acting agent for New Hampshire, did Elisha Cooke and his friends the great favor of obtaining the removal of Surveyor-General John

---

[26] Dummer's correspondence in this affair is printed in *Publ. C.S.M.* xiv, 360–72. Lieutenant-Governor Tailer's Letter Book (M. H. S.) discredits this story.

[27] *Diary*, iii, 78–9.

[28] Cotton Mather, *Diary*, ii, 418.

[29] *Id.* 420.

Bridger, who had been disagreeable about Cooke's cutting the King's mast trees into staves. Bridger complained to the Council of Trade and Plantations in picturesque terms, but without success.[30] While Dummer was thus assisting the vested interest of plundering the King's woods, he was lobbying for bounties upon the production of naval stores in the colonies. So ardent was he in this cause that he was temporarily excluded from the Board for intimidating an advocate of the Baltic trade by ill language publicly administered.[31]

Despite such petty bickerings, Dummer was the greatest colonial agent before Franklin. On the one hand he was constantly seeking to moderate the actions of the colonial legislatures, for example opposing the publication of the Massachusetts House Journals lest Englishmen think the colonies too independent. "People here are very apt to read these things with jealous eyes . . . they fancy us to be a little kind of sovereign state, and conclude for certain that we shall be so in time to come. . . ."[32] On the other hand he was constantly rising to defend the colonies, his supreme service being a famous pamphlet, *The Defence of the New England Charters*. This essay, written in 1715 when the helplessness of Carolina before the Indian attack had given rise to a plan to recall the charters and reorganize the colonies, was first published in 1721 to ward off a similar suggestion.[33] This essay frustrated sundry efforts to meet the colonial situation by recalling the charters, and closed that discussion for a generation. When it was renewed on the eve of the Revolution, James Otis and John Adams made Dummer's pamphlet a text-book.

Dummer wrote bitterly: "I expect no thanks from the assembly for this service. . . . It is a hard fate upon me when, I am doing the province and the gentlemen in it all the honour and justice that is in my power, that some persons in the lower house should take equal pains to lessen and expose me";[34] and no thanks he got. To Edmund Quincy he frankly avowed his preference for the conservative party: "I am of your mind that

[30] *Cal. St. Pap., Am. & W. I.*, 1717–1718, p. 368.

[31] *Journal of the Commissioners for Trade and Plantations*, 1718–1722, p. 21.

[32] 3 *Coll. M.H.S.* I, 145.

[33] Louise P. Kellogg, "The American Colonial Charter," in *Annual Report of the American Historical Association for 1903*, I, 310–13.

[34] Thomas Hutchinson, *History of Massachusetts Bay* (Boston, 1767), II, 255 n.

some Gentlemen who set up for the onely Patriots, are far from being so, and it is my firm beleif that in the present Situation of Our Affairs the Governour's friends are the true friends of the Countrey. You'l see by my Publick letter that I have not conceal'd these sentiments, notwithstanding I beleive that my free expressing them will procure my Quietus this next session; especially if I may beleive Col. Tayler, that Mr Cook is the great Darling of his Countrey, and carries what points he pleases in the Assembly. . . ."[35]   He informed the General Court that the Ministry was displeased by the obstructive tactics of the House, and he had the temerity to oppose the charges which Elisha Cooke was sent to England to press against Shute. The 'popular' party then set out to ruin him. The House on June 23, 1721, voted that it was not for their interest to continue his agency any longer, and that he should be dismissed.[36]   The Council unanimously non-concurred and began an acrimonious struggle over its constitutional powers, a struggle which ended a year later when the House admitted that it could not act without the consent of the Council. Dummer continued to perform the duties of agent, appearing in behalf of Massachusetts in legal cases and boundary disputes, giving the contract for printing bills of credit, buying "some tyn kittles" on the province account, and doing all of the hundred and one thankless tasks of the colonial agent.

This experience with the House did not intimidate Dummer, for a few years later he was supporting the authority of admiralty officials over the common courts. In 1728 his ill health gave the House an excuse to appoint Francis Wilks to assist him. The next year an attempt to exclude Dummer by appointing his classmate Jonathan Belcher to serve with Wilks precipitated a quarrel between the two houses which subsided when Dummer's illness incapacitated him. In 1730 Connecticut, alleging the state of his health and his absence from London, dismissed him after eighteen years of service. He apparently tried to regain his position, for five years later Belcher wrote to Wilks that Dummer was "very angry" with him, "and a gentleman lately from your side the water says he can't brook the thought of your succeeding him in Connecticut and in this

---

[35] *Publ. C.S.M.* XVII, 93 n.
[36] Most historians assume that this was the end of Dummer's service as agent.

Province. You know him, and your prudence will make you cautious." [37] Whatever Dummer might have been in private life, he was faithful in public affairs. After leaving office, he never ceased to serve the colonies whenever opportunity offered. In the meantime he was active in the South Sea Company.[38]

It was largely thanks to Dummer that the Collegiate School of Connecticut became Yale College. About 1711 James Pierpont (A.B. 1681) asked Dummer to look up Elihu Yale in London. Dummer investigated, discovered that the nabob had "a prodigious estate," and set to work on him. It took several years of patient labor before there was a show of color in the pan, and much longer before pay dirt was yielded. A letter from Dummer to Gurdon Saltonstall (A.B. 1684) illustrates the difficulties encountered:

Mr. Yale is . . . more than a little pleas'd with his being Patron of such a Seat of the Muses, saying that he expresst at first some kind of concern, whether it was well in him, being a Church man, to promote an Academy of Dissenters. But when we had discourst that point freely, he appear'd convinc't that the buisness of good men is to spread religion and learning among mankind without being too fondly attach't to particular Tenets, about which the World never was, nor ever will be agreed.[39]

Elihu Yale was not Dummer's only victim; no Englishman with a book or a guinea was safe from his importunities, although the library of one Dr. William Salmon was saved: "An Apoplexy took him off before he had time to make a new will." [40] Dummer induced Dick Steele to present complete sets of the *Tatler* and *Spectator*, persuaded Sir Isaac Newton and Blackmore, the poet laureate, to give some of their works, procured a choice volume from the library of Halley, and so on. It was a most remarkable collection of 700 volumes which Dummer sent to Saybrook for the Collegiate School.

In 1720 Thomas Hollis wrote to Benjamin Colman of an attempt to intercept the Hollis bounty in the interest of Yale: "Mr Dummer, I think your agent, sent for me yesterday to a coffee-house, said he had a letter to show me, mentioning thanks

[37] 6 *Coll. M.H.S.* VIII, 181 (Belcher Papers).
[38] *Gentleman's Magazine*, II, 927.
[39] Franklin B. Dexter, *Documentary History of Yale University* (New Haven, 1916), p. 193.
[40] Half of this library was to go to Harvard, Dummer explained later.

for my bounty to your College, but had mislaid it, and acquainted me about a College, building at New Haven, which he proposed to my bounty. I could only answer, I had not heard of it before." [41] Dummer apparently heard from his Cambridge friends, for he wrote Timothy Woodbridge (A.B. 1675): "Mr Hollis has given me Some hopes that he will think of you when he has finish't what he intends to do for Harvard Colledge, which I'le do every thing in my power to promote, though I've receiv'd very Severe reprimands from some of my friends in Boston for having made application to him." [42] To Benjamin Colman he explained his position:

I have your repremand for Speaking to Mr Hollis in favour of Yale Colledge, which however I thought my Self oblig'd to do, and I fancy you your selfe will acquit me, when I tell you what obligations I am under to Serve that Colledge. When the Noble Legacy of Mr Hopkins was to be paid here, Their Trustees sent me a power of Attorney to assert and maintain at law their right to at least part of it. . . . I consider'd that a Suit in Chancery would Swallow up the Whole legacy before it could be ended . . . therefore I prevail'd with the Trustees of Yale Colledge to drop the Suit, upon a promise that I would endeavour to make up the loss to them Some Other way. I'm sure this was a good Service to Harvard Colledge and therefore their Sons should not grudge the little Services which I am Capable of doing the Collidge in Connecticut. . . . Tis true the Library at New Haven is something better than that at Cambridge, but then you are supported and your buildings enlarg'd by a powerful province and a great Patron [Hollis].[43]

Hollis was disgusted, suspected a "snake in the grass," and wrote Colman that Dummer seemed to boast that in a little time that nursery (Yale) would exceed Cambridge: "he has such a vollible tongue, which takes with some People, it is not fitt to disgust him, but rather to treat him civilly, because he may be of some use to you." [44]

Dummer's catholic tastes in literature had an unexpected result. He had selected the library for Yale College with a regard for quality and a disregard of sect, naturally including the works of the great Anglicans. From these seeds the great Anglican revival of Connecticut sprouted. Malignant motives were then discerned in Dummer's conduct, for, it was said, he had

[41] Josiah Quincy, *History of Harvard University* (Cambridge, 1840), I, 527.
[42] *Publ. C.S.M.* VI, 189.   [43] Colman Mss., M. H. S., Jan. 16, 1719/20.
[44] *Publ. C.S.M.* VI, 211; XXV, 7.

early conformed to the Church of England.[45] He was in later years restored to Yale's Pantheon and his benefactions sung in what was appropriately called a poetical "Attempt."

> By *Dummer* nurs'd as by a Patron's care,
> Still *Science* grows and grows divinely fair:
> His opening hand her num'rous wants supplies
> And next to Heav'n on *that* her hope relies.
> *Yalensia's* Sons in gen'rous *Dummer* find
> *Mecænas's* bounty and great Tully's mind.[46]

One curious happening throws light upon Dummer's character. On one occasion when attending service in the great church of St. Sulpice at Paris he arose and engaged in a Latin debate with the Rector of the Sorbonne on the subject of the invocation of saints:

I told the Jesuit, before I propos'd my Arguments, that I was sensible of the *Impar Congressus* between him, a profound Doctor in Theology at the head of the Learnedest University in Europe, and my Selfe an Itinerant Layman, who had receiv'd my birth and Education in the wilds of America; But that I was firmly perswaded of the goodness of my cause, which alone gave me the Courage to enter the lists with him. Nor should I have done it nevertheless, if he had not from the Pulpit invited any person in the Audience who was dissatisfy'd with his doctrine to oppose him. Nor perhaps then neither if Sir Biby Lake, who sate on the one side of me, and a Learned Swede of my Acquaintance, who Sate on the Other side of me, had not forc'd me up, and then I did not know how to sit down again; for as soon as I rose The Jesuit fix't his eye upon me, and the whole Audience Seem'd to expect Something.[47]

The Jesuit, who probably guessed that this backwoods theologian was concealing something like a Utrecht Ph.D., expressed his desire to conclude the controversy in private.

We have another and very different glimpse of Dummer. The Massachusetts Historical Society has a copy of his *Discourse on the Holiness of the Sabbath Day* upon which is a manuscript note in the hand of Jeremy Belknap (A.B. 1762). Referring to the author he writes:

In his latter days he grew a Libertine and kept a Seraglio of Misses around him to whom he was lavish of his favours.[48] Col. S — who was

---

[45] *Id.* VI, 195.   [46] *Doc. Hist. Yale*, p. 301.

[47] *Publ. C.S.M.* VI, 192–3. There is a different account in the *Boston News-Letter*, May 1–8, 1721.

[48] Are these the "nymphs of Plaistow fields" of whom he sings in the *Gentleman's Magazine*, III (1733), 490–1, 546?

in England in 1738 went to wait upon him at his Seat in Plaistow on a Sunday after Church and found him with his Ladies sitting round a Table after dinner drinking Raspberry Punch. As he entered the Room he observed a confusion in Mr Dummer's countenance and the Girls fled out of the Door like Sheep — almost over one another's back. At another Time, I think on Saturday Evening the same Gentleman was in company with Mr D. and a number of other New England Gentlemen. After much gay Discourse and when the Company were grown merry with Wine, Jeremiah Allen, who had an excellent Memory begged Leave to entertain the Company with part of a New England Sermon which he had formerly read and then repeated *verbatim* some of the most striking Passages of this Discourse on the Sabbath beginning at p. 24, Mr D. was struck dumb with astonishment and unfitted for any further enjoyment that Evening and the recollection of it worried him ever after. . . . The above I had from Col. S — own Mouth Aug. 27, 1776.[49]

John Eliot (A.B. 1772), to whom similar sources of information must have been open, said that Dummer "could only bring his views to a state of forlorn scepticism, and was never able to fix his mind in infidelity. Amidst scenes of dissipation, he had some reflections which prevented him from enjoying what *commonly gives delight to the sons of men*, and confessed to a friend that he wished to feel what he once experienced, when he was a pious man in New England, without any great expectations, and had not other desire than to settle in the ministry of the gospel." [50] In spite of the fact that Dummer so far departed from the tenets of New England Puritanism, it was generally admitted that he was the most brilliant man whom the English Colonies produced in that generation. His English friends said that "He had an elegant Taste both in Men and Books, and was a Person of excellent Learning, solid Judgment, and polite Conversation, without the least Tincture of Political or Religious Bigotry." [51] He died at Plaistow on May 19, 1739, and was buried at West Ham, Essex, under a monument which announced that he was "distinguished by his excellent life probity and humanity." [52]

[49] Also printed in 6 *Coll. M.H.S.* I, 305 n. Mr. D. H. Mugridge, who has corrected many errors in this sketch, doubts the story.

[50] *A Biographical Dictionary* (Salem and Boston, 1809), p. 165.

[51] *Gentleman's Magazine*, IX, 273. An obituary notice is printed in the *Boston News-Letter*, Aug. 23, 1739.

[52] *Publ. C.S.M.* VI, 174 n. The Massachusetts Historical Society owns an invitation to his funeral, a curious document 10 × 8¾ inches, with the particulars type-set in a ready-engraved print.

Dummer's will reflects the man:

I do on this solemn occasion commend my soul to Almighty God
... firmly confiding in the Benignity of his Nature that he wont
afflict me in another World for some follys I have committed in this,
in common with the rest of mankind, but rather that he will graciously
consider the frail and weak frame which he gave me and remember
that I was but Dust.

As to the Interment of my body I should think it a trifle not worth
mentioning but only desire my executors kindly to invite to my funeral
all such New England gentlemen as shall be in London at the time of
my decease and to give each of them a twenty shilling ring without
any name upon it but only this motto which I think affords a good
deal of reflection — *Nulla retro via*.

I bequeath ... to my brother Dummer of Newbury twenty pounds
New England money to distribute among the poor Indian Squaws
that may come a begging at his door in the country.[53]

A portrait of Dummer formerly attributed to Kneller, here
reproduced, is owned by the Yale University Art Gallery.

THOMAS WELLS

THOMAS WELLS, minister of Amesbury, was born at Ipswich
January 11, 1646/7,[1] the youngest son of Deacon Thomas
Wells, a leading inhabitant of Ipswich and a member of the
Ancient and Honourable Artillery Company. His mother was
Abigail, the daughter of William Warner. Young Tom had a
normal boyhood. He constructed a "litle chist & table" and
cast a longing eye at his father's "chist planks, to make him a
chist on."[2] But he was fitted for a gentler profession; his
father bought books for him and sent him to sit under the rod
of Thomas Andrews at the Ipswich grammar school. Perhaps
he was not a pious youth, for his father thought of sending him
to one "Mr. Allcocke" to learn the profession of pills and
plasters. Thomas may have attended the College for a time;

---

[53] *N. E. Hist. Gen. Reg.* XLI, 57–8.

[1] A later date is sometimes given, but it can not be reconciled with his age as given in
affidavits.

[2] *Probate Records of Essex County* (Essex Institute, 1916–20), II, 67.

the death of his father in 1666 would explain his leaving without a degree. His inheritance included £250 in property, payable at various future dates, his father's "three phisicke bookes & the booke called the orthodox evangelist, the great sermon booke, & Hyelings Geogripha . . . sworde & scabitt . . . fire locke musket, with a square barrell." All of this was "toward his chardges, of his goeinge to the Colledge." [3]

Thus Thomas Wells Jr. was launched upon the world equipped to combat Indians, disease, or the devil. He chose the last of these professions. He may have begun preaching at Newbury [4] shortly after his father's death; but his steps are hard to trace. The various Thomas Wellses of the time showed a very annoying predilection for having brothers named John and wives named Mary. Our Thomas may or may not have been one of the individuals of that name who kept the quarterly courts busy by such activities as calling a neighbor a "dammed wrech & a lim of the Divell," or wiping their muddy feet on the neighbors' clean linen.[5] But he was certainly the Thomas Wells who in 1667 bought a plot of land in Wells, Maine, of a Newbury man, and another of a Cape Porpoise man two years later. His father had owned considerable property in Wells, which brother John had inherited.

On January 10, 1669, Thomas Wells took to wife Mary Perkins, daughter of John and Elizabeth of Ipswich.[6] The young couple immediately went to the Eastward, where Mr. Wells may have preached on the Isles of Shoals. He was mentioned as minister of Kittery in October, 1670, when he drew up the will of a fisherman from the Isles and received a bequest of "one pound tenn shillings . . . one halfe In money & the other halfe In fish." [7] After preaching at Kittery for about a year he was forced out by the activities of a gentleman who disliked parsons.[8]

Mr. Wells's next step is not known, but in October, 1671,

---

[3] *Id.* 65–8.

[4] Joseph Merrill, *History of Amesbury* (Haverhill, 1880), p. 196.

[5] *Records and Files of the Quarterly Courts of Essex County* (E. I., 1911–19), IV, *passim.*

[6] *Vital Records of Ipswich* (E. I., 1910–19), II, 450.

[7] *Probate Records of the Province of New Hampshire* (*State Papers Series*, XXXI–XXXIII), I, 121–2.

[8] *Documentary History of Maine* (2 Coll. Me. Hist. Soc.), IV, 340.

he was sued in the Hampton quarterly court "for withholding 50s pay due for work done about a frame, with divers years' forbearance." [9] He gave bond. The following year an opportunity opened in the recently incorporated town of Amesbury, which was trying to settle one "Mr. Hoberd," perhaps one of the Hobart boys of the Class of 1667. The candidate rejected the proffered salary of 40l a year, and the town would offer no more. On May 11, 1672, the town appointed a committee "to see if they can obtain Mr Weels to be helpfull . . . in the work of the Ministry." [10] Mr Wells had independent means; so he could afford to accept. The town voted to furnish his firewood and to pay his widow 50l if he died in office. In spite of some opposition it was voted to build him a house "fower and forty foot long." The following year the town repented this generosity, and, remembering a house which it had purchased for a parsonage some time before, offered to give him that and to move it to a site that suited him. He rejected the offer and built for himself. The house is now gone, but within the memory of living men lilacs still bloomed around the cellar hole.

In due time Mr. Wells was ordained and admitted a commoner —"resaived a towns man amongst us." [11] John Weed gave him three acres of land, "Always provided that the said Weells doe not voluntarily desert the work of the ministry in Amsbury." [12] The town was less generous, and arguments over the parsonage land resulted. Perhaps the parson had ideas of leaving, for he bought more land in Wells. However, in 1681 he began to buy land in Amesbury, and a few years later he sold his Maine property. He accumulated horses as well as land, but from the description of them in the records, the town historian does not think that the parson's judgment was of the best.

The Weed family was generous with other things than property, according to a complaint which the parson lodged with the quarterly court in October, 1677. It seems that one evening in his absence there was a gathering of townspeople in the parsonage. A quantity of liquor was staked on the question of

9 *Essex Quarterly Courts*, IV, 428.
10 Charles K. Wells, *Genealogy of the Wells Family* (Milwaukee, 1874), p. 39.
11 Merrill, p. 100.
12 *E. I. Hist. Coll.* LX, 232.

whether or not Samuel Weed was bold enough to kiss the parson's wife, who was asleep in bed with her latest baby. Mr. Weed took the bet, and John Colby took a light and went along to make sure that the liquor was earned. Mrs. Wells, who was awakened, said that it was. The court fined Mr. Weed and adjudged the others present "to be either abettors or countinansers of whatt was then acted," and fined them accordingly.[13]

In 1680 Mr. Wells's salary was increased 10*l*, but as it was paid in provisions and goods and the Boston market was far away, he found it very difficult to get the few imported objects which the family needed. When approached on the subject several years later, the town refused to add anything "tow Mr Wells is Mayntainance," but allowed him "fifty shillings in money out of the fifty pounds which they formerly granted him." [14]

Once a colonial minister was ordained, it was felt to be a disgrace to him and to the church if he ever dissolved the connection. This security of office tended to make some parsons lazy. Apparently Amesbury resolved to keep Mr. Wells in doubt and working hard, for it adopted the amazing procedure of voting him into office annually. The town meeting would vote that "the towne was cleare from Mr. Wells, and Mr. Wells from the towne," and would send to him "to know his mind whether he would still continue in the work of the ministry." It then would solemnly resolve that he should be the minister. In 1693 the townspeople voted that they "ware willing to have Mr Wells," which was a little too insulting. He replied tartly that he would a great deal rather that they supply themselves with a better parson. He offered to take a salary cut because of the burdens of war, but he insisted that his contract stand without annual renewal: "iff you are not disposed to comply with the terms above propounded I only request you never to trouble mee more with proposals upon this account Just as you please." The chastened town voted to "comply with Mr. Wells in the severall pertickelers," and that was the end of the matter.[15]

13 *Essex Quarterly Courts*, VI, 428; VII, 283.
14 Merrill, p. 111.
15 *Id.* 118–9, 124–5.

Amesbury was a frontier town, and despite the garrison houses which studded it, many members of the congregation fell under the tomahawk. Violent death was so commonplace that when a leading man of the town was captured and put to death in the lingering fashion favored by the Indians, Mr. Wells entered in his book of births and deaths only the date of his taking-off. Still the people held stubbornly to their farms. The enemy and bad roads made coming to church a burden for those living in outlying districts; so the town voted that the parson go out to preach to them every third Sunday. Considering the danger one wonders if they had in mind the possibility of a less expensive preacher.

Indian warfare was not the only weapon which the devil used against the Amesbury flock; a woman went to the scaffold for witchcraft, following the path laid out by her own shrewish tongue. There is no record that Mr. Wells had any more to do with this affair than to appoint a day of fasting for one of the witch's victims.

During the war the parson undertook to increase his burdens considerably and his income by 20l a year by teaching the town school, contracting "to teach all persons that belong to the town that shall attend the school at any time, except such lettell ons as cannot say there a b c." [16]

By the close of the long struggle the Wells family fortunes seem to have suffered considerably. Several of the minister's relatives had been killed and much of their property destroyed. In 1716 he asked the town for a piece of land and had some dispute with it in regard to his firewood. The next year it gave him the old meetinghouse.

Even in the midst of these difficulties the population of the town increased until Mr. Wells had to ask permission to build his family a pew outside the meetinghouse: "These request you to grant mee the liberty to build a place fer the use of my family on the outside of the meeting house betwext the south doore & south wast corner thereof & to make a convenient opening to the congregation not damnifying the meeting house nor any room or seat or place therein. . . ." [17] When the new meetinghouse was completed and another projected for Jamaco

---

[16] *Id.* 137.　　　　　　　　　　　[17] *Id.* 141.

parish, he asked permission to build his family pews in such places and dimensions as the church directed "provided it may not appear to be a ridiculous mockery." In return for this permission he gave the town the arrears of his salary "from the beginning of the world" to 1714.[18]

At Commencement, 1703, Mr. Wells was created Master of Arts. His name does not appear on the *quaestio* sheet with those bachelors who qualified for the second degree by presenting a dissertation, but "Mr. Belcher[19] of Newbury Testified his Education under Mr. Andros at Ipswich, that he was a good Latin and Greek Scholar."[20]

Four years later Mr. Wells joined the younger and more liberal members of the clergy in supporting John Leverett for the Presidency of the College. But he himself was passing into the class of venerable clergymen as years of obscurity flowed by. Finally the town called a meeting to "make choyce of men to procure an orthodox Schooller to assist our reverent Mr Wells Minister in the work of the Ministry."[21] No assistant was procured at the time, but the western part of the town was set up as a separate parish and Paine Wingate, A.B. 1723, ordained over it in 1726. As was almost always the case, the two parishes fell to quarrelling over the division of the parsonage land and the burden of supporting the elder minister in his declining years. The east parish sued the west, which appealed to the General Court.[22] Finally the west parish paid Mr. Wells some 23*l*, but precisely from what obligation this freed them we do not know.

There is no indication that "the ancient Minister" was interested in these bickerings. He lost his wife on January 26, 1726/7. A little over a year later, unable now to carry the burden of even the east parish, he was given Edmund March, A.B. 1722, as a colleague. On July 10, 1734, he died. The inscription on his gravestone with (we trust) unconscious skepti-

---

[18] *Id.* 171.

[19] Samuel, A.B. 1659, who may have known him when preaching on the Isles of Shoals.

[20] Samuel Sewall, *Diary* (5 *Coll. M. H. S.* v–vII), II, 81. "Mr. Andros" was Thomas Andrews, his old schoolmaster.

[21] Merrill, p. 178.

[22] *Acts and Resolves of Province of Massachusetts Bay* (Boston, 1869–1922), xI, 500.

cism reads that "having served his Generation by the will of GOD, he fell on sleep, and (we trust) enjoyt a Prophets reward."[23] None of his sons attended the College.[24]

TIMOTHY CUTLER

TIMOTHY CUTLER, Rector of Yale and minister of Christ Church in Boston, was born at Charlestown May 31, 1684, and baptized in the Congregational church the following day. He was the fifth child of Major John Cutler, who was an anchorsmith, an Andros adherent, and one of the larger taxpayers of Charlestown. His mother's maiden name was Martha Wiswall. At the College Timothy was either frugal or frequently absent from commons, but he paid a few more fines than the average student. After commencing A.B. he returned, rented a study and a cellar, and remained in residence for six months. Two of his college texts survive — Heereboord's *Meletemata* at the Boston Public Library and Keckermann's *Logica* at the American Antiquarian Society. Sir Cutler reappeared to take his second degree in course, holding the negative of the question, "An Detur in Deo Scientia Media?"

Joseph Parsons, A.B. 1697, told Increase Mather some scandalous things about Timothy Cutler, but the dean of the New England clergy refused to believe them,[1] and the College indicated a good opinion of him when it forgave him a bill of 16s. He was admitted to the Charlestown church on Septem-

---

[23] Charles K. Wells, p. 40. *The New-England Weekly Journal* for July 22, 1734, says that he died on the 11th.

[24] For Wells's children see David W. Hoyt, *Old Families of Salisbury and Amesbury* (Providence, 1897–1916), *passim*.

[1] *Proc. Am. Antiq. Soc.* n.s. XIV, 313.

ber 30, 1705. "Under the advantage of a good natural capacity, joined with singular diligence and application, he made an early proficiency in useful literature and was accordingly soon distinguished as likely to become eminent in his profession." [2] Consequently the Governor and Council offered him the pulpit at Dartmouth, Massachusetts, an offer which he promptly and wisely rejected, for the idea was to force a Congregational minister upon an unwilling town.

The following year Mr. Cutler was chosen for a more delicate missionary task. The Church of England had obtained a small following in the town of Stratford, Connecticut, and had corrupted the theological principles of the local Congregational minister, John Read, A.B. 1697. Worried by this breach in the dike of New England orthodoxy, the ministers of the two colonies consulted together and "determined that one of the best preachers that both colonies could afford should be sought out and sent there," and Mr. Cutler "was accordingly pitched upon." [3] He accepted the call, sold the Charlestown property which he had inherited from his father, and set out for the front. The town of Stratford, which for some time had been unable to agree upon a minister, was charmed by him and on September 16, 1709, voted "103 in favor and none against" "for the continuance of Mr. Cutler . . . in the work of the ministry in order for a settlement," and offered him a salary of 80*l*. [4] He accepted and terms were arranged, "All, provided his disciplining be agreeable with the way of the Colony or country at present or future." [5] On January 11, 1709/10, he was ordained. Of course a wife was essential for a parson; so on March 21, 1710/11, he took to himself Elizabeth, daughter of Samuel Andrew (A.B. 1675). [6] Of these experiences he wrote to his classmate, George Curwin of Salem:

Twice have I been neglected by you for some Months and altho I had reason to think I was Antiquated by you, because forsooth, I

[2] Henry Caner, *The Firm Belief* (Boston, 1765), p. 17.

[3] Col. Heathcote to Gen. Nicholson, Apr. 19, 1714, in *Documentary History of the Protestant Episcopal Church in . . . Connecticut* (New York, 1863), I, 49–50.

[4] Samuel Orcutt, *History of Stratford* (Fairfield County Hist. Soc., 1886), I, 229.

[5] *Ibid.*

[6] In a genealogical publication he is given Mary Dimond and Mary Gedney as wives. This is an error arising from the use of an annotated Triennial in which the wives of his classmate Cotton were carelessly entered on the wrong line.

TIMOTHY CUTLER

wear a Connecticut Wigg and have an Agreable Bow and Scrape, yet
I passed by these things, and as soon as ever I could hear that you was
alive wrote down to you, and when I could say nothing else I Entertained you with the tagg end of an old sermon: But notwithstanding
that of Late I have been Entering the Dangerous Depths of Matri-
mony, and Since Marryage have had the usuall troubles of it, as that
of the Pewking of my Spouse, care to get the Baby Clowts ready, to
get the chimnys Swept, to buy cradles and Closestools, etc. . . . And if
you please you may take this last Paragraph as an answer to the
Question you put to me about the conveniences & Inconveniences of
Marryage. . . . It is Pure Sport. I never thought there could be such
Intimate Fellowship and noble Happiness found in the world. May be
you may next Commencement See my wife down in a Black Silk
Gown Longing for Plum Cake, Canary wine, Silver Tankards, etc.
But however pray take my word at present that she hath the Softest
Lips in the world next to your wife or Dear. Pray Sir, if you are not
Marryed be not Discouraged by any thing that I may have said; tho'
you may be sure that if any man hath fallen into the Mire he would
have as many others fall in as he could. You tell me Nat. [Gookin,
A.B. 1703] is Marryed, I think he hath been Rude in pretending to such
things sooner than I, pray tell him of it and that if now he doth not
give way to his betters and let me get a Boy before him, I shall not
Easily be reconciled to him. (P. S.) Sir I begg a little more of your
Patience about your Books, and desire you the next time you write to
treat me with the Gravity you should use to a Marryed man.[7]

Although he did not get the boy that he expected, he found
plenty of use for the numerous purchases which he mentions,
for his wife presented him with twin girls.

Mr. Cutler probably got to Commencement in 1714, for two
weeks later he preached in Boston and prayed well enough to
please even Judge Sewall. It was presumably on this trip or on
one of John Checkley's visits to the Episcopalians in Stratford
that the two men met. Probably Mr. Cutler was attracted by
the brilliant mind and European polish of the older man. The
fact that Major John Cutler had been an Andros man may
have made it easier for his son to become friendly with a
Jacobite like Checkley. With Timothy Cutler as with William
Vesey (A.B. 1693), the son of another Andros adherent, there
was probably a certain predilection toward the Church of
England. Checkley undoubtedly fanned any smoldering
embers of doubt to be detected in the young parson's mind and

encouraged him to read widely. Perhaps a like part was played by Tutor Henry Flynt, who was suspected of Episcopalian leanings, and who in May, 1715, "lent Mr Timothy Cutler of Stratfield Dr Morrice [Sir William Morice?] his discourse of primitive Episcopacy." [8]

Still another assault on the poor minister's peace of mind came as a result of his friendship for the brilliant young Yale tutor, Samuel Johnson. Jeremiah Dummer (A.B. 1699) had gathered for Connecticut's infant college the best library to be obtained in England. It included, naturally, the works of the great Anglicans as well as those of the well-known dissenters. Johnson's mind, having consumed all that the New England theological course ordinarily contained, fell eagerly upon this rich feast. He invited his friend down from Stratford to enjoy it with him. So it happened that the omnivorous appetite for learning that caused Timothy Cutler to stand out from his puritan fellows and to be chosen to hold the critical parish against the forces of Episcopacy led him into the enemy's camp.

In 1717 the Connecticut government did the young Stratford minister the honor of inviting him to preach the election sermon. He gave a long and practical discourse, one paragraph of which reflects the doubts in his mind.[9] A few months after the delivery of this sermon, Yale College held its first Commencement at New Haven. Mr. Cutler was interested in the vicissitudes of the institution if for no other reason than that his brother-in-law, Jonathan Law (A.B. 1695), was one of its chief backers, while his father-in-law, Samuel Andrew, was acting Rector. In the spring of 1719 some of the scholars remained at Wethersfield under Elisha Williams (A.B. 1711), and only seventeen were studying under Johnson at New Haven. Rector Andrew was anxious to be released from his duties, and it was perhaps at his suggestion that the trustees voted, on March 11, to ask Mr. Cutler to assume the post until their June meeting. They then informed the Council that they thought him "a person of those qualifications that they could not but think him very proper to take charge of the tui-

---

[8] Henry Flynt, Ms. Diary, II (M. H. S.), 392.
[9] *The Firm Union of a People Represented* (New London, 1717), p. 52. Note also the longing for rest from Protestant searchings, in his *Depth of Divine Thoughts* (New London, 1720), p. 35.

tion and government of the students." The Council replied "that if the trustees could by any means obtain the Rev. Mr. Cutler . . . it would prove an expedient universally acceptable to the Colony, as it was to this board, and to all persons who have been under any uneasiness respecting the state of that college. . . ." [10]

Perhaps it was the lure of the Dummer books; at any rate Mr. Cutler had taken over the instruction of one of the classes before the month was out. There was no regular meeting of the trustees in June, but the few who convened asked him to continue, and later some of the absentees expressed their approval. The Hartford ministers, who had opposed settling the college at New Haven, had no reason to dislike him, and reconciled themselves to the situation. The students studying at Wethersfield came down, increasing the student body to thirty-five or forty. Jonathan Edwards, who was one of these new arrivals, found the situation quite satisfactory:

> Mr. Cutler is extraordinarily courteous to us, has a very good spirit of government, keeps the school in excellent order, seems to increase in learning, is loved and respected by all who are under him, and when he is spoken of in the school or town, he generally has the title of President.[11]

At the commencement meeting of trustees, September 9, 1719, his appointment was formally approved, although not by unanimous vote. When he was inaugurated, the founders of Yale sat back, weary but happy after their twenty-year struggle to bring the institution through a sickly infancy and adolescence. The College was now united, the warring factions had laid by their disputes, the students were housed in the magnificent new building, and a library unequalled in New England was ready for their use. And not least was the promise of the brilliant young man who was now Rector. Ezra Stiles said of him:

> Rector Cutler was an excellent Linguist — he was a great Hebrician & Orientalist. He had more Knowlege of the Arabic than I believe any man ever in New England before him, except President Chauncy and his Disciple the first Mr. Thatcher. Dr Cutler was a good Logi-

---

[10] *Public Records of Colony of Conn.* (Hartford, 1850–90), 1717–25, p. 101.
[11] Jonathan Edwards, *Works* (New York, 1830), I, 31; for Cutler's opinion of his brilliant student see *id.* 30–1.

cian, Geographer, & Rhetorician. In the Philosophy & Metaphysics & Ethics of his Day or juvenile Education he was great. He spoke Latin with Fluency & Dignity & with great Propriety of Pronunciation. He was a noble Latin Orator. . . . He was of a commanding Presence & Dignity in Government. He was a man of extensive Reading in the academic Sciences, Divinity & Ecclesiastical History. He was of an high, lofty, & despotic mien. He made a grand Figure as the Head of a College.[12]

Tutor Samuel Johnson shared the general attitude toward Mr. Cutler and was happy in studying and teaching with him. Thanks to the Dummer books these men effected a revolution in the Yale curriculum, lecturing on Newton and Locke, offering instruction in the new science and philosophy better than anything Mother Harvard could afford.

So eager was Rector Cutler to put this intellectual feast before the undergraduates that he failed to watch their physical diet. The food became so bad that the students signed an agreement to eat in commons no longer, and indulged in serious excesses.[13] Even worse, however, was in store for the College. Early in 1722 Cotton Mather heard that "Arminian books are cried up in Yale College for eloquence and the learning, and Calvinists despised for the contrary." [14] If this was surprising, the rumors that followed on its heels were incredible. Obadiah Ayers (A.B. 1710) took Judge Sewall aside, whispered in his ear, and pointed out a piece of property which, it was said, had been bought for the site of an Episcopal church in which Rector Cutler was to preach![15] And there were good grounds for the stories, for on August 20, the Reverend Mr. Pigot, a Church of England missionary, wrote home that he had lately had a conference with Mr. Cutler and five of the six ministers in the New Haven association who were determined to declare themselves professors of the Anglican doctrines as soon as they were assured of support from the mother country.[16]

Under these circumstances it was inevitable that the air should be charged with suspicion when the clergy gathered for

[12] Ezra Stiles, *Literary Diary* (Scribners, 1901), II, 339–40.

[13] Franklin B. Dexter, *Documentary History of Yale University* (Yale, 1916), pp. 210, 211.

[14] Franklin B. Dexter, *Biographical Sketches of the Graduates of Yale College* (Henry Holt and Yale, 1885–1912), I, 260.

[15] 1 *Proc. M. H. S.* XII, 378.

[16] *Doc. Hist. Ch. Conn.* I, 57.

Yale Commencement on September 12, 1722. Rector Cutler confirmed the rumors by ending his prayer with the Episcopal sentence, "And let all the people say, Amen." The following day he, the two tutors, and four ministers (including Jared Eliot, Class of 1703, and John Hart, Class of 1704) appeared in the library and announced to the trustees that they doubted the validity of presbyterian ordination and were considering seeking orders in the Church of England.

The news of the defection swept the colonies.[17] With one voice the orthodox majority "bewail'd the Connecticut Apostacie." Joseph Webb (A.B. 1684) expressed the general reaction: "I apprehend the axe is hereby laid to the root of our civil and sacred enjoyments; and a doleful gap opened for trouble and confusion in our Churches. . . . It is a very dark day with us; and we need pity, prayers and counsil." [18] The agony of the moment even drove one man to this:

> Oh! now alas, alas, what's come to pass
> In our Horizon?
> 'Tis strange for to tell, five Stars are now fell
> And a very great one.
> The famous great Rector, a fine Director,
> To prevent Schisms and Heresy:
> However was devout, is now turn'd about
> To Episcopacy. [19]

Mere conversion might have been taken less seriously, but Rector Cutler's case seemed to include duplicity and treachery of the worst sort. A reasonable interpretation of his statement to the trustees was that he had deliberately given up a pastorate in which his conversion, honestly confessed, would have done little harm to the New England system, for a position from which he might corrupt the Congregational church at its source and make a breach in the very foundations.[20] Over his own signature he gave Hollis, the benefactor of Harvard, the worst possible statement of his case: "I was never in judgment heartily with the Dissenters, but bore it patiently until a favorable opportunity offered. This has opened at Boston, and

[17] See particularly *New-England Courant*, Oct. 8, 1722; *Boston News-Letter*, Oct. 15, 1722.
[18] 2 *Coll. M. H. S.* II, 131.
[19] *New-England Courant*, Mar. 11, 1723 (No. 84).
[20] 2 *Coll. M. H. S.* IV, 297–301.

I now declare publicly what I before believed privately." [21]
Since Cutler's choice of the Church over Yale has been cited to
prove his sincerity, it should be pointed out, without any
question as to his motives, that the Christ Church berth
promised to be much more comfortable and remunerative; at
least three Harvard men had refused the New Haven rectorship
in recent years.

Governor Gurdon Saltonstall (A.B. 1684) proposed that the
converts and the orthodox clergy hold a friendly argument in
the college library in an effort to settle the affair. Of the famous
meeting which took place on October 16 our best account is
that by Samuel Johnson. From this it is evident that the
young rebels, armed with arguments that had been whetted
by generations of scholars of the Church of England, easily
sliced through the rusty armor of puritanism which had hung
untested in New England parsonages for nearly a century.
Routed by logic, the Congregationalists took refuge in a cloud
of meaningless rhetoric. Seeing this, Governor Saltonstall
quietly called off the contest. Later New England tradition
asserted that, seeing his side worsted, he came down from the
moderator's chair and put Cutler and his friends to the sword
of invincible logic. However, the important point is that
all the waverers but the Rector and the two tutors were brought
back into the fold by the combined influence of gentle persuasion and the uncertainties of the path to the Church of
England.

The next day after the debate, the trustees "excused" Mr.
Cutler from the rectorship.

In the meantime matters in Boston were progressing with
long strides. Two weeks before the debate a group of King's
Chapel, men had written Mr. Cutler promising that they would
care for his family and further his voyage to England to obtain
ordination.[22]

The Reverend Mr. Henry Harris, Assistant Rector of King's
Chapel, was by no means favorable to the project. English born
and educated, a Whig, a liberal in theology, and a friend of the

[21] Josiah Quincy, *History of Harvard University* (Cambridge, 1840), 1, 365. There is
other damning material in Hollis Letters and Papers (Harvard College Library), No.
35, p. 6; Hollis' Letters to Leverett (H. C. L.), Aug. 16, 1723.
[22] Calvin R. Batchelder, *History of the Eastern Diocese* (Claremont, 1876), 1, 507-9.

dissenters, whom he hoped to convert by kindness, his particular enemy was John Checkley, who not only refused to take the oath of allegiance to King George, but persisted in throwing theological vinegar on the dissenters. At the news of the affair at Yale, Mr. Harris wrote to the Bishop of London that Checkley was responsible for the Rector's conversion, that he had "plyed him with such irresistible arguments as compelled him to declare for the Church of England upon Jacobite principles namely, the invalidity or nullity of the Baptism & other ordinances administered by the Dissenters." Mr. Harris added that he "had a great deal of reason to believe that the chief motive of this person's conversion was the prospect of a new Church in this Town." [23] The rank and file of the Church of England in Boston were prepared to welcome the converts and took up a subscription to pay their passage.[24]

A few days after the Rector and faculty of Yale had been released, they set out for Boston. At Bristol they attended a Church of England service for the first time. The Rhode Island Episcopalians were pleasantly impressed by them, and the wardens and vestry of Newport wrote to the Society hailing their defection as portending "a grand revolution, if not a general revolt." [25] At Boston the pilgrims were given a hearty welcome. On November 4 they partook of holy communion, and the following day they sailed. In these affairs John Checkley appears only once, and then in the uncertain light of a doggerel verse:

> But shall not we poor Checklies cause lament,
> Where in Newhaven and five Towns three Weeks spent;
> Who with the Devil five Churches near rent;
> Yet his Master Cutler when on board he went,
> No Thank-offering to his faithful Servant sent;
> But let Checklie yet a while wait and be content,
> His great Master after a while will give him Cent per Cent.
> Unless the Godly pray that God will give him to repent.
> When Rector Cutler aboard going so fast,
> Unto his Friend he said, Oh that it were with me as in Months past! [26]

[23] *Papers Relating to the History of the Church in Massachusetts* (n.p., 1873), p. 157.
[24] Henry W. Foote, *Annals of King's Chapel* (Boston, 1882–96), I, 317.
[25] *Doc. Hist. Ch. Conn.* I, 90–1.
[26] *New-England Courant*, Mar. 11, 1723 (No. 84).

After a rough passage Mr. Cutler and his friends landed at Ramsgate on December 15 and proceeded to Canterbury. Finding that they would have to wait three days for a London coach, they visited the cathedral, where both architecture and service awed them. They had no letters of introduction to Dean Stanhope; so it was with trepidation that they presented themselves at the deanery. It so happened that the cathedral chapter at that very moment had gathered to hear the news-paper accounts of the happenings at Yale. From that point the New England pilgrims' progress became a triumphal procession in which they were both the victors and the spoils.[27] One man in England who did not greet Mr. Cutler and his friends with open arms was Jeremiah Dummer, agent of Massachusetts and Connecticut, who had collected that fatal library for Yale. He had been volubly explaining to the New Englanders that his intentions had been entirely honorable, and he resolved not to color their suspicions by introducing the pilgrims to bishops or other people of influence.[28]

The mission was delayed by smallpox, which brought Mr. Cutler low and carried off Mr. Browne. At the end of March, 1723, the two survivors were ordained in the parish church of St. Martins-in-the-Fields by Dr. Green, Bishop of Norwich. Mr. Johnson recorded in his autobiography:

In the beginning of May Mr. Cutler and he pursued their visit to Oxford. Their great friend Dr. Astry, who had long resided and had much influence at Oxford had of his own accord procured their degrees to be passed and diplomas prepared ... which were presented to them by the Vice Chancellor, Dr. Shippen, upon their arrival with much respect; and they were treated by the heads and fellows of the house in the kindest manner and shown every thing curious in the Bodleian Library and each college. . . . After a fortnight most agreeably spent they returned to London and in the beginning of June they visited Cambridge . . . where they were admitted to the same degrees and treated in a like respectful manner as at Oxford. . . .[29]

Twenty-five years before when William Vesey (A.B. 1693) received a degree by diploma from Oxford, it was that of Master of Arts, but on this occasion the ancient university signalized

[27] Samuel Johnson, *Works* (Columbia Univ., 1929), I, 16–17.
[28] *Publications Col. Soc. Mass.* VI, 194–6.
[29] Johnson, I, 18.

her pleasure by conferring on Mr. Cutler the highest degree within her power to grant, *Sacrae Theologiae Doctor*, commonly Englished as D.D. Cutler was the first Harvard man to obtain an English Doctorate of Divinity. Later in the month he was licensed by the Bishop of London to officiate in Massachusetts.

Supplied with the D.D. and a chair which had belonged to Dean Berkeley (and is now in the possession of the Massachusetts Historical Society), Dr. Cutler set sail for home on July 26, 1723, with his friends Johnson and Checkley. After a voyage of two months they made land at Piscataqua,[30] whence Dr. Cutler rode to Boston "in the great Rain." [31]

The Cutlers took up their residence at the corner of Salem Street and Love Lane (Tileston Street). The Doctor found that the corner stone of Christ Church had been laid some months before. On December 29 he was inducted at the first service in this, the "Old North" of Longfellow's poem, now the oldest church building in Boston. The congregation allowed him 3*l* (Massachusetts paper money) a week from the collection, and the Society for Propagating the Gospel paid him £60 (sterling) a year, which was much more than his Congregational rivals usually enjoyed. Nevertheless he early began to petition Christ Church and the S. P. G. for more, urging the high cost of living and the fact that for appearances' sake he had not asked for 90*l* due him from Yale.[32]

The congregation of Christ Church included some notable and pleasant people, such as Thomas Greaves, A.B. 1703, who had been the childhood playmate of Rector Cutler.[33] To all Episcopal churches, however, as to the Quaker and Baptist meetings, there flocked the atheists, deists, rate-dodgers, and those who manfully objected to the moral scrutiny of the dissenting churches or longed for the various fleshpots of the less rigid society of Europe. The Rector's flock was described as "a Little sorry, Scandalous Drove, which have Little but Baseness and Impiety and Jacobism to distinguish them." [34]

[30] *Boston News-Letter*, Sept. 26, 1723.
[31] Samuel Sewall, *Diary* (5 *Coll. M. H. S.* v–vii), iii, 326.
[32] *Doc. Hist. Ch. Conn.* i, 60–1.
[33] Timothy Cutler, *The Good and Faithful Servant* (Boston, 1747), p. (5).
[34] Cotton Mather, *Diary* (7 *Coll. M. H. S.* vii–viii), ii, 804.

It was distinctly more Jacobite than that at King's Chapel. A lady wrote to Rector Cutler:

I pray preach more on true conversion, and the life of Christianity, and not so much on passive obedience and non-resistance. Pray in your little prayer before the Sermon for King George and Royal Family, and for the Governor, as our Ministers do, and I will come often. I know many others of my mind; and I am sure your Church will be full.[35]

Cotton Mather was pleased to observe that the Rector's "high Flights" were discrediting the Church of England.[36]

Half of the Episcopal clergy of the vicinity regarded Rector Cutler with hostility. Henry Harris, David Mossom, Matthias Plant, and John Usher (A.B. 1719) could usually be relied upon to be whole-heartedly uncooperative in any project which he favored.[37] They opposed even his project for the appointment of a bishop for America, as unwise and an insult to the Bishop of London to whose see the colonies belonged.[38] Mr. Harris summed up his objections in a letter to the Bishop of London dated June 22, 1724:

My present opinion of him is that his behavior is so imprudent his notions so wild & extravagant & his principles so uncharitable that I may venture to affirm that the Church will never flourish under his care, the affections of the dissenters being entirely alienated from him, & there is not so much as one person of tolerable note & distinction whom he has brought off from the congregational persuasion. This is what I foresaw would be the issue of his management, & to shew my dislike of it I declined having any intimate conversation with him, lest his principles should be thought to be espoused by all of our communon & so the whole Church suffer thro' the indiscretion of one man.[39]

Three weeks after this letter was written, Dr. Cutler preached for the Reverend Mr. Myles (A.B. 1684) of King's Chapel in the morning service. In the afternoon sermon Mr. Harris took the Doctor to task for popery, plainly stating that the Church had changed its opinions in the course of a century or so, and

[35] John Nichols, *Illustrations of the Literary History of the Eighteenth Century* (London, 1817–58), IV, 279.

[36] Cotton Mather, *Diary*, II, 797.

[37] The letter quoted in Foote, I, 318, to illustrate their mercenary motives, was apparently written twenty years after this date. Cf. Joshua Coffin, *History of Newbury* (Boston, 1845), p. 381.

[38] *Paps. Rel. Hist. Ch. Mass.* pp. 142–4 ff.

[39] *Id.* 160–1.

that no Protestant could hold to the old doctrines.[40] This sent the factions again flying to the wearied Bishop of London. It happened that the see of London at that time was occupied by Edmund Gibson, an ardent Whig, Hanoverian, and latitudinarian. He let it be understood that the doctrines attributed to the Checkley-Cutler faction could not be tolerated.[41] It may have been this that led Dr. Cutler to point out that the Anglican clergy were not being led by Mr. Checkley.[42] The Doctor's friends came to his assistance, Samuel Johnson writing to Bishop Gibson that the whole reason for Mr. Harris's malicious persecution of him was the fact that Mr. Harris had coveted the Christ Church rectorship, and that his accusations of popery were quite unfounded.[43] According to Dr. Cutler, Mr. Harris was the chief fly in his ointment.

My Church grows faster than I expected, and while it doth so I won't be mortifyed by all the Lyes & Affronts they pelt me with. My great difficulty ariseth from another Quarter, and is owing to the covetous & malicious Spirit of a Clergyman in this Town, who in Lying & Villainy is a perfect over match for any dissenter that I know, and after all the Odium that He contracted heretofore among them, is fully reconciled & endeared to them by his Falsehood to the Church & Spite at Me. I have a clear Conscience towards Him & have tryed to gain Him, and for the peace of the Church have passt over many affronts that every body would not have thought supportable, & have stirred till he gave such a Vent to his Furious Malice that none but an ass would bear.[44]

This passage is a good example of the venom and scurrility in Cutler's letters (judiciously deleted by his Victorian editor) which explains why he could not live in peace and amity with the Congregational clergy, as did some of his Anglican brothers. He appeared on the scene when the distance between the New England churches and that of old England had appreciably narrowed. The Mathers praised the bishops of their day, and younger Congregational ministers like Benjamin Colman (A.B. 1692) were friendly with the liberal Anglican leaders. With the convert's zeal, however, Dr. Cutler attacked

[40] *Id.* 162-4.
[41] *Id.* 167; Cutler denied being a Jacobite, and tried to prove to the Bishop of London that Harris was one. Fulham Mss., Mass. Book II, No. 37 (L. of C. transcripts).
[42] Nichols, IV, 281.
[43] *Doc. Hist. Ch. Conn.* I, 97.
[44] Ms. Boston Public Library, G.380.27.

the New England system and leaded his whip with words of contempt and scorn:

> We are told Col. Shute is to come again over to us, which surely will be much better than if our fanatick Country prevail'd in their cause against Him. And certainly if He hath any sence in him, He will not think our Fanaticks worthy of any more of his Favours.[45]

He was reported to have declared, as no bishop would have done, "that, ordinarily, there was no salvation out of the communion of the episcopal church; and that none but an episcopally ordained minister could perform any religious offices, with validity and effect." [46]

The relations between Rector Cutler and the Congregational clergy were restricted to bad feelings and bad language until, in the spring of 1725, the Puritan ministers proposed to the General Court that a synod be called to consider the very real moral and religious difficulties with which the colonies were faced. Quite unexpectedly, Dr. Cutler and Rector Samuel Myles of King's Chapel protested to the General Court that such a synod would be illegal, and insinuated that its purpose was to formulate oppressive measures against the Anglicans.[47] The legislature rejected the Cutler-Myles petition as "an indecent reflection" on the Congregational ministers, and Benjamin Colman protested to Bishop Kennett of Peterborough that nothing was further from their minds than discrimination against the Episcopalians.[48] Cutler won a complete victory, however, when the royal law officers ruled that the calling for the synod would be a violation of the King's prerogative.

Having thus given the Congregational majority reason for anxiety lest their liberties go the way of those of the dissenters of England, Rector Cutler promptly gathered a convention of Episcopal ministers in Rhode Island, and in their name petitioned that a bishop might be appointed to reside in the colonies. Harris and his friend Mossom bitterly protested to the Bishop of London that the whole affair was simply Checkley-Cutler Jacobitism, and that such an appointment

---

[45] *Ibid.*
[46] Alden Bradford, *Jonathan Mayhew* (Boston, 1838), p. 382.
[47] *New-England Courant*, July 3, July 10, 1725.
[48] Ebenezer Turell, *Benjamin Colman* (Boston, 1749), pp. 136–41.

would destroy all hopes of reconciliation between dissenter and Anglican.[49]

Dr. Cutler's ideas of effecting a revolution in New England included the capture of Harvard for the Anglican minority. Of the College he wrote to a friend in the English Cambridge: "Here is a Snotty Town of the Same name where there are near 300 Scholars among whom a Churchman durst hardly say that his soul is his own, and I think it will never be well till that College become an Episcopal College, or we have one founded with us."[50] The Doctor did not propose to liberalize the College, but to restrict it; for its students had access to the conservative Anglican writers whose books he approved, and read under tutors who thought highly of such latitudinarian Episopalians as Bishop Tillotson, whose works were anathema to him. As early as December, 1725, his offensive movement against the established order had caused the Congregationalists to close their ranks and tighten their grip on the College. Benjamin Colman wrote to his friend, Bishop Kennett:

The Catholic Spirit . . . is the very Spirit of our College and has been so these forty Years past. . . . And such it has continued till of late a Parcel of High-flyers have poisoned and stagnated it, by leading us into a Course of angry Controversy which has alarmed and narrowed us, who before received the Writings and Gentlemen of the Church of England with the most open Reverence and Affection.[51]

According to the college charter, the "teaching elders" of six neighboring towns were automatically Overseers of Harvard College, and soon after Dr. Cutler's arrival, he was invited to sit with the other clergymen. But in 1725, after his hostility to the Congregational order had become apparent, the authorities of the College ceased to notify him of meetings. Neither he nor Rector Myles had taken the trouble to attend the Overseers' meetings when invited, but now he expressed a determination to die rather than surrender his right.[52] The Overseers, perhaps remembering that Oxford and Cambridge excluded dissenters from even the student body, now boldly declared that Anglican ministers had no right to sit because they had never even

[49] *Paps. Rel. Hist. Ch. Mass.* pp. 176–8, 200–2.
[50] Ms. B. P. L., G.380.27; Nichols, IV, 282–3.
[51] Turell, pp. 136–7.
[52] Nichols, IV, 284.

claimed to be "teaching elders." Dr. Cutler hired the best lawyer available, John Read (A.B. 1697), to take the matter to the General Court, where the decision was dictated by the Congregational affiliations of the majority. The angry Rector then tried to bring pressure to bear from the Bishop of London and the S. P. G., to whom he accused the College of using "all possible art consistent with safety & secrecy . . . to suppress any good inclinations in the Students towards our excellent Church," and he even attacked the sacred college charter.[53] He made efforts to get Dean Berkeley to set up his proposed college in Boston as a rival to Harvard, but in the end he was compelled to accept the favors of the Puritan institution. "Notwithstanding my struggles about it," he wrote to an English friend, "I have been forced to put my son under Dissenting tuition; but I must do them the justice to say that I know not that he suffers for my sake." [54] In 1730 he renewed his efforts to get a seat on the board of overseers and failed again.[55]

These experiences naturally made Dr. Cutler feel that Harvard and Yale were much inferior to the Anglican colleges, and in 1734 he wrote to a friend at the English Cambridge: "Excuse my sending you the paltry Theses & performances of our Commencement: In one of which you will see my Youngest Son's Name, and that of my Eldest in the Catalogue." [56] And again, "The Episcopal Students are much discouraged from coming to our Worship on Sundays, and it is to be feared will be restrained on this only pretence, That they are hindered in their Studies by it for which there appears no great Zeal and Exactness in other Particulars." [57] It was with pleasure that he reported that his old friend Johnson was sapping away at the Congregational foundations of Yale.[58] He hoped that an Anglican missionary would establish his church within a quarter of a mile of Yale College and ferret schism out of that nest of it.[59]

[53] *Paps. Rel. Hist. Ch. Mass.* pp. 210–1.
[54] Nichols, IV, 289.
[55] Ms. Overseers' Records (Harvard College Library), I, 92–112. Most of the documents relating to the case are printed in Quincy, I, 560–87.
[56] Cutler to Z. Gray, Ms. New York Public Library, Nov. 8, 1734; Ph. M. H. S.
[57] *Paps. Rel. Hist. Ch. Mass.* p. 330.
[58] Nichols, IV, 292–4.
[59] *Id.* 299.

If Rector Cutler was not successful in converting Harvard and Yale, he found a compensation in bringing into the world the Episcopal churches at Canton, Dedham, Stoughton, Braintree, and Scituate, and probably found no little satisfaction in the anxiety caused the Congregational churches in those towns.[60] Of his own congregation in these years, he reported: "My people do constantly and reverently attend the public worship, making some small deduction for common sailors, whereof I have great numbers and who are too much of the unthinking kind." [61] It was less the presence of unthinking sailors than his own domineering attitude which caused the internal difficulties from which Christ Church suffered. The wardens had been accustomed to call vestry meetings without previously consulting him, and were indignant when he denied their right to do so. They wrote their statement of the case into the church records with the provision that it must be read to each new minister.[62] Apparently within the pale of Christ Church there was some of that Independency against which the Rector was contending. Indeed it is a wonder that he got on as well as he did with his parishioners considering the egotistical and otherwise unpleasant attitude which he showed even toward his friends.[63]

In spite, however, of what Jonathan Edwards, who admired him, called his "haughty, stiff and morose" spirit, the colonial Anglicans had to recognize that he was their leader in the struggle against the New England system. If they were not concerned in his opposition to a synod or his efforts to obtain a seat on the board of overseers, they were concerned in his long battle to free them from taxation for the Congregational ministers.[64] In the midst of these struggles Mr. Myles suddenly died and in so doing created a problem which again split the Boston Anglicans. Mr. Harris naturally expected to succeed to the King's Chapel pulpit, and he complained to the Bishop of London that even before the old Rector's death the Cutler-

---

[60] For an amusing account of the resulting difficulties at Scituate, see the *Boston News-Letter*, Aug. 19, 1725. For his version see Nichols, IV, 275–8.

[61] *Paps. Rel. Hist. Ch. Mass.* p. 205.

[62] Foote, I, 425–6 n.

[63] See, for example, his *The Final Peace* (Boston, 1735), p. (i).

[64] For a detailed account of this struggle see Susan M. Reed, *Church and State in Massachusetts* (Univ. of Ill., 1914).

Checkley faction had been plotting to get a Roman-educated man for the place.[65]

Benjamin Colman, who so far tolerated the opinions of others that it was sometimes said he had none of his own, arose in Mr. Harris's defence and wrote to the Bishop of London that Mr. Cutler's faction was "a Jacobite . . . party in the Church, as inimical to the Kingdom of Christ, as to the reign of King George & the Protestant succession."[66] Even Rector William Vesey (A.B. 1693), of New York, whose story is such a curious parallel to that of Dr. Cutler, may have supported Mr. Harris, for the Doctor called the New Yorker "a very useless mortal," and said that his catechist was "very scurvily treated."[67] The Bishop finally gave the King's Chapel place to Roger Price, a neutral stranger.

Nothing better shows Dr. Cutler's "Jacobite" and un-New England attitude than the affixing of his signature to a certificate supporting Henry Phillips (A.B. 1724) after he had killed Benjamin Woodbridge in a duel on Boston Common. The Rector thus aligned himself with the royal officials and the English element against the Congregational clergy, who adhered to the more civilized attitude of the old puritanism.[68]

In his attitude toward theology, the Doctor was more uncompromisingly conservative than the majority of the puritan ministers. He was horrified as any at the religious laxity of Governor Burnet.[69] It seemed to him that the writings of Dr. John Clarke, the liberal Dean of Salisbury, were doing as much evil here as the works of the deist Henry Chubb. Thomas Woolston was as corrupting an influence among the Congregationalists as the *Independent Whig*. To the S. P. G. he reported: "Some have lately asserted that Hell torments will have an end, and that wicked men and devils will at last be saved thro' the goodness of God and the merits of Christ. I have privately borne witness against this corrupt doctrine."[70] But he had the consolation that "the state of

[65] *Paps. Rel. Hist. Ch. Mass.* pp. 246–7.
[66] *Id.* 249.
[67] Nichols, IV, 280.
[68] *Publications Col. Soc. Mass.* XXVI, 374.
[69] Nichols, IV, 287.
[70] *Paps. Rel. Hist. Ch. Mass.* p. 263.

New England" was "too near that of the barbarous ages to distinguish itself much in Infidelity." [71]

At the time of popular agitation against inoculation for smallpox, Dr. Cutler joined the puritan clergy in approving the practice. His letters to England told how he was howled at and called no Christian and no Churchman for having his family inoculated. The fact that the only notice that the town government ever took of him was to ask him to go on board ships suspected of carrying smallpox probably reflects a faith in the immunity which he had acquired in England rather than any hope arising from the religious difference between them. [72]

To a man who longed for the annulment of the Massachusetts charter as did Dr. Cutler — he more than once said that the vacating of it would be one of the most eminent blessings that the King could confer — the appointment of Governor Belcher (A.B. 1699), a native of the province, was a blow. A curious accident changed the situation soon after Belcher arrived. Rector Cutler had been at swords' points with Commissary Roger Price of King's Chapel, much to the distress of the Anglicans and the amusement of the dissenters. So when the Commissary rankly insulted the Governor, the incident naturally improved the latter's opinion of the Doctor. A year after his arrival, Belcher wrote to the Bishop of London that Mr. Cutler had acquired so much respect and reputation, not only by preaching but by his virtuous and regular life, that he had the good will of all who had the pleasure of his acquaintance. [73] Three years later when it seemed likely that Mr. Price would vacate the post of commissary, the Governor urged the Bishop of London to confer it upon the Doctor, praising him as "a gentleman of good learning, and of a good life," although somewhat given to "hierarchycal principles." [74]

Rector Cutler's temper improved with the passing years. He took little part in the brawl between the Anglicans and the Congregationalists in the thirties, although he referred to the champion of the latter as "Tom Foxcroft [A.B. 1714], a bitter creature." [75] Jonathan Edwards regarded him as "a man of

[71] Nichols, IV, 293.
[72] Id. 290.
[73] Paps. Rel. Hist. Ch. Mass. p. 271.
[74] Belcher Papers (6 Coll. M. H. S. VI–VII), II, 175–6.
[75] Paps. Rel. Hist. Ch. Mass. p. 673.

much sobriety and gravity and of more decent language" than the Mayhew faction.[76]

The Christ Church pastor was far from using sober and grave language when he described to his English friends the Great Awakening [77] and the subsequent "spread of Infidelity." [78] To the Bishop of London he bewailed that Whitefield had corrupted Benjamin Colman and many other leading dissenters. When Whitefield visited Boston in September, 1740, he met the Episcopal clergy at the home of Mr. Price, where Dr. Cutler argued with him, holding as a first premise "that the Church of England was the only true Apostalic Church," and that the Prayer Book was no less inspired Scripture than the Bible.[79] He described Gilbert Tennent as "a Monster! impudent & noisy," who "told them all they were, *damn'd, damn'd, damn'd*. This charm'd them!" [80]

The one good thing about the Awakening, in Dr. Cutler's opinion, was the comforting thought that its excesses would make the dissenters appreciate the solidity of the Church of England. And so it happened. On Christmas day, 1744, his church was thronged with hundreds of dissenters, and more followed as the coals of the Awakening cooled. Happy in his victory, he called the attention of the S. P. G. to the fact that Congregationalists were treating him kindly and respectfully, and even approving of his missionary preaching at Billerica. His last years would have been ones of peaceful contentment had not Methodism begun to multiply its "very awful & deformed" shape.[81] According to Rector Cutler so many of the dissenters were discouraged by the "Growth of Principles very unfriendly to Revelation" that they favored the appointment of a resident bishop, who would be a bulwark against heresy and excessive enthusiasm.[82] This encouraged him to circulate a petition in favor of such an appointment, a petition which fairly states the objections of the Congregationalists and lays down a

[76] Johnson, IV, 318.

[77] Nichols, IV, 298, *et passim*.

[78] *Paps. Rel. Hist. Ch. Mass.* p. 336.

[79] George Whitefield, *A Continuation of the Reverend Mr. Whitefield's Journal from . . . Savannah . . . to . . . his Departure from Stanford for New York . . .* (Boston, 1741), p. 49.

[80] Chamberlain Mss. (B. P. L.), A3.1-5.

[81] *Paps. Rel. Hist. Ch. Mass.* p. 439, *et passim*.

[82] *Id.* 433.

series of principles which, had they been acted upon, would
have fully safeguarded the interests of the non-Anglicans.[83]
Curiously enough the Doctor, who got on famously with that
rank dissenter, Governor Belcher, was not in the good graces
of the Anglican Governor Shirley, apparently because the latter
suspected him of disingenuous conduct.[84]

There is not room to discuss the more personal side of Dr.
Cutler's life, his befriending Nathan Prince (A.B. 1718) after
he had lost his tutorship, his unending efforts to place his sons
in good Anglican pulpits, or his bitter disappointment and
grief at the premature death of Timothy Jr. William Samuel
Johnson spoke of this side of him in a letter to his father, Dr.
Johnson:

> I am thankful to Dr. Cutler for his kindness, and am glad of the
> good opinion you entertain of him. He is certainly a gentleman of
> great capacity, learning and integrity and of a generous benevolent
> hospitality and you are secure of his friendship and the best good
> offices he can do if you behave with innocence and great modesty, and
> say nothing that savors of self-sufficiency and latitudinarianism.[85]

Rector Cutler gave himself to Christ Church without stint.
He obtained for it an organ, silver, bells, and a library which
it yet retains. But his last years were saddened by the state
of his congregation, which was never able to sustain the building and the minister without aid from the S. P. G. The neighborhood did not grow because it was chiefly divided into
large estates, whose owners frequently preferred to attend
King's Chapel. The wars, he complained, cut down the number
of his hearers, and an admonitory smallpox epidemic failed to
encourage attendance. To the Rector's account of the state of
his church in the S. P. G. report of 1756 is affixed this note:

> This, it is too probable is the last Account the Society will receive
> from this very worthy Divine, who was struck with the Palsy on his
> right Side in the latter End of the April following, and for some time
> his Death was daily expected; but by the last Accounts from Boston,
> he appears to be yet living, and somewhat better, and his Church is
> taken Care of by the Neighbouring Clergy.[86]

[83] Johnson, III, 240–1.
[84] *Paps. Rel. Hist. Ch. Mass.* pp. 400–1.
[85] Johnson, I, 122.
[86] Batchelder, I, 528.

The Rector lingered on until August 17, 1765.[87] The Anglican ministers who gathered to hear Henry Caner preach his
funeral sermon decided to institute an annual convention in
Boston. Through this convention Timothy Cutler in his
death as in his life contributed to the religious unrest of his
country, for the sight of "priests" in their robes in the streets
of Boston excited fears which helped drive the dissenting
clergy into the rebel ranks in the years that followed. Henry
Caner said little of the Rector's character, but John Eliot
(A.B. 1772), who knew many of his parishioners, observed:

He was haughty and overbearing in his manners; and to a stranger,
in the pulpit, appeared as a man fraught with pride. He never could
win the rising generation, because he found it so difficult to be condescending: nor had he intimates of his own age and flock. But people
of every denomination looked upon him with a kind of veneration, and
his extensive learning excited esteem and respect where there was
nothing to move, or hold the affections of the heart.[88]

In spite of the Doctor's unusually large salary, his estate
came to only 210*l*, of which almost a third was his great library
of 1130 volumes. The portrait of the Rector which is reproduced here is from a mezzotint which Peter Pelham made from
one of his own paintings.

The Cutlers seem to have had eight children: (1) Martha,
b. Dec. 30, 1711; m. John Gibbs, Feb. 24, 1730. (2) Elizabeth,
b. Dec. 30, 1711; d. young. (3) John, b. June 9, 1713; A.B. 1732;
d. Jan. 1771. (4) Elizabeth, b. Aug. 1715; d. unmarried Mar.
26, 1795. (5) Timothy, b. Oct. 22, 1718; A.B. 1734; d. 1739.
(6) Sarah, b. Aug. 25, 1720. (7) Abigail. (8) Ruth, b. Apr. 15,
1722; m. Thomas Potts, July 18, 1758.

---

[87] *Boston News-Letter*, Aug. 22, 1765, p. 3.
[88] John Eliot, *Biographical Dictionary* (Salem and Boston, 1809), p. 144.

ISRAEL LORING

The writings of James Truslow Adams, Charles Beard, and many other historians of the last generation describe the clergy of colonial New England as intolerant bigots whose minds were firmly closed to every new idea. According to this thesis, colonial history is the story of the liberation of society from these tyrants. Israel Loring, although a conservative among his fellows and in his day the oldest minister in New England, is a good illustration of the typical clergyman as he actually was, far more tolerant because better educated than his parishoners. He was often the only inhabitant who subscribed to a newspaper and read new books, which he obtained on his annual visit to Boston at Commencement time. There was sometimes complaint that he took his sermon material from these instead of sticking to the Bible and theology.

ISRAEL LORING, minister of Sudbury, was born at Hull on April 6, 1682. His father was John, who once sat for Hull in the General Court, and his mother was Rachel, daughter of John Wheatley of Braintree and widow of one Buckland or Bucklin of Rehoboth. The Loring household was imbued with the religious spirit, and Israel contracted it from his brother Jacob, who "himself was Under awakening impressions by the Spirit of God."[1] Increase Mather, who admitted him to the College and kept an eye on him, observed that "he was studious, blameless, and serious."[2] According to the college records he was obscure except for some rather large bills for window-glass. He owned a copy of the *Graecae Linguae Fundamenta* of Matthias Martini, which is now preserved at the American Antiquarian Society. When he had taken his first degree, he was appointed

[1] Israel Loring, Ms. Account of the Life and Death of Mr. John Loring of Hull (M. H. S.); Israel Loring, Ms. Diary xvi (M. H. S.), 94. The date of birth is taken from his diary, which does not agree with the town records.

[2] Increase Mather, introduction to Israel Loring, *Duty and Interest of Young Persons* (Boston, 1718).

a Scholar of the House on a stipend of 4*l*; so he kept his chamber and his cellar and remained in residence. In August, 1703, he preached his first sermon at Scituate lower parish and probably filled the pulpit several Sundays, for the College cut his fellowship to three-quarters. He passed the winter in Cambridge and took his second degree at Commencement, 1704, arguing the affirmative of the question, "An Mores animi sequantur Temperamentum Corporis?" Two Sundays later he preached at Braintree. But the autumn found Mr. Loring at the College again, and there he remained, enjoying his fellowship, until Commencement, 1705.

On July 29 of that year Mr. Loring preached at Sudbury, and three weeks later he was called back, probably from Falmouth, where he was for a time. For a while the town paid him 20*s* a Sunday, then engaged him for a quarter, and finally for six months. This caution is to be explained by the fact that the town had been unfortunate in his predecessor, an unworthy man who had recently been removed from the pulpit by a council of churches. But on October 22, 1707, the town felt sufficiently assured of the young parson's character to vote him a salary of 70*l*, the use of the ministerial lands, and 50*l* settlement.[3] He accepted, although the salary and settlement were small and, to judge by the experience of his predecessor, prompt payment could not be relied upon. On November 20, 1706, he was ordained.[4]

Three years taught Mr. Loring that a minister needed a helpmeet, and on May 25, 1709, he was married, probably at Hingham, to Mary Hayman or Heman, daughter of Nathan of Charlestown. The bride effected a revolution in her husband's religious conceptions. He, like many others, had believed that the birth of a child on the Sabbath was proof that it had been conceived on the Sabbath, and hence in wickedness. Such children he had refused to baptize despite the bitter protests of their parents. Then, to his horror, his wife presented him one Sunday with a daughter.[5] There was nothing to do but to

---

[3] "Wayland Local History," notes by James S. Draper in the *Waltham Free Press*, 1867–68.   [4] *Boston News-Letter*, Nov. 25, 1706.

[5] The story was told by Dr. Thomas Stearns, M.D. 1812, who owned now lost portions of the Loring diary. The date of the birth of the child is given as Nov. 16, 1712, that of his daughter Elizabeth. Another version makes the embarrassing offspring his twin daughters, but they were not born on a Sunday.

admit his error and call in the other little ones for their delayed baptisms.

The old Sudbury meetinghouse was on the east side of the Sudbury River in the present town of Wayland, but by Mr. Loring's day a majority of the inhabitants lived on the west side of the river and were frequently kept from church by high water. Consequently, the people on the west side obtained permission from the General Court to set up another parish, the present Sudbury. The Court's insistence that the town settle accounts with Mr. Loring indicates that his salary was in arrears, and when the new parish offered him a settlement of 100*l*, he crossed over to the new meetinghouse, taking the church books with him. This removal was in July, 1723. William Cooke, A.B. 1716, took the old pulpit and proved to be an obliging neighbor.

As a young man Mr. Loring showed that he belonged in the camp of the religious liberals by supporting Leverett for President of Harvard, but he was no latitudinarian. On one occasion he informed his parishioners: "It is my design by God's assistance, to entertain you at this time, with a solemn and awakening Discourse of Hell." He then described luridly the torments of the damned, worse than those inflicted by "The Tyrants of Japonia," and lasting to all eternity. And what was eternity? "Suppose a little bird should once in a thousand years, carry in her bill a drop of water out of the Sea, when would the Sea be emptied? Yet this would infinitely sooner be done, than the Misery of a Sinner shall be ended." [6] Yet he dodged the horror of infant damnation by saying only that the Lord had given us no assurance that those who died young should not go to Hell. [7] In like manner, when he maintained the doctrine of justification by faith alone, he dodged the damnation of the man without faith but of perfect works by saying that there could not be perfect works without faith. [8] He regretted the religious laxness of his times and gave "a Solemn Testimony against the vain, vile Sinful Frolicks" of the younger generation. [9]

[6] *Serious Thoughts on the Miseries of Hell* (Boston, 1732), p. 20, *et passim*.
[7] *The Service of the Lord* (Boston, 1738), pp. 34–5.
[8] *Justification not by Works* (Boston, 1749).
[9] Diary, XVI (M. H. S.), 55.

Mr. Loring was popular as a preacher. There was a demand for his sermons in print, and he was frequently called to preach to neighboring congregations and in the conventions of ministers. In 1737 he preached the election sermon before the General Court and took the opportunity to urge legislation to remedy the general negligence in church attendance, to keep retail liquor-licenses in the hands of godly, church-going people, and to restore the estates of those condemned of witchcraft in Salem Village a generation before. In regard to the last he stood with the clergy of 1692, holding that the judges had acted honorably and according to the law, but that the rules of evidence accepted had led to the death of innocent people.[10] At his suggestion the General Court initiated such legislation.

In this sermon Mr. Loring concisely expressed the spirit of religious toleration that was prevalent among the puritan clergy: "Unity of the Faith is not to be expected, till we get to Heaven. . . . By all Means, let us espouse generous Principles; let us breathe a catholick Spirit; let us be one with every one, that is one with Jesus Christ; whether they be Lutherans, or Calvinists, Episcopalians, or Presbyterians, Congregationalists, or Antipœdobaptists; or whatever other Denomination they may be of." [11]

There was one religious group which Mr. Loring would not tolerate: the illiterate preachers who intruded themselves into established parishes, attacked the ordained ministers, and broke up the congregations. To the convention of ministers of 1742 he denounced them:

As for any outward call to authorize them to this work, this is what they can't pretend to. They never were regularly introduced into this, were never selected thereunto by that order that God hath appointed in his Church. And as for an inward call enabling them to teach and exhort, it may justly be feared that they are utterly destitute of it. . . . Such as set up to be teachers and exhorters of others, should doubtless be men of superior understanding themselves; but are the persons that I am now speaking of such? How should they come to an eminency of knowledge in divine things? Knowledge in the liberal arts and original tongues is an handmaid to Divinity, and a great help to attain it; but this our exhorters are destitute of.[12]

---

[10] *The Duty of an Apostatizing People* (Boston, 1737), pp. 50–2.
[11] *Id.* 67.
[12] William B. Sprague, *Annals of the American Pulpit* (New York, 1857–69), I, 258.

Naturally he was to be found opposing Whitefield in the later days of the Awakening and calling him erroneous in doctrine and disorderly in conduct. In his ninetieth year he was enraged to hear that an itinerant had the effrontery to invade his parish and pray for him, and in retaliation he went down on his knees and prayed furiously for the conversion of his enemy.

In spite of this the Sudbury parson remained a friend and active correspondent of Thomas Prince, a chief supporter of the Awakening. He subscribed to Prince's *Chronology*, a compliment which was not returned when Mr. Loring published a sermon by subscription. But when he preached a Thursday lecture sermon, Prince said of him:

He was so plain and easy in his expression and method, so familiar and moving in his delivery, so affected by himself with the momentous truths he would inculcate on us, that we must have harts of adamant to resist the impressions, or continue indifferent whether we pass through so great a change as he clearly explained and earnestly urged as of the last necessity.[13]

Although the Great Awakening did not, as in many instances, break up Mr. Loring's church, it gave rise to serious difficulties. There had always been a large minority who differed with him as to the admission of candidates, and generally kept affairs in a turmoil. Encouraged by the majority, he steadfastly refused to call church meetings for the particular purpose of permitting the minority to air their grievances. On one occasion, when he was attending a meeting of ministers at Westboro, he was amazed to have the leaders of the opposition walk in with complaints against him to the effect that he was the cause of the dissension in the church and the decline of religion in Sudbury. Ignoring the request of his colleagues, he refused to listen and stalked out.

The ministers then authorized the rebels to call for a council of churches if he would not face their attacks in church meeting. When the dissenters acted on this advice and sent out letters for such a council, he galloped off to Boston and hamstrung the affair by getting the support of Dr. Sewall (A.B. 1707) and the leaders of the Old South, although he was repulsed by his friend Josiah Willard (A.B. 1698), the Province Secretary.

[13] *Id.* 260.

There were other ways to make trouble for him, however. On one occasion when the howling of an infant disturbed the service he directed the mother to take it home, but she carried it instead to a place near the pulpit and permitted it to distract the congregation for the rest of the service.[14]

Mr. Loring's salary averaged no more than 50*l* lawful money plus something for firewood. Unlike most underpaid parsons of the time, he was humbly thankful for what he got and did not make a great to-do about finances. While others were writing sermons on their wrongs he preached on the subject of the support of the ministry by lay Christians without once mentioning salaries.[15] Needless to say he was not one of the ministers who neglected their parishes for business; he limited such worldly activities to one suit to recover on a bond which he held, and one summons to the Reverend William Brattle to come into court about some land in Princeton.[16]

Outliving the ministers of his generation, he naturally became depressed: "Tis a Dying World that we live in our turn is Coming. With Me it Can't be far off Considering my very advanced age. . . ." He always gladly looked forward, however, to a world of "light, Love & Joy above." [17] Mrs. Loring died on December 24, 1769, after sixty years of harmonious married life:

There were a few Things to be observed concerning her; for 45 Years past she eat but one Meal in 24 Hours, and that was ordinarily only a little Bread and Cheese at Night, a little before she went to Bed, yet her Health was such that she was at the Head of her Family Affairs and Business till about ten Days before her Death.[18]

When growing age finally prevented Father Loring from going down to Commencement, he sat at home on Commencement day and wrote in his diary, "May God bless the College and continue it a blessing from Generation to Generation." [19]

[14] Diary, *passim*, esp. 1748–50 (Sudbury); *N. E. Hist. Gen. Reg.* VII, 326. His experience in Boston is reflected by a remark some years later that Secretary Willard, "tho' Such a Child of light, often Walked in Darkness." — Diary, 1753–57 (Sudbury), p. 174.
[15] *Private Christians Helpers of their Ministers in Christ Jesus* (Boston, 1735).
[16] Early Files in the Office of the Clerk of the Supreme Court of Suffolk County, 165076, 165243; Misc. Mss. M. H. S. Aug. 6, 1760.
[17] Chamberlain Mss. (B. P. L.), A 1.66.
[18] *Boston News-Letter*, Jan. 4, 1770, p. 2.
[19] Diary, 1770–72 (Conn. Hist. Soc.), p. 31.

In 1770 he made one trip more to Boston and on June 4 with other parsons he was dined at Faneuil Hall by the Sons of Liberty, and venerated as the oldest minister in America.[20] He agreed with the popular party, for he recorded in his diary "The horrible Slaughter of the Men at Boston by regular Soldiers."[21]

Parson Loring was never given an assistant, but carried the full burden of his work until the end. On March 1, 1772, he preached all day, but was taken sick at a town meeting two days later and had to be hurried home on a sleigh. He died on the ninth. Ebenezer Parkman (A.B. 1721) of Westboro was called to preach the funeral sermon and had great difficulty in reaching Sudbury because of the depth of the snow. "I arrivd seasonably for Dinner at Mr. Lorings. Messers Goss [Thomas, A.B. 1737], Smith [Aaron, A.B. 1735], two Bridges [Ebenezer, A.B. 1736, and Matthew, A.B. 1741], Woodward [Samuel, A.B. 1748] & Cushing [Jacob, A.B. 1748] of Waltham there. The Corps was carryd to the Meeting House. I could not be excused from praying. The Corps were interrd in Col. Browns Tomb."[22]

A woman who remembered Father Loring described him as tall and thin,[23] but Ezra Stiles, who knew him well, said that he was of small stature.[24] All agreed that his manner had the apostolic dignity which they imagined resembled that of the early puritan divines. We know from his diary that he refrained from the appearance of levity, although risking a little "innocent and Civil Mirth" in order to overcome his natural melancholy.[25]

The Lorings had seven children: (1) John, b. Apr. 27, 1710; A.B. 1729; m. Elizabeth Vryling, Dec. 9, 1736; a physician of Boston; d. Apr. 11, 1744. (2) Elizabeth, b. Nov. 16, 1712; m. Richard Manson, June 6, 1746. (3) Mary, b. Sept. 14, 1716; m. Elisha Wheeler. (4) Jonathan, b. Aug. 29, 1719; A.B. 1738; m. Elizabeth Woods, Jan. 21, 1740. (5) Nathan, b. Nov. 27, 1721; m. Keziah Woodward, Dec. 31, 1747; d. Apr. 25, 1803. (6) Sarah, b. Nov. 10, 1724; m. Hopestill Brown, Dec. 30, 1746.

[20] Ezra Stiles, *Literary Diary* (New York, 1901), I, 54.
[21] Diary, 1770–72, p. 12.
[22] Ebenezer Parkman, Ms. Diary (Am. Antiq. Soc.).
[23] *N. E. H. G. R.* VII, 328.
[24] *Literary Diary*, I, 218.
[25] There is an obituary and character in the *Boston News-Letter*, Mar. 19, 1772, p. 2.

(7) Susannah, b. Nov. 10, 1724; m. William Moulton, Apr. 24, 1770. Parson Loring also brought up his nephew Nicholas, A.B. 1732, and minister at North Yarmouth. The estate was not large, Jonathan getting fifty acres in Rutland, Nathan the homestead, and the others small sums of money and shares of a farm in Princeton.[26] Jacob Bigelow, A.B. 1766, succeeded to the Sudbury pulpit.

JARED ELIOT

The Puritans had a passion for science, as the history of Restoration England and the Royal Society show. Many of the colonial ministers devoted much of their time to reading and amateurish exploration of this part of the handiwork of God. The introduction of inoculation against smallpox by Cotton Mather is only the most famous example of the tendency of the clergy to be more advanced than the professional practitioners of medicine. In Jared Eliot this took the form of seeking for useful and practical applications of the new discoveries. Yale is properly very proud of him, but she still ignores the fact that he began his collegiate education and earned his M.A. at Harvard.

JARED ELIOT, the physician, political philosopher, parson, and scientist, was born at Guilford, Connecticut, on November 7, 1685, the eldest son of the Reverend Joseph (A.B. 1658) and Mary (Wyllys) Eliot. One of his grandfathers was the Apostle to the Indians, and the other Governor Wyllys of Connecticut. When Jared was eight, his father died, leaving provision in his will that one or both of his sons be brought up to the ministry. So Abial was permitted to become a farmer, while

---

[26] Middlesex Probate Records, LIII, 159–63.

Jared was shipped off to Cambridge, where he was placed third in his Class — an honor not surprising for a boy with two distinguished grandfathers. It was said that he was intellectually a little slow in his younger days. For the first half of freshman year he did not live or board at the College, and in the second half he was fined small sums, after which he disappears from the books.

There is no great mystery involved in the fact that Jared did not return to Cambridge in the summer of 1700. In the parsonage of Abraham Pierson at Killingworth, where he, as the son of the neighboring minister, must have spent many days, there was talk of founding a "Collegiate School" which would be more convenient than Harvard for Connecticut boys and, perhaps, more orthodox. In the fall of 1702, it is said, Jared went to study with Mr. Pierson in the Killingworth house where Yale University was having its beginning. Perhaps health was a factor in the transfer, for he did not take his first degree until 1706.

Jobs were plentiful for young graduates in those days, and on September 27, 1706, Sir Eliot was made schoolmaster in his native town, receiving a salary of 35*l* in country pay. He had been a favorite pupil in the Killingworth parsonage, and when Mr. Pierson received the call which no man can reject, he told his mournful parishioners that young Eliot was the man for his pulpit.

Consequently Sir Eliot took up the ministry in Killingworth, in what is now the First Church of Clinton, Connecticut, on June 1, 1707. After eighteen months the people of the town decided that they liked him well enough to sit under his ministry for life, and they encouraged him to settle by promising him sixty loads of firewood a year when he should "marry, or have a family." In September, 1709, the town "did by their vote conclude to Indever that Mr. Jared Eliot be settled a monst us in office, (or ordayned) on the Last Wednesday of October next insueing," the charge of the ceremony to "be boren or paid by a Rate Levied upon the Estates of the subscribers to the Covenant in the Town, allwaye provided that any other person is not bared from doing what they shall see cause." [1] The ordina-

---

[1] *Two Hundredth Anniversary of the Clinton Congregational Church . . . .* (New Haven, 1868), p. 22.

tion took place on October 26, 1709, and exactly a year later the parson qualified for the firewood by taking to wife Hannah Smithson, of Guilford.

In the meantime the parson had earned his title of "Mr." It was an ancient academic custom for universities to admit graduates of other institutions *ad eundem gradem* ("to the same degree") as a symbol of the fellowship of all learned men. Harvard had not as yet taken up this practice, and when Sir Eliot and two other young Yale graduates asked to be admitted to the degree of M.A. at Commencement, 1709, the Corporation voted that it was not advisable. But five days before Commencement the Corporation, having heard from Governor Joseph Dudley, an old friend of the Eliot family, changed its mind.

Upon his Excellencys desire that Mr Jared Eliot should be admitted to the degree of Master of Arts Agreed and Voted that the said Jared Eliot be allowed to proceed Master of Arts the approaching Commincement he performing the usual Exercise in the Colledg Hall and paying the Stated fees of the Colledg and 20s Detriments as is Usual for those that take their second degree after their discontinuance.[2]

The *quæstio* sheet indicates that to meet the requirements for the degree he presented a negative answer to the question, "An Duellum sit Licitum." Thus his degree was an ordinary M.A., and he was admitted to it as a strayed son of Harvard. His name should not stand, as it does in the *Quinquennial*, among the holders of honorary degrees.

For a while the years flowed quietly by the Killingworth parsonage, betraying nothing of the storm that was brewing for the minister. His father-in-law, Samuel Smithson, was an English immigrant and strongly attached to the Book of Common Prayer. And in the great library which Jeremiah Dummer (A.B. 1699) had gathered in England for Yale, Samuel Johnson, his friend and former pupil in the Guilford school, now a tutor in the College, had found many wonderful and, from the Congregational point of view, unorthodox volumes. In consequence Eliot participated with Rector Cutler (*q. v.*) and Samuel Johnson in that "Great Apostasy" in the Yale library one September day in 1722. Although missionary

[2] *Publications Col. Soc. Mass.* xv, 389 (Harvard College Records).

George Pigot described him as being one of those who were fully persuaded of the invalidity of their Presbyterian ordinations,[3] at the critical moment he was unable to bear the anger of the great majority of New Englanders, who with some reason denounced Rector Cutler and his fellows as guilty of flagrant treachery in betraying the trust which had been imposed on them when they had been given control of Yale College. Knowing the Jared Eliot of later years, one feels that he weighed a point of dogma against the fact that he and his family had to live, and allowed himself to be convinced by the "learned reasonings" of Governor Saltonstall.[4] Like the majority of the Puritan clergy of his day, he felt that the world would never agree on all points of religion and that there was no use in making a martyr of oneself over a pinpoint of ceremony. So he shrank from the pillory of public opinion and fled to the quiet security of the Killingworth parsonage, where by distinguished service he lived down the story of his weakness. In later days, however, when the Church of England and her bishops loomed as a menace to the established order in Puritan New England, he did not join in vilifying her, and refused "to act or say or insinuate" anything to her disadvantage.[5]

After eight years of quarantine, Yale chose Mr. Eliot to be one of her trustees on September 9, 1730. He was the first graduate of the College to fill that office. He served for more than thirty years, during many of which the records are in his hand. With his old friend Samuel Johnson, now an Anglican clergyman, he was instrumental in securing Dean Berkeley's donation for Yale, and he was sent by the trustees to take charge of the good bishop's gift of Rhode Island real estate. As a theological liberal he is credited with opposing the grilling which the trustees gave Naphtali Daggett, the first professor of divinity. Benjamin Franklin complimented him on his "Catholick Divinity,"[6] and his old friend Thomas Ruggles praised his quality of Christian charity:

As he thought and judged freely for himself; he was persuaded that every man had the same right. . . . Hence he was an enemy to all

[3] *Documentary History of the Protestant Episcopal Church in . . . Connecticut* (New York, 1863), I, 58.   [4] *2 Coll. M. H. S.* VI, 162.

[5] E. Edwards Beardsley, *History of the Episcopal Church in Connecticut* (New York, 1869), I, 30.   [6] Benjamin Franklin, *Writings* (Smyth ed.), III, 281.

imposition, and arbitrary dominion over other men's faith. . . . Hence he was free from all bitter words, or reproachful reflections; but spake, judged and acted freely, without fear or restraint, but from the great law of prudence, which he ever exemplarily practised. By this upright conduct he gained the esteem, confidence and good-will of persons of every denomination, who were fond of his company, and valued his friendship highly.[7]

Like many theological liberals he was opposed to the excesses of the Great Awakening and the insistence of some of its adherents on the narrowest of orthodoxy. In a petition to the General Assembly dated October 8, 1741, he speaks of "several persons [who] under Religious Pretenses have taken up severall new Practices which tend to hinder those good Ends and Effects and to make Divisions and Convulsions, in our churches," and proposes that the civil government bear the expense of a general consociation of churches.[8] The New-Lights did not succeed in capturing his church, and in the midst of the Awakening the town cheerfully paid him a salary of 120*l* "in Bills of publick credit, or in provision, at the Current market price, exclusive of the use of parsonage and his wood," and in addition voted him firewood valued at 40*l*.[9] This may partly be explained by the fact that "He was truly a good preacher, in a proper sense: Though he never studied to shine in rhetorick, and the enticing words of men's wisdom; yet his discourses were always instructive and entertaining; and from the peculiar manner in which he communicated his ideas, were animated, entertaining, and always engaging the attention."[10] Against this must be balanced the statements that before the Awakening, of the 130 inhabitants of Killingworth, 44 were communicants in his church;[11] while thirty years later there were only two families in his church and 140 in the rival, New-Light, Presbyterian congregation.[12]

No doubt there were those who thought his unrhetorical preaching dull, and were angry when he signed a petition that "various Incroachments and Infringements having been made

---

[7] Thomas Ruggles, *The Death of Great, Good, and Useful Men Lamented* (New Haven, 1763), p. 18.
[8] Conn. Archives, Ecclesiastical Records, VII, 243.
[9] *Two Hundredth Anniversary of the Clinton Congregational Church*, p. 23.
[10] Ruggles, p. 17.
[11] *Coll. Conn. Hist. Soc.* III, 291.
[12] Ezra Stiles, *Itineraries* (Yale, 1916), p. 216.

upon our Ecclesiastical Constitution . . . by young men's taking upon them to preach without licence and contrary to order, by Minister's Entring into other parishes besides their own and preaching in a disorderly manner," the civil government should intervene.[13] This was the attitude of the majority of the clergy, and not unnaturally Mr. Eliot was elected moderator of the general association of ministers at Middletown North in 1755.

President Thomas Clap (A.B. Harvard 1722) of Yale, somewhat less liberal in matters of theology than Eliot, provoked a series of clashes among the trustees which once drove the Killingworth pastor to the point of resigning his trusteeship.[14] When the President separated the college church from that of New Haven, Mr. Eliot led the Old-Light opposition, evidently thinking that it was unwise to permit Yale to become too independent of the moderating influence of the neighboring ministers.[15] In 1758 he and his neighbor, Thomas Ruggles, complained to the Assembly that the college church was "a glaring Evidence of an undue aim at exorbitant power, as loudly calls for the timely check of the Legislature." [16] He never lost his interest in Yale, expressing concern when the students beat a tutor with clubs, and writing to enquire what had been done at Harvard, where the students had figuratively "thrust their Tutors through the fires." [17]

It would be a mistake to make Jared Eliot the hero of the development of religious thought in eighteenth-century Connecticut, for he was a man of his times, if a liberal one. His sermon on the taking of Louisbourg is as full of the wonderful providences of God as are the writings of Increase Mather, and his letters are crammed with stifling theology. He had no use for Deists and their assertion that such biblical stories as that of the Fall were mere fables. On the other hand he admitted that these stories had grown much in the retelling, and he defended the study of Egyptian mythology as a means of understanding the Bible.[18]

[13] Conn. Archives, Ecc. Rec. VII, 250.
[14] Letter of Sept. 25, 1756 — Hobart to Eliot, in Yale archives.
[15] Conn. Archives, College and Schools, I, 338 ab, 339 a.
[16] Franklin B. Dexter, *Biographical Sketches of the Graduates of Yale College* (New York, 1885-1912), II, 507.
[17] Ezra Stiles, Mss. Letters (Yale Archives), May 25, 1761; July 28, 1761.
[18] *Ibid.*, Apr. 12, 1757; Aug. 21, 1761.

Jared Eliot is best known today, not for his struggles in religion, but for the election sermon which he preached in 1738. After its droning predecessors, this one sounds like a sudden crash of drums in the yet distant Revolution. To the historian of political thought his direct references to Locke and his assertion that the colonial assemblies were little parliaments represent a milestone on the road to John Adams and Thomas Jefferson. Perhaps the people of Connecticut realized the significance of this sermon, for when the Freemen of New London held their annual meeting two years later they were "Generally for mr Eliot Governour & Thomas Fitch not above 6 or 7 . . . for the old Governour." [19] Samuel Johnson wrote George Berkeley that "Upon a considerable struggle last election for a new Governor . . . Mr. Eliot had a vast many more [votes] than all other competitors put together, and will doubtless succeed whenever there is a new choice." [20] But when Governor Talcott died the following year, it was Deputy Governor Jonathan Law (A.B. 1695), Mr. Eliot's brother-in-law, who gained the succession.

Office interested Jared Eliot far less than did bottles and pills, and he was fast becoming a famous physician. At the outset of his practice he received a cruel disappointment, for when Joshua Hobart (A.B. 1650), the eminent physician and divine of Southold, Long Island, was on his death-bed, he sent for Mr. Eliot "to come over to Long Island, that he might impart to him some Secrets in Physic & Chemistry which he had never communicated; but it being deep Winter Dr Eliot could not pass the Sound, & so saw Mr. Hobart no more." [21] He had a great advantage over the ordinary uneducated barber-surgeon of his day, for he could study in the original the then still standard works of Hippocrates, Celsus, Galen, and Aretæus, as well as read the Latin works of the contemporary European school, such as the volumes of Boerhaave. One may judge his prescriptions by this which he gave to the Reverend Daniel Wadsworth for a "Catarrhous humour": "take a vomit once in awhile of Ipacacuania, and also bitters, as gentian, camamile &c in powder or steeped in wine, and also the yolk of an egg in

[19] Joshua Hempstead, *Diary (Coll. New London County Hist. Soc.* 1), p. 362.
[20] Samuel Johnson, *Works* (Columbia, 1929), 1, 101.
[21] *Itineraries*, p. 364.

cyder sweetened with honey, once twice or 3 times a day." [22]
In regard to a famous prescription he remarked, "Pearls are
prescribed in Medicine for great People: but is not of Use but as
a testacous Powder; and for that Use an Oyster Shell will do as
well." [23] His fame as a practitioner was wide:

> As his principal natural talent was for physic; so he by study and
> reflection . . . became, at least, one of the ablest phisicians in his day;
> by a quick discernment, he was seldom at a loss what the disease was,
> where he came, though before a secret; by a penetrating judgment
> what was the person's constitution. . . . But that part of physic wherein
> he conspicuously excelled was the method of treating and the remedies
> he invented, for the cure of that afflictive disease, the dropsy; whereby
> great numbers were released of their distresses, their lives preserved,
> and their health and comfort restored to them. Hence his practice
> became very extensive in this and other colonies; taking long journies
> of hundreds of miles, to relieve the distressed; by which the blessing of
> many ready to perish came upon him. He was thus a physician for
> the bodies as well as spirits of his fellow men; for he never put off the
> Christian or minister, when he undertook to practice as a phisi-
> cian. . . .[24]

In 1766 one of Jared Eliot's medical students, a Dr. Ayers
of Newport, laid before the New Jersey Medical Society this
wonderful cure for dropsy, only to have it rejected as purely
sympathetic medicine.[25] In order to avoid neglecting his
clerical duties (he was careful to be in his pulpit at every
service), Eliot formed the habit of reading on his long horseback
rides to visit his patients, and it is said that he too frequently
aroused himself from his book to find himself facing a hay-
stack far from his destination. After he had trained many
students in physic, and so many went out from the Killingworth
parsonage that he has been called "the father of regular medical
practice in Connecticut," he laid aside his pills to take up the
study of chemistry, in which he had become interested while
compounding medicines. He also turned his attention to
metallurgy, which was an old love, for as early as 1716 he held
a considerable interest in the Simsbury copper works.[26] With

[22] Daniel Wadsworth, *Diary* (Hartford, 1894), pp. 39–40.

[23] Jared Eliot, *Essay on Field-Husbandry* (Boston, 1760), p. 97. The M. H. S. copy
of this volume has Ms. comments in the hand of Ezra Stiles.

[24] Ruggles, pp. 18, 19.

[25] For a critical account of this remedy see Gurdon W. Russell, *Early Medicine, and
Early Medical Men in Connecticut* (Proc. Conn. Med. Soc., 1892), pp. 65–6.

[26] *Wyllys Papers* (*Coll. Conn. Hist. Soc.* xxi), 389.

Elisha Williams (A.B. 1711) he was an original proprietor of the Salisbury iron ore bed, and by gift from the General Assembly and careful purchase he became perhaps the chief holder of Connecticut ore lands, particularly in New Milford. In 1744 he erected at Killingworth the first, and for many years the only, steel furnace in Connecticut. Two years later he added a forge with a tilt-hammer. In 1750 Governor Law reported to the enquiring royal authorities that the works had not been in use of late, but Governor Wolcott a year later said that they were active.[27]

Having sold his Salisbury ore holdings, Mr. Eliot was attracted to the possibility of making iron of a black sand which he found on the beaches. After applying much argument and a bottle of rum he induced his furnace man to make the attempt, which resulted in a small quantity of metal which he pronounced to be as good as Spanish iron, and much better than the brittle American product made from bog ore.[28] Naturally elated, he informed the Society Established at London for the Encouragement of Arts, Manufactures, and Commerce, which promptly awarded him a gold medal for his "publick spiritedness, in promoting their views for the advantage of the british colonies."[29] It was Mr. Eliot's connection with this institution which caused the statement to be made that he was elected to the Royal Society of London, and the letters F.R.S. to be affixed to his name in the Yale catalogues. There is no evidence that he ever received this greater honor. The sequel to the black-sand discovery is disappointing. Colonel Aaron Eliot reported that he had successfully smelted a large quantity of it and produced such fine iron that he intended to use little else in his furnace,[30] but neither he nor any one else ever claimed the bounties which the Society of Arts offered for the importation of such iron into England.[31]

[27] *Law Papers* (*Coll. Conn. Hist. Soc.* XI, XIII, XV), III, 427; *Wolcott Papers* (*Coll. Conn. Hist. Soc.* XVI), pp. 74-5.
[28] *An Essay on the Invention, or Art of Making Very Good, if not the Best Iron, from Black Sea Sand* (New York, 1762), p. 6; *Philosophical Transactions of the Royal Society*, LIII, 56. The M. H. S. copy of the *Essay* has Ezra Stiles's Ms. comments.
[29] *Itineraries*, pp. 188-9.
[30] Ruggles, p. 21 n.
[31] Robert Dossie, *Memoirs of Agriculture* (London, 1768), I, 298. The black-sand iron cropped up again in 1781 when Ezra Stiles communicated the essay to the American Academy of Arts and Sciences — *Continental Journal*, Feb. 22, 1781, p. 3.

Mr. Eliot's writings are an amusingly magpie collection of speculations and observations, typical of that age of intellectual enthusiasm. On one occasion he wrote Thomas Prince (A.B. 1707) a letter urging the codification of Indian law, arguing that, applied to frontier conditions, it would be as useful "as the old laws of Rhodes and Oleron in maritime affairs."[32] He applauded when Professor John Winthrop "laid Mr Prince flat on his back" for using earthquakes for purposes of religious terrorism. Earthquakes, he wrote, were natural phenomena: "the fermentation which will arise upon putting water to Iron and Sulphur I know to be fact, that two cold Liquors upon my putting them together have kindled into a flame in an Instant. Such fluids and Solids encountering each other in strait places confined and pent up will produce dismall Effects."[33] But Mr. Eliot's famous book on scientific agriculture perpetuated an idea much more unreasonable than that of the hand of God in earthquakes. Having heard that it was best to kill trees and brush in the old of the moon, he tried it (although the auspicious moment annoyingly came on the Sabbath) and found that it was true:

> To shew such a Regard to the Signs, may incur the Imputation of Ignorance or Superstition; for the Learned know well enough, that the Division of the Zodiac into Twelve Signs, and the appropriating these to the several Parts of the animal Body, is not the Work of Nature, but of Art, contrived by Astronomers for Convenience. It is also as well known, that the Moon's Attraction hath great Influence on all Fluids.
>
> It is also well known to Farmers, that there are Times when Bushes, if cut at such a Time, will universally die. A Regard to the Sign, as it serveth to point out and direct to the proper Time, so it becomes worthy of Observation.[34]

Mr. Eliot defended the fact that he, a parson, should write essays on field-husbandry on the ground that a monk had invented gunpowder and a soldier, printing. He gives an amusing picture of himself directing the work of his men, pleased with their amazement at his ideas, and edified by his apt proverbs. He thought the new English seed drills too com-

[32] *Coll. Conn. Hist. Soc.* III, 291–2.
[33] *Itineraries*, p. 480.
[34] *Field-Husbandry*, pp. 16–17, 123, 124.

plex and induced his mechanically-minded friends to simplify
them. The *Essay on Field-Husbandry* is as significant in
American agricultural history as his election sermon is in the
field of political thought, for in it he points out that peculiar
American conditions make European customs unprofitable, but
argues against extensive in favor of intensive agriculture.
In spite of this preoccupation with colonial conditions, the book
was better received in England, wrote Benjamin Franklin,
than any other contemporary work on agriculture, because it
was based upon actual observation.[35] Its popularity abroad was
attested by the complimentary letters and packages of seed
which poured into the Killingworth parsonage. Peter Collin-
son, F.R.S., sent him wheat from Thessaly and flax from Siberia,
and corresponded with him on biological as well as botanical
subjects.[36] At home the graying sage of Killingworth was
recognized as a Connecticut Benjamin Franklin, and he
numbered among his correspondents such men as Ezra Stiles,
Peter Oliver (A.B. 1730), and Edmund Quincy (A.B. 1722).[37]

The most famous of Mr. Eliot's friends was Benjamin
Franklin himself, who had at least one of the essays printed
and corrected the Yankeeisms in the parson's style. Of one of
his frequent visits to the Killingworth parsonage he wrote:

> I remember with Pleasure the cheerful Hours I enjoy'd last Winter
> in your Company, and would with all my heart give any ten of the
> thick old Folios that stand on the Shelves before me, for a *little book*
> of the Stories you then told with so much Propriety and Humor.
> Adieu, my dear friend, and believe me ever yours affectionately.[38]

We have evidence that other people found the old minister a
charming host:

> In his house he was liberal, courteous and generous, in a gentleman
> like hospitality. . . . His whole deportment was engaging and agree-
> able; nothing stiff, unsociable, assuming or imperious had a place in
> his soul or actions; but he was always humble and condescending,
> plain and manly in his whole behaviour. . . . His person was well pro-
> portioned: The dignity and gravity and openness of his countenance,

---

[35] Franklin, *Writings*, III, 53; see also the account by Rodney H. True affixed to
Harry J. Carman and Rexford G. Tugwell, *Essays Upon Field Husbandry* (Columbia
Univ. Studies in Hist. of Am. Agriculture, 1934).
[36] *American Museum* (Carey ed.), IV (1788), 504.
[37] Belknap Mss. (M. H. S.), 161. A. 45.
[38] Franklin, *Writings*, III, 78, 128, 268, 281.

were plain indications of the penetration and greatness of his mind, and the agreeable turn of his conversation. He was favoured with an excellent bodily constitution, capable of enduring all the fatigues of hunger and thirst, heat and cold, without sensible relaxation or weariness. . . . He had a turn of mind peculiarly adapted for conversation, and happily accomodated to the pleasures of a social life. . . .[39]

From Mr. Eliot's casual correspondence we know the sort of thing which drew men to the parsonage at Killingworth. He told with a chuckle the story of the misprint in the Bible which substituted the word "printers" for the first word in the text "Princes have persecuted me." He discussed the probable state of education in Abyssinia, cut off from the sea by Turks and harassed by the Gallas, and he watched with excited expectation for the unearthing of manuscripts at Herculaneum.[40] At times his agile mind projected itself into the future:

I wonder very much that in all the projected Expeditions there is nothing said of Massasippy which would be the greatest accquisition, and in time of the greatest importance to the Crown and North America. . . .[41]

Even in his last days the old gentleman was no armchair philosopher. Besides writing on agriculture, he and his son exported live stock to Halifax in their own sloop.[42] With President Clap he acted as the New England representative of the Society of Arts in paying bounties for silkworm cocoons raised in that area.[43] He tried to raise cotton at Killingworth and would have succeeded, it has been remarked, if frost had not killed the plants. He is accredited with having introduced the white mulberry tree and rhubarb. With all his fame and attention, he was not too great to pause in the midst of a learned discussion to express delight at the gift of a pair of garters, or to twinkle at an invention of his to mend runs in black silk stockings — the application of ink.

Nine days before he died, Mr. Eliot was made happy by the letter from the Society of Arts telling him of the award of the gold medal, which had been entrusted to Benjamin Franklin for

[39] Ruggles, pp. 16, 17, 22–3.
[40] Stiles, Ms. Letters, June 28, 1752; July 14, 1757; May 3, 1762.
[41] *Itineraries*, p. 481.
[42] Hempstead, *Diary*, p. 593.
[43] Dossie, I, 25, 235.

delivery. He died on April 22, 1763, having survived all of his generation in the colleges and the church. His wife had preceded him on February 18, 1761. Three of his sons were graduated at Yale, and may be followed in the Eliot genealogies. His estate came to 1800*l*, of which 10*l* went to Yale to begin the library fund. Some of his books went to President Stiles, the heirs agreeing that he, the President, "was the fittest person on Earth to be Sentenced to Read them." [44] A portrait of Jared Eliot is owned by his descendants in Clinton, Connecticut, and reproduced in the *Century Magazine*, v (1884), 437, the *Genealogy of the Descendants of John Eliot* (New Haven, 1905), p. 44, and elsewhere.

WILLIAM SHURTLEFF

Because of the diversity of its output, Harvard College had among its graduates a number of New Light revivalists whose often incredible antics made it a temptation to load this volume with such individuals as Solomon Prentice (A.B. 1727). True charity moved the non-Harvard majority on the board of editors to pass up the opportunity to suggest that, in colonial times at least, a majority of the graduates were crackpots. Shurtleff was chosen as a New Light because he dreaded the effects of heterodox theology and did not go to excess in using emotion to bring his hearers back into the orthodox path.

WILLIAM SHURTLEFF, minister at New Castle and Portsmouth, was born at Plymouth, April 4, 1689, the third child of Captain William Shurtleff and Suzannah, the daughter of the Hon. Barnabus Lothrop. The first Shurtleff child came too soon, so the Captain "was called forth before the church in the open

---

[44] *Itineraries*, 490.

Assembly," where he "shewed little sence of sin"; so "the church voted, & the Elder laid him under Admonition, for his sin, & for the pride & hardnesse of his heart." [1] But when little William was two years old, his father had a change of heart, took the covenant, had the children baptized in one batch, and was chosen one of the board of psalm readers. And by the time little William was ready for college, his father had been selectman, town treasurer, town clerk, surveyor, and even Representative.

At Harvard, William showed his father's later and better qualities. A book, now at the Massachusetts Historical Society, in which he summarized the sermons he heard, shows that he remained in the vicinity of Boston for several years. Part of the time he was keeping school at Beverly. At the Commencement of 1710 he qualified for the second degree by presenting the negative of the question, "An Patres veteris Testamenti in Limbum detrusi fuerint?" In 1712 he was chaplain at Castle William in Boston harbor [2] and one of the unsuccessful candidates for the pulpit at Salem middle parish.

From the Castle, Mr. Shurtleff was called to New Castle, New Hampshire, where he was ordained the successor to John Emerson, A.B. 1689, on December 24, 1712.[3] The town voted him 65*l* a year while single and 80*l* when he should marry, and paid 90*l* for a house for his use. Thus encouraged, the young parson sued and won the belle and heiress of the town, Mary, the daughter of the elder Theodore Atkinson and the sister of Theodore, A.B. 1718. This proved a miscalculation that would have broken most men, but it served to make of him a saint, and to make his parishioners regard with wondering awe "his uncommon meekness and patience under great trials." [4] For his "great trial," as a contemporary called his lady, was a woman of most whimsical ways, delighting in such tricks as making him cook his own supper and then suddenly seasoning it with his snuff before he could eat it. One Sabbath morning she locked him in his study (this is the later, Victorian, version) and went and sat demurely in their pew while the congregation

[1] *Publications Col. Soc. Mass.* XXII, 256, 279 (Plymouth Church Records).
[2] Samuel Sewall, *Diary* (5 Coll. M. H. S. v–VII), II, 339.
[3] *Boston News-Letter*, Dec. 29, 1712.
[4] 1 Coll. M. H. S. x, 54.

waited in vain. In later years she reformed and took to reading such good and sober works as *Pilgrim's Progress*.[5]

Mr. Shurtleff's meekness is illustrated by a letter requesting his classmate Thomas Prince to see one of his sermons through the press.

When you are so much at Leisure as to inspect the following Sermon, you will see upon what Occasion it was preach'd & how I came to consent to its being made more Publick. I objected all I could against it, but it was so insisted on that I could not fairly avoid it. Though since I promised a copy of it, I have a great many Times censor'd & condemn'd my Self for it, especially when I have considered how many there are who refuse to print those Discourses that are really worthy of it. . . . It is not long since I sent a Sermon to Boston in order to be publish'd, & now upon my sending another, It may possibly look as if I intended to make a Practice of it. But I can assure you I have no such Design.[6]

When Mr. Prince delayed and the Lieutenant-Governor pressed Mr. Shurtleff for a copy, the latter wrote to Mr. Prince asking him to lend his strength in resisting the demand for the publication of "so mean a discourse."[7] The two ministers were friends, Mr. Shurtleff subscribing to Prince's *Chronology* and discussing scientific questions with him:

I should be glad, if you would favour me with some of your thoughts as to the late terrible Earthquake and let me know whither the great noise that attended it be not uncommon in other parts of the world, whither it being throughout the countery at the same instant does not seem in some measure to jarr with the causes, that are usually assigned for such events in nature. & if it proceeded from any subterraneous passage, it be not reasonable to think there should be some difference of time at several hundred miles distance. It would like wise be very gratefull to me to know your opinion with respect to the consequent rumblings, & shakings, that by times, till very lately have been plainly perceiv'd here. . . .[8]

Unlike his Boston friend and other ministers of the sort, however, Mr. Shurtleff did not preach sermons that were the latest European scientific thought only thinly clothed in theology. Once, in speaking of stars, he said:

[5] *Provincial Papers . . . of New Hampshire* (Concord, etc., 1867–1915), xviii, 177.
[6] Letters and Papers (M. H. S.), 71. J. 58.
[7] *Ibid.*, 85.
[8] Misc. Mss. (M. H. S.), Nov. 20, 1727.

And here you will none of you imagine, that I am going to amuse you with any new and curious Speculations, concerning these Celestial Luminaries. Should I attempt to give you a nice and philosophical Account of the Nature, Influences and Motions of the Stars, it would argue a great deal of Vanity, especially, unless I was more a Master of the Science; and would perhaps be altogether as impertinent and unprofitable to the greater Part of those with whom I have to do, as it would be vain and arrogant in my self. I shall not therefore soar so high, nor wander so far beyond my proper Orb.[9]

The Assembly of New Hampshire thought well of Mr. Shurtleff, commissioned him to pass upon the amazing "Theosophical Thesis" of Hugh Adams (A.B. 1697), watched out for his interest when Sandy Beach (Rye) was set off as a separate parish, and had him accompany the New Hampshire delegation to the intercolonial conference with the Penobscot Indians in the summer of 1726. In these matters and others he enjoyed the company of the Reverend Jabez Fitch, A.B. 1694, of the north parish of Portsmouth, a fact which no doubt influenced him to take the most unusual step of leaving a poor parish for a richer one. Perhaps he was also influenced by the fact that a small minority of his congregation had protested the raising of his salary to 80 and later to 100*l.* However that may be, he was dismissed from the church at New Castle on January 21, 1731/2, and installed as the successor of his old friend John Emerson in the south parish of Portsmouth just one month later.[10]

The south parish was happy in its choice, although tradition says he used the training-field and the burial-ground for a horse pasture.[11] He, on the other hand, was dissatisfied by the prevailing religious attitude.

How did not only Pelagianism, but Arianism, Socinianism, and even Deism itself, and what is falsely call'd by the Name of Freethinking, here and there prevail? How much was it grown into Fashion to throw off all Manner of Regard to strict and serious Godliness? . . . a horrid Contempt was put up on the Ministry of the Word. . . . Indeed upon the Lord's-Day, when the Season was inviting, and there

---

[9] *Gospel Ministers* (Boston, 1739), pp. 4–5.

[10] *Boston News-Letter*, Mar. 2, 1732. It was said at the time that he was leaving New Castle to take orders and become curate of a proposed Anglican Church at Portsmouth, — Robert Hale, Ms. Journal of an Expedition to Nova Scotia (Am. Antiq. Soc.), p. 10.

[11] Pearson and Bennet, *Vignettes of Portsmouth* (Portsmouth, 1913), p. 28.

was nothing in the Way, there would (it may be) be what some call a handsome Appearance: That is, there would be a Number of Persons of both Sexes, especially in some Congregations, richly and curiously dress'd, and making as fine and glittering a Shew as if this was the Thing they chiefly aim'd at. . . .

Even the parsons erred, he thought, in failing to preach "some weighty points, such as that of Original Sin, Regeneration and Conversion, Justification by Faith only," and the like. They erred as well in unbecoming levity.

Our Association Meetings had not always that Seriousness in them that might be expected from Persons of our sacred Character: Insomuch that some have since told me, that being occasionally present, it was Matter of Stumbling to them to see us behave as if we had nothing further in View than to smoke and eat together, to tell a pleasant Story, and to talk of the common and ordinary Affairs of Life.[12]

Feeling thus about the general situation, Mr. Shurtleff naturally welcomed Whitefield to his pulpit when he arrived at Portsmouth in October, 1740. Of this service, the evangelist reported:

Instead of preaching to dead Stocks, I had now Reason to believe I was preaching to living Men. People began to melt soon after I began. . . . Mr. Shutlif the Minister, when he afterwards sent me 97*l* collected at this Time for the Orphans, wrote thus: 'You have left great Numbers under deep Impressions, and I trust in God they will not wear off.' . . .[13]

Far from dying out, the religious feeling in Portsmouth reached a point where Fitch and Shurtleff needed no outside assistance to work up religious frenzy in the congregation. Late one dark November day in the meetinghouse, "the Chimney of an House that stood near to it happening to take Fire and blaze out to an uncommon Degree: upon the sudden Appearance of the Light breaking in at the several Windows, there was a Cry made, that Christ was coming to Judgment. . . ." [14]

[12] *A Letter* (Boston, 1745), pp. 4–6.
[13] *A Continuation of the Reverend Mr. Whitefield's Journal from a few Days after his Return to Georgia to his Arrival at Falmouth on the 11th of March, 1741* (London, 1741), p. 35.
[14] *Christian History*, I, 385. For his encouragement of the revival, see Misc. Mss. (M. H. S.), 81. F. 149.

The excesses of such revivalists as Gilbert Tennent, whom Mr. Shurtleff called "that faithful Servant of Christ," [15] now split the ministry into warring camps, the Portsmouth pastor taking the side of his classmates Sewall and Prince and thus opposing the great body of educated clergy. Not that he favored the uneducated itinerants, for he had long maintained "that the Knowledge, not only of the Learned Languages, but of the whole Circle of Arts and Sciences, may be of use to a Gospel-Minister," and sound divinity "is what Men must not expect to be inspired with, in the present Age." [16] Neither was he one of John Wise's democrats, seeking to place the lay elders in control of the churches, for he had the highest opinion of the place of the clergy.[17] His moderate position is well defined in a letter to the convention of ministers summoned to meet at Boston on July 7, 1743.

I should be for using a becoming Care that the Disorders complain'd of might not be magnified in an undue Measure, and that nothing might come under that Character and Denomination that is not worthy of it. Whilst I should be for guarding our Pulpits and Parishes against bold, and ignorant Intruders, and such as may unjustly pretend to an extraordinary Call & Warrant from God, I should be careful that none of the zealous and faithful Preachers of the everlasting Gospel . . . might be excluded.[18]

He welcomed and enjoyed Whitefield's second visit, writing to Sewall and Prince:

I have frequent opportunities of being with him, and there always appears in him such a concern for the advancement of the Redeemer's Kingdom and the good of souls, such a care to employ his whole time to these purposes . . . that . . . I find my heart further drawn out towards him.[19]

The following year, 1745, he published an open letter to the clergy defending Whitefield.

Mr. Shurtleff was frequently asked to open the House of Representatives with prayer and was given the sinecure of the chaplaincy of Fort William and Mary. He was on intimate

---

15 *The Obligations upon all Christians* (Boston, 1741), p. 25.
16 *The Labour* (Boston, 1727), pp. 8, 9.
17 *Gospel Ministry* (Boston, 1739).
18 *Christian History*, I, 173.
19 Hamilton A. Hill, *History of the Old South Church of Boston* (Boston, 1890), I, 551.

terms with such distinguished men as Sir William Pepperrell.[20] He died on May 9, 1747, and was buried under the communion-table of the meetinghouse, from which his remains were re-moved to the Auburn cemetery when the building was taken down in 1863. His estate came to only 433*l* 3*s* 9*d*, which caused Brother Theodore to send to England for some goods to set "Sister Shurtleff into Some Little business."[21] She died on April 9, 1760.[22] They had no children.

The Massachusetts Historical Society owns portraits of Mr. and Mrs. Shurtleff from which those belonging to the South Church of Portsmouth were copied. No original painting can be found corresponding to the engravings reproduced in the *Century Magazine* for 1893 (p. 388) and in the family gene-alogy (1, 8); they are apparently pious improvements on the rather unflattering originals at the Historical Society.

JOSHUA PARKER

In contrast with eighteenth century Europe, or the modern United States, crime was very rare in colonial New England. Observant foreign visitors said that there was no crime, and, rightly, said that the reason was that there was almost no poverty. The poorest man could expect to become comfortable by industry. What public punishments there were seem brutal to us because imprisonment was used chiefly for detention while awaiting trial. There is some-thing to be said for the system which punished by a day in the pillory or even the loss of an ear instead of by long jail sentences.

JOSHUA PARKER, who has the distinction of being the first Har-vard man to lose an ear in the pillory, was one of the older children of Captain Josiah and Elizabeth (Saxon) Parker. He was born about 1690,[1] after the captain and his family had left Groton and before they appeared at Woburn. In 1698 the family moved to Cambridge, had Joshua baptized, and bought an inn standing at what is now the northwest corner of Brattle

[20] Pepperrell Mss. (M. H. S.), 71. B. 193.
[21] *Provincial Papers*, xviii, 310–11.
[22] *New-Hampshire Gazette*, Apr. 11, 1760, p. 2.
[1] Middlesex Probate Records, xii, 329.

Square. The captain was a respected citizen and a holder of town offices. Joshua's college career was cut off in the third quarter of his junior year, very likely by invitation of the faculty, for fines were one of the chief items on his quarter-bills.

On June 15, 1712, Parker was married to Mary Fessenden, the sister of his brother-in-law, Nicholas Fessenden, A.B. 1701. Their daughter Mary was baptized in Cambridge on October 12, 1712. A year later they had removed to Sudbury, where Mary Parker died on March 16, 1714/15. A year or so later the widower had acquired to care for his two small daughters a second wife, whose given name was Parnall or Parnel.

There was a family named Parnall in Boston, and thither Parker removed in July, 1717. There, to reconstruct from the court records, he paid Richard Pullen 40*l* for a quarter interest in the White Horse tavern.[2] The selectmen, however, turned a deaf ear to his petitions to be allowed to sell strong drink or set up as an inn-holder, and on October 28 warned him to leave town. This he did, to the inconvenience of certain Boston and Dorchester gentlemen to whom he was indebted for buttons, mohair, shalloon, snakeroot, and cloves.[3] Their suits against him died when the sheriff reported that he could find neither goods nor body; but the widow of the keeper of the White Horse, who asserted that he had forged a receipt, was more persistent and kept after him. He was found in Cambridge, where he, "pointing to Cambridge prison Said, but if you will put me to trouble about what I owe you, there is a house provided for mee."[4] His father had offered to pay the bill, and probably did so.

Parker then settled in Sudbury as a yeoman, finding variety and a little ready money in serving as an enlisted man in the militia. When he was stationed at Rutland in 1722, Colonel Joseph Buckminster discovered that he had an interest in the plantation, and thought best to discharge him, giving him a 20*s* note to buy a substitute. When this note came back, it had strangely grown to 30*s*, and for this "and other Enormities" Buckminster brought Parker before the superior court, where

[2] Early Files in the Office of the Clerk of the Supreme Court of Suffolk County, 163812.

[3] *Ibid.*, 13150, 14868, 16447.

[4] *Ibid.*, 17382.

the jury found for the defendant.[5] Parker then haled his colonel into court for padding the muster rolls, but failed to prove his point.[6] In his disappointment he swore that "he would be Revenged" on Buckminster and his friends, particularly Captain Thomas Smith, whose "Divilish doing" it was that he was sent home from Rutland. Smith successfully sued Parker for slander.[7] The latter got revenge by going before the General Court, in which Buckminster sat for Framingham, and getting the colonel unseated for having juggled his rolls "for the Lucre of Gain." The sweetest part of the victory was that Buckminster's pay was docked and given to his accuser for expenses.[8]

For two years the three men battled in the courts, and on one occasion Parker was "taken on a Surprise at Boston distant from his Place of Residence and Friends and clapt into Goal." He frankly admitted to the court that, having observed how Buckminster had padded his bills when awarded costs by listing witnesses who never came, he himself resolved to profit in the same manner in his suit against Smith. But Smith, when he had costs assessed against him, objected to paying for six witnesses for Parker when there were none, and to submitting to subpœnas which Parker blandly swore had been issued but which had not.[9] Again Captain Parker was called upon to aid.

In 1725 Parker was back in Rutland with the Sudbury company which "scouted and guarded the meadow, for the people in their getting of hay."[10] It was profitable service until the General Court discovered that they were drawing pay to "March into the Enemy's Country," "Altho' they were in the Service but one Week, and did not March Twenty Miles out."[11]

The following year, on July 31, a jury found that Joshua Parker, gent., "of his own head and false Imagination Contrivance and Covin, Wittingly subtilly and falsly forged and Counterfeited A Certain false Deed . . . purporting that One

---

[5] Ms. Records of the Superior Court of Judicature of Suffolk (in the office of the clerk of the supreme court of Suffolk), 1721–1725, p. 97.

[6] *Ibid.*, 132.

[7] Suffolk Files, 17022, 17038, 17052.

[8] *Journals of the House of Representatives of Massachusetts* (M. H. S., 1919–), IV, 161, 162, 168, 175, 216.

[9] Records of the Superior Court, 1721–1725, p. 246; Suffolk Files, 17907, 18101.

[10] Alfred S. Hudson, *History of Sudbury* (Sudbury, 1889), p. 301.

[11] *House Journals*, VI, 334–5.

Joseph Reed of Sudbury did Bargan & Sell all his Upland and Meadow in Wethersfield . . . to the said Joshua Parker."[12] Consequently he was "Sentenced to be set upon the pillory in the Town of Cambridge aforesaid and there have one of His Ears Cut off, and have & Suffer Imprisonment by the Space of one whole Year." The condemned man petitioned pitifully against execution of the sentence, pleading his respectable mother and father in Cambridge, his brother Thomas (A.B. 1718) in the ministry, and "his Wife who descended from very Credible freinds by whome he has seven small Children." Sheriff Gookin complained of "Trouble and Affronts in the execution of the Writ against Joshua Parker" and threatened to prosecute Tutor Sever (A.B. 1701) for refusing to assist: "It is Intollerable that the Law should be so Abused by such villianous actions." The sentence was executed "October 14th 1726 between the Hours of ten and four."[13] From the Cambridge jail the unfortunate man in vain petitioned the General Court that his wife might "be allowed to Retail Strong Liquors in Sudbury for her support and her Families."[14]

Once more at liberty Parker managed to collect from the province for some express-riding done two years before,[15] and, still described in the writs as "gentleman," began to pursue Colonel Edmund Goffe (A.B. 1690) through the courts with some success.[16] July, 1731, found him once more in the Cambridge jail, convicted of "selling strong drink contrary to Law."[17] A year later he was at liberty and again being sued.[18]

That is the last that we hear of Joshua Parker. His widow married Uriah Moore at Sudbury on January 2, 1742. Of Parker's seven children, two were by his first wife, and all but the first were born at Sudbury: (1) Mary, bap. Oct. 12, 1712. (2) Elizabeth, b. Feb. 23, 1713/14. (3) Josiah, b. Apr. 19, 1717. (4) John, b. Mar. 15, 1718/19. (5) Susanna, b. Dec. 23, 1723; d. July 31, 1740. (6) Sarah, b. July 14, 1726. (7) Jerusha, b. Dec. 31, 1728; d. in infancy.

[12] Suffolk Files, 164231.
[13] *Ibid.*, 19763.
[14] *House Journals*, VII, 186–7.
[15] *Ibid.*, 295, 314.
[16] Suffolk Files, 20518.
[17] *Ibid.*, 32097, 32990; Records of the Superior Court, 1730–1733, p. 9.
[18] Misc. Mss. (M. H. S.), III, 81. F. 37.

SAMUEL PHILLIPS

The mariners are the least adequately recorded of our graduates because they sailed over the horizon to unrecorded adventures among Turks and Moors, or in Africa or the Indies, to lay their bones in unknown places. Phillips was more fortunate in coming from a bookish family and in coming to his end near home.

SAMUEL PHILLIPS, mariner, was the son of the book-seller, Samuel Phillips, by his "pretty and obliging Wife," Hannah. The father was a business friend of John Dunton, who described him as "young, witty, and the most Beautiful Man in the whole Town of Boston; He's very Just, and (as an effect of that) Thriving."[1] More than that, he was a member of the Old South, "a vertuous Man, an exemplary Christian, an Indulgent Husband, a kind Father, and a true Friend."[2] To these remarkable parents little Samuel was born on May 26, 1693, probably in the family home on what is now Washington St. near Court St. He must have perplexed them sadly, for he was fined in thirteen of the sixteen quarters of his college career. The fines were incredible: one of 10*s* being coupled with a bill for 11*s* 6*d* worth of window-glass. Such activities probably cost his father more than his formal education.

Obviously Samuel was not cut out for the ministry; so a year or two after graduation he sailed from Marblehead in the *Province Galley* with a cargo of bad fish. In September, 1714, he wrote from Barcelona that they had tried peddling their powerful cargo at 'Cales, Allicant, Mayorca, Minorca, and Salee,' but that even the strong-stomached Moors were not eager to buy. Barcelona being in a state of siege, they lay outside waiting for it to surrender; but when it did, even the starved inhabitants rejected their offering.[3]

[1] John Dunton, *Letters from New England* (Prince Society, 1867), p. 79.
[2] *Boston Gazette*, Oct. 31, 1720, p. 4.
[3] Early Files in the Office of the Clerk of the Supreme Court of Suffolk County, 9821.

Phillips made the most of his opportunity for research and mailed back to the College a *quæstio* entitled, "An Papa potius quam Turca habendus sit pro Antichristo? Aff." On the strength of this the President and Fellows at Commencement, 1715, declared him a Master of Arts, "Gratia Curatorum."[4]

When the *Province Galley* arrived from London in 1716, lawsuits began at once.[5] In June, 1717, Mr. Phillips was testifying in regard to the fish.[6] In the fall he made a short voyage which came to disaster on October 28:

On Monday last a very Awful & Lamentable thing fell out, A Vessel from Newfoundland for this Place, having Sixteen Persons on board struck upon Cony-Hasset Rocks, who was split & lost in a few Minutes, Six of those on board got a-Shoar in the Boat, Four were got a-Shoar next Day, who were alive on the Rocks, Six were Drowned, viz. Mr. Patrick Ogilvie, Mr. Samuel Phillips Jun. Mr. Shippard Mariner, the other three know not their Names; and of those Drowned, none yet found but Mr. Phillips, who was Decently Interr'd here on Friday last.[7]

The funeral was on November 1, and among the mourners was Major General Winthrop.[8]

---

[4] John Leverett, Ms. Books Relating to College Affairs (Harvard College Library), p. 103. The *Quinquennial* is in error in assigning this degree to Samuel Phillips, A.B. 1708.

[5] *Essex Institute Hist Coll.* LVIII, 169–70.

[6] Suffolk Files, 163288.

[7] *Boston News-Letter*, Nov. 4, 1717.

[8] Samuel Sewall, *Diary* (5 *Coll. M. H. S.* v–VII), III, 145.

HUGH HALL

HUGH HALL, a Boston merchant, was the son of a Barbados gentleman of the same name and occupation. His mother was Lydia, the daughter of Benjamin Gibbs of Boston. Mrs. Hall died in Philadelphia in 1699, and Hugh, then about six years old, was sent to Boston and placed under the care of Grandmother Gibbs, who was then enjoying as her third husband William Colman, the father of Benjamin (A.B. 1692). Hugh was the kind of small boy who took notes on the sermons he heard,[1] and at the college was much more orderly than the boy at the head of the class was expected to be. His father entrusted him with business errands and the disposal of goods, some of which were turned over to the steward's wife in payment of term bills.[2] Although quiet and practical, he was not drab, for he had three of his father's silk coats (one of them with peach bloom lining) cut down for his own use. Judge Sewall records that at the Commencement of 1713 "Sir Hall made the Oration very well."[3] He remained in residence three years longer, and for the second degree offered an affirmative answer to the *Quaestio*, "An Dogma illud Jesuiticum, Fides Hæreticis non est Servanda, sit maxime Abominandum." At that Commencement likewise he made the masters' oration.[4]

The Reverend Benjamin Colman and President Leverett urged the elder Hall to permit Hugh to enter the ministry, to which he inclined, but without success. Hugh obediently sailed for home in October, 1716, and on the way encountered a storm, which, however bad it may have been, could not have been so bad as the verse in which he celebrated it:

> To Those that Travell through the Seas,
> the Lord is their Defence;
> His Wisdom always is their Guide,
> their Help Omnipotence.

*   *   *   *

[1] Ms. volume of sermon notes at Mass. Hist. Soc.

[2] Hugh Hall, Sr., to Hugh Hall, Jr., Apr. 22, May 28, 1712 (copies in the possession of Mrs. Jerome W. Coombs of Scarsdale, N. Y.).

[3] Samuel Sewall, *Diary* (5 *Coll. M. H. S.* v–vii), ii, 390.

[4] Joseph Sewall, Diary, 1711–16 (in the possession of Miss Edith Woolsey of New Haven), p. 165.

Confusion dwelt in every Face,
    & Fear in Every Heart,
When the Unyeilded Waves did Rage
    & Quash the Pilotts Art.

*   *   *   *

The Storm Allayed the Winds Retir'd,
    Obedient to thy Will;
The Sea that Roar'd at thy Command,
    At thy Command was still.[5]

Hugh was greeted with a "joyful & Affectionate Reception," and his proud father, amazed at the amount which he had saved out of his college allowance, gave him an additional sum to set up in business.[6] The new merchant promptly wrote back to Boston to establish commercial connections:

I am now from the Assistance my Father affords me, & the Encouraging Offers he yet Tenders, fixing my Self in a Merchantile life & have very promising Views of an Answerable Success to the Schem laid; however if any of Your or my Friends are for Adventuring this way, I am Conscious there are none here in a better Capacity to Serve them than my Self, which You may hint at every Oportunity that shall propitiously Occur.[7]

To his friend Elisha Callender (A.B. 1710), who was studying for the Baptist ministry, he wrote that Barbados was not so wicked as its reputation,[8] and to President Leverett he wrote saying that he intended to go to London, and asking for recommendations "to some of the Royal Society, or South Sea Company; either of which I am well assured will be very Serviceable to some Views I have upon Initiating my Self into a Correspondency with them." [9]

Hall sailed for London in April, 1717, and after a voyage of six weeks, landed in Sussex and proceeded to London. Here he was coldly received by his father's representative, who had urged against his being taken into the business, and he was insulted by being told about the "new" books, which it was assumed that he had not seen in the provinces.[10] He wrote

---

[5] Hugh Hall, Jr., Letter Book (Mrs. Coombs), pp. 73–4. Seasickness may be an explanation. For a full account of the letter book see *Publ. Col. Soc. Mass.* XXXII, 514–21.

[6] Coombs copy, Mar. 6, 1716/17.

[7] Letter Book, p. 1.

[8] *Ibid.*, p. 5.

[9] *Ibid.*, p. 9.

[10] *Ibid.*, p. 21.

HUGH HALL

amusingly of the absurdities of London society and reported
that English students were rustic and unpolished.[11] Leverett's
letters of recommendation had introduced him into "the Company of several Members of Parliament, & an Intimacy with
Men of bright Character & good Figure," and as a small testimony of his debt to his "Beautiful Mother," Harvard, he "dedicated to the Use of her Sons, a Compleat System of Divinity,
by that Noble Prelate, Bishop Beveridge." [12] He travelled
widely in England, visiting Oxford and spending some time
at the country seats of Sir William Humfreys and Sir Peter
Meyer.[13] By way of business he negotiated for the shipment of
slaves from Guinea to his father. In December he sailed for
home.

During the next few years Hall was alternately in Barbados
and Boston, fighting at law for the estate of his grandfather,
Benjamin Gibbs, which Major John Richards had acquired on
mortgage from Grandmother Gibbs's second husband, Anthony
Checkley. At the same time he was active in the slave trade,
although the blacks he sold had a way of dying before he could
get them off his hands. Once he unsuccessfully petitioned the
Massachusetts House "praying that the Duty of Four Pounds
on a Negro Boy he lately Imported . . . be abated; Inasmuch
as the Boy Died in Ten Weeks after his Arrival here." [14] His
letters from Barbados give a nauseating account of the trade
and suggest efforts on his part to conceal from prospective buy-
ers the fact that smallpox, flux, and eye infection raged in his
stock.[15] Of the small merchants who accused him of cheating,
he wrote, "Though Fortune had Canted them a little above
their Mechanick Fraternity, yet they had much better been
Improving themselves in their Old Philosophy of the Spade,
Trowell, & Anvil; than Vilely Treating their Superiours." For
his part, he thanked God for his Harvard education which had

[11] An example of what Hall expected of a "polished" scholar may be culled from
one of his letters to a college friend: "It is now again my good Fortune to be excluded
the Stride of Colossus and no Longer to be in danger of Submarine Mountains, where
the echo's of the Ominous Pitteril Resound, & the Corbosants are Capering most high
before poor mortals tost from Pillow to Post, Cooing out their fearfull Apprehensions
of an Elementary Change." *New England Hist. Gen. Reg.* XLII, 300.

[12] Letter Book, p. 23.

[13] Hall gives his itinerary in his Ms. almanac (Mrs. Coombs).

[14] *Journals of the House of Representatives of Massachusetts* (M. H. S., 1919—), II, 67.

[15] Letter Book, p. 57.

"laid a Foundation for the Strictest Morality." "I can Say I always Act from the Golden Law of Equity of doing as I would be done by." [16] To Benjamin Colman he wrote indignantly of the treatment of the "Poor Slaves, as if they had no more Souls than Brutes, & were really a Species below Us." [17] As a side line he and Benning Wentworth (A.B. 1715), future governor of New Hampshire, engaged in a little venture in contraband "B[rand]y" which he shipped "in a Barbados Terse, as Rum, to Prevent any Suspition or Trouble about it" on the part of the customs officers.[18]

The Hall family was doing well in Barbados, and in 1719 the father was appointed judge of vice admiralty. Hugh served as his deputy [19] and was a member of the commission of the peace, but was impatient to make a sufficient fortune to get married and settle in Boston. To a friend there he wrote: "I shall be pretty Difficult under my present [bachelor] Circumstances, & much more if I Arrive to be a Ten Thousand pound Man, which I doubt not a few Years will Effect . . . upon which I shall Spend the Remainder of my Dayes in Your Place." [20] Puritan Boston, he thought, better suited his inclinations and character:

It's a singular satisfaction . . . that I have spent nigh twenty years of my Life in that Metropolis. I never was Taxt with any thing Dishonorable, Unmanly or Unchristian, & I . . . may say . . . I have by my Education laid a good Foundation for Strict Piety as well as undisguised Probity, & that I have always lived up to that Lunate Principle of doing as I would be done by. [21]

On one visit to Boston, he wrote to his father:

It will be high time for me upon my Return with You, to think of the Conjugal State; & I Presume it may not be Amiss, to get a Private Information of the Quantum of Mrs P- - -y (if Mr H- - -r dont Address her) & how the Pulse of the Family would beat at my Application.

But if You should upon my Arrival Advise me better I shall wholly Submit to You, & beleive I can heartily Close with Your Choice, knowing You will Study with all Propper Deliberation my highest Felicity. . . .[22]

[16] Letter Book, p. 95.
[17] Colman Mss. (M. H. S.), Mar. 30, 1720.
[18] *New England H. G. R.* XLII, 303.
[19] *Belcher Papers* (6 *Coll. M. H. S.* VI–VII), II, 370–1.
[20] Letter Book, p. 176.
[21] Hall Mss. (Mrs. Coombs), Oct. 20, 1718.
[22] Letter Book, p. 92.

No doubt the elder Hall approved when his son was married, on October 31, 1722, to Elizabeth Pitts of Boston. She was a daughter of John and Elizabeth (Lindall) Pitts, and her father was, one of Hugh's college friends said approvingly, "a Gentleman of a large Estate." [23] The couple settled in Boston, where Hall bought a house on the south side of Tileston Street and in the next fifteen years acquired much additional real estate. He prospered as a commission merchant, at times doing business as a member of the house of Bowdoin, Hall, and Pitts. His stock in trade included "blew & Gold Japan Clocks," "Good Barbados Sugars, Bag Hollands, Garlets, Druggets, Mulmuls, and Caps of sundry prices," but above all, slaves.[24] He inherited property in Pennsylvania and bought some in Sutton, but his interest was always trade.

Except for the fact that Hall subscribed to the *Chronological History* of Thomas Prince (A.B. 1707), he showed very little interest in literature or the college, nor was he inclined to politics. When he was first married he was elected constable, but was excused on the grounds that he had held higher office in Barbados. In later years he served on many town committees, particularly those to visit and warn undesirable inhabitants and to trouble harmless Huguenots at the time of the French spy scare. In 1732 he served as a special justice on the court of common pleas, and sat on the case of a judge of vice admiralty who had been making more money from his office than the law allowed. Perhaps it was this that excited his interest in obtaining an admiralty judgeship for himself. He offered Jonathan Belcher, Jr., (A.B. 1728) twenty-five guineas for urging his appointment at court, and he had the hearty support of his "very good friend," Governor Belcher (A.B. 1699), who recommended him to the Admiralty as "a gentleman of good integrity & capacity, of a liberal education, & of a plentiful fortune." [25] The governor said that the appointment would be a personal favor to him, which was a fatal recommendation, for the block and axe were already prepared for the Belcher administration. The friendship between Hall and Belcher may have been largely an

[23] *Publ. C. S. M.* xxvi, 394.

[24] *New-England Weekly Journal*, Aug. 14, Sept. 11, 1732; Early Files in the Office of the Clerk of the Supreme Court of Suffolk County, 19,662; customs statements, etc., in Mass. Archives, *passim*.

[25] *Belcher Papers*, ii, 370–1, *et passim*.

outgrowth of seven years of labor which the former put into the battle against inflation. His signature appears on the notes issued by the hard-money party in 1733 and 1740.[26]

Hall's first children were baptized at Benjamin Colman's church in Brattle Square, but in 1736 he took a leading part in the foundation of the West Church. Ten years later he was back with Colman. Although ardently religious, he was not narrowly sectarian, for he subscribed liberally for the purchase of bells for Christ Church and an organ for King's Chapel.

In these years Hall needed the consolation of religion. He appears to have suffered from the business depression, for he sold a great part of his Boston property, including his mansion house, which was described as "near the Orange-Tree, three Story high, four Rooms on the Floor, besides a Kitchen; with a Barn, Garden, & other Accomodations." [27] His wife died "after a long Indisposition" on May 16, 1747.[28] In his bereavement he appears to have turned to the poets, for after the death of his daughter Lydia [29] he wrote in an almanac:

> Oh! Lead me where my Lydia lies
> Cold as the Marble Stone
> I will recall her with my Cryes
> And wake her with my Moan.
>
>   * * * *
>
> Since then my Love, my Souls delight,
> Thou canst not come to me;
> Rather than live without thy Sight
> I'le find the way to Thee.[30]

These losses appear to have embittered Hall, for thereafter he was frequently in trouble for the way in which he carried out his duties as a justice of the peace and quorum. Once two angry farmers "in the King's highway with force and arms did assault Hugh Hall Esq" and would have damaged him had not "Pitt Hall his son distracted & hindered them." [31] Later he seized a supposed deserter from the royal navy and committed him "to

[26] *New England H. G. R.* LVII, 280, 388; *New-England Weekly Journal*, Sept. 16, Nov. 11, 1740.
[27] *Boston Gazette*, Jan. 1, 1754, p. 3.
[28] *Ibid.*, May 19, 1747, p. 3.
[29] *Ibid.*, Jan. 4, 1762, p. 3.
[30] Ms. almanac (Mrs. Coombs).
[31] Suffolk Files, 63,546.

his Majesty's Goal . . . upon Suspicion of Disertion, without Trial or Conviction," where he lay illegally confined for eight months. The House of Representatives took up the case and after some difficulty compelled the Governor and Council to set April 14, 1748, for a hearing of its charges against Hall for this action. The Council Chamber proved too small for the crowd which gathered, so the hearing was adjourned to Faneuil Hall, where the defendant attempted to have the case dismissed on the grounds that the House had no authority to prosecute complaints "for Injuries supposed to be done to private Persons." The public hearings continued for a week (ignored by the press) until the Council voted unanimously that he "had been guilty of illegal & unwarrantable Practices . . . & thereby rendered himself unworthy any longer to sustain the . . . Office" of justice of the peace. On May 24 the Governor publicly struck his name from the list of justices.[32]

Hall was also very much of a stormy petrel in town affairs. The selectmen prosecuted him for renting his houses without first giving them the opportunity to pass on the tenants, as the law provided, and for ten years tried to get him to pay over 8*l* which he had fined an offender on their behalf.[33] Small disputes grew into great ones because of the "haughty Airs" which he showed to even his fellow Harvard men.[34]

The weight of years made Hall's last decade more quiet. He died on June 15, 1773, and was interred in the Old Granary burying ground. The pastel portrait by Copley which is here reproduced is owned by a descendant of Hall's, Michael C. Janeway (A.B. 1962). The estate was troubled by the needs of a *non compos mentis* son [35] and by the exile of the Loyalist administrator, Foster Hutchinson (A.B. 1743). There were eleven children: (1) Elizabeth, b. Oct. 2, 1723; d. in infancy. (2) Lydia, b. Apr. 20, 1725; int. to m. Daniel Fowle of Portsmouth, Jan. 31, 1750; d. Dec. 31, 1761. (3) Elizabeth, b. Sept. 4, 1727; m. John Welch, 1753; d. after 1788. (4) Pitts, b. Feb. 8, 1728/9; A.B. 1747;

[32] *House Journals* (Boston, 1747–48), June 22, 23, 25, Aug. 25, Sept. 4, 1747, Apr. 7, 9, 20, 1748; Executive Records of the Province Council (Mass. Archives), 1744–57, pp. 32–41.

[33] *Selectmen's Minutes from 1754 through 1763* (Boston Records Commissioners, xix), pp. 62, 157.

[34] Paine Mss. (M. H. S.), I, 103.

[35] *Massachusetts Spy*, May 5, 1774, p. 3, May 19, p. 1; *Boston Post-Boy*, May 9, p. 4.

d. May 25, 1752. (5) Hugh, b. Feb. 4, 1731; d. in infancy.
(6) Gibbs, b. Jan. 18, 1733; d. in infancy. (7) Mary, b. Oct. 3,
1736; d. May 13, 1757. (8) Hugh, b. Sept. 2, 1737; d. in infancy.
(9) Sarah, b. Feb. 2, 1738/9; m. Elisha Clarke, Oct. 21, 1766, and
Wensley Hobby, May 19, 1785; d. Aug. 3, 1801. (10) Benjamin, b. Apr. 17, 1740; m. Sarah Marrow, Apr. 22, 1774, and
Margaret – – – –. (11) Susanna, b. Mar. 10, 1741; d. in childhood.[36]

Mrs. Jerome W. Coombs of Scarsdale, New York, owns Hall's
Ms. diary for 1714–17, and his letter book for 1716–20. The
New England Historic Genealogical Society has an interleaved almanac for 1723 and formerly had a volume of notes on
the Boston Thursday lectures of 1706–8; the Massachusetts
Historical Society has the companion volume of sermon notes
for 1709, a book of business accounts for 1728–30, and a few
scattered letters.

THOMAS WALTER

THOMAS WALTER, musician and minister of Roxbury, was born
in that town on December 13, 1696, the second son of the Reverend Nehemiah Walter (A.B. 1684), and a grandson of Increase
Mather. When he entered college he was placed first in his
class, but was later forced to give way to Hugh Hall. The college records show that his first year was a lively one; but he
soon settled down, and left no record of the wickedness with
which tradition has credited him. The college gave him liberal
financial aid. His scholarship may be judged by the fact that
Charles Chauncy (A.B. 1721), who knew him at this period, in
later years reckoned him as one of "the three first for extent
and strength of genius and power New-England has ever yet
produced":

I . . . often had occasion to admire [him] for the superlative excellence of his natural and acquired accomplishments. His genius was
universal and yet surprisingly strong. He seemed to have almost an
intuitive knowledge of every thing. There was no subject but he was
perfectly acquainted with it; and such was the power he had over his
thoughts and words, that he could readily and without any pains,
write or speak just what he would. He loved company and diversion,

[36] Compiled from the family Mss., supplemented, and corrected where obviously
wrong.

which prevented his being the greatest student; and he had no need to study much: for his powers were so quick and retentive that he heard nothing but it became his own. . . . He made himself master of almost all Dr. Cotton Mather's learning, by taking frequent opportunities of conversing with him. I suppose he gained more learning this way than most others could have done by a whole life's hard study.[1]

Some months before graduation Walter began to keep the Dedham school, and remained there for a year or two. He then turned his attention to theology, for he was "designed for the evangelical Ministry."[2] At the Commencement of 1716 he took his second degree, giving the traditional negative answer to the *Quaestio*, "An ad bonitatem Actus sufficiat bonitas Intentionis?" A few days later Cotton Mather entered in his diary the fact that he was grievously disappointed in his brilliant young nephew: "I will once more admonish him; and if no Impressions be made, I will then cast him off."[3] The difficulty appears to have been that the young man had been attracted into "fatal Entanglements" with the gay and anti-Puritan group of which the magnetic John Checkley was a leader.[4] Mather's admonitions were disregarded, and he was enraged at "miserable T. W., abandoned by the Wrath of God, into unaccountable Stupidity!"[5]

For two months the angry Mather excluded his nephew from his household, and then resolved to make "Yett one Essay more to recover" him.[6] This succeeded, for in December, 1716, Walter began to preach publicly and was invited to occupy the Needham pulpit. In the spring he was "lying dangerously and dubiously sick, at Dedham,"[7] and in the fall he was admitted into full communion in his father's church at Roxbury. He preached from various pulpits, including that of the Old South where he "pray'd well, and made a very good Discourse." On another occasion at the Old South he "made a very good Sermon" and showed unusual musical ability in setting the

---

[1] *Coll. Mass. Hist. Soc.* x, 155–6.
[2] Cotton Mather, *Diary* (7 *Coll. M. H. S.* vii–viii), ii, 232.
[3] *Ibid.*, ii, 359.
[4] John Eliot (A.B. 1772), *Biographical Dictionary* (Salem and Boston, 1809).
[5] Mather, *Diary*, ii, 363–4.
[6] *Ibid.*, ii, 376.
[7] *Ibid.*, ii, 459. He was "taken sick of the measels at Dedham, they turned well, but a feaver sets in & he is very dangerously ill, but Little hopes of him." Nathaniel Cotton (A.B. 1717) to Roland Cotton, Misc. Mss. (M. H. S.), June 12, 1717.

psalm.[8] Mather eagerly moved to have the young man made his colleague, but in this was blocked by certain leading parishioners:

All the Brethren of the Church, except four or five Gentlemen, who must always be the Rulers of all, are fond of Inviting Mr. Walter unto the Assistence and Succession in the Ministry. . . . There is now a more general Desire, and a very vehement One, for this Person, who is one of rich and rare Accomplishments, and such another cannot presently be hoped for. But from I know not what Principle, these Gentlemen clog all our Motions; and Roxbury is like to sieze upon him.[9]

The First Church at Roxbury had already voted unanimously to call the young man as his father's colleague, and on October 29, 1718,[10] he was ordained, Mather presiding. On Christmas Day he was married to Rebecca, the daughter of the Reverend Joseph Belcher (A.B. 1690) of Dedham.[11]

The return of Walter to the Mather fold irritated John Checkley, who, in a pamphlet [12] directed against the Calvinistic leanings of the Congregational church, took the opportunity to make slighting comments on Mather's remarks at the Roxbury ordination. The Puritan clergy were in an uncomfortable position because, while logic compelled them to accept the disagreeable facts of election and predestination, they were, at this time, accustomed to dismiss them as inscrutable mysteries of God, and to preach a happier religion. To bring forth these skeletons from their theological closet for a sober anatomical dispute with Checkley would have been to play into his hand, for his chief purpose was to call attention to the existence of them. Here Walter's wit came to the aid of his party, and he produced a pamphlet which, while accepting election and predestination, distracted attention from theological argument by quips: "Now stand clear, the Prefacer has took Voyage over to Holland. Huzzah! The Dutch are routed! Jack has broke down the Dikes, and drowned them." [13]

[8] Samuel Sewall, *Diary* (5 *Coll. M. H. S.* v–vii), iii, 142, 200–1.

[9] Mather, *Diary*, ii, 519.

[10] This date is frequently given as the 19th, but contemporary diaries agree that it was the 29th.

[11] His classmate and successor in the Dedham school, Perez Bradford, married a sister of Rebecca.

[12] *Choice Dialogues Between a Godly Minister & an Honest Countryman* (Boston, 1719).

[13] *A Choice Dialogue Between John Faustus a Conjurer and Jack Tory his Friend* (Boston, 1720), pp. xiv–xv.

Checkley described this pamphlet to the Reverend James McSparran as "being the joint Labours of the grand Committee, but taggd together by Mr Walter and by him adorned with those many Billingsgate Flowers which have so delicately perfum'd the whole Piece." [14] In private Walter was "particularly insulted and abused by the scurrilous and scandalous Libels" of the Checkley faction,[15] and publicly slurred in their organ, the *New-England Courant*. Walter replied, in a vein that then passed for wit, with a broadside entitled *The Little-Compton Scourge: or, The Anti-Courant*. In the next edition of the *Courant* John Gibbins (A.B. 1706) replied by hinting that Walter drank too much and threatening to publish scandals about him. But Checkley, in the same edition, produced the scandals. He called Walter an "obscene and fuddling Merry-Andrew," the "Tom-Bully" of the Mather faction, and printed what he claimed to be a letter from Walter promising him "a Kick on the Arse, a Slap on the Chops":

The very same Night . . . that he wrote this Letter, he was with another Debauchee, at a Lodging with two Sisters, of not the best Reputation in the World, upon the Bed with them several Hours, and this Spark sent for Punch to treat them with, and would have had the Candle put out, but they not having a Conveniency to light it again, it was lock'd in a Closet, and ---- etc." [16]

Too fond of the bottle Walter may have been, but to accuse the young husband of being a rake was too much. Public opinion reacted in his favor, the *Courant* published its regrets, and Checkley's letters no longer appeared in its columns. He had been led into the trap he had laid for the Congregational clergy, and had impaled himself on his own pen.

The religious dispute drifted into the controversy over inoculation against smallpox, where the party lines were much the same. Thomas Walter, Samuel Aspinwall (A.B. 1714), and Richard Dana (A.B. 1718) were inoculated together, lay in the same chamber, and made a picnic of the whole affair.[17] One night, while Walter was recovering at his Uncle Cotton's, some

[14] *John Checkley* (*Prince Soc. Publ.*, 1897), II, 154-5.

[15] Mather, *Diary*, II, 605.

[16] *New-England Courant*, Aug. 21, 1721, pp. 2-3.

[17] Zabdiel Boylston, *An Historical Account of the Small-Pox Inoculated in New England* (Boston, 1730), p. 18.

unrestrained member of the Checkley-Douglass faction threw a bomb into his room:

The Weight of the Iron Ball alone, had it fallen upon his Head, would have been enough to have done Part of the Business designed. . . . The Granado was charged . . . in such a Manner, that upon its going off, it must have splitt, and have probably killed the Persons in the Room, and certainly fired the Chamber. . . . But . . . the Granado passing thro' the Window, had by the Iron in the Middle of the Casement, such a Turn given to it, that . . . the Fuse was violently shaken out. . . .[18]

Again the popular reaction was in favor of the Mather faction, and others, including the Reverend Nehemiah Walter and his wife, flocked to be inoculated.[19]

Walter, like Mather, was eager to rescue New England church music from the slough into which it had fallen, and in 1721 he published a little book in which he explained the rules for reading music and printed the favorite hymn tunes in modern musical notation, the first time, apparently, that the modern system was used in the colonies. The introduction explained the difficulties he faced:

Singing is reducible to the Rules of Art; and he who has made himself Master of a few of these Rules, is able at first Sight to sing Hundreds of new Tunes, which he never saw or heard of before, and this by the bare Inspection of the Notes, without hearing them from the Mouth of a Singer. Just as a Person who has learned all the Rules of Reading, is able to read any new Book, without any further Help or Instruction. This is a Truth, although known to, and proved by many of us, yet very hardly to be received and credited in the Country.[20]

Although the little book sold for 4s, it was very popular, and singing societies sprang up throughout the colony, aided and directed by the clergy:

On Thursday last in the Afternoon, a Lecture was held at the New Brick Church, by the Society for promoting Regular Singing in the Worship of God. The Reverend Mr. Thomas Walter of Roxbury preach'd an excellent Sermon of that Occasion . . . *The Sweet Psalmist of Israel* — The Singing was perform'd in Three Parts (according to Rule) by about Ninety Persons skill'd in that Science, to the great Satisfaction of a numerous Assembly there present.[21]

[18] Mather, *Diary*, ii, 657–8.
[19] Boylston, p. 24.
[20] *Grounds and Rules of Musick Explained* (Boston, 1746), p. 2.
[21] *New-England Courant*, Mar. 5, 1722. Samuel Dexter (A.B. 1720) said that Walter "preached a Charming sermon." Diary (Dedham Hist. Soc.).

No bequest that wealth could have made would have given the New Englanders of the next two generations the pleasure that came from the opening of the world of music to them by this little volume.

The young Roxbury minister was already "Languishing of that English Disease, a Consumption." [22] His last fire went into an answer which he and his Uncle Cotton prepared to a new Checkley tract, in which they demolished one of their opponent's authorities by calling him "that sorry superstitious Fellow, Austin the Monk." [23] His last months were spent in bed, sometimes at Mather's and sometimes at home, dictating the arguments he was too weak to pen. On Sunday, January 10, 1724/5, he failed so rapidly that his father, leaving the house to conduct the afternoon service in their pulpit, said his final farewell. "He continued in Heaven-ward Aspirations, till towards the close of the Afternoon-Service, when he fell a smiling: And one asking him the Reason of it, he only made an Unintelligible Answer, which intimated as if an admirable Consort, in Singing the Praises of his Redeemer now entertained him. So he expired. . . ." [24] His funeral was the kind he would have liked to attend, if we may judge by the bill:

To a coffin	2*l*	10*s*	0*d*
" the pall	0	12	0
" opening the tomb	0	10	0
" 5 dozen and 3 payrs of gloves, at 45*s*	12	0	0
" 6 rings	6	12	0
" a barrel of wine	9	01	6
" tolling the bell	0	01	6
" a box to put the bones of old Mr Eliot and others in	0	06	0
" pipes and tobacco	0	03	0
" three payres of women's mourning gloves, allowed to this accomnt by the town, att 36*s*	1	16	0
	33*l*	12	0[25]

Walter had dedicated the *Sweet Psalmist* to Paul Dudley (A.B. 1690), who now left a bequest to the church with the

[22] Cotton Mather, *Christodulus* (Boston, 1725), pp. 27–8.

[23] *An Essay upon that Paradox* (Boston, 1724), p. 41. This was a reply to Checkley's *Discourse Shewing Who is a True Pastor.*

[24] Mather, *Christodulus*, p. 31. This was Walter's funeral sermon. There is an account of his character by Thomas Prince and Joseph Sewall in Nehemiah Walter, *Discourses* (Boston, 1755), p. xx.

[25] Walter E. Thwing, *History of the First Church in Roxbury* (Boston, 1908), p. 115.

proviso that Mrs. Walter receive from the income 20*s* per annum as long as she remained a widow, which she did until death claimed her on April 30, 1790. The Walters had a daughter Rebecca, who died unmarried on January 11, 1780, and they may have had a daughter Sarah, about whom nothing is known.

WORKS

Walter may have written "An Account of Mrs. Katharin Mather," pp. 47–82, in Cotton Mather, *Victorina* (Boston, 1717). See Mather, *Diary*, II, 391.

A CHOICE DIALOGUE Between John Faustus a Conjurer, and Jack Tory his Friend. . . . By a Young Strippling. Boston, 1720. (2), xxi, (1), 79, (3) pages. BPL, H, MHS.

THE GROUNDS AND RULES OF MUSICK EXPLAINED. . . . ["A Recommendatory Preface" signed by Increase Mather, Cotton Mather, Nehemiah Walter, Joseph Sewall, Thomas Prince, John Webb, William Cooper, Thomas Foxcroft, Samuel Checkley, Joseph Belcher, Benjamin Wadsworth, Benjamin Colman, Nathaniel Williams, Nathaniel Hunting, and Peter Thacher. For a bibliographical account of the different editions of this work, with locations, see *Proc. Am. Antiq. Soc.* XLII, 235–46.] Boston, 1721, and five subsequent editions.

THE LITTLE-COMPTON SCOURGE: or, The Anti-Courant [Anon.]. Boston, [1721]. Broadside. LC.

Appendix to Cotton Mather, *Sober Sentiments* (Boston, 1722).

THE SWEET PSALMIST OF ISRAEL. A Sermon Preach'd at the Lecture Held in Boston, by the Society for Promoting Regular & Good Singing. . . . Boston, 1722. (8), 28 pages. AAS, BA, BPL, H, JCB, LC, MHS, NYH, NYP.

THE SCRIPTURES THE ONLY RULE OF FAITH & PRACTICE. As it was Delivered in a Sermon at the Lecture in Boston, September. 5. 1723. Boston, 1723. (4), ii, 45 pages [last page numbered 54 by error]. AAS, BA, BPL, CHS, CL, HEH, LC, MHS, Y.

AN ESSAY UPON THAT PARADOX, Infallibility may Sometimes Mistake. Or a Reply to a Discourse Concerning Episcopacy. . . . By a Son of Martin-Mar-Prelate. . . . [Errata vary.] Boston, 1724. (2), 120, (2) pages. AAS, BA, BPL, CL, H, MHS, Y.

BENNING WENTWORTH

BENNING WENTWORTH, Governor of New Hampshire, was born in Portsmouth on July 24, 1696, probably in the house which his grandfather Samuel Wentworth had built near the dock as a convenient place to "sell and brew beare." His father, John, was a wealthy merchant who was appointed to the Council when Benning was an undergraduate, and made lieutenant governor shortly thereafter. In the summer vacation of 1714 Benning began his political life by putting his signature to the treaty concluded with the Indians at Portsmouth.[1] At Cambridge he set a new high record for fines and broken windows. Thence he entered the countinghouse of a Boston relative, where he pondered the merchant's *Quaestio* which he answered in the affirmative at the Commencement of 1718: "An Mare æque ac Terra subjiciatur Legibus Proprietatis, et Dynastarum particularium Jurisdictioni?"

Wentworth had by this time established his own business in Boston, his stepbrother Hugh Hall (*q. v.*) acting as his Barbados agent and joining him in some brandy imports which would not have borne the scrutiny of the customs office. On December 31, 1719, he married Abigail, daughter of John Ruck, a Boston merchant. Increase Mather performed the ceremony, and Wentworth's first two children were baptized in the Mather church on the strength of their mother's membership. Although most of his business papers of the period [2] describe him as of Boston, he never owned real estate there and declined the only town office, that of constable, to which he was elected. During these years he was importing wine through Portsmouth (where the customs collector was more lax) and selling munitions to the New Hampshire government. Much of the family business was with Cadiz, where the Wentworths sold what they said was ordinary lumber and what their enemies alleged to be mast timber from the king's woods. Returning from Cadiz, late in 1730, Benning nearly ended his career. At the Old South of Boston, where some of his creditors worshiped, thanks were "return'd for the wonderfull Preservation of 5 of the Went-

[1] *Coll. Maine Hist. Soc.* vi, 258.
[2] In the Jeffries Mss. (Mass. Hist. Soc.).

BENNING WENTWORTH

worths (3 Brothers). The Ship was foundred, they were taken up in their Boat by a Ship that had Spent her Mast or masts." [3]

Later in that same year Lieutenant Governor John Wentworth died, leaving Benning a rich estate:

One moity of my farm in Cascobay Caled mare point & Two thousand pounds in any Part of my real Estate at a reasonable Price and where he Inclines also all my armory as guns Pistolls Swords etc and after his mothers deceas my Large Silver Punch Bowl.[4]

Although Benning continued for some time to date his business correspondence from Boston, he now assumed his place in the Portsmouth oligarchy, whose peculiar customs so interested Robert Hale (A.B. 1721) when he visited "the Bank":

Their Manner of Living here is very different from many other places. The Gentlemen treat at their own houses & seldom go to the Taverns. Their treats are Splendid, they drink Excessively all Sorts of Wine & Punch — their Women come not into Company, no not so much as at Dinner.

In fact in five dinners, three of them at Wentworth houses, Hale "saw not one woman except a serving Girl." [5]

When Jonathan Belcher (A.B. 1699) was appointed governor of Massachusetts and New Hampshire in 1730, John Wentworth wrote him humbly from Portsmouth asking to be continued as lieutenant governor. Benning wrote "as handsomly" from Spain,[6] but on his return "behav'd with a great deal of insolence & ill manners" toward the governor,[7] who in surprise and indignation sent a word of warning to Portsmouth: "Notwithstanding the young gentleman's pertness, you may let him know I will be his master & every body's else in the Province." [8] The trouble was, Belcher surmised, that he had not tossed enough of the political plums to the oligarchy: "I cou'd not give the bread from my own family to theirs." [9] David Dunbar, who had become lieutenant governor on the death of John Wentworth, was determined "to have a sett of his creatures" on the Council,[10] and to that end asked the Board of Trade to appoint

---

[3] Hamilton A. Hill, *History of the Old South Church* (Boston, 1890), I, 456.
[4] *Probate Records of New Hampshire (Provincial Papers Series)*, II, 379.
[5] Hale, Journal of an Expedition to Nova Scotia (Am. Antiq. Soc.), pp. 6–7.
[6] *Belcher Papers* (6 Coll. Mass. Hist. Soc. VI–VII), I, 325.
[7] *Ibid.*, p. 58.   [9] *Ibid.*, p. 325.
[8] *Ibid.*, p. 455.   [10] *Ibid.*, p. 84.

Benning Wentworth, Joshua Pierce, and Theodore Atkinson (A.B. 1718). When Belcher heard of it he wrote to his New Hampshire supporters that he despised "the contemptible simpleton" Wentworth, who had "neither head nor interest to do good or hurt" to the administration.[11] He took care, however, to keep track of Wentworth's sailings to Spain and England, and to send to the Board of Trade a set of affidavits that they might "see how unreasonable it is that such a rascal shou'd sit at the Council Board with the Governour." [12] Wentworth and Atkinson in turn sent to England a complaint against Belcher's military negligence, magnifying a few canoeloads of peaceful Indians intent on gathering berries into a menace against Fort William and Mary, twenty leagues away.[13]

In the summer of 1732 Wentworth wound up his affairs in Boston and obtained election to the House of Representatives from Portsmouth, where he entrenched for the battle. About the same time word reached Belcher that the order in council appointing Wentworth and his friends had passed before the arrival of the protest against them, and in his disappointment he made up his mind to withhold the mandamus to seat them when it should arrive.[14]

The documents arrived in Boston in December, 1732. The governor maintained that they were directed to him; be that as it may, the first that he heard of them was that they were in the hands of Wentworth, Atkinson, and Pierce. Angrily he ordered Richard Waldron (A.B. 1712), secretary of New Hampshire, to advertise for the mandamuses with a clear statement of the penalties of the law. Pierce, in a fright, brought in his, and was confirmed and sent back to Portsmouth with the gubernatorial blessing. But "Bin & At," as Belcher called them, did not present theirs until they "had done all the mischeif they cou'd in the House of Representatives," blocking the appropriation bills and generally making "every thing in the government as uneasy as possible." [15] Then they showed the mandamuses to the Council, but they refused to surrender them to be sent to Belcher in Boston.[16] Dunbar, acting governor in the absence of

---

[11] *Ibid.*, pp. 46, 98.
[12] *Ibid.*, p. 476.
[13] *Ibid.*, p. 154.
[14] *Ibid.*, p. 161.
[15] *Ibid.*, p. 295.
[16] Jonathan Belcher, Letter-Book (M. H. S.), Jan. 15, 1732/3.

Belcher, hesitated to admit the trouble-makers to the Council. Perhaps he, like his superior, suspected Wentworth of trying to buy the office of lieutenant governor.[17] From Boston Belcher wrote to Dunbar ordering him not to admit the two: "Had they acted with that insolence & rudeness to you as they did to me, I believe you wou'd 'a' thought a jayl a proper place to teach 'em more modesty & manners."[18] When Belcher visited New Hampshire, the offenders appeared, after long delay, with the missing mandamuses and demanded admission to the Council. "But after such an insolent behaviour," wrote the governor, "I could not think it for his Majesty's honour to admit them."[19] He was privately determined to make short work of "Pilgarlic," as he called Wentworth, and gathered affidavits showing "how vile & impudent" he had been.[20] The Board of Trade, however, pointed out that the governor did not have the power to disallow councillors in New Hampshire.[21] Reluctantly he called Wentworth before the Council at the meeting of January 1, 1733/4, and ordered the secretary to administer the oath, "to which Mr. Wentworth replied, 'I should have been glad to have known it sooner Sir — for I am now Engaged to serve in the assembly for this town and therefore cannot accept now but when the session is over I may be ready.'"[22] During that session of the Assembly Wentworth and Atkinson made all the trouble they could for Belcher, and in the autumn they were admitted to the Council, "tho'," complained the governor, "they have been and are the greatest opposers of the publick safety and justice."[23]

"Toby, the Cadix pedlar," as Wentworth's enemies called him, may have accepted the seat on the Council because he was too busy with his private affairs to continue to lead the resistance to Belcher in the House of Representatives. When relations between England and Spain became strained in 1733, the latter government refused payment on an enormous shipment of oak timber which he had, upon order, laid down at Cadiz. He hastened to Madrid and pleaded his case in vain.

---

[17] *Belcher Papers*, I, 261.
[18] *Ibid.*, p. 319.
[19] *Provincial Papers . . . of New Hampshire* (Concord, etc., 1867–1915), IV, 665.
[20] *Belcher Papers*, I, 378.
[21] *Journal of the Commissioners for Trade and Plantations*, 1728–34, pp. 391, 392.
[22] *Provincial Papers*, IV, 794.   [23] *Belcher Papers*, II, 158.

By 1735 his creditors in Boston were pressing.[24] For a time he kept his head above water by means of drafts on London, but in January, 1737/8, these were being protested.[25] The British government did what it could for him and London merchants like situated, but the growing feeling that war was inevitable put an end to the efforts of the diplomats. John Thomlinson, the New Hampshire agent in London, induced Wentworth's creditors to sign an agreement not to jail him if he appeared there, and summoned the "Spanish bankrupt," as Belcher jeeringly called him, to come and see if something might not "be done for him." [26] The bankrupt arrived in London in the late summer of 1738 and found that Thomlinson's "something" was a plot to have him made governor of New Hampshire. The Wentworth creditors joined the "she Sh-r-ly," who was in London to obtain the Massachusetts governorship for her husband (that "mean, false, ungrateful beggar," William Shirley), and other of Belcher's enemies in a jangling "triple confederacy" which agreed on but one thing, "Delenda est Carthago." [27] Belcher swiftly moved to get "the royal creditor's" place on the New Hampshire Council vacated on the grounds of his long absence and accused him of being a party to a "villanious forg'd letter." [28] Wentworth might have been, as Belcher assured himself, "very poor & insignificant," but Wentworth's creditors had political influence, and on February 6, 1740/1, Thomlinson wrote gleefully to Atkinson that "His Grace the Duke of New Castle has now actually promised him your Government so soon as it shall be seperated." [29] The £300 which the commission would cost was no barrier to Thomlinson, who passed the hat among the Wentworth relatives and friends, and on June 4, 1741, the document was signed. There was, as the new governor's nephew and successor later remarked, nothing honorary about the appointment.[30]

There was no prospect that the Portsmouth friends and relatives who had bought the commission for Wentworth would tolerate his administration of the government in such a way as

[24] Jeffries Mss. VII, 118, 119.
[25] Thomlinson to Wentworth (Library of Congress transcripts), Jan. 21, 1737/8.
[26] Ibid., July 14, 1738.
[27] Belcher Papers, II, 222, 314, 385.   [28] Ibid., p. 204.
[29] Provincial Papers, XVIII, 169.
[30] John Wentworth to Rockingham (L. C. trans.), Mar. 10, 1765.

to permit him to pay off his creditors. But the ever efficient Thomlinson had in mind "something more" to be "done for him." [31] The "something more" was the purchase for him of the post of surveyor of the king's woods in North America, which was finally consummated in 1743. The arithmetic involved was peculiar. On the credit side, Wentworth received a salary of £800. On the debit side he paid the salaries of four deputies (one of whom remained comfortably at home in Ireland); he paid Dunbar, the former surveyor, £2000 to resign; and he dropped claims of about £11,000 against the government, thus assuming the liability of meeting the claims of his personal creditors for something like the same amount. [32] Yet Wentworth ended his term of office possessed of a great fortune and popular applause, although his predecessor, Dunbar, had ended in the debtor's prison, despised of men.

At two in the morning, November 30, 1741, ex-Governor Belcher lay in his bed in the "lowly Cottage" at Milton and listened to the hoofs of the horses of Governor Benning Wentworth and his party as they rode triumphantly northward along the King's Highway from Cape Cod, where they had landed. [33] On December 10 the new governor and Thomas Hutchinson (A.B. 1727) stopped at Salem to consult with the Lynde clan, Harvard men all, high in the government of Massachusetts. [34] Three days later Wentworth arrived in New Hampshire, welcomed by his creditors.

The people of New Hampshire knew their new governor as an honest business man (according to provincial lights) of good average ability, a generous and kindly friend, but a furious and vindictive enemy who nursed his hatreds. Partly because he aped the haughty manners and governmental maxims of the Spanish grandees he had known, and partly because of his oft-repeated story of the dreadful mountain passes between Granada and Madrid, he was known, behind his back, as "Don Granada." [35]

The day after his arrival at Portsmouth he met the Assembly

---

[31] Thomlinson to Atkinson, *ibid.*, May 9, 1741.
[32] John Wentworth to Rockingham, *ibid.*, Mar. 10, 1765.
[33] *Belcher Papers*, II, 549.
[34] *Diaries of Benjamin Lynde and Benjamin Lynde Jr.* (Boston, 1880), p. 123.
[35] Jeremy Belknap (A.B. 1762), *History of New Hampshire*, II (Boston, 1791), 339; *Belcher Papers*, II, 515.

and asked for an adequate salary, which his friends defined as
1500*l* (provincial currency). The House, instead, voted to give
him, out of the funds raised for the West Indies expedition,
500*l* "towards the charge he has been at in coming to the Government, etc." In the end he accepted an annual salary of 500*l*
and although this was supplemented by what would appear to
be very generous extra grants, he complained that his expenses
were double his compensation and asked the royal government
to coerce the colony into generosity.[36] His former friends in the
House of Representatives were indignant at his Spanish manners and scale of values:

Your Excellencys manner of treating this House, the Representatives
of a Free People, is intirely new unparliamentory & without Precedent
of which your Excellency's verbal & written Messages of the 3rd Current are not the first Instances. Your Excellencys Reflections on the
House are so Gross & Coarse that should we answer them in equal
Terms and Language it might be justly stiled unworthy and unbecoming.[37]

It should be said that in the black and white of print the gov-
ernor's messages are not so astonishing as the resolutions of the
House; and that in time he learned moderation, which the
House never did.

Part of the general dislike of Wentworth was due to the fact
that he, like the rest of the oligarchy, flaunted his Anglican
religion in the face of a people who were mostly Congregationalists or Ulster Presbyterians. In 1743 he was elected to the
Corporation of the Society for Propagation of the Gospel, the
famous S. P. G. In accepting, he spoke of his stand against
"that flood of errors & Enthusiasm, propogated & Spread by
Whitefield," and rejoiced at "the Over-Ruleing Providence of
God in favour of our holy Church, in the removal of Mr Belcher." "This happy Event," he added, "makes me hope that
it will allways be his Majesty's pleasure to Appoint for Succeeding Governors Such only as are true sons of the Church." [38]
No doubt Belcher, who had been an honest governor, would
have been interested to hear Wentworth's creditors called "the
Over-Ruleing Providence of God." The governor served the

[36] Wentworth to Bedford (L. C. trans.), Oct. 24, 1748; *Provincial Papers*, v, 136–46.
[37] *Provincial Papers*, XVIII, 220.
[38] Bearcroft-Wentworth Correspondence, S. P. G. Mss. (L. C. trans.), *passim*.

S. P. G. assiduously, distributing its tracts among the savages
rather than among the Presbyterians, and pointing out that
missionaries might be sent with advantage to Massachusetts.
To aid in the establishing of an American bishop he offered to
bear the expense of laying out and obtaining a patent for an
episcopal land grant if the S. P. G. would move the king to
order it. He also promised to set aside in each new township a
lot for a glebe and one for the S. P. G., paying the fees himself.[39]

The unpopularity which Wentworth's war policies brought
upon him was all to his credit. He showed in these matters a
grasp of the greater issue and a determination to press it which
the members of the legislature never comprehended. In 1744
he reported French encroachments to the Board of Trade and
sent it a plan of the fort at Crown Point.[40] When the Louisbourg expedition was planned, Shirley wrote to him: "It would
have been an infinate satisfaction to me, and done great honor
to the expedition, if your limbs would have permitted you to
take the chief command." Wentworth promptly replied that
his gout could not keep him from his duty, and Shirley had to
retract the offer.[41] In spite of this the two men cooperated well
and relied upon one another's judgment, although they quarrelled about the payment of the New Hampshire soldiers by
Massachusetts, and Shirley complained to the Board of Trade
that Wentworth had not been frank about the intention of the
New Hampshire Assembly to acquire and neglect Fort Dummer.[42] Wentworth urged Shirley to send the victorious Louisbourg army against Montreal as a surer way of taking Crown
Point than by frontal attack,[43] and with him planned an attack
on St. Francis.[44] Nor was he a quill and ink soldier. He personally directed the strengthening of Fort William and Mary,[45]
sent out scalping parties, and paid them 50*l* for each of their
trophies.[46] There was no intercolonial diplomatic dawdling
when Indian raids struck Maine.[47] Of course there was con-

---

[39] Bearcroft-Wentworth Correspondence, *passim*.
[40] *Journal Comm. Trade and Plantations*, 1741/2–49, p. 122.
[41] William Shirley, *Correspondence* (New York, 1912), I, 187–8.
[42] 1 *Coll. M. H. S.* III, 108.
[43] Shirley, I, 206–8.
[44] *Provincial Papers*, XVIII, 299.
[45] Wentworth to Newcastle (L. C. trans.), Nov. 14, 1746.
[46] *Provincial Papers*, V, *passim*.
[47] Dartmouth College Archives, 746,271.

fusion at which the political faction out of power and responsibility could laugh:

Our Don Diego has acted such a part that he is quite entangled and hardly knows what to do about commissions, besides is embarrass'd by the Assembly, and moreover much plagued in his mind. . . . He said lately it was better to be a porter than a Governor.[48]

The Assembly was controlled by "patriots" and inflationists to whom the war was a God-sent opportunity. They compelled Wentworth to violate his instructions against issuing paper money by refusing to vote reenforcements to the Louisbourg garrison until he signed an inflation bill,[49] and they tried to compel him to abandon various elements of the royal prerogative by such practices as threatening to repudiate the engagements he had concluded with the Indians.[50] His complaints to the Board of Trade were bitter.[51]

Wentworth, at the outset of his administration, tried to choke the opposition with political plums, and to this end created a great swarm of justices of the peace, of whom a contemporary remarked:

> This was the happy silver age
> When magistrates, profoundly sage,
> O'erspread the land; and made, it seems,
> Justice run down the streets in streams.[52]

In general this plan succeeded, although it was not applied in the case of Belcher's old friend, Secretary Waldron, who was ignominiously ejected from the Council. The two exiles from office plotted "a bold push on Granada," the former governor setting forth the advantages of scandal:

A well cookt bill of parcels may be of great service in private conversation with those near the candle; and his illicit trade with Spain in naval stores, inabling that crown to build their powerful fleet, & now therewith to carry on a warr with Great Britain, I say, this should be one grand article, & that the late great v-ll-n [Sir Robert Walpole] as a reward for Granada's villany (I had almost said treason) gave him a government.[53]

[48] *Pepperrell Papers* (6 *Coll. M. H. S.* x), p. 113.
[49] *Ibid.*, p. 463; Shirley, I, 301.
[50] *Provincial Papers*, v, 221.
[51] Wentworth to Bedford (L. C. trans.), Oct. 24, 1748; *Journal Comm. Trade and Plantations*, 1741/2–49, p. 282.
[52] Belknap, II, 341.   [53] *Belcher Papers*, II, 431.

If the Don and his friends were "carefully watcht, & a journal kept of their proceedings," sooner or later they could be trapped.[54] The stigma of inflation could not be used as an argument, for Belcher and Waldron, like Wentworth, were hardmoney men; but there were endless possibilities in patriotism.

The first opportunity offered itself early in 1745. By the adjustment of the Massachusetts boundary, several towns and fragments of several others had come under the jurisdiction of New Hampshire. Wentworth issued writs for the election of representatives by these towns; the House denied his right to do so. Like the excellent politician he was, the governor dropped the issue until he could obtain specific instructions and thus shift the onus of the prerogative to the Board of Trade. Waldron, meanwhile, was getting restless under "Diego's" "persevering malice" and "ungent'm and cruel Partiality" against his sons.[55] He concocted a plot by which Colonel Isaac Royal of Medford was to finance an attempt to have himself made governor in Wentworth's place. They were encouraged by Colonel John Vassall (A.B. 1732), who had been after the same plum and reported that he could have obtained it easily "if he had had but the shadow of a complaint, for Mr. W------th had no interest at Court worth naming, nor any Friend save Capt. Tomlinson, but that the Ministers, when they made a removal chose to have some color or Pretence for it, tho' it was ever so insignificant." [56] Nothing was easier to produce than "color or Pretence," so Waldron promptly dispatched a complaint against Wentworth "exhibiting some parts of his male Administration," particularly "an invasion of the civil and religious rights of the People, superseding a Law, arbitrary removals of Judges, partially bestowing most of the places of Power and Profit on his relations, endeavoring to destroy the Independency of the General Assembly, taking money out of the Treasury without an Act for it, and cashiering and reducing worthy and brave officers, who served at the siege of Louisbourg." [57] The governor sniffed the wind, guessed that Waldron might be saying something about selling naval stores to Spain, and promptly wrote to the Duke of Bedford describing his "abhorrence of every thing that might have the least tendency to encourage an Illicit

54 *Belcher Papers*, II, 434.
55 *Provincial Papers*, VI, 40.
56 *Ibid.*, pp. 43–4.
57 *Ibid.*, p. 58.

Trade," which had "the good effect to prevent the Merchants of this Province from making any attempt to supply his Majesties Enemies." [58]

With the war drawing to a close, Wentworth prepared to reassert his right under the prerogative to issue writs for the election of representatives by the annexed towns. The royal government had not only approved his claim, but had suggested that his troubles were partly due to being too yielding. As early as May, 1748, Waldron had determined to attack along this line at the next session of the General Assembly.[59] It was, in the opinion of the modern authority on colonial government, the most serious challenge to the prerogative which had arisen in America.[60]

When the new Assembly met on January 5, 1748/9, it refused to seat the men chosen by the annexed towns and, with but a single dissenting vote, chose Waldron speaker. By a strange metamorphosis Belcher's former henchman was become the champion of popular rights. Wentworth disallowed the speaker on the grounds that the members from the annexed towns had been excluded from the voting. The House promptly denied his right to disallow a speaker and reasserted its right to reject representatives elected on no "other authority . . . than the King's writ . . . contrary . . . to the usage and custom of the Province." [61] With an eye to the veterans' vote it announced that "the Soldiery who scouted & Guarded the frontiers last Summer are crying aloud for their wages" which were being held up by the governor's obstinate stand, and it made insulting insinuations about the safety of the money parliament had contributed for the Louisbourg expenses, which was in the governor's hands.[62]

After a month of this Wentworth showed his instructions. The Assembly calmly demanded that he ignore them and submit until "his Majestys pleasure" should "be farther known." [63] To the king the Assembly addressed a silly petition quibbling about such points as the governor's right to combine annexed fragments of towns into a single district for election purposes.

[58] Wentworth to Bedford (L. C. trans.), Oct. 24, 1748.

[59] Provincial Papers, VI, 49.

[60] Leonard W. Labaree, Royal Government in America (New Haven, 1930), pp. 180–4.

[61] Provincial Papers, VI, 69–79.

[62] Ibid., p. 80.

[63] Ibid., p. 82.

The logic was inferior to the oratory: "The Governours arbitrary proceedings set the Government on fire. If he had consulted the good of the people & not private Interest, none of these difficulties would have happened." [64] The Assembly instructed the colony agent in London to work for the recall of the governor, "whose arbitrary attempts on our civil and religious Privileges, together with his unskilfull conduct, in the Administration of Government" had "bro't us into the most distressing circumstances, and rendered himself quite disagreeable to the Generality of our People." Particularly he had reserved to Harvard College or to the two senior ministers in New Hampshire the nomination of the pastor for a new town. [65] Indeed the Assembly was anxious to appeal to an even higher Authority by a fast for "the various Tokens of the Divine Displeasure," particularly the "unhappy circumstances of our Provincial Civil affairs." [66]

The New Hampshire Council and the Board of Trade stood firmly by the governor, and the Lincoln's Inn lawyer who had been consulted in behalf of the Assembly reported that its case was bad. If, he warned the Assembly, it continued to press this case against the governor, the effect would be to "fix him firmer in his seat of Government & procure him the Royal sanction and protection." All that the Assembly could do was to ask for a change of instructions to the governor and charge "him with such acts of male-administration as they shall be able to make out." [67] The difficulty here was that they were as deeply concerned as he in his chief act of "male-administration," the theft of the king's mast trees, but Captain Joseph Sherburn did appear on their behalf before the Board of Trade and complain that the governor had replaced him and another good militia officer by the Wentworth boys, "who had not done service." [68] In January, 1749/50, the heir apparent of the opposition, Colonel Royall, gave up the struggle, which had cost him 1000*l.*[69]

The Assembly then changed its tactics and begged the governor to recognize it as a legal body in order that appropriations might be voted for defence:

[64] *Provincial Papers*, VI, 89–100.
[65] *Ibid.*, p. 63.
[66] *Ibid.*, p. 106.
[67] *Ibid.*, p. 66.
[68] *Journal Comm. Trade and Plantations*, 1741/2–49, pp. 470–1.
[69] *Provincial Papers*, VI, 67–8.

At least for mercy's sake and in compassion to our poor Naked Brethren in the frontiers we earnestly Beseech your Excellency to allow us
to do what is our Duty for the Relief of as many of them as we can,
for by the accounts we have from all Quarters they are in great Danger
of the Indians.[70]

He answered by sending them a section of a letter from the home
authorities describing their stand as a "notorious and unjusti-
fiable . . . Act of disobedience." They replied by demanding
his "sorrowfull Reflections" on "his Majestys Exposed De-
fenceless subjects in the frontiers," and, secondly, asking for the
whole of the letter that they might take it into consideration
and reply to it. He naturally retorted that it was not addressed
to them, pointed out that they were not a legally constituted
body, and added some reflections on their veracity and literacy.[71]

For three long years the deadlock continued. The provincial
currency declined fifty per cent, the taxes remained uncollected,
the Louisbourg bounty money lay idle, the garrisons went
unpaid, and the registry office remained closed. Grimly Wentworth kept the weary representatives in session until the triennial precedent put an end to their misery on January 4, 1752.
The new Assembly without question admitted the members
from the annexed towns. It has been said, with reason, that
this was the most important political victory in the prerogative
field in the history of the colonies. Had there been more royal
governors with Wentworth's political courage, American history
would have been different.

A liberal distribution of jobs and commissions broke the back
of the "popular" party, but scraps from its "well cookt bill of
parcels" against the governor were magnified by the Revolution, became folklore, and still appear in popular novels.[72]
Among these is the tale of Molly Pitman, to whom the governor
is supposed to have made advances after the death of Mrs.
Wentworth on November 8, 1755. According to the story,
Molly rejected him and married a shipwright. The enraged
governor then had his lowly rival pressed into the navy, and for
seven years besieged the young wife, who nevertheless remained
loyal to her absent husband. Had stories of this sort been true,
Wentworth's enemies would have presented them to the Board

---

[70] *Ibid.*, p. 117.
[71] *Ibid.*, pp. 118–24.
[72] As in Kenneth Roberts, *Northwest Passage* (New York, 1937).

of Trade instead of the feeble and silly accusations they did make.

In these years Wentworth acquired an old house at Little Harbor, two miles from Portsmouth, and added to it an imposing series of rooms, one of which served as the Council Chamber. But it was a lonely great house, for the governor's son Benning, who had entered Harvard with the Class of 1741, died about the time that his mother did, to be followed on November 8, 1759, by the only surviving child, John, "whose benevolent and charitable Disposition, inoffensive Life and Conversation, had justly rendered him very agreeable to all." [73] Tradition, charmingly improved by Longfellow,[74] tells how the lonely widower cast his eye upon his housekeeper, forty years his junior, and a former tavern serving wench who had scandalized the ladies by going inadequately attired to the public pump. The governor, the story runs, invited the heads of the oligarchy to dinner, and before them ordered the dumbfounded Parson Browne to perform the ceremony. In London Wentworth's enemies tried to obtain his removal on the ground that he had "married a dirty slute of a maid." [75] Martha Hilton, to whom Wentworth was married on March 15, 1760,[76] was indeed his housekeeper. Her father, William, died at Louisbourg in 1745, which may well explain her poverty. Her grandfather was the Honorable Richard Hilton of Newmarket,[77] connected by marriage with Governor Dudley of Massachusetts. So much for provincial slander.

At the outbreak of the French and Indian War the conflict between the governor and the Assembly was renewed, but with more moderation and good sense on both sides. Although the Assembly was stubborn and short-sighted, it had better reason for its actions than the old family contest for office. It refused to legislate against the worthless Rhode Island paper money or to prepare for war. It insisted on mismanaging the Canada expedition and disposed of the munitions which Wentworth had been ordered by the home authorities to turn over to Shir-

[73] *Boston Weekly Post-Boy*, Nov. 26, 1759, p. 2.
[74] "Lady Wentworth," in *Tales of a Wayside Inn*.
[75] James Nevin to —— (L. C. trans.), Nov. 14, 1761.
[76] *Boston News-Letter*, Mar. 27, 1760.
[77] *New England Hist. Gen. Reg.* xix, 65. Mrs. Marcia Hilton, Mss. on the Hilton family, at the N. H. Hist. Soc. These were kindly called to our attention by Miss Vaughan of the Portsmouth Public Library.

ley. For ten years Wentworth had tried in vain to persuade the Assembly to build roads north and west; nor could he now, backed by the urgent pleas of British commanders, induce them to cut a road to the Connecticut, build a fort at Coos, or even provide uniforms for the New Hampshire regiment. Braddock's defeat on the Monongahela was followed by a no less furious and much more prolonged political battle on the Piscataqua before Wentworth could prevail upon the Assembly to vote a hundred additional troops.[78] Massachusetts was bitter about his failure to do more,[79] but he was obliged to pawn his personal credit to feed the New Hampshire troops.[80]

When the flower of the New Hampshire army was captured at Fort William Henry and when the forces on the Connecticut deserted, it was Wentworth who set about to rebuild the defences while mobs attacked the sailors of the royal navy in Portsmouth and recruiting officers in Exeter.[81] With imperial vision Wentworth grasped the importance of distant operations, but never was he able to convert the Assembly. When he sent down to that body the orders of Lord Halifax to raise forces for service in the southern colonies, it refused to obey, saying that Halifax had misinterpreted His Majesty's pleasure in the matter.[82] Wentworth's war messages were eloquent enough to have produced better results:

Therefore let it not hereafter be told in Gath, or ever published in the streets of Askelon, that so many populous Colonies of Protestants should tamely submitt to entail irretrievable misery and bondage on the Generations yet to be born, (a burthen which our fathers could not bear) without makeing our strongest efforts to Repel the threatening danger.[83]

The hour seems to be approaching when the Inhabitants of this Continent must universally unite in puting a stop to the progress of the French King's army, in failure of which, we shall soon become Provinces and Subjects of the French King, Subjected to a Government whose civil Polity is Tyrany, and to a religion, teaching superstition and the worship of wood & stone, instead of that pure and uncorrupted adoration due only to the Supreme Being.[84]

[78] Wentworth Correspondence (L. C. trans.), 1748–64, *passim*; *Provincial Papers*, VI, *passim*.

[79] Mass. Archives, v, *passim*.

[80] Wentworth to Robinson (L. C. trans.), Sept. 19, 1755.

[81] *Provincial Papers*, VI, 635, *et passim*.

[82] *Ibid.*, VII, 28–30.

[83] *Ibid.*, VI, 357.

[84] *Ibid.*, p. 534.

This imperial sweep of vision and the business sense which appeared in his administration of the office of surveyor of the king's woods show in his land policy. He discreetly avoided participation in the purchase of the Mason claims by the oligarchy just as the Assembly was about to buy them for the public; but his son John was involved. Somewhat more popular was his policy of granting freely lands west of the Connecticut which he claimed came to New Hampshire when the Massachusetts boundary was fixed at the beginning of his administration. In 1749 he granted the site of the present town of Bennington, Vermont. For twelve years thereafter he corresponded in a most gentlemanly fashion with the governors of New York about the disputed territory, always continuing while he did so to dispose of the lands in question. When in 1761 he actually granted sixty-eight new townships in this area, New York protested bitterly to the Privy Council that he had violated all and sundry regulations relating to land grants. Three years later the Board of Trade decided in favor of New York, placing the boundary at the western bank of the Connecticut. But by this time Wentworth had built up vast holdings for himself and his friends. He had erected three tiers of townships on either side of the Connecticut, in some of them giving the greater part of the land to members of the oligarchy or to leading Bostonians.[85] His own profits were enormous. By making a practice of reserving 500 acres in each township for himself, he accumulated between 65,000 and 100,000 acres. Traditionally he exacted a fee of 100l for each of the 130 or more town charters which he granted, and pocketed "presents" of unknown amounts.[86] By 1764 a considerable degree of popular resentment had been generated by these practices.[87]

About the same time the home government began to question other aspects of Wentworth's administration. He thought it wise to take ostentatious steps toward the enforcement of the forest regulations,[88] no doubt confident that no jury would convict the thieves. But he did suggest the appointment of a colonial judge of vice-admiralty to take over the task of preserving

[85] George P. Anderson, in *Publ. Col. Soc. Mass.* xxv, 33–8.
[86] Otis G. Hammond, in *ibid.*, xxiv, 33; Matt B. Jones, *Vermont in the Making* (Cambridge, 1939).   [87] *New Hampshire Gazette*, Aug. 29, 1764; Belknap, ii, 337.
[88] *Boston Evening-Post*, Mar. 29, 1756, p. 2.

the royal mast trees.[89] Not many were left in New Hampshire. To Pitt's angry enquiry about trade with the enemy under color of flags of truce Wentworth replied:

Applications were made to me for Commissions for Flaggs of Truce, for Issueing of which, I might have received Considerable Sums of Mony, yet I ever treated the Applications with the greatest Contempt and disdain, which when known, freed me from future trouble, and was the means of keeping the port pure.[90]

To the accusations that he had connived with smuggling he replied to the Board of Trade in like vein:

The place of my Residence is within a Mile of His Majesty's Fort at the entrance of the Harbour, & no Vessel can come into the Port, without coming within my sight, which I believe has contributed in a great measure to the Chastity of the Port.[91]

Perhaps he was here misleading the Board of Trade as he did in the matter of the judges' salaries. All judicial cases of any importance had to come before a single court sitting at Portsmouth. The Assembly wished, for convenience, to divide the colony into three counties, each with its court, and refused to pay the salaries of the judges until such a division was made. The governor, for no apparent reason, refused and ostentatiously paid out of his own pocket the small salaries which supplemented the fees received by the judges. The Board of Trade, no doubt confusing this situation with the serious one in other colonies, warmly commended the governor and ordered the Assembly to pay the salaries and reimburse him. To this the Assembly calmly replied that it would pay the salaries "as soon as the Province is divided into three Counties & not before." [92] The case was still deadlocked when Wentworth left office six years later. His relations with the Assembly at this period were not bad-tempered but very frank, as some of the resolutions show:

That very large sums of money have been put into the Treasury in the four years past (& not yet accounted for) most of the members of this Assembly are sensible; but as your Excellency has the Keys of the Treasury and no money can be paid out without your warrant, wee would hope that there has been no misapplication of any.[93]

[89] *Journal Comm. Trade and Plantations*, 1759–62, p. 420.
[90] William Pitt, *Correspondence* (New York, 1906), II, 363.
[91] Sparks Mss. 43 (Harvard College Library), British Transcripts IV, 2.
[92] *Provincial Papers*, VI, 721 ff.   [93] *Ibid.*, p. 695.

Education was not a political issue in New Hampshire because the leaders of both political factions were Harvard men. But an attempt in 1758 to establish a new college failed because the movers of the plan were the Congregational clergy, who could not accept the pro-Anglican provisions which the governor wished to insert in the charter.[94] Wentworth was interested enough in Eleazar Wheelock's Indian school at Lebanon, Connecticut, to write and ask for particulars, and later offered land for a site if the school were removed to New Hampshire.[95] Wheelock tried to get the governor to establish scholarships to be known by his name, and considered the possibility of agreeing that the holders be under the Anglican missionary for the district.[96] Wentworth, however, wished to have the college placed under the care of the Bishop of London. Consequently the charter of Dartmouth College was not issued until John Wentworth (A.B. 1755) had become governor. Then the old gentleman gave the site on which Dartmouth was built.

After the burning of Harvard Hall, Wentworth obtained an appropriation of 300*l* with which the college bought 743 volumes, some of which may still be identified by bookplates with the manuscript note, "Prov. Neo-Hanton. Auspice B. Wentworth Praefect." [97]

As early as 1758 the gout had so mastered the governor that he was unable to go to Boston for a conference with General Lord Loudoun:

It is a great Mortification to me That I have it not in my power to meet your Lordship at Boston, I am such a Cripple, that I cannot go out of my doors without the help of Sticks, added to this, The danger of being siezed with a fit of the Gout on the road, where I might be Confined Three or four Months.[98]

It was apparent that the old gentleman must soon be removed from office, and no doubt this knowledge, coupled with his physical incapacity, discouraged him from defending the inter-

[94] *Coll. New Hampshire Hist. Soc.* IX, 36–8.

[95] Wheelock to Wentworth (Dartmouth College Archives), Sept. 21, 1762; Wheelock to Sir William Johnson, in *Documentary History of New York* (Albany, 1850–54), IV, 324.

[96] Wheelock to Wentworth (D. C. A.), Oct. 29, 1764. This is a draft in which the offer in regard to the Anglican missionary is scratched out.

[97] *Publ. C. S. M.* XXV, 33.

[98] Wentworth to Loudoun (Henry E. Huntington Library), Jan 12, 1758.

ests of the home government in 1765 with the same energy that he had shown twenty years earlier. He never received an official copy of the Stamp Act or any instructions regarding the enforcement of it.[99] When the distributor took fright at the popular clamor, he turned over his warrant and instructions to the governor, who found them too hot to hold:

It soon took Air, that the Distributor, had placed his Warrant, & Instructions, under my care, I soon had Notice that two thousand men determined to march out of the Country to demand the papers out of my hands, but to prevent this Insult on The Kings Governor I thought it most prudent, to return Mr Meserve his Warrant and Instructions. . . . I am [unable] to Enforce a due Obedience to the Stamp Act, unless I could withstand thousands with my own hands.[100]

The stamps themselves he lodged in Fort William and Mary where they were guarded by five men, all that the Assembly would allow. He had, he reported, done his best to reconcile his people to the Act, but he had no "Remedy to suggest for the Madness that prevails all over the Continent." [101] He did, however, prevent the representation of New Hampshire in the Stamp Act Congress by proroguing the Assembly. In announcing to the Assembly the repeal of the Stamp Act he pointed out that the prosperity of the mother country and of the colony were inseparable, and that if the colony did not faithfully observe the new resolutions of Parliament it need expect no more favors.[102]

By this time the home government had resolved to remove Wentworth because of his neglect of correspondence with the boards, his failure to submit acts for royal approval, his failure to protect the mast reservations, his venal land grants, and his simony.[103] His nephew, John Wentworth, who was in England at the time, was able to cushion the fall:

I wrote a hasty explanation & defence of the good old gentleman for the information of my noble friend & patron, thro' whom I prevailed to obtain time for him to resign, which saved all the disgrace which might have attended his removal, especially as it appeared he resigned in favor of his nephew.[104]

[99] Belknap, II, 330.
[100] Wentworth to Charles Lowndes (L. C. trans.), Jan. 10, 1766.
[101] Wentworth to ——, *ibid.*, Oct. 5, 1765.
[102] *Provincial Papers*, VII, 99, 100.
[103] *Ibid.*, XVIII, 560–7.   [104] 6 *Coll. M. H. S.* IV, 498.

This defence was a curious thing, excusing the governor's failure to keep up his correspondence on the ground of the gout, but asserting that it did not prevent his attending to his other duties, and arguing that he could not have taken bribes for land grants because the colonists were too impecunious to be able to tempt him.[105]

The resolution of the Assembly on Wentworth's retirement is an amusing contrast to their earlier estimates of him:

The House . . . would take the occasion to express their gratitude and give you their hearty thanks for all the signal services you have done this Province in the course of your administration and during the long time you have with such Reputation & Honor fill'd the Chair; for the steady Administration of Justice, the quiet enjoyment of Property, the Civil and Religious Liberties and Priviledges his Majesty's good subjects of this Province have experienced and Possess'd during this Period. That mildness and moderation with which you have conducted the Publick affairs justly Demand our acknowledgements.[106]

The twenty-five years of service, the longest continuous administration in American colonial history, had shown Benning Wentworth to be an able man. But unlike Belcher and Shirley, he failed to rise above the prevailing public morality of European government officials. Belcher left office poorer than he entered it. Wentworth built up a great fortune based largely on peculation from the king's woods and upon monopolization of the province lands. In these practices he had the company of most English government officials and many "patriot" leaders of the province.

On June 13, 1767, John Wentworth arrived in Portsmouth and took over the government. The old governor retired into obscurity, to emerge in astonished indignation two years later when a widely published letter attacking the Church of England was ascribed to him.[107] He "Dy'd on Sunday about 6 o Clock Evening," October 14, 1770,[108] and was buried with the highest honors:

The Regiment was under Arms on this Occasion, the Honorable, Free, and Accepted Masons walking before the Corpse, to Queen's-Chappel.

[105] *Provincial Papers*, XVIII, 560–7.
[106] *Ibid.*, VII, 116.
[107] See the Boston and New York newspapers for Jan.–Mar. 1769. It was Charles Chauncy (A.B. 1721) who signed Wentworth's initials to the letter, perhaps as a joke. — *Boston Gazette*, June 30, 1769.
[108] *New England H. G. R.* LXXIV, 128.

. . . During the Procession, Minute Guns were fired from . . . Castle-William and Mary.[109]

Wentworth's five children had died before him, and his relatives expected to profit greatly by his will; but to the amazement of all he left the entire estate to his widow, "Whereby the said Lady was tho't to be the richest Widow in New-England, as . . . the old Gentleman died possessed of more than Ten Thousand Guineas in the Hands of Mr. Alderman Trecothick in London — besides a large paternal Estate in New-Hampshire." [110] This prize promptly fell into the hands of one Michael Wentworth of Yorkshire, who married the governor's widow on December 19, 1770. She survived him and died at Portsmouth on December 28, 1805.[111] The family, angered at the loss of the estate, precipitated a quarrel in which the government became involved.[112]

The portrait of Wentworth by Blackburn which is reproduced in this volume is owned by Mrs. Strafford Wentworth of Brookline, Massachusetts. His private manuscripts are not now to be found, and his official ones are widely scattered. There is a group of fifty letters at the Henry E. Huntington Library.

[109] *New Hampshire Gazette*, Nov. 2, 1770, p. 1. This issue contains a highly complimentary character of Wentworth.

[110] *Boston News-Letter*, Dec. 27, 1770, p. 1.

[111] *New England H. G. R.* xix, 66.

[112] Peter Livius memorial, in *Essex Journal*, Feb. 16, 1774, *et seq.*

ISAAC GREENWOOD

The many amateur scientists among the clergy were most respect-
able men, but Harvard was unfortunate in her two first Hollis
professors of mathematics, Greenwood and Prince (A.B. 1718).
Fortunately the College did not conclude from their careers that
professors of science ought to be clergymen.

ISAAC GREENWOOD, first Hollis Professor of Mathematics, was
born at Boston on May 11, 1702, and baptized a week later at
the Old North. His father, Samuel, was a Boston shipwright;
the Samuel Greenwood who graduated in 1709 was an older
brother, and Isaac, who graduated in 1685, an uncle. The wife
of Samuel the shipwright was Elizabeth Bronsdon. An account
of Isaac's early years from the pen of a friend has come down
to us:

In his Childhood he is said to have been very negligent and an Enemy
to his Book, insomuch that I have often heard him say, that he was
Old before he could read Letters. But by a closer Application afterwards, and the Advantage of an uncommon Memory, he soon made
so prodigious a Progress in his Studies as is scarce credible. At 15 he
was thought fit for an Admission into College, having run thro' the
usual Exercises in about 5 or 6 Years, which generally take up a much
longer Time.[1]

Isaac did not live at the college or take much part in student
activities other than to read a paper on the progress of philosophy before a student society.

At the University he was distinguish'd by a sprightly Wit and Genius,
temper'd with a studious Diligence & Application. Here he devoted
himself to the Muses and the Mathematicks; always making the gayer
Scenes of the one his Recreation and Amusement after the severe
Studies of the other. . . . We scarce ever met with the Poet & Philosopher more agreably united than in him. This Love to Poetry led
him naturally into an Esteem for the Classics . . . and by an early
and intimate Acquaintance with them, he attain'd to that Elegance
& Gracefulness, that set so easy upon him ever after. At this place
too he had the Happiness to ingratiate himself into the Affections of
his Tutor Mr. Roby, a Gentleman at that Time famous for his great
Skill in the Mathematicks.[2]

[1] *Boston Gazette*, Nov. 26, 1745.   [2] *Ibid.*

ISAAC GREENWOOD

During the smallpox epidemic of 1721, Isaac, his brother Samuel, and Obadiah Ayers (A.B. 1710) were inoculated together, had light cases with few pocks, and were soon well.[3] This experience encouraged Isaac to defend his pastor, Cotton Mather, who had encouraged the introduction of inoculation, against the criticisms of the conservative Dr. William Douglass and his friends. For this purpose he wrote a little tract called *A Friendly Debate; or, a Dialogue between Academicus; and Sawny & Mundungus*, in which he called Douglass "a Credulous and Whimsical Blade, a Madman and a Fool," and ridiculed his Scotch dialect. The performance was so dully sophomoric that Mather, to whom the tract has sometimes been attributed, must have been distressed; his own invective was far more effective. Douglass replied with an even and mild *Postscript to Abuses*, in which he quite demolished "Academicus" Greenwood and mercifully spared Harvard the thrusts to which the *Friendly Debate* had laid it open.[4]

After taking his first degree Greenwood returned to Boston, where he joined a pious club organized by some of his classmates.[5] In December, 1722, he was admitted into full communion at the Old North:

His Friends had an Inclination to have him a Clergyman, and in Compliance with their Desires he gave himself . . . to the study of Divinity, under the Conduct of the late learned Dr. Cotton Mather, and not without Success, as we may well conclude from that Encouragement and Applause which always accompanied his publick Dispensations. But whether this Kind of Life did not so well suit his philosophical Genius, or his Inclination to travel had got the entire mastery of him, in 1723, he took a Voyage to England, hoping there to satisfy his Curiosity in every Thing he could wish for.[6]

He carried with him a letter from Mather to Dr. James Jurin, in which he was described as "an Ingenious Young Gentleman . . . who proposes to make some Addition to his Accomplishments, by Visiting Europe, and particularly by visiting of Dr. Jurin."[7] As an "Inoculate," the young man was an object of interest to the learned gentlemen of England.

   [3] Zabdiel Boylston, *An Historical Account of the Small-Pox Inoculated in New England* (Boston, 1730), p. 20.
   [4] Another party replied to Greenwood with *A Friendly Debate; or, A Dialogue between Rusticus and Academicus* (Boston, 1722).
   [5] Edmund Quincy, Letters, 1722–81 (Mass. Hist. Soc.), Sept. 24, 1722.
   [6] *Boston Gazette*, Nov. 26, 1745.
   [7] Cotton Mather, Letters to the Royal Society (Gay transcripts, M. H. S.), p. 210.

It is not altogether clear how Greenwood passed his first two years in England, but the *Quaestio* which he mailed home for his second degree in 1724 suggests that he may have been studying medicine.[8] Thomas Hollis, to whom he had delivered a parcel on his first arrival, reported in May, 1725, that he had "begun to preach at London with approbation." "Mr. Greenwood dined with me to day, says he has preached with courage 7 times, which I am glad of, I feard he could have been daunted at his first setting out." [9]

He was much admir'd as a Divine by many of the dissenting Party, & had several advantageous Offers made him of spending the remainder of his Life among 'em: but either thro' too great a Fondness for his native Country and Friends, or a much more prevalent Passion to Philosophy, he chose rather to postpone his Interest for that Time. His Genius leading him chiefly to the Mathematics and Philosophy, he applied himself mostly to these, and under the Direction of Dr. Disaguliers [John Theophilus Desaguliers] that great Mecanic & accurate Master of experimental Philosophy, he made such strange and surprizing Advances in these his darling Studies, as not only gained him the entire Love and Esteem of his Tutor, but wro't so much upon that Gentleman, as that he condescended to make him an Assistant in his Lectures, and wou'd even venture to leave him to officiate in his Stead if he was ill himself, or oblig'd to be otherwise absent. By this Gentleman and Dr. [William] Derham he was introduced to the Royal Society, and had the Honour of answering several Questions to their Satisfaction propos'd to him by their President the great Sir Isaac Newton. At this Time too he contracted an Acquaintance with many of that Society, with whom he maintained a good Correspondence ever after.[10]

In June, 1725, Hollis wrote to the Harvard authorities that Greenwood intended to return in the fall with a quantity of apparatus with which to teach "the Mathematicks." "I hope he will prove an usefull instructor in your college ... [he] apears to me sober, and dilligent to acquire knowlege." [11] Apparently Greenwood had already approached the college authorities when, in December, Hollis sent for him and opened his plan of founding a professorship of science. At Hollis' request

[8] "An Variolarum in corpora humana Transplantatio, licita sit et tuta? Aff."

[9] Hollis Letters and Papers (Harvard University Archives), pp. 61–2.

[10] *Boston Gazette*, Nov. 26, 1745. This is printed at some length because it corrects the suppositions in Foster Watson, *The Beginning of the Teaching of Modern Subjects in England* (London, 1909), pp. 251–2.

[11] Hollis Letters and Papers, p. 62.

he drew up and sent to Boston a set of proposed regulations for the professorship, a set which no doubt much better fitted the situation than did those drawn up by various learned English gentlemen to whom Hollis also turned. Greenwood asked that the Brattle funds be added to the Hollis endowment because of the scientific interests of the Brattles. He asked that the students who elected to take his courses pay him special fees, and promised to go through the motions even if no students appeared. It is evident that he hoped to be, as master of the philosophical apparatus, somewhat independent of the Corporation.[12] Lack of funds with which to purchase the necessary apparatus kept him in London through the winter of 1725–26, but he employed the time to gain the respect and sympathy of Hollis, who definitely nominated him for the professorship:

I have discorst him many times, & had him examined by Mr Hunt Mr Watts Mr Neale Mr Ingram — and am incouraged to hope he may come over to you in July or August next, well qualified for an instructor of youth in those Sciences. I have obtaind of Mr Professor Emmes, Mr Hunt, Mr Watts, Mr Neale, papers mentioning what his work should be, and method of his publick and privat Lectures. . . . Mr. Greenwood's brothers have put him here this winter to great straites, for want of Remitting him in due time moneys for his necessary expences . . . but as the mans heart is in the pursuit of these Studyes, it is pitty he should return before he is Master of what he designs and aimes after.[13]

By July he was no nearer having the apparatus and was getting very much out at the elbows. This did not prevent his beginning to show vices which caused Hollis to warn the college to make no promises about the professorship until it had given him a thorough trial. But on July 5 Greenwood suddenly slipped out of England in Captain Prince, bound for Boston via Lisbon, "without paying his debts or taking leave of Dr Desaguliers, his Landlord or Tutor," or of Hollis, which gave "great Scandal and reflection on him . . . at the New England coffee house." A little investigation confirmed Hollis in his bad impression of the situation:

I am told he owes about £300. . . . Mr Dummer tells me he is bound to him with Peirpont for £150 or thereabouts, which money was all spent in a ramble of a few weeks — he owes for his bord etc £- - - to

Mr Desaguliers, and to sundry tradesmen — one demands money for 3 pair perle colour silk stockings.[14]

Hollis was very much at a loss where to turn for another mathematics professor. The only likely candidate was, like himself, a Baptist and for that reason less palatable to the Harvard Corporation than the spotted Greenwood. In view of Hollis' recent letters, the president and fellows did not dare to nominate Greenwood for the professorship when he arrived, but put the gentle Benjamin Colman (A.B. 1692) to the task of placating the prospective donor.[15] Meanwhile Greenwood, who had arrived in Boston about October 20, 1726,[16] preached a few times at the Old North and, in January, 1726/7, announced what was hailed as the first lecture course in science in New England. After delays necessitated by the failure of sufficient subscribers, at 4*l* the head, he opened at "Mr. Howard's in King-Street." According to his description, it was a course

Whereby such a competent Skill in Natural Knowledge may be attained to, (by means of various Instruments, and Machines, with which there are above Three Hundred Curious, and useful Experiments performed) that such Persons as are desirous thereof, may, in a few Weeks Time, make Themselves better acquainted with the Principles of Nature, and the wonderful Discoveries of the incomparable Sir Isaac Newton, than by a Years Application to Books, and Schemes.[17]

This carried Greenwood through the winter, and in the spring came the grudging approval of Hollis:

I take notice what you and others write concerning Mr Greenwood . . . and some of you admitting of him to preach, which I think was very hasty. However I shall forbear telling heare says . . . wishing his future cariage may be sober, Religious, dilligent, and becoming his Profession and calling. And if the Corporation shall be unanimous in electing of him and recomending of him to me, I think I shall accept him as my Professor.[18]

In spite of Hollis' uncertainty, the Corporation immediately elected Greenwood "Professor of Mathematicks & Natural &

[14] *Ibid.*, pp. 72, 74.

[15] Greenwood to Colman, Colman Mss., Feb. 7, 1726/7.

[16] Ebenezer Parkman, Diary (Am. Antiq. Soc.), Oct. 27, 1726.

[17] *An Experimental Course* (Boston, 1726), title-page. Thomas Prince took the course and noted the dates of the lectures in the copy now at the M. H. S. For newspaper advertisements for Greenwood's school, see the list in *Publ. Col. Soc. Mass.* xxvii, 142–3.   [18] Hollis Letters and Papers, p. 76.

Experimental Philosophy," and ordered him to "read his publick weekly Lecture in the Hall on Every wednesday at two of
the Clock afternoon." Only upperclassmen could attend, and
they must first obtain the "Consent and allowance of their
parents" and pay 10*s* the quarter. Pending Hollis' definite approval, the inaugural ceremony was put off for nearly a year.
It finally took place on February 13, 1727/8:

The President being ill Mr Flynt was desired by the Corporation to
direct the Affair of the inauguration to begin with Prayer and make
an introductory Speech which Speech (in Latin) being finished Mr
Professor Wigglesworth was desired to read Mr Hollis's Rules &
Statutes respecting the Professor of Mathematicks & Natural & Experimental Philosophy. Then the Oaths to the Civil Government
were read by Mr Sever and repeated verbatim by Mr Greenwood and
. . . the printed copies of them were Signed by Mr Greenwood. . . .
After this Mr Greenwood was desired by Mr Flynt to Express his
declarations and promises agreable to the . . . articles. . . . Then
Mr Flynt called for Mr Greenwoods Inaugural oration which oration
being finished Mr Flynt asked Leave of the Overseers and Corporation to declare Mr Greenwood Hollisian Professor of Mathematicks.
. . . After this Mr Appleton was desired to make the Last Prayer then
the two first with the last Staff of the 104 Psalm were sung and the
whole Company went to Diner in the College Hall.[19]

Settled at last in the only position in America where a man
could live by science, Greenwood went back to Boston and
married Sarah, daughter of John Clark (A.B. 1687), on July 31,
1729. She was "a young Lady of Character & Fortune . . . no
less esteem'd for her acquir'd than natural Accomplishments,
which were not small nor common." [20] Thomas Robie, who had
first interested Greenwood in science, had, during his tutorship,
gathered meteorological data for the Royal Society. The new
Hollis professor forwarded the last of Robie's notes and offered
to continue the observations:

I shall have all the Opportunities imaginable [for weather observa-
tions] being chosen by the College their Hollisian Professor of the
Mathematicks & Expirimental Philosophy which place has some
peculiar Advantages for Observation, above Most of the same Nature
in the World, being accommodated with a very large Apparatus of
Glasses, and other Instruments, besides by it's Institution Furnish'd

19 Overseers' Records (H. U. A.), I, 101, 104–6.
20 *Boston Gazette*, Nov. 26, 1745.

with 10 Pensionary Scholars of the 2 upper Classes who will always be ready to continue on the Observations in case of Sickness.[21]

He urged upon the Royal Society the codification of ships' registers on vast daily charts: "hereby We may be able to judge in what Place such a Wind has its Origin, how long a Time it continues, with what Velocity it moves, where its greatest Strength is, and how great a Part of the Earth it passes over." [22] This proposal was published in the *Philosophical Transactions*,[23] with an account "of some of the Effects and Properties of Damps" in wells [24] and "an Account of an Aurora Borealis." [25] At the request of English archaeologists he made careful drawings of the inscription on Dighton Rock which are now of considerable importance.[26] In his spare time he prepared and published an arithmetic, presumably the first from an American author.[27] In vacations he carried on public lecture series, for which he issued an enticing printed prospectus. In 1734 his public lectures on the orrery attracted attention as far away as New York.[28] On the death of Hollis he delivered a *Philosophical Discourse*, which was purely mechanistic in its reasoning about immortality and completely ignored the function of revelation in religious consolation.

Professor Greenwood's unorthodox religious philosophy went unchallenged, but his other weaknesses gradually began to make trouble for him. Originally granted "the same authority with the Tutors in Governing the Schollars," he exercised it in such a way that the faculty had to remit some of the fines which he laid on the students. He quarrelled with Judah Monis over the classroom which they shared, and became so intemperate that in April, 1737, a visiting committee of the overseers presented a report which resulted in his being called before the Board. There "he confessed the Charge of intemperance . . . and cast

[21] Greenwood to Jurin (Library of Congress transcripts), May 1, 1727.

[22] Royal Society Record Books (L. of C. trans.), 1727–30.

[23] *Philosophical Transactions of the Royal Society*, xxxv, 390–402.

[24] *Ibid.*, xxxvi, 184–91.

[25] *Ibid.*, xxxvii, 55–69.

[26] A full account of the Dighton Rock work, with reproductions of Greenwood's drawings, may be seen in *Publ. C. S. M.* xviii, 240–96.

[27] A long prospectus of this volume appeared on the front page of the *Boston News, Letter*, July 3, 1735.

[28] *New-York Gazette*, July 8, 1734, p. 1. *New-England Weekly Journal*, June 24-1734.

himself on the Lenity of the Overseers & professed his resolution of reformation." [29] Let off with a formal admonition drawn by his old friend Thomas Prince, he soon lapsed into "Various Acts of gross Intemperance, by excessive drinking." In November, irritated by his disregard of "repeated warnings & Admonitions," the Corporation called him up for a formal examination, in which "he made a free Acknowledgment" of his failings. He was ordered to exhibit "an humble confession, and at the Same time receive an Admonition publicly in the Hall," and warned that if he did not remain sober he would be removed from the professorship within five months.[30] The professor promptly appealed to the overseers, got drunk again, and refused to answer the repeated summons of the Corporation, which thereupon, on December 7, "Voted, That He be remov'd from his Office." [31] The overseers ordered a committee to take over the apparatus, authorizing them to force the door of the philosophical chamber if necessary. A humble confession and nearly four months of sobriety induced the Corporation to intrust Greenwood with the apparatus once again, and to increase his salary. But this was too much prosperity; within two weeks he had relapsed, and when, with difficulty, he was dragged before the Corporation, he was too sick and befuddled even to offer one of his habitual confessions. The president and fellows had had enough, but the overseers three times postponed action. Finally, on July 13, 1738, they gave up hope and removed him.

According to the professor's eulogist, Harvard proving a "Sphere of Action . . . too small and confin'd for the enterprizing Genius of Mr. Greenwood, in Imitation of his Master Dr. Disaguliers, he set up a School of experimental Philosophy." [32] This new institution appeared in Boston in the fall of 1738, invitingly advertised in the press:

Such as are desirous of learning any Part of Practical or Theoretical Mathematics may be taught by Isaac Greenwood, A.M. etc. in Clark's-Square, near the North Meeting-House; where Attendance will be given between the Hours of 9 and 12 in the Forenoon, and 2 and 5 in the Afternoons. N. B. Instructions may also be had in any Brand of Natural Philosophy, when there is a Number sufficient to attend.[33]

---

[29] Faculty Records (H. U. A.), i, 80; Overseers' Records, i, 152–3.
[30] *Publ. C. S. M.* xvi, 669.   [32] *Boston Gazette*, Nov. 26, 1745.
[31] *Ibid.*, p. 671.   [33] *Boston News-Letter*, Nov. 9, 1738.

The best Gentlemen of all Orders became his Scholars, and he had the happiness to perform his Lectures to the Satisfaction of all that heard them. . . . He had a happy Talent of adapting himself in such a Manner to the capacity of his Hearers, of representing the most obscure and difficult Things in such a plain and easy Light, as it could not fail to satisfy the most ignorant, at the same Time that it would please the most learned.[34]

For all this success, prosperity eluded Greenwood. In the year in which he established the school he sold several houses on what is now Harris Street and a piece of pasture on Hanover, but by 1740 the receipts from this source were gone. In the spring of that year he tried his luck in Philadelphia, where he advertized a course of lectures and by the intercession of Benjamin Franklin obtained the loan of the experimental equipment of the Library Company.[35] He had no better success there, so returned to Boston where, in July, 1742, the Second Church voted that whereas he had "for a Long time bin ill reputed of as giving Scandal by a Course of excessive Drinking, & those Effects of it which bring reproach upon our holy Profession," he should be called before the deacons. He met them and "with an appearing seriousness & concern of mind acknowledged himself to be guilty, profess'd his Submission to the Censure of the Church, and affectionately asked our Prayers for his Reformation." [36] The prayers of others, however, were not enough:

Not seeing any Likelihood to succeed better on Shore, by the Advice of several of his good Friends, he was content to try his Chance once more at Sea, and accordingly ship'd himself on board the Rose Man of War as a Companion and Instructor to Capt. Frankland. With him he was in the Engagement of the Conception, the rich Prize that they took after a smart Fight, and behaved to the Acceptance of that brave Commander. But Capt. Frankland soon after returning to England, and Mr. Greenwood not being willing to go so far from home, he exchang'd the Rose for the Alborough, finish'd one Cruise in her and dy'd [in South Carolina, on October 12, 1745].[37]

Mrs. Greenwood died at Falmouth, Maine, on May 23, 1776. They had five children, of whom Isaac became a famous mathe-

---

[34] *Boston Gazette*, Nov. 26, 1745.

[35] *Pennsylvania Gazette*, June 5, 1740. Information from Mr. I. Bernard Cohen.

[36] Records of the Second Church of Boston.

[37] *Boston Gazette*, Nov. 26, 1745.

matical instrument maker in Boston.[38] The original of the portrait here reproduced shows the yellowish brown gown faced with red which appears in other academic portraits of the period.

Mr. Frederick G. Kilgour of the Harvard College Library provides this evaluation of Greenwood's place in the history of science:

Isaac Greenwood and American Science

When Isaac Greenwood returned to Boston from London in 1726, he was a promising young scientist who had received a good training in England. He was an accurate observer and he possessed a capable intellect, but for some reason — it may have been the 'demon rum' — he never became a productive scientist.

Greenwood did not invent any scientific theories, and he recorded but few observations. He revealed his appreciation of the value of observation when he wrote that he was 'perswaded there is no better way to arrive at the true cause of this extraordinary phenomenon [an aurora borealis], than by attending to the minutest particulars and circumstances thereof.' His description of the aurora is good and an excellent example of his powers of observation. So also is his paper on damps, in which he describes the lowering of a lighted candle into a well that had damp in it: 'In about 6 Feet below the Top of the Well, the Flame would grow dim, and if not immediately raised, would change to a bluish colour, and become more and more contracted or diminished, till in about a Minute's Time it would be totally extinguished, without any Remains or Stench accompanying the Wick.' And when John Eames of London asked Greenwood for a transcription of the figures on Dighton Rock, Greenwood, unlike Cotton Mather before him, was not content with a copy of someone else's work: Greenwood went to Taunton and made his drawing directly from the rock.

But although Greenwood made no contribution to the history of science, he did play an important rôle in the tradition of science in New England. If we can judge from the prospectus of

---

[38] For the children, see *New England Hist. Gen. Reg.* XIV, 172.

Greenwood's *Experimental Course of Mechanical Philosophy*, his 1727 lectures were remarkably good. And he apparently made an effort to keep his scientific knowledge up to date, for in his *Philosophical Discourse* (1731) he quoted Stephen Hales' *Vegetable Staticks* (1727), an excellent work on plant physiology that was published in London the year after Greenwood returned to Boston. Another instance of his continued pursuit of scientific learning is his inclusion in the undated prospectus of his *Course of Philosophical Lectures* of "an Account of Mr. Professor Bradley's new discovered Motion of the Fixt Stars" and a promise that a "particular consideration will be taken of Dr. Desaguliers, late Theory of the Rise of Vapours and Formation of Clouds, and Meteors, with his Experiments concerning them." Both of these papers were published in the *Philosophical Transactions* of the Royal Society, Bradley's in December, 1728, and Desaguliers' in January and February, 1729. From internal evidence it appears that the little four-page announcement was published around 1735 or shortly thereafter.

In philosophy Greenwood was a materialist, as Dr. Shipton has pointed out. His materialism was based on "Sir Isaac Newton's Laws of Matter and Motion," and he was supported in his views by the eminent success of Sir Isaac's laws. Greenwood's public lectures were almost entirely discourses based on Newton's work. In this connection it is interesting to note that Greenwood possessed the third copy of Newton's *Principia* to come to the colonies.

It is difficult to evaluate accurately Greenwood's position in the scientific tradition in America. When he returned to Boston in 1726, science was rapidly becoming a living part of New England culture. Greenwood's public lectures and his teaching at Harvard were undoubtedly potent factors in the continuation of the interest in science. Greenwood was much more learned in science than Cotton Mather, Paul Dudley, Zabdiel Boylston, Thomas Robie, and other scientifically curious New Englanders of the 1720's, and was able to transmit this learning to his pupils. John Winthrop, later Hollis Professor of Mathematics, was one of his students. Greenwood not only instilled in Winthrop a deep interest in science, but also undoubtedly gave Winthrop a good basic scientific training. Thomas Robie was probably the first New Englander to whom science was a systematic body of knowledge and not just something "curious." Greenwood is the logical step after Robie, and Greenwood's pupil, Winthrop, brought the position of science in colonial New England culture to its greatest height.

EBENEZER PARKMAN

EBENEZER PARKMAN, the first minister of Westborough, was born at Boston on September 5, 1703, the eleventh child of William and Elizabeth (Adams) Parkman. The family was solid, humble, and pious. William was by trade a shipwright, and when his son was in college, served Boston as a hogreeve. Later he became a ruling elder of the New North. When the North Latin School was opened in April, 1713, little Ebenezer was sent to study at the feet of John Barnard (A.B. 1709), whom he always remembered with affection.[1] The combined influence of a religious father, a saintly schoolmaster, and a nearly fatal siege of measles made him a very pious lad and marked him for the ministry.[2] At the official placing of the Class of 1721 the college authorities dropped him from tenth place to twenty-eighth, probably in part at least because of his humble social status; but this brought no rebellion, perhaps because, as he said, he found President Leverett "great and awful in Government."[3] As an undergraduate he eschewed student disorders and sat quietly in his chamber, copying bits from Shakespeare's Tempest, Dryden's Virgil, and contemporary English poetry.[4] But in taking his second degree he showed his fundamental interest in religion by presenting a denial that ad Annihilationem, Actus positivus requiritur.

[1] Parkman, Diary (Mass. Hist. Soc.), Apr. 20, 1781.
[2] Parkman Mss. (Am. Antiq. Soc.).
[3] Ezra Stiles, Literary Diary (New York, 1901), II, 231.
[4] Parkman Mss. (A. A. S.). His college Homer is at the Harvard College Library, and his Wollebius at the M. H. S.

In his sophomore year Ebenezer took a journey to Nantucket,[5] the only sea voyage of his long life. When he graduated he was invited to teach at Newton, but President Leverett refused to permit him to accept until it was agreed that the school be kept near the meetinghouse.[6] After a year at Newton, Ebenezer went to live with his brother Elias in the North End of Boston, where he made his home until his settlement in the ministry. Here he became the secretary of the "Friday-Evening Association of Batchelors," an organization of pious young men who met in his chamber for "the Promotion of Good Morals and Good Citizenship."[7] On one occasion he read a paper on the mind of man.[8] As a step in the preparation for the ministry he joined the New North. His "first Attempt at Sermonizing was delivered . . . Sept. 1. 1722 . . . occasioned by the happy Return of Brother John from the Pyrates."[9] The following April he preached his first public sermon at Wrentham, and in the course of the next few months he supplied the pulpits of Westborough, Newton, Hopkinton, and Worcester. The itinerant candidate of those days was everywhere welcomed with respect and sometimes with Lucullan feasts at which "roast Goose, roast Pea hen Bak'd, stuff'd Venison, Pork, etc.," graced the board together.[10] At Worcester, Adam Winthrop (A.B. 1694) took the young parson fishing for "Salmon Trouts." Other entertainment is suggested in Parkman's note that "We Stopd at Sundry Taverns (which particularly I do not remember)."[11] On February 28, 1723/4, the frontier town of Westborough recalled Parkman to become its first minister. The twenty-eight families could promise only a salary of 80*l* and a settlement of 150*l*, but in June he accepted. On July 7, 1724,[12] he married Mary Champney, a daughter of Samuel and Hannah Champney of Cambridge. Although she was more than four years his senior, she won from him affection of unusual depth and dura-

[5] Parkman, Commonplace Book (M. H. S.), June 15, 1719.

[6] Parkman, Book for Sundry Collections (A. A. S.), July 10, 1721.

[7] Lincoln Papers (A. A. S.), Feb. 1, 1722/3. The minutes of the organization for Sept. 14–Dec. 7, 1722, are in Edmund Quincy, Letters, 1722–81 (M. H. S.).

[8] The original is at the A. A. S.

[9] Flyleaf to vol. VII (the oldest surviving) of his diary (A. A. S.).

[10] Diary (A. A. S.), Jan. 10, 1724.

[11] *Ibid.*, Aug. 3, 1723.

[12] The date is from Parkman's Natalitia (A. A. S.) and the Cambridge Church Records; the other dates which have been given are evidently wrong.

bility. Together they returned to Westborough, where on October 28, 1724, he was ordained:

The Rev. Mr. Dorr [A.B. 1711] of Mendon, opend the Solemnity with Prayer. The Rev. Mr. Prentice [A.B. 1700] of Lancaster preachd a suitable sermon. . . . Afterwards The Rev. Mr. Williams [A.B. 1705] of Weston . . . prayd, & gatherd the Church. There were 12, besides the Pastor Elect, who signd the Covenant & answerd to their Names in the Assembly. The Rev. Mr. Prentice gave the Solomn Charge. The Rev. Mr. Loring [A.B. 1701] of Sudbury the Right Hand of Fellowship.[13]

The colonial minister regarded all knowledge as his province, and the gaining of it one of his duties to his parish. Parkman read from cover to cover every encyclopedia and biographical dictionary which came his way, and devoured the many-volumed political histories of his day. He subscribed to the *Chronological History* of Thomas Prince, recorded the Indian captivities of his parishioners, and wrote a history of Westborough.[14]

The scandalous travels of Ned Ward he read without comment, but the Jesuit missionary letters he found "edifying."[15] Although political theory interested him but little, he read works as varied as Montesquieu [16] and a "Pamphlet, the Oeconomy of human Life, by an Oriental Bramin." [17] After the manner of his time, Parkman recorded the reading of "Romances & Tales Poems & Plays" in his list of his personal vanities and impieties,[18] but it is evident that he did not take seriously this traditional indictment of light literature. Addison, Pope, the *Spectator,* and "the incomparable writings of Mrs. Rowe" were among his favorites.[19] For years the Westborough parson tried to induce the house of Kneeland to print a collection of English poetry.[20] He enjoyed the "unbounded" wit of

---

13 Diary (A. A. S.), Oct. 28, 1724.

14 His account of an Indian captivity is printed in *Coll. Wor. Soc. Antiq.,* XIII, 425–8. His sketch of Westborough is printed in Heman P. Deforest, *History of Westborough* (Westborough, 1891), pp. 479–81. Parkman is treated at some length in this sketch, not because he was important among his contemporaries, but because the survival of the greater portion of his diary gives a unique opportunity to describe an ordinary country parson.

15 Diary (A. A. S.), Jan. 3, 1724.

16 *Ibid.,* Aug. 14, 1765.   17 *Ibid.,* Feb. 28, 1756.

18 Natalitia, Sept. 5, 1729.

19 Diary (A. A. S.), Nov. 29, 1756.

20 *Ibid.,* May 30, 1751, *et passim*; Commonplace Book, p. 44.

Swift but wished that "we might have the Instruction & Entertainment without so great Expence of Decency." [21] When a neighbor lent *Pamela* to Molly Parkman, her father picked it up and without regret spent one of his precious days upon it.[22] Of *Clarissa Harlowe* he remarked:

I think these sorts of Books are indeed to be read Sparingly; and others to be preferrd far before them but yet Such as are bred in the Country & cant be afforded to live at Boarding schools, may by those Means come to some Taste of brilliant sense, when they cant be polishd by Conversation. But this indulgence had need be kept under a Strict Guard, & Caution.[23]

The Fair Circassian he regarded as "an unhappy Mixture of heavenly, Seraphic Ardors, with odious, infernal pollutions." [24] In general, his discrimination between good and bad literature accorded with modern taste, and his fear of the effects of the glorification of sexual excesses was that of parents of all times.[25] Young Sam Haven (A.B. 1749) found a stay at the Westborough parsonage anything but dull rustication:

I enjoy not only the Advantage of his most Learned as well as Christian Conversation; but also the pleasure & profit of perusing a Collection of Books not inferior to the best in the hands of Any Country Minister nor do I imagine my priviledges less noble than those to be enjoyed at College.[26]

Science interested Parkman less than literature, although he took tea with his "dear friend" Professor Greenwood (*q. v.*), carefully recorded earthquake phenomena, and observed eclipses with a telescope.[27] As there was no lawyer in Westborough, he grew skilled in drawing deeds and wills which were tight against the quibbles of the practitioners in the Worcester courts. Medicine was a part of the function of every country parson, and in this field his life bridged the gap between medieval lore and the Massachusetts Medical Society. Early in his

[21] Diary (A. A. S.), Oct. 2, 1765.
[22] *Ibid.*, June 23, 1746.
[23] *Ibid.*, Jan. 9, 1756.
[24] Commonplace Book, p. 41.
[25] For a lengthy, sound, and reasonable criticism of contemporary romances, see the Diary (A. A. S.), Nov. 9, 1773. For his criticism of Chesterfield's Letters, see Harriette M. Forbes, *Diary of Ebenezer Parkman* (Westborough, 1899), pp. 210, 213.
[26] Haven to Robert Treat Paine in Paine Mss. (M. H. S.), VII, 36.
[27] Diary (A. A. S.), Feb. 13, 1728, Jan. 23, 1758, *et passim*.

ministry he received from his neighbor, Parson Swift (A.B. 1697), an old-fashioned system of uroscopy which provided forty-one means of diagnosis such as the rules that "If you see your Face in her Water If She hath not a Fever, she is with Child," and that if the urine is "Bright as Gold," the trouble with the patient is "Lust or Desire to marry." [28] When son Ebenezer fell into the fire, the minister seized and killed the parsonage cat to bathe the burned hands in its blood.[29] He took garlic steeped in rum for his rheumatism and on the advice of Parson Gee (A.B. 1717) took "a Tea of Mullen, Cullenbine and Sage" for a sore throat.[30] In his old age he submitted to electric shocks to cure his cramps.[31] He was always keenly aware of the importance of studying and recording the symptoms of disease, but on one occasion he received some unfavorable publicity because of an error resulting from his ignorance of botany.[32]

Parkman once borrowed from William Nurse of Salem a copy of Calef's book on witchcraft, and thus left the only near-contemporary record of the reading of that controversial tract which has ever come to the notice of the editor.[33] On the other hand, his life is a striking example of the great and good influence of the Mathers. In young manhood he watched the funeral procession of Cotton Mather and noticed that it seemed "almost as if it were the funeral of the Country," so general was the mourning.[34] In his old age he read his family the elder Mather's books on witchcraft and remarked, "How exceeding plain yet solemn!" [35] His superstitions, if such they can be called, consisted of crediting stories of contemporary occurrences which were parallels of Biblical accounts. Thus he records the case of one Robert Woodberry of Beverly who during his life had "been used to profane Cursing, & wishing persons might go to Hell & blare like a Calf and would often curse his worthy Mother." In due time this Robert "sickend and dyd, & when the Corps had lain a suitable Time to be laid out, a horrible Sound broke forth from the Corps, blaring like a Calf." [36]

[28] Commonplace Book, 1729.
[29] Diary (A. A. S.), Dec. 2, 1747.
[30] *Ibid.*, Nov. 28, 1745.
[31] Forbes, *Parkman*, p. 136.
[32] *Boston Gazette*, Feb. 18, Mar. 24, 1752; Ezra Stiles, *Literary Diary*, I, 75–6.
[33] Diary (A. A. S.), May 14, 1740.
[34] *Ibid.*, Feb. 19, 1728.
[35] Diary (M. H. S.), Feb. 4, 12, 1781.   [36] Diary (A. A. S.), June 25, 1750.

But in times when death walked everywhere in Westborough, and diphtheria swept away whole families, and agonized people saw coffins in the air, the minister did his best to allay the excitement.[37]

It is unfair to judge Parkman and his colleagues by their attitude toward social problems without taking the problems themselves into consideration. Westborough was a frontier town. Once when Parkman was preaching there as a candidate he walked from the meetinghouse to his boarding place with pistol in hand as protection against hostile Indians. Thus engaged he was "much affrighted with the Sight of an Indian" ahead of him, but on approach found it to be his landlord.[38] As his eyesight was remarkably good, this evidence as to the personal appearance of a well-to-do inhabitant is interesting. In later years the minister had his firelock cleaned at the news of Indian hostilities, but he put up savages for the night if they were too drunk to continue their journey safely. One of his guests was a Thomas Rice who had been carried into captivity as a child and now returned to his old home brave in his finery as "one of the principal men of the Cagnawaga Tribe." [39] Frontier manners were sometimes distressing to the Boston-bred parson. Occasionally a parishioner called him in to help control a "well-liquored" wife. Once he complained that tobacco spittle drooled down from the meetinghouse gallery upon daughter Molly's head. Frequently he reproved the "frolicks" and "ungodly Mirth" of the "giddy Youth," who were generally, in his opinion, allowed too much liberty by their parents. Sometimes he "requested Young women to dismiss Seasonably the Young Men that wait upon them." [40] The problem was not peculiar to Westborough; at the raising of the new Framingham meetinghouse he complained that "the Impudence of Young Men with the Young Women was with them very Shameless." [41] Crime was rare, although growing; but the steady increase in fornication with each period of warfare justified the parson's attitude. It finally reached a point where he could preach a sermon against fornication with the knowledge that it applied

[37] Diary (A. A. S.), Sept. 16, 1745.
[38] Ibid., Aug. 26, 1723, footnote.
[39] Executive Records of the Province Council (Mass. Archives), x, 434.
[40] Diary (A. A. S.), May 27, 1751.
[41] Diary (M. H. S.), May 3, 1749.

to a number of his hearers.[42] On one occasion a deacon who had recently been churched for this offence was called upon to defend a night frolic at his house, which he justified on the grounds that "a great Number of Wrestlers from Hopkinton came swaggering & challenging." [43]

Parkman was no enemy of innocent fun. He went fishing down Boston Harbor and got back at 2 A.M., having had "a very Pleasant Time." [44] At the age of fifty-six he put on skates with a small son. With the neighboring ministers he picnicked on Mt. Wachusett on "Bacon, Bread & Cheese, Rum," and in descending "Sang a Stanza to the praise of the Great Creator." [45] All through his life he encouraged group singing among the young people, teaching "The Rules of Musick" himself and offering board and lodging to squads of young ladies who came from a distance to attend the singing schools set up under his guidance.[46] The manuscript singing book from which he introduced written music to Westborough began as a pious collection but later acquired such tunes as "The Jolly Young Swain," "Cheshire Rounds," "Lord Biron's Jigg," and "Love Triumphant." The words of this last were:

> When I beheld Clarinda's Eyes,
> Love did my trembling Heart Surprise;
> & long have I hugg'd my fond amorus Chain,
> & long have I mourned the fair Tyrant's disdain;
> Stil Whining & Sighing & Pining & dying;
> Nor once bravely trying relief to obtain.[47]

Against those who would celebrate Christmas, he set his face, however:

Were any of them rationally & Sincerely Enquiring & Examining into the Grounds of the Controversie between the prelatists & the Dissenters it were a far different Case; but they manifest only a Spirit of Unsteddiness. May God grant 'em a Sight & Sense of their Folly and Childishness! [48]

When the Parkmans first moved to Westborough they built a house on the hill where the Lyman School later stood. Before

---

[42] *Ibid.*, Aug. 12, 1781.
[43] Diary (A. A. S.), Aug. 30, 1768.
[44] *Ibid.*, July 1, 1740.
[45] Diary (M. H. S.), Sept. 3, 1755. [47] The Ms. singing book is at the M. H. S.
[46] Diary (A. A. S.), Mar. 19, 1778. [48] Diary (A. A. S.), Dec. 25, 1750.

the door they set tulips and "Honey Suckle," gifts from the Roxbury Dudleys.[49] The people of the parish flocked to the garden to enjoy cherries and currants in season. In the back yard chips, muck, and manure accumulated for years until the problem became one of excavation. In 1751, when the church was moved, a new parsonage was begun near-by. It was a relatively elegant structure, with windows so large that a parishioner made remarks about the "pride of Ministers." The exterior was "coloured yellow" [50] and shaded by elms planted for the purpose. Inside hung framed pictures of the city of Hamburg and the River Elbe.[51] The fact that sand for the floors was drawn by the load suggests that it may have been used to cover them in the old manner.

In 1728 Parkman bought a raw Guinea boy of his father for 74*l*. The slave, Maro, trotted to Westborough behind the parson's horse. He died during the first winter, mourned, not as a financial loss but as "The First Death in my Family." [52] The next experience of the sort was the purchase of an Irish boy, John Kidney, of Solomon Lombard (A.B. 1723) for 7*l* 15*s*. Although John had given his consent to the indenture, he soon exhibited such "Rude and Vile Conduct" that Mrs. Parkman was "afraid to be left alone with so brutish a Creature." As a climax he tried to rape Molly and cut her arms badly. Although he begged for forgiveness from his knees, he was soon again impudent, lying, and stealing. After long conferences with the neighboring ministers and justices, the parson finally sold him, at a handsome profit, to two farmers who were quietly confident that they knew how to care for unruly live stock.[53]

Mary Champney Parkman died on January 29, 1735/6. For more than forty years her husband recorded the agony of that day when its anniversary came around; neither time nor his comfortable second marriage erased the pain. But since there were four small children in the parsonage, it is not surprising to find him courting Hannah, daughter of the Reverend Robert Breck (A.B. 1700) of Marlborough, within the year. Parkman's motives were not entirely utilitarian, however, for he sent

49 Diary (A. A. S.), Mar. 28, Apr. 26, 1739.
50 *Ibid.*, Sept. 24, 1771. Great difficulty was experienced with the paint.
51 Diary (A. A. S.), Feb. 2, 1760.
52 *Ibid.*, 1728, *passim*; Natalitia, Dec. 5, 1729.
53 Diary (A. A. S.), 1738–39, *passim*.

Hannah "Letters, Poems, etc.," which on second thought he asked her to burn. "Her Conversation was very Friendly, and with divers expressions of Singular and Peculiar Regard. *Memorandum Oscul:* But she cannot yield to being a step mother." [54] But Hannah, who was then twenty-one, thirteen years his junior, did finally yield and marry him at Boston on September 1, 1737, Joshua Gee presiding. She was a woman who "derived lustre from her parantage . . . but more from her own conduct and character, being a real, but Cheerful Christian, whose religion was free from the odious extremes of bigotry, superstition, or enthusiasm." [55] Her portrait survived to frighten little girls of a later generation; the painter was either no artist or a very brave man. [56] However that may be, she made the minister a good wife; in his later years he remembered "How Ardent and United" were their first years together. This marriage meant no break with the Champneys of Cambridge, who took the surplus sons and daughters of the Westborough parsonage for years at a time. Little Ebenezer, by the way, thought the smells of Boston shocking. [57]

It was a disappointment to the minister when Ebenezer, his oldest son, refused to study for college; but probably the family could not have met the expense. They were so poor and the daughters had to work so hard that what little education they obtained came from copying words with chalk on the barn floor in odd minutes. Parkman was depressed because he could not do better by his children, but they seemed content. Neighbors were kind, and one gave little Molly "an Irish or Foot Wheel" which must have made her the envy of her playmates. [58] When the north parish, now Northborough, was set off in 1744, the financial situation became so difficult that the minister considered resigning. [59] The Marlborough association, of which he was a member, argued him out of suing the town for his back salary in 1747, but the depreciation of the paper "Soldier Money" further aggravated the situation. Nor was the matter im-

---

[54] Forbes, *Parkman*, p. 29.

[55] *Columbian Centinel*, Aug. 29, 1801, p. 2.

[56] *Some Old Houses in Westborough* (Westborough, 1906), p. 24. The portrait is reproduced in Forbes, *Parkman*, op. p. 143.

[57] Diary (A. A. S.), June 1, 1738.

[58] *Ibid.*, June 30, 1738. A flax wheel?

[59] Mass. Archives, cxvi, 208.

proved by the pressing of his hired men into the army over his protests. In 1755 his salary hit a hopeless low of 32*l.* As an excuse his people pointed out that he sometimes repeated old sermons. But with a neighborly kindness which is as characteristic of Yankees as penuriousness, they showered "above an hundred Valuable presents" upon him on one occasion when he was confined with rheumatism. In temporary moments of solvency he contributed to the support of Josiah Cotton (A.B. 1722) as a missionary among the Baptists of Providence,[60] and bought a share in a town in the Berkshires, a futile speculation.

The needy and unfortunate always found welcome at the parsonage no matter how low the larder might be. The Acadian exiles from neighboring towns visited him, partly perhaps because he could read and write French and make some stabs at speaking it. Some of them attended his services, but when he tried gently to point out the errors of popery, they retorted that they were more religious than the English, whose young people were so wicked that they could not permit their children to associate with them.[61] These Catholic strangers were no less welcome than were the great men of the colony who made a practice of stopping at the Parkman home as they travelled the main highway which passed the door. How the mistresses of the parsonage cared for the constant throng of guests is a mystery. The minister, in his striped cotton gown, entertained piously and graciously, inwardly fuming because the presence of his guests prevented proper Sabbath preparations.

Parkman was himself an inveterate visitor, although always with a pious end in view. Wearing in season a "valuable wig" which had belonged to his old friend Judah Monis (A.M. 1723) and a muff which had belonged to his classmate Hubbard, he frequently travelled the road to Boston. His farthest venture was New Haven, which he visited in 1738. He attended the Yale Commencement and found the company "not very numerous" but the "Exercises and Entertainment handsome & agreeable." "The Custom of giving Diploma at the time of giving the Degree" he thought worthy of adoption at Har-

[60] Colman Mss. (M. H. S.), May 27, 1731.
[61] Diary (A. A. S.), 1756, *passim.*

vard.[62] He attended the Harvard Commencement whenever he could and did not miss a single election day in Boston until 1745. The convention of the clergy of the province, which met at the same time, was an opportunity to meet old friends. He was not averse to being called out of the meeting by his classmates Lowell and Chauncy to have a pipe and a word of gossip. In 1761 he delivered the convention sermon.

If we may judge by what has survived, Parkman's sermons were dull and scholarly. He was so dependent upon his notes that if he left them at home he could not preach. His preaching, praying, and introspection were of the type of the seventeenth century, although some of his pious resolutions have a practical ring: "To Return or pay for the Books I have some Time agoe borrowed and negligently & unjustly detained for Some years from the owners." [63] Although for him God's hand was not to be discerned in every phenomenon, there were sermons in stones, and even in a jackal which was exhibited in Westborough: "An Entertaining Sight. How wondrously are the works of God diversifyd. How Manifold are thy Works, O Lord in wisdom hast thou made them all!" [64]

In theology Parkman was little touched by his age. He read Tillotson on his texts but disagreed heartily with him. He clung firmly to such fundamental doctrines as justification by faith, "it being of vastest Consequence & weight in the Christian Religion." [65] When his friend Jonathan Edwards brought out the famous tract on Original Sin, he thanked God and prayed for its happy success. Such a man naturally favored the Awakening. He affectionately invited Gilbert Tennent to preach at Westborough [66] and entertained George Whitefield at the parsonage on October 14, 1740.[67] In the course of the next three years he was greatly troubled by uneducated preachers who invaded his parish, so while generally approving the manifesto issued by the provincial convention of the clergy on July 7, 1743, he regretted that it had not sufficiently damned itinerancy.[68] But when the Marlborough association, of which

[62] *Ibid.*, Sept. 13, 1738.
[63] Natalitia, Mar. 13, 1740/1.
[64] Diary (A. A. S.), Apr. 25, 1750.
[65] Commonplace Book, p. 114.
[66] *Ibid.*, p. 106.
[67] Diary (A. A. S.), Oct. 14, 1740.   [68] Commonplace Book, pp. 110–11.

he had been clerk for twenty years, passed resolutions against Whitefield, he attempted to resign rather than enter them in the records. His associates, however, affectionately held him to his post.[69] Among his parishioners there were many who condemned what he himself called his "Credulousness and Abounding Charity towards the New Lights, who in many Instances have (its said Commonly) been found too forgetfull where they ought to Remember both words & Facts." [70] When members of the congregation, infected with the general religious enthusiasm, cried out during the sermon, he begged them to compose themselves, but could not prevent a "great crying out in the woods where a Number were retired" during the noon hour.[71] Forty years after the Awakening he was again denounced as a tyrant for refusing to permit an itinerant to preach erroneous principles in the town.[72]

In religious forms Westborough, like most frontier towns, lagged far behind the general practice. It was not until 1748 that Parkman dared institute the public reading of the Scriptures, and even then he met with considerable opposition. Our natural New England cantankerousness magnified these small differences. The new meetinghouse was built so far from the old that the minister could not go home for dinner, but no one invited him in, and the parish refused to pay for his meal if he bought it at the tavern. Still, when he offered to resign in the midst of these difficulties, the parish unanimously invited him to remain. A public aversion to "Triple Time Tunes" unexpectedly caused trouble in 1750. At the close of one service the minister "appointed Mear Tune to be sung," but the man who set the tunes "Sat Canterbury" instead. The parson thought that this was "thro either Mistake, or because he could not strike upon Mear at the Time," so he set "Mear Tune" himself and thereby raised a teapot tempest.[73] He could not even make such an obvious statement as "There is a God" without having some theological hair-splitter in the congregation take him to task for preaching "damnable Doctrine." [74]

[69] Diary (A. A. S.), Jan. 22; June 11, 1745.
[70] *Ibid.*, Mar. 9, 1743/4.
[71] *Ibid.*, June 3, 1744; Aug. 24, 1746.
[72] Diary (M. H. S.), Nov. 12, 1781.
[73] Diary (A. A. S.), Apr. 30, 1750.
[74] *Ibid.*, Feb. 26, 1761.

It was fortunate for Parkman, who was accustomed to being dined and entertained by such aristocrats as the Dudleys, that the American Revolution had no social implications in West-borough. Like his fathers he denounced the wearing of "Velvet Whoods" by "young persons of low Rank in the Congrega-tion." [75] When a young Scotch blacksmith offered bags of hard English coin for the old parsonage, the minister refused to sell to a man so far below him socially. A wealthy farmer bought the property and turned it over to the Scot, whose solid virtues eventually won the minister's firm friendship. [76] He regarded the Stamp Act riots as "A melancholly Occurrance! much to be deplord" [77] and heartily pitied his old friend Governor Hutchin-son, whom he continued to visit on his now rarer trips to Boston. Naturally he was not one of those to whom the Boston agi-tators sent the propaganda resolutions:

I askd publicly to be gratifyd with the pamphlet which the antient & respectible Town of Boston had sent hither, for I have neither seen it nor can find where it is. I would Seek the Peace and Welfare of this Place, of this Province, & of the British Realm, on which we are de-pendant. — But I could not hear of Said Pamphlet. [78]

Never himself in any way discontented with the British con-nection, Parkman did not awaken to the political situation until 1773, when he began to take stock of his ideas by reading such classics as Jeremiah Dummer's *Defence of the Charters*. In July, 1774, he opposed the passing of patriotic resolutions by the town on the grounds that they were illegal and untimely and that there was "not such an Alternative as . . . Suffering Bloodshed or slavery" as the radicals made out. [79] When, in spite of his opposition, the resolutions were passed, he refused to read the document:

If I heard it, I must either approve or condemn it — but do which I will, I must of necessity be blamd. If I approvd of their Draught, I must have exposd my self to the resentments of Authority which I must teach all Men to avoid: for I must teach & enjoin that "every soul be subject to the higher powers" — "to obey Magistrates. . . ." If I should dislike it, I was aware that they would not be easily turned aside. . . . I was not 'o mind to render my Self Obnoxious either way. [80]

---

[75] *Ibid.*, Dec. 29, 1738.
[76] Forbes, *Parkman*, p. 206.
[77] Diary (A. A. S.), Aug. 28, 1765.
[78] Diary (M. H. S.), Mar. 1, 1773.

[79] Diary (A. A. S.), July 4, 1774.
[80] *Ibid.*, July 7, 1774.

In August the people became "very fierce about our Rulers," and Parkman, to avoid being forced to sign their resolutions, drew up an "agreement" of his own which he proposed to sign instead "Rather than have such an Hubbub & Uproar." [81] This was probably the document in which he published the fact that he was "heartily Set against Despotism & Oppression," a statement in which the hottest Tory would agree.[82] This did not save him from "the Sore Frown of Heaven by the Displeasure of very many of" his "dear people, for . . . refusing to intermeddle with their civil Transactions when they Signd a Covenant." [83] He regarded the situation of the colony as tragic: "Mobs & Riots, Whigs and Torys — as if our Happiness were nigh to an end! O God save us!" [84] When contributions were solicited for the people of Boston, suffering under the Port Bill, he headed the list, but with the proviso, intended to avoid political implications, that his money go to certain poor people of his acquaintance who were ordinarily recipients of his charity.[85]

On April 19, 1775, Parkman rode over to preach for his neighbor at Northborough, where he found that the outbreak of civil war, as he called it, prevented the meeting. During the next week it seemed as though the whole western world was pouring by the door of the parsonage. To the Westborough company at Cambridge he wrote:

May abundant Grace be given you to prepare & qualifie you for the most Memorable Action — Not in Rebellion, not in Transgression but in just & unavoidable Defence of our invaluable Rights, Laws, Libertys & Privileges.[86]

His situation now became more difficult, and in May he was forced to sign a political covenant "containing 3 Articles — all of which were exceptionable, but as the present Torrent of Liberty" was "irresistable," he "was obligd to condescend, for Peace Sake, & to Avoid a Rupture among us." [87] He visited with the recruits who poured through Westborough to join in the siege of Boston and was particularly interested in one company of a hundred and thirty strange men who said that they

[81] Diary (A. A. S.), Aug. 4, 29, 1774.
[82] *Ibid.*, Dec. 30, 1774.
[83] *Ibid.*, Jan. 1, 1775.
[84] *Ibid.*, Dec. 31, 1774.
[85] *Ibid.*, Feb. 21, 1775.

[86] Commonplace Book, p. 40.
[87] Diary (A. A. S.), May 23, 1775.

had "come from Countrys some hundred Miles beyond Pittsburg on the Ohio." [88]

Parkman's second son, Thomas, had died fighting for King George in 1759. After his first son, Ebenezer, joined the Continental army, his sympathies for the American cause mounted, but his attitude was always one of sorrow. Once more war inflation made things difficult in the parsonage. On the day that the minister completed his fiftieth year of service, his people cut his salary to 55*l* without wood. Such adjustment as was made for inflation lagged far behind soaring prices which compelled him to pay $50 for a silk handkerchief and made the quarterbills of his son Elias at Harvard look like pages of Professor Winthrop's astronomical calculations. As he sank further and further into debt, his quarrels with the town over the arrears of his salary increased.

The ills of gathering years, long overdue, now added to Parkman's difficulties. One day in 1772, when he was reading the manuscript diary of Israel Loring (A.B. 1701), his eyes suddenly failed and compelled him to go out and borrow a parishioner's spectacles.[89] In April, 1782, his sight suddenly failed him in the pulpit, compelling him to finish with extemporaneous remarks. Still he continued to carry on all of his ministerial duties until a shock brought him down early in November. On December 5, 1782, he made the last feeble entry in the great diary which he had begun in 1720. Four days later he died, honored by a long obituary in newspapers, which were crowded with news of events utterly foreign to the days of his youth.[90] He lies in the Memorial Cemetery of Westborough. Hannah Breck Parkman died on August 20, 1801.

Any extended account of Parkman's sixteen children and the equally distinguished people whom they married would be a history of their century: (1) Mary, b. Sept. 14, 1725; m. Eli Forbes (A.B. 1751), Aug. 6, 1752; d. Jan. 16, 1776. (2) Ebenezer, b. Aug. 20, 1727; m. Elizabeth Harrington, Sept. 21, 1752; d. July 5, 1811. (3) Thomas, b. July 3, 1729; d. in the army, Oct. 23, 1759. (4) Lydia, b. Sept. 20, 1731; d. in childhood.

---

[88] *Ibid.*, Aug. 26, 1775.

[89] *Ibid.*, Oct. 11, 1772. Parkman has the sympathy of the editor, who in the preparation of the previous volume in this series strained his eyes over the same Ms.

[90] *Mass. Spy*, Dec. 26, 1782, p. 3; *Boston Gazette*, Jan. 6, 1783.

(5) Lucy, b. Sept. 23, 1734; m. 1st Col. Jeduthan Baldwin, int. Feb. 12, 1757, 2nd Eli Forbes (A.B. 1751). (6) Elizabeth, b. Dec. 28, 1738; d. in infancy. (7) William, b. Feb. 19, 1740/1; m. Lydia Adams, Sept. 10, 1766. (8) Sarah, b. Mar. 20, 1742/3, m. John Cushing (A.B. 1764), Sept. 28, 1769; d. 1825. (9) Susanna, b. Mar. 13, 1744/5, m. Jonathan Moore (A.B. 1761), Oct. 13, 1768; d. Nov. 30, 1772. (10) Alexander, b. Feb. 17, 1746/7; m. Keziah Brown. (11) Breck, b. Jan. 27, 1748/9; m. Susanna Brigham, int. Nov. 14, 1776; d. Feb. 3, 1825. (12) Samuel, b. Aug. 22, 1751; m. Sally Shaw Feb. 11, 1773. (13) John, b. July 21, 1753; d. Sept. 10, 1775. (14) Anna Sophia, b. Oct. 18, 1755; m. Elijah Brigham (A.M. 1794), Sept. 21, 1780; d. Nov. 26, 1783. (15) Hannah, b. Feb. 9, 1758; d. Oct. 14, 1777. (16) Elias, b. Jan. 6, 1761; A.B. 1780.

Parkman left but a small estate,[91] and for a portrait nothing but a drawing made from memory in later years.[92] More important by far was his diary, of which we can account for the following parts: Feb. 1719/20 through Dec. 1722, burned by Parkman himself;[93] Jan. 1723 through Sept. 1728, at the American Antiquarian Society; Feb. through Nov. 1737, printed by the Westborough Historical Society; Jan. 1738 through Dec. 1740, A. A. S.; Dec. 21-31, 1742, at the Massachusetts Historical Society; Jan. 1743 through Dec. 1748, A. A. S.; Jan. 1749 through Dec. 1749, M. H. S.; Jan. 1750 through Dec. 1754, A. A. S.; Jan. 1755 through Dec. 1755, M. H. S.; Jan. 1756 through May, 1761, A. A. S.; June, 1764 through Dec. 1767, A. A. S.; Jan. 1768 through June, 1769, A. A. S.; Aug. 1771 through Nov. 1772, M. H. S.; Nov. 10-21, 1772, A. A. S.; Nov. 1772 through June, 1773, M. H. S.; June, 1773, through Oct. 1778, A. A. S.; Nov. 1778 through Dec. 1780, printed by W. H. S.; Jan. 1781 through Dec. 1782, M. H. S. Photostats of the portions now at the Antiquarian Society may be consulted at the Massachusetts Historical Society. The gaps may be partly filled from a volume of "Natalitia," or birthday reflections, preserved at the American Antiquarian Society along with two bundles of manuscripts. In the *Proceedings* of that Society the full text of the diary and the natalitia is now being

---

[91] The inventory of the estate is printed in Forbes, pp. 298–300.
[92] This picture is reproduced in Deforest, *Westborough*, op. p. 66.
[93] Diary (A. A. S.), Jan. 16, 1752.

published under the editorship of Dr. Francis G. Walett of Worcester State College. The Massachusetts Historical Society has his commonplace book. Manuscript sermons are preserved at the above institutions, the Congregational Library in Boston, and the New York Public Library.

WILLIAM BRATTLE

GENERAL WILLIAM BRATTLE was born in Cambridge on April 18, 1706, the son and namesake of the famous Parson Brattle (A.B. 1680) of that town. As the minister lay on his deathbed in the spring of 1717, he realized that his small boy would inherit, along with his wealth and that of college treasurer Thomas Brattle, an intellectual tradition second in New England only to that of the Mathers and, in one important way, less provincial. So he patiently wrote out a long letter of advice which concluded:

Acquaint thy Self with History; know something of the Mathematicks, and Physick; be able to keep Accompts Merchant like in some measure; but let Divinity be thy main Study. Accomplish thy Self for the worke of the Ministry. . . . My dear Child, be of a Catholick Spirit.[1]

Had the younger William died before reaching the age of twenty-one, the bulk of his father's large estate would have gone to the college;[2] but instead he turned out to be a considerable financial drain upon that institution, for the £165 which his ten years in residence cost were charged against the gifts of his father and uncle. Tutor Flynt found him "as quick as any" of his class[3] and lent him chocolate and brandy. Although the youngest and, by

[1] *New England Hist. Gen. Reg.* I, 285.
[2] Middlesex Probate Records, XIV, 539.
[3] Henry Flynt, Diary (Mass. Hist. Soc.), p. 420.

WILLIAM BRATTLE

favor and promise, the head of his class, William was no grind, for
at the tender age of fourteen he "debauched" with Thomas Smith
(A.B. 1720) and throughout his undergraduate years incurred heavy
fines for violations of college rules.⁴ In 1725, as the "Head of the
Masters Class," he "made a gratulatory oration." On this occasion
his *Quaestio* was a negative reply to the question, "An, Humanæ
Naturæ Christi, per Unionem personalem, formaliter communi-
catæ sint, quædam Proprietates Naturæ Divinæ?"

As this topic would suggest, Brattle intended to enter the ministry. He owned the covenant in his father's church, preached
where occasion offered, and in the spring of 1725 joined in the
competition for the Ipswich pulpit.⁵ Evidently his talents did not
lie in that direction, for when he once turned for confirmation of
his preaching ability to a friend who used to weep in his pew
when Brattle preached, he got the unexpected answer: "When I
saw the miserable figure which you made in the pulpit, I burst into
tears, nor could I refrain whenever I heard you preach, and was
rejoiced when you attempted it no more." ⁶

Discouraged, Brattle turned to the practice of medicine. In
December, 1726, he accompanied Lieutenant Governor William
Tailer's expedition to make peace with the Indians at Casco in the
capacity of physician and later he began to practise on the soldiers
at Castle William.⁷ Soon such distinguished families as the Jeffreys
were employing him for his "superior skill in physic," ⁸ and for
many college generations he was called to treat the ills which
afflict students. The duties of the many civil offices which he held
in his later years prevented his keeping abreast of the improvements in the practice of medicine, and the time came when people
complained that he came to their bedsides with filthy hands, reek-
ing with the odor and plastered with the refuse of the stable.⁹

Brattle finally left his college chamber to marry Katherine
Saltonstall, the daughter of Governor Gurdon Saltonstall (A.B.

----

⁴ His Ms. copy of Charles Morton's "Physics," with Hebrew notes at the end,
is in the Harvard University Archives.

⁵ William Waldron to Richard Waldron (Library of Congress), May 3, 1725.

⁶ An undated clipping in J. L. Sibley, Collectanea Biographica Harvardiana
(H. U. A.), I, 7.

⁷ *Journals of the House of Representatives of Massachusetts* (Boston, 1919–),
VII, 138; Executive Records of the Province Council, VIII, 486, XI, 22.

⁸ Jeffries Mss. (Mass. Hist. Soc.), June 1753.

⁹ Clipping, *op. cit.*

1684), on November 23, 1727. He plunged immediately into public affairs, for in the course of his twenty-third year he was elected to the first of his twenty-one terms as selectman of Cambridge, was elected to the Ancient and Honourable Artillery Company, in which he soon became a captain, was chosen a major of militia, was appointed a justice of the peace, and was sent to the House of Representatives. In the General Court, in spite of his youth, Major Brattle leaped into the thick of the battle which the popular party was waging against Governor Burnet. He was on the committee which welcomed Governor Belcher in 1730 and that which ran the Rhode Island boundary line in 1733. With this experience fresh upon him he refused to join in running the Connecticut line. But making political speeches in the House also had its discomforts, as the Major once learned after remarking on the floor that one Mr. Giles Dulake Tidmarsh was "a begerly fellow." Stepping over to the Bunch of Grapes tavern to refresh himself after his speech, he unexpectedly encountered Mr. Tidmarsh, who informed him that he was "a Damned Villin," "and took him said Brattle by the Nose and Wrung it." Tidmarsh then ran from the tavern hotly pursued by that hard-bitten Indian fighter, Colonel John Stoddard (A.B. 1701). Later, when Stoddard was testifying in behalf of his friend, the judge suddenly asked him if he had overtaken Tidmarsh, and the surprised Colonel could only reply that he could not remember; by which we may conclude that the Indian fighter had stepped behind the first water barrel to have his laugh in private. But the injured Major went before the House and "Complained that he was much hurt and . . . did then spit Blood, and . . . did Bleed at his Nose." The General Court in great indignation voted that Tidmarsh's act was "a very high Indignity and Affront offered as well to this Court, as to any Members thereof" and posted a £50 reward for the apprehension of the offender.[10]

By this time Brattle had developed a private law practice and in 1736 he was designated as one of the justices of the peace who sat with the judges in the Middlesex County Court. In that same year, and on several other occasions, the House elected him At-

[10] *Acts and Resolves of the General Court of Massachusetts* (Boston, 1868–1922), XI, 750, 761; *House Journals*, XI, 309, 310–1, 333; Early Files in the Office of the Clerk of the Suffolk Supreme Court, 38,536, 38,873, 166,311.

torney General of the province, which was an empty honor because
the Governors insisted that the office could be filled only by appointment by the Governor and Council. There is no evidence that
he ever prosecuted as Attorney General, but he did sit as a special
justice of the Superior Court in 1749. It is evident that Brattle's
ability in law was not much greater than in medicine and preaching. When he was employed to represent the college he showed
little decisiveness.[11]

There is no question, however, of Brattle's willingness to serve
Harvard. In later years he was one of the most faithful members
of the Board of Overseers and served on the committees to build
Hollis and to rebuild Harvard Hall. He was also a trustee of the
Hopkins foundation. Once the college asked him to complain to
the Cambridge selectmen "that a Billiard Table" had been "set up
in Cambridge not far from the College viz at the house of Capt.
Samuel Gookin," which, they apprehended, would be "of pernicious
Consequence to the Society." [12] On the other hand, the Major once
complained to the faculty that a student "had grossly insulted his
train'd Company when under arms, by fireing a Squib or Serpent
among their firelocks when loaded & primed & all grounded,
whereby he greatly endangered the limbs at least of the Souldiers
& Spectators." [13]

The people of Cambridge knew Brattle as an active proprietor
of the town common lands and a faithful church member who
sometimes found time to represent them in ecclesiastical councils.
He was a charter member of the abortive Society for Propagating
Christian Knowledge among the Indians of North America, and
was at one time agent for the Indian proprietors of the island of
Chapaquidic. An old friend and patron of Thomas Prince (A.B.
1707), the historian, Brattle parted with him over the issue of the
Awakening and moved into the camp of Charles Chauncy (A.B.
1721). His chief contribution to the religious controversy was his
defense of the college against the attacks of George Whitefield. In
this he refuted the charges of irreligion and Arminianism by muster-
ing a quantity of misinformation about the early days of the college
and by quoting the library charging lists to show how little such

[11] *Publ. Col. Soc. Mass.* XVI, 702 (Harvard College Records).
[12] Faculty Records (H. U. A.), III, 51.
[13] *Ibid.*, II, 115.

liberals as Tillotson were read.[14] Chief Justice Paul Dudley (A.B. 1690) was so impressed by the Major's defense that he made a speech about it to a jury.[15]

On the other hand Tutor Nathan Prince (A.B. 1718) at this time was informing the Harvard authorities that the Major was "a Devilish Lyar" and that "what he said was no more true than that the Devil was in heaven, & as cursed a Lye as ever was hatch'd in the infernal Pit." [16] There is no surviving evidence to support Prince's characterization but a great deal to corroborate a description which Governor Belcher had recently penned:

Old Br-ttle [he was thirty-five, and much younger than the Governor] is honest, Let us be thankfull if we think we know more, than he does . . . God alone is able to make Creatures . . . free from Oddities, of mind, and yet when they are odd, they wont generally be so acceptable, in the world . . . I really think my far distant Kinsman to be a queer fellow. . . .[17]

These characteristic oddities grew on Brattle with the passing years and made him something of a joke to those who disagreed with him on political matters:

He entertained company at his table in a handsome style, and selected many of his guests from a class, who could return his attentions in the same way. Having kept much good company he was a gentleman in his manners and his politeness was that of the old school. He was clumsy in his person and though his dress was expensive, it was always in a fashion that had gone by. To the pleasures of the table he was a devotee. . . . He used to travel to Marblehead, to feast upon cod taken fresh from the water. . . . With an affectation of humility he entered at the kitchen door of his friends when it was convenient — kindly inquired of the servants respecting their health — cast a sheep's eye at the fire, and if what was cooking pleased him he would contrive to tarry to dinner.[18]

In much this manner the jolly Major rolls his great bulk through the diaries of John Rowe and his contemporaries, betting, fishing, and, particularly, dining well. But if he enjoyed leisurely pleasures unknown to us, he also cringed under scourges which we have

[14] *Boston Gazette*, Apr. 20, May 25, June 22, 29, 1741; William Hobby, *An Inquiry* (Boston, 1745), p. 22.

[15] *Diaries of Benjamin Lynde and Benjamin Lynde, Jr.* (Boston, 1880), p. 113.

[16] Nathan Prince Case Mss. (H. U. A.).

[17] Jonathan Belcher to Richard Waldron (M. H. S. photostat files), May 25, 1741.

[18] Clipping, Coll. Biog. Harv. I, 7.

learned to avoid. When he was quarreling with George Whitefield, his wife and daughter lay sick of throat distemper,[19] probably
in this instance diphtheria which he had brought into his home from
the bedside of some patient. Five small daughters and a son were
carried from the mansion to the burial ground. Katherine Saltonstall Brattle survived these blows only to succumb to the smallpox
on April 28, 1752. The Major was married again on November 2,
1755, this time to Martha Fitch Allen, widow of James Allen
(A.B. 1717). She died on August 26, 1763.

While Martha Brattle was alive, the Major took a house on
King Street, in Boston, drawn, no doubt, by her friends and by the
demands of the Boston property which he had inherited from his
Uncle Thomas, as well as by the convenience of being near the General Court. In 1765 he bought a house on the east side of Tremont
Street, between Court and Bromfield streets, and for some years he
served as one of the visitors of the Boston schools. By purchase
and inheritance he acquired considerable land on the frontiers and
was an active proprietor under the old Plymouth grant at Sheep-
scott. Brattleboro, Vermont, was named for him because, as a leading proprietor, he had chosen the site. For short-term investments
he took so many mortgages and personal notes that he had special
forms printed.

The contemporary who described Brattle as "a man of universal
superficial knowledge" [20] was as right as witty, but perhaps he did
not do justice to a real knack for things military which contrasts
with the mediocre record in matters of religion, law, medicine, and
business. At the age of twenty-seven Brattle had published a set of
"Rules for Drawing up a Regiment" which was widely used. At
the time of the French invasion scare of 1745 Governor Shirley
appointed him to the command of the auxiliary forces at Castle
William, where his main function was that of drillmaster.

Military matters were then closely associated with politics, for
there were no professional soldiers in the province, and the General
Court insisted on making even the minor decisions. This made the
presence of the militia officers in the General Court highly desirable, which was no doubt one reason for the return of Brattle to the

---

[19] *Lynde Diaries*, p. 111.

[20] John Foxcroft (A.B. 1758), quoted in Emory Washburn, *Sketches of the
Judicial History of Massachusetts* (Boston, 1840), p. 210.

House of Representatives in 1754 after an absence of nearly twenty years.[21] Almost at once he attracted popular disfavor by his support of an excise bill and was lampooned by his opponents as "Mrs. Biddy":

Mrs. Biddy. . . . is the Wife of a certain valiant C-ll-n-l; and, in Imitation of him, wears a great Tuft of Feathers on the top of her Head; so weighty, that it is thought to have overstrain'd, and weaken'd, that tender Part; and made her a little giddy. She pretends to a great deal of Religion; tho' but few think she has more Virtue than her Neighbours. She has the Misfortune to have a very sly, cunning Jewish Look; and not being of the fairest Complexion, nor able to look a good Christian in the Face; some have suspected her to be of that Religion; and hinted, that the only Reason why she is not conformed to 'its whole Ritual, is her Sex. Yea, many are very confident, that she is an Isr-l-te ind-d-bating this single Branch of the Character, that in such Persons there is no Guile. . . . There are others who have suspected Mrs. Biddy of being a follower of Mahomet, instead of Moses: But, in my Opinion, with far less Reason. For whereas the Mahometan Religion forbids its Votaries to use the Fruit of the Vine, she is a great Admirer of it: And is known to have spent the whole Afternoon of the Christian Sabbath, over her Bottle. . . . If she is really a Jewess . . . it is to be hoped, that, since her Sex will not allow of the most distinguishing Mark and Characteristick of that Perswasion, she will not always want another, which is very compatible thereto; I mean that she will not be long unc-rc-sed in Ears! [22]

Brattle's other activities in that session were more popular than wise, for he sought to delay action on the proposed colonial union until the constituents could be consulted.[23]

The following year the Colonel was raised to the Council, where he began his long years of administrative service. Faithfully he labored on committees to care for the Acadians, to burn Indian scalps,[24] to raise stores, and to keep Loudoun advised and in good humor. Perhaps His Lordship was edified by the sermons which Colonel Brattle enclosed with his dispatches. In 1758 the Colonel was named Adjutant General and in this capacity he raised 7,000 troops in one year. Two years later he was appointed Brigadier

---

[21] For an enlightening account of his election methods, see the *Boston News-Letter*, May 15, 1755, and the *Boston Gazette*, May 12, 1755.

[22] *Monster of Monsters* ([Boston], 1754), pp. 13–4. The identification is confirmed in a contemporary hand in the Jeremy Belknap copy at the M. H. S.

[23] *House Journal* (Boston, 1754–1755), Dec. 14, 27, 1754.

[24] Executive Records of the Province Council, 1755–1759, p. 319.

General and Commissary. For the rest of his life he was "Brigadier Brattle" to the whole province, although he was appointed Major General in 1773. During the war he represented the province in negotiations with Connecticut for common defense, and after that, in 1767 and 1772–1773, he served with Thomas Hutchinson and John Hancock in the New York boundary negotiations. In 1770 he was engaged in the survey of the northern boundary of Maine.

It was commonly said that Brattle, like the elder James Otis, hoped to succeed Stephen Sewall (A.B. 1721) as Chief Justice and was bitterly disappointed at the appointment of Thomas Hutchinson. There is no question that after Hutchinson's appointment he plunged into politics with new vigor and headed the anti-government party in the Council,[25] fully as ardent a Whig as the younger Otis,[26] who commanded the corresponding faction in the House. While serving as moderator of the Cambridge town meeting on October 14, 1765, he publicly read an "outrageous & indecent" "libell against the Government of Great Britain," and, although he expressed innocent surprise at the resulting furor, he was regarded by Governor Bernard as having dictated the whole proceeding.[27] Two weeks later the Governor reported to the Privy Council that the General had openly walked arm-in-arm with Mackintosh, the leader of the Boston mob, in a public demonstration.[28] About the same time Brattle had difficulties with Colonel Murra of the regulars not unlike his earlier encounter with Mr. Tidmarsh.[29]

On Election Day, 1769, the General Court as usual chose Brattle to the Council, but Governor Bernard promptly disallowed both him and James Bowdoin (A.B. 1745) as "the Managers of all the late Opposition in the Council to the Kings Government."[30] Bowdoin replied to the Governor with a nasty speech, but Brattle contented himself with pointing out that his character was vindicated by the fact that he had been elected by the unanimous vote of

[25] Edmund Trowbridge to William Bollan, Dana Mss. (M. H. S.), July 15, 1764; *Coll. M. H. S.* LXXIV, 32, 66; John Murray, the future Loyalist, once "grossly insulted" him at a meeting of the Council and was obliged to apologize. Executive Records of the Province Council, 1761–1765, pp. 255, 256–7.

[26] John Adams, *Works* (Boston, 1850–56), x, 193.

[27] Bernard to Board of Trade, Sparks Mss. (Harvard College Library), Bernard Papers, IV, 166–7; *Acts of the Privy Council. Colonial. Unbound Papers*, p. 412.

[28] *Acts of the Privy Council. Colonial. Unbound Papers*, p. 414.

[29] John Rowe, Diary (M. H. S.), June 5, 1766.

[30] Bernard to Pownall, Sparks Mss., Bernard Papers, IV, 295.

both houses and leaving the Governor to his conscience and to God.[31] The word went around that Bernard, after consulting his conscience, at least, had deprived the General of his position as Colonel of the First Regiment.[32] Brattle privately sent Lord Dartmouth a loyal explanation of the situation [33] and then went forth to dine in public with the Sons of Liberty.[34] In the spring of 1770 he was elected to the House, where he was appointed to a committee for building province magazines which were certainly not called for by any foreign menace.

As late as the end of November, 1772, Brattle was "chatty" with John Adams and was regarded by Samuel Adams as being sound in his political principles; [35] but in a special Cambridge town meeting called at the end of December to consider the question of fixed salaries for judges, the General argued that it was desirable to make the judiciary financially independent of both the Governor and the popularly controlled House. As a result he became engaged in a newspaper controversy with John Adams which brought that rising young lawyer for the first time into the political limelight. Adams's version of the affair has become the classic account:

He [Brattle] was so elated with that applause which this inane harangue [in the town meeting] procured him from the enemies of this country, that in the next Thursday's Gazette he roundly advanced the same doctrine in print, and, the Thursday after, invited any gentleman to dispute with him upon the points of law.

These vain and frothy harangues and scribblings would have had no effect upon me, if I had not seen that his ignorant doctrines were taking root in the minds of the people, many of whom were, in appearance, if not in reality, taking it for granted that the Judges held their places during good behavior. . . . Brattle's rude, indecent, and unmeaning challenge, of me in particular, laid me under peculiar obligations to undeceive the people.[36]

The literature of the controversy, when reexamined, does not indicate any intention on Brattle's part to blow up a storm. His letters, if somewhat pompous and superficial as to points of constitutional law, were courteous and sensible, and his arguments soundly based on the fact that the courts of the province had been for a century

[31] *Essex Gazette,* June 6, 1769.
[32] Rowe, Diary, June 21, 1769.
[33] Dartmouth Mss., II (*Hist. Mss. Comm., Report* 14, Pt. X), p. 43.
[34] 1 *Proc. M. H. S.* XI, 140.
[35] *Warren-Adams Letters* (*Coll. M. H. S.* LXXII–LXXIII), I, 13.
[36] Adams, *Works,* II, 315–6.

terrorized by wine smugglers, timber thieves, and their allies, the popular politicians. Adams, for his part, was smart and ill-tempered and tried to split legal hairs the existence of which is doubtful.[37]

Writing of the affair forty years later, Adams gave an account of the conversion of Brattle to the Tory party which has, unfortunately, become fixed in American history:

> Brattle was a divine, a lawyer, and a physician, and, however superficial in each character, had acquired great popularity by his zeal, and I must say, by his indiscreet and indecorous ostentation of it, against the measures of the British government. The two subtle spirits, Hutchinson and Sewall, saw his character, as well as Trowbridge, who had been his rival at the bar, for many years. Sewall was the chosen spirit to convert Brattle. Sewall became all at once intimate with Brattle. Brattle was soon converted and soon announced a brigadier-general in the militia. From this moment, the Tories pronounced Brattle a convert, and the Whigs an apostate. This rank in the militia in time of peace was an innovation, and it was instantly perceived to have been invented to take in gudgeons.[38]

A review of the twelve years after Brattle's appointment as Brigadier in 1760 shows, to the contrary, that the Adamses and the royal governors regarded him as one of the popular leaders, and both treated him accordingly. Nor is it at all likely that the appointment of Brattle as a Major General in 1773 reflects such a bribe, for there is little likelihood that the man who for a dozen years had commanded the militia of the province as Brigadier General would have thought such a promotion to be worth the danger involved in changing his party.

By the winter of 1773-1774 "Old Brattle," like a majority of the men of his class, had decided that the political agitators and legal metaphysicians were driving the province into civil strife likely to produce much more evil than good, but by his public position he was compelled to declare his views while the more fortunate majority could confine their doubts to their diaries and private correspondence. He entertained the Governor and loyal Councillors at his house,[39] and it was there that the Council finally met after the repeated efforts of the Governor to obtain a quorum of Councillors, intimidated by the Tea Party, at the Council

---

[37] *Boston Gazette*, Jan. 4, 11, 18, 25, 1773; *Boston News-Letter*, Dec. 31, 1772; *Boston Post-Boy*, Jan. 25, 1773; Adams, *Works*, II, 315, III, 513-74.

[38] Adams, *Works*, X, 194.   [39] *Ibid.*, IX, 334.

Chamber.[40] When Brattle signed an honest testimonial supporting Hutchinson against silly accusations of treason, popular writers promptly attacked him for "meanly stooping to this servility" and speculated as to whether he was more odious and contemptible than the Governor.[41] He joined Hutchinson in his last reviews of the militia, and, in the absence of Governor Gage, presided over the Artillery Company election of June 6, 1774, for the Whigs refused to tolerate the ministrations of Lieutenant Governor Thomas Oliver (A.B. 1753). A small boy long remembered his appearance on that day:

Dressed in a superb suit of scarlet, trimmed with broad gold lace, with a campaign wig, gold laced hat, and a very handsome sword, he presented a most gorgeous spectacle. He performed his part with great propriety, though accompanied with some degree of pomposity.[42]

When, two months later, the list of Mandamus Councillors was announced, it was remarked that Brattle's name did not appear, although two less distinguished fellow townsmen were honored.

It was obvious to every one of the least discernment [remarked a contemporary] that the Brigadier was much chagrin'd at not receiving a mandamus, though he has since declar'd, in a scoffing way, that he was exceeding glad he was not appointed, and would only wish to have been, that he might have had an opportunity to shew that he had the good of his country at heart, by resigning with contempt, or rather refusing.[43]

As it turned out, not even an opportunity to decline royal favors would have saved Brattle's reputation from the ill winds of chance which were already gathering. In July, 1774, the royal officers were complaining of the enormous quantities of ammunition being withdrawn from the Medford powder house by the towns which had deposited their stocks there.[44] Late in August, Gage, as each of his predecessors had done, sent Brattle, who had charge of the stores, a routine request for a report of the quantity on hand. With the report the Brigadier sent a fatal letter:

Mr. Brattle presents his duty to his Excellency Gov. Gage; he apprehends it is his duty to acquaint his Ex-cy from time to time with every

[40] Executive Records of the Province Council, XVI, 749.

[41] Boston Gazette, Jan. 31, 1774.

[42] Clipping, Coll. Biog. Harv. I, 7.

[43] 1 Proc. M. H. S. VIII, 351.

[44] John Andrews to William Barrell, Andrews and Eliot Mss. (M. H. S.), p. 34.

thing he hears and knows to be true and is of importance in these troublesome times, which is the apology Mr. Brattle makes for troubling the
General with this Letter. — Capt. Minot of Concord a very worthy
man this minute inform'd Mr. Brattle that there had been repeatedly
made pressing applications to him to warn his company to meet at one
minute's warning equipt with arms and amunition according to law he
had constantly denied them. Adding, if he did not gratify them he
should be constrained to quit his farms and town. Mr. Brattle told him
he had better do that than lose his life and be hang'd for a rebel. He
observed that many Captains had done it. . . .

Mr. Brattle begs leave humbly to quere whether it would not be best
that there should not be one Commission Officer of the Militia in the
Province. This morning the Selectmen of Medford came and received
their Town stock of powder which was in the Arsenal, on Quarry hill.
So there is now therein the King's powder only which shall remain there,
as a sacred depositum, till ordered out by the Capt. General.[45]

Gage made no reply but three days later, on the evening of September 1, sent Sheriff David Phips (A.B. 1741) to Brattle with an
order for the delivery of the powder and guns. Without discussion
the General turned over the keys. The supplies were quietly removed by the regulars to the Castle. It happened that that morning
General Gage, while walking down the street, had, in pulling out
his handkerchief, dropped the Brigadier's letter.[46] The popular
leaders, into whose hands it came, took it as proof of Brattle's responsibility for the removal of the supplies and at once published
it by manuscript copy and broadside. So swiftly did the propaganda
machine work that within a few hours the Brigadier was forced to
saddle and ride for Boston; and just in time, for shots were fired at
him before he reached the Brighton bridge. Even Boston did not
seem safe, so he took refuge at the Castle while Gage quickly rein-
forced the guard and placed guns at Boston Neck against the threatening mob of country people.[47] These, finding that Brattle had
escaped, dispersed after a threatening visit to his Cambridge house.

Brattle's letter to Gage was published in the newspapers far and
wide with invidious remarks which induced him to answer with a
public statement:

It is assumed, I advised the Governor to remove the powder; this I
positively deny, because it is absolutely false. . . . As I would not have

[45] From a contemporary Ms. copy in the Artemas Ward Mss. (M. H. S.), II,
Aug. 29, 1774.
[46] Rowe, Diary, Sept. 1, 1774; 1 *Proc. M. H. S.* VIII, 350–1.
[47] 1 *Proc. M. H. S.* VIII, 352; Rowe, Diary, Sept. 2, 1774.

delievered the provincial powder to any one but to his Excellency, or order, so the towns stocks I would have delivered to none but the select-men. . . . My . . . grief is much lessened by the pleasure arising in my mind, from a consciousness that I am a friend to my country; and . . . that I really acted according to my best judgment for its true in-terest. I am extremely sorry for what has taken place; I hope I may be forgiven, and desire it of all that are offended, since I acted in an honest, friendly principle, though it might be a mistaken one.[48]

It was hopeless for this "deputy serpent" to that "vile serpent," Hutchinson, as John Adams called them,[49] to try to regain public favor. He remained a joke to the Whigs, caricatured as "Proteus" and "Brigadier Paunch" in their farces.[50] In November a story that his house had been leveled by a mob [51] induced him to get rid of what property he could sell. His Cambridge estate, which he never saw after his flight to the Castle, he deeded to his only son, Major Thomas Brattle (A.B. 1760), who fled to England where he posed as a refugee and whence he later returned to prove, with great difficulty, that he had always been an ardent Whig. After the Battle of Lexington the mob plundered the cellars of the Cambridge house of their liquors, and the Provincial Congress took over the remaining stores.[52]

The Brigadier remained in Boston, wringing his hands and groaning, "We shall lose the day. Good God! what will become of us?" [53] Stricken by fever and flux in July, he tried to put the war out of his mind and wrote of pears, grapes, and rabbits.[54] His only surviving daughter, the widow of John Mico Wendell (A.B. 1747), remained in the Cambridge mansion in constant fear lest the bombs exchanged by the armies end her life and that of her father. General Washington, hearing of her fears, once pulled up his horse before the window in which she sat and called out: "Madam! there is no reason for your apprehension of danger to your life here or to that of your father. . . . You may rest in quiet repose night and day, for ought I know to the contrary at present." [55]

[48] *Massachusetts Spy*, Sept. 8, 1774; also published in broadside form.
[49] Adams, *Works*, I, 133.
[50] *American Gazette* (annotated American Antiquarian Society copy), July 2, 1776; *Massachusetts Spy* (A. A. S. copy), Jan. 26, 1775; *Proc. M. H. S.* LXII, 20.
[51] *Boston News-Letter*, Dec. 1, 1774.
[52] *Journals of Each Provincial Congress* (Boston, 1838), p. 532.
[53] 1 *Proc. M. H. S.* VIII, 399.
[54] *Massachusetts Spy*, Aug. 16, 1775; Dixon's *Virginia Gazette*, Sept. 9, 1775.
[55] *Proc. M. H. S.* LII, 147.

In October, Brattle plucked up courage to sign the farewell address to Gage, thus burning what little was left of his last bridge to popular favor.[56] When the army evacuated Boston he went with it, still under the fire of the rebel wits:

The Rev. General Brattle, Attorney at Law, and Doctor of Physic, went from Boston to Halifax, in the character of Commissary's Cook. It seems that in the hurry and timidity of the flight, this complication of excellencies, notwithstanding his eminent services, particularly in feeding the Rabbits, and singing that beautiful elegy to their memory, was entirely forgotten, and had no birth provided for him; although he was always allowed to have a singular talent at running away.[57]

At Halifax the Brigadier and Simon Tufts (A.B. 1767) "Mess'd together in a little chamber over a grog shop." [58] After the short rations of besieged Boston, General Brattle made the most of the fresh fish and vegetables of Halifax and so brought himself to the grave there on October 25, 1776:

He was always a great feeder, and being at dinner at a gentleman's table, having his plate filled with fish, one who was at table took notice of his countenance, and said to him — "You are not well, General," — but he went on eating, until it was observed that his mouth was drawn on one side, and he was advised to get up, or some thing to that purpose, which he agreed to, and had just time to say to the servant, — "Set the plate by for supper." These were the last words he spake.[59]

What with inflated currency and conflicting claims, Brattle's estate was insolvent.[60] His Boston house and his Oakham lands were confiscated by the province and sold, but the Cambridge property was left in the hands of his children. The original of his portrait by Copley, which is reproduced in this volume, is the property of Thomas Brattle Gannett of Wayland, Massachusetts. Mrs. Lovel Hodge of Kittery Point owns an old copy.

[56] *Boston Gazette*, Oct. 30, 1775.
[57] *New England Chronicle*, May 30, 1776.
[58] *Connecticut Courant*, July 1, 1776.
[59] Thomas Hutchinson, *Diary and Letters* (London, 1886), II, 221.
[60] Clipping, Coll. Biog. Harv. 1, 7.

JONATHAN FRYE

JONATHAN FRYE, famous in New England song and story as the chaplain of Pigwacket Fight, was the third but only surviving son of Captain James and Lydia (Osgood) Frye of Andover. He entered Harvard at the then normal age of fifteen and distinguished himself as a leader in the student riot of 1722. If one may trust posthumous statements, "he was a very worthy and promising young gentleman." [1]

Frye fell in love with Susanna Rogers, the thirteen-year-old daughter of the Reverend John Rogers (A.B. 1705) of Boxford, and pressed his attentions with success as far as she was concerned, but against the bitter opposition of his father, who could well object on the ground of the girl's age and the boy's inability to support her. In the spring of 1725 parental obstinacy drove the lad[2] into enlisting in Captain Lovewell's famous company of rangers as a common soldier;[3] chaplains were appointed only through favor and by the General Court. His last act before leaving home was the planting of an elm which he charged his friends to guard and cherish until his return; and cherish it they and their sons and grandsons did until it died a century and a half later.[4]

The young lover was a great favorite with the company of rangers, which he served as chaplain. While he was praying before the company at daybreak on May 9, 1725, they saw a lone Indian some distance away. After some consultation as to the likelihood that the savage was a lure to draw them into ambush, they set out after him. The chase was successful, and young Frye joined in the scalping. Returning to their packs, they were ambushed by a much superior force of Indians. There followed one of the two bitterest fights in New England history, a battle so hot that it was declared that during the ten hours it lasted some of the rangers discharged their guns "more than 20 times." "About Five hours" after the bloody struggle began, young Frye, "having fought with undaunted

---

[1] Samuel Penhallow, *History of the Wars of New England* (Cincinnati, 1859), p. 112.

[2] William Waldron to Richard Waldron, Dana Mss. (Mass. Hist. Soc.), May 24, 1725.

[3] *Acts and Resolves of the Province of Massachusetts Bay* (Boston, 1869–1922), X, 720.

[4] *Columbian Centinel*, May 25, 1825.

Courage, and scalp'd one of them In the Heat of the Engagement," fell wounded.[5] Unable to use his gun, the young chaplain resorted to prayer and encouraged his companions by his loud petitions for divine aid. At nightfall the Indians withdrew, and the surviving English set out for a neighboring fort. After a mile or two Frye and three other wounded men collapsed and told the others to go on.

When they'd waited some Days for the Return of the Men from the Fort, & at length despair'd of their coming, tho' their Wounds Stank & were Corrupt, & they were ready to Dy with Famine; yet they all Travell'd several Miles together, till Mr. Frie desired Davis & Farwell not to hinder themselves any longer for his sake, for that he found himself Dying, & so lay down.[6]

His last murmured request was that the survivors tell his father that he was not afraid of the eternity which lay just before him. And so they left him on the third day after the fight. His body was never found.[7]

New England was shaken like Rome at the news of Teutoberg Forest. Cotton Mather publicly addressed himself to his "Worthy Friends, the Parents of Mr. Jonathan Frie, an Only Son, who after a Liberal Education, and a Temper and Conduct, which made him universally Beloved, and raised considerable Expectations of him . . . with Admirable Expressions of Piety, and Magnanimity and Resignation, Sacrificed his Life." [8] For generations New England sang a ballad which recited how

> Our Worthy Captain Lovewell among them there did die;
> They killed Lieutenant Robbins, and wounded good young Frye,
> Who was our English chaplain: he many Indians slew,
> And some of them he scalped when bullets round him flew.[9]

To assuage the sorrow of Jonathan's stricken mother, Susanna poured her grief into appreciative verse:

[5] *Boston News-Letter*, May 27, 1725.

[6] Thomas Symmes (A.B. 1698), quoted in Frederic Kidder, *The Expeditions of Capt. John Lovewell* (Boston, 1865), p. 36.

[7] Certain traditional detail regarding Frye's last hours is probably a confusion of events which took place after his death. Cf. Caleb Butler, *History of Groton* (Boston, 1848), p. 107 *n.*, and Penhallow, p. 113.

[8] Cotton Mather, *Edulcorator* (Boston, 1725), p. 38.

[9] The long verse epic first published in J. Farmer and J. B. Moore, *New Hampshire Historical Collections*, III (1824), pp. 94–7, was a nineteenth-century production. See Kidder, p. 119. There is a broadside copy of the eighteenth-century version of the ballad at the Harvard College Library.

And there they left him in the wood,
Some scores of miles from any food;
Wounded and famishing all alone,
None to relieve or hear his moan,
And there without all doubt did die.
    *   *   *   *   *
Not from mine eyes alone, but all
That hears the sad and dolefull fall
Of that young student, Mr. Frye,
Who in his blooming youth did die.
Fighting for his dear country's good,
He lost his life and precious blood.
His father's only sone was he
His mother loved him tenderly
And all that knew him loved him well,
For in bright parts he did excell
Most of his age, for he was young,
Just entering on twenty-one.
A comely youth, and pious too,
This I affirm for him I knew.
He served the Lord when he was young,
And ripe for Heaven was Jonathan.[10]

Susanna remained loyal to her first love until she married at the ripe age of twenty-three.

Mrs. Fannie H. Eckstorm, the present authority on this subject, has recently reviewed the Pigwacket Fight material and arrived at new conclusions. According to these, Frye joined the army to obtain scalp money to enable him to marry, and in his eagerness he stirred up the rangers against the better judgment of Lovewell to go off on the chase which resulted in the ambush. Mrs. Eckstorm also points out that the fight apparently took place on Sunday, May 9, although the contemporary historian, Thomas Symmes, explicitly places it on the day before. This was, she concludes, a deliberate plot concocted by good Parson Symmes to conceal the fact that poor Frye had been driven into the sin of scalp hunting on the Sabbath, and had been driven to his death by the obstinacy of "his purse-proud family and a miserly, arrogant, besotted old father." [11] To arrive at this thesis requires, as Mrs. Eckstorm says, "a little imagination."

[10] *New England Hist. Gen. Reg.* XV, 91.
[11] *New England Quarterly,* IX, 381–2.

SAMUEL COOLIDGE

SAMUEL COOLIDGE, schoolmaster of Watertown, was born there on August 16, 1703, the fifth son of Richard and Susanna Coolidge. Richard was a lieutenant in the militia and a member of the General Court. Samuel, who was a quiet and orderly student, returned to Watertown after graduation and for several years kept the town school. In 1726 he joined the church with Josiah Convers, who had been graduated a year ahead of him. It was a simple matter for him to ride over to Cambridge to qualify for the M.A. by commonplacing in the college hall from the text, "The wisdom of God is a mystery." According to the *Quaestio* sheet he stood ready at Commencement, 1727, to defend the negative of the question, "An Gratia in Renatis, sit suâ Naturâ interminabilis?" After that he began riding out to preach in frontier pulpits; he was at Southborough in March, 1728.[1] Later in the year he agreed to go to Lyme, Connecticut, as assistant to the aged minister, Moses Noyes (A.B. 1659), and gave up the Watertown school. The winter of 1728–29 found him preaching in the North Parish of Killingly, now the town of Thompson. Here he rejected a call to settle as the first minister on a salary of £80,[2] perhaps because he had an eye on the far better pulpit at Dorchester, Massachusetts. However, better men eliminated him in the semi-finals of the contest for that famous church.

After four more years of such wanderings Coolidge returned to the college to live in November, 1733. He was "free and ready in Classical Learning, expert in Tullys orations,"[3] and had shown enough interest in contemporary scholarship to subscribe for the *Chronological History* of Thomas Prince. Accordingly in November, 1734, he was appointed keeper of the college library. As such he drew his minute salary from the Mary Saltonstall donation and was assigned the task of preparing for the press the continuation of the library catalogue. He lasted only a year as librarian, at least partly because of his bad relations with the students, who showed him "very gross ill manners," breaking into his study, taking away his freshman, and in his hearing and before witnesses saying that

---

[1] Ebenezer Parkman, Diary (Am. Antiq. Soc.), Mar. 22, 1727/8.
[2] Thompson Church Records (Connecticut State Library).
[3] Parkman, Diary, July 19, 1744.

he was a preacher who "would curse and swear." [4] Still he remained at the college until January, 1737/8, when a place was found for him as chaplain of the garrison at Castle William. While he was there, an attempt to advertise him was made by publishing a sermon which he had preached on the death of Queen Caroline, but this document was a little too queer to interest parishes looking for ministers. His manner may be judged by a letter which he wrote to a Boston publisher who had reflected upon the ministry of Watertown: "For Dust we all are and by a Just and irrevocable Decree of the allmighty unto Dust we must return there is Scripture for you you Dog." [5] At Castle William he lasted only a year, and from there he went to the Leicester school, where he lasted for six months.

Once more he wound up at the college, where he plagued the authorities until, on February 4, 1742/3, they were obliged to take action against him:

Whereas Samuel Coolidge M. A. who has been (within this half Year or more) much about in Town & more especially at College, in which Time he hath behav'd himself as a Vagabond but hath of late (especially) carried himself in an insolent & outragious Manner, indulging himself in Cursing & Swearing profanely, in drinking to Excess, in a rude & indecent Behaviour at Divine Worship, particularly in the time of Prayer in the College-Hall, Insulting & reproaching the Governors of this House, Hindring the Students from their Business & abusively endevoring to force himself into their Chambers, On all which Accounts, It is hereby agreed & order'd That every Member of this house be & hereby is strictly forbidden to Recieve him the Said Coolidge into their Chambers, either by Day or by Night, or to associate with Him on any Pretence whatsoever, & That upon their Peril. [6]

Exiled from Cambridge, he became a familiar and pitiful sight in Watertown, where in November, 1743, a collection was taken so that the deacons might buy clothing to carry him through the winter. To their great relief a place was soon thereafter found for him as schoolmaster in Westborough. There he dined Sundays with the minister, Ebenezer Parkman (A.B. 1721), who was greatly distressed because he could not, as courtesy demanded, risk asking him to say grace. The parson likewise firmly rejected the pleas of Lieutenant Simon Tainter, one of the town fathers, that the school-

[4] Faculty Records (Harvard University Archives), I, 66.
[5] Wyman Mss. (Woburn Public Library).
[6] Faculty Records, I, 177–8.

master be admitted to the church communion. This was wise, for in August, 1744, Coolidge was "in great Horrors & Despaire," and in September he gave up the school, "being far gone in Despair, sordidness and viciousness (viz Idleness and sloth, Smoaking & Drinking)" and feeling himself "utterly without Hope." [7] The Westborough people kindly fed and cared for him, but finally he wandered off again.

At Commencement, 1745, "Poor Mr. Samuel Coolledge" appeared "in his Destractions and Delirium" and was "plucked out of the presidents Chair in the Meeting House & draggd out on the Ground by a Negro like a Dead Dogg in presence of all the Assembly," a "Most pitious Sight." [8] Winter found him again brought to the attention of the Watertown selectmen:

Complaint being made to the Selectmen that Mr Samuel Coollidge is come again to town and is under Such Circumstances that he is in Danger of Suffering if Some thing be not Emediately done for him (Altho his Conduct has been Such as is very unaccountable whereby he has forfitted both Credit & Kindness) The Selectmen out of Humanitie towards him under his Necessitous Circumstances think themselves Obliged to Do Some thing for him at the Towns Cost to releive him for the present untill further Measures may be taken Concerning him. Accordingly Agreed to procure for him a pair of Shoos a pair of Stockins a pair of Britches and as the Selectmen are Informed there are Some old cloths belonging to him Left at Certain places Agreed to make inquirey and if any be to get the Same for him that he might not Suffer till further care can be taken of him. [9]

The warrant for the next town meeting carried an article, next to that dealing with the swine, regarding the care of Mr. Samuel Coolidge. The town voted to leave the matter to the selectmen, who cared for the vagrant but could not keep him from wandering to Cambridge, where he made himself "a Reproach & Scandal to the College" by "Cursing & Swearing," "drinking to Excess," and "a rude & indecent behaviour at the Divine Worship, particularly, in time of religious Exercises in the College Chapell, & on the last Lords Day in the Scholars Gallery in the Meetinghouse." [10] The

[7] Parkman, Sept. 8, 1744, *et passim.*
[8] *Ibid.*, July 3, 1745.
[9] *Watertown Records*, III (Boston, 1904), Pt. I, p. 317.
[10] Faculty Records, I, 243-4.

rereading of his excommunication in the college chapel immediately after morning prayers had only a temporary effect, for in the winter of 1747–48 he was back again, so troublesome that the faculty had to ask the justices of the peace to "take effectual Care to keep them, from being insulted by Samuel Coolidge."

The winter of 1748–49 found the vagrant back in Watertown staying alternately with his brothers John and Nathaniel. The selectmen assumed the expense of his board and supplied him with "a Coat a Shurt a hat a pair of Stockens & a pair of Shoois which came to 17*l* 13*s* with the Making." They likewise urged him "to git in to Some way of business" and in the spring bought him a gun "to Cary to the Eastward with him." They gave him much good advice, but it is not recorded that they warned him not to shoot himself with the gun. As usual November found him once more in Watertown "in a Suffering Condition . . . being Destitute of every thing of Clothing fit to cover his Nackedness" and terrifying the villagers lest he fire the barns in which he slept. The selectmen in vain tried to find a boarding place for him and in January, 1749/50, were informed "that the Town of Watertown was in Danger of A Presentment on Account of Samuel Cooledge his Runing about & being Mischeivs." They sent word to the grand jury that they were doing all that they could to keep him out of trouble and that they would try to get him into the Boston or Charlestown workhouse. In August, 1750, he was arrested in Boston and passed back to Watertown by way of the constables of Charlestown and Cambridge. For lack of any other place to keep him, he was confined in the town jail, from which he promptly escaped to resume his "Strowling about Vagabone Like." The selectmen asked that he be apprehended wherever found and enlisted the aid of Colonel William Brattle (A.B. 1722) to get him into the Boston workhouse, agreeing "to Reimbust the Charge of his keeping there what his Labour will not Answer." Finally he was taken up as "a Common disturber of the peace" in a neighboring town, and two selectmen were sent with a horse and chaise to bring him home.

In February, 1751, Coolidge showed such a "Considerable Alteration (for the better)" that the town voted to try him at keeping the school. For a year he did so well that the selectmen

agreed to permit him to board himself, but, left without supervision, he soon became "so disorderly as not fit to keep the School."
In August, 1752, he was placed in the care of Samuel Whitney,
who was ordered to restrain him "from Runing about from House
to House" by chaining him up, if need be, until he "be brought to
his right Mind so as to be Serviceable again." This succeeded, for
by December he was so well "Composed again" that the selectmen placed him in charge of the school after giving him "a thorough
talk relating to his past Conduct and what He might Expect if he
did not behave well in the School for the future." To compensate
for the means of internal warmth which he was denied they had
"a Bare Skin Coat" made for him. By keeping money out of his
hands, buying his clothing, and paying Captain John Brown to
board him they kept him in serviceable condition until the school
was closed on October 5, 1754. The selectmen continued to board
him around while making efforts to get him "into Some way of
Business whereby he" might "in some measure Earn his Living."
It was probably through Captain John Tainter, with whom he had
been living, that his old friend, Lieutenant Simon Tainter of Westborough, came down to get him to try the school in that town in
April, 1755. This proved a false hope, for once in Westborough
he relapsed so that Tainter had to remain at home to keep a hand
on him.[11] After a month Watertown took him back and paid
Lieutenant Tainter for his effort.

For the next two years he kept the Watertown school on and off
under strict supervision, but the old story was repeated. The various
Coolidge families and others tried boarding him at town expense
but found his ways intolerable. In April, 1763, Joseph Wellington
of Cambridge undertook to keep him but found him so "very much
Disordered" that within two weeks he had to ask to be relieved of
his contract. He now required a physician's care. In 1763 the town
bought a lock "for Mr Samuel Coolidge" and thus put an end to his
wanderings. On January 12, 1767, the selectmen, "being Informed
that Samuel Coollidge was Dead . . . Agreed that his Funeral
Should be the next day and that mr Moses Stone provide Nine paire
of Gloves one for the Minister and a pair for each of the Bearers &
two pair for the two women that Assisted in Laying him out also
that mr Stone Speak to the Saxton to Dig a grave and Tole the

[11] Ebenezer Parkman, Diary (Mass. Hist. Soc.), Apr. 28, 1755.

Bell." [12] A year later the selectmen inventoried his "weareing Apparril" and closed his account by selling what remained of the things which the town had bought for him.

This dreary story has been given in detail here in order to show how a New England community cared for one of its unfortunate members.

HENRY PHILLIPS

HENRY PHILLIPS, the duelist, was born in Boston on February 21, 1704, a son of Samuel and Hannah (Gillam) Phillips. Samuel was a wealthy publisher and importer of books and had the reputation of being the handsomest man in Boston. Henry inherited his father's appearance and charm and, in 1720, enough of his fortune to make him one of the richest boys in college. In the college slang of the day "Blubber Phillips" was "a clean fellow." [1] For joining Winthrop and Belcher in one of their disorders he was fined no less than 10*s*. There was probably a sense of humor behind the fact that when he took his M.A., he was assigned the negative side of the *Quaestio*, "An Generis Humani Fælicitas, in Sensuum Voluptatibus consistat?"

Henry joined his brother Gillam in the family book business but on the side exported fish and boards to Spain. He belonged to the gay set and frequented Luke Vardy's Royal Exchange Tavern on King Street, where there was more drinking and gambling than was good for the town. There he had differences with Benjamin Woodbridge, "a pretty young man," [2] a son of Judge Dudley Woodbridge (A.B. 1696) of Barbados and at the age of nineteen a partner of Jonathan Sewall, a busy merchant. It is evident from the attitude of the clergy at the time of the duel that the falling out was a result of gambling. A "Vile Fellow" named Robert Handy for weeks nursed along the quarrel and urged Woodbridge to challenge Phillips.[3] Perhaps Handy taught fencing, for he had Woodbridge's sword in his possession. According to his story he

---

[12] *Watertown Records*, v (Newton, 1928), p. 320.
[1] An inscription in a text used by his classmate Oliver and now owned by Mr. W. H. P. Oliver of Morristown, N. J.
[2] Benjamin Walker, Diary (Mass. Hist. Soc.), July 3, 1728.
[3] Phillips to his mother, 2 *Proc. M. H. S.* XVIII, 241.

was aware of the bad blood which existed between the young men, and only with reluctance and upon assurance that no violence was intended did he surrender the weapon to its owner on the evening of July 3, 1728.

The two young men met alone near the water edge of the Common about dusk. Before they began to fight, Handy, who had followed Woodbridge to enjoy the affair or to stop it, joined them. Phillips gave him a tongue lashing which sent him away. Then the two young men fell to with a bitterness which no minor blood-letting would assuage. Phillips was wounded slightly in his belly, leg, and hand,[4] and Woodbridge was struck under the right arm. Apparently the final, fatal, stroke came when the Barbadian fell forward, grasping his foe's blade with his left hand as it ran him through from his right breast to the small of his back. He dropped his sword, turned, and walked toward the Powder House where Handy was standing. Phillips, now greatly distressed at what he had done, picked up the sword, followed, and replaced it in Woodbridge's scabbard. When the wounded man, who had been staggering along with his left hand clutched to his breast, collapsed, Phillips hurried after Handy and told him for God's sake to go back and look after Woodbridge while he went for a surgeon. But Handy walked off, leaving his friend to die alone while Phillips went to the Sun Tavern and called out Dr. George Pemberton. They went to the Doctor's home, where Phillips showed his wounds and begged him to come to the Common to find Woodbridge. They called out a surgeon, John Cutler, and went to the place where Phillips had left Woodbridge, but they searched for him in vain. Cutler told Phillips to go walk in Bromfield Lane, and the two doctors continued the search without success. When they returned to Cutler's house empty-handed, they found Phillips there and bound his wounds.

In spite of the fact that Woodbridge had not been found, Phillips was sure that he had mortally wounded him, so he went from Cutler's to the home of Peter Faneuil, who was Gillam's brother-in-law. In order to delay any possible searching party Faneuil took Phillips to the home of Colonel Estes Hatch and concealed him while Gillam set out to find a small boat to take him to H. M. S. *Sheerness*, which lay between the Castle and Spectacle Island. He

---

[4] Walker, Diary, July 3, 1728.

enlisted the services of Captain John Winslow, who had come ashore from his pink *Molly;* but Winslow's boat was tied to Long Wharf, so Peter took Henry to a secluded dock while Gillam and the Captain rowed out to the pink to get hands to row them off to the man-of-war. This accomplished, they returned to the hiding place, where Peter handed Henry over to Gillam. The fog was so thick that night that the men from the *Molly* ran the boat aground on Dorchester Neck, where they wandered around for some time before putting off again. Not much of the night remained when the sophisticated officers of the *Sheerness* welcomed them aboard. Before dawn Gillam returned to Boston, and the man-of-war dropped down the bay with the fugitive aboard.

Meantime the search for Woodbridge continued, but it was not until about three in the morning that he was found where he had died, alone, in the rain, near the Powder House. The body was taken to Jonathan Sewall's house. The fact that the victim's sword was in its sheath gave rise to talk of murder, but to the orderly Bostonians dueling, hitherto unknown among the native stock, was an equal crime. In spite of the fact that Boston was a busy seaport, violent crime was almost unknown, so the town was convulsed with horror at this killing. The province Council met in the morning and "Ordered that the Sheriff of the County of Suffolk forthwith send men to the Places following Viz, Nantasket, Long Island, Peeling Point, Deer Island, Spectacle Island, or any other of the Islands, and that they watch any Boats that may carry Henry Philip to or from the shoar . . . [and] That Orders be sent to the Keeper of the Light House, that he watch all Boats & use his utmost Endeavours to seize the said Henry Phillips." [5]

Woodbridge "was decently and handsomely interr'd, his Funeral being attended by the Commander in Chief, several of the Council, and most of the Merchants and Gentlemen of the Town." [6] His stone is still to be seen in the Granary Burying Ground.

The Congregational clergy did all that they could to comfort Hannah Phillips, but for her son they had no kind word. Joseph Sewall (A.B. 1707) preached a vigorous public lecture against gambling and dueling, and to the printed version his fellows contributed a preface, saying:

[5] Executive Records of the Province Council, IX, 74-5.
[6] *New-England Weekly Journal*, July 8, 1728.

The ensuing Sermon was preach'd upon as lamentable an occasion, (of a private nature) as has been known among us. That any of the sons of New-England, who have been born and educated in this land of light, should be so forsaken of God, and given up to their lusts and passions, as to engage in a bloody and fatal Duel, deserves to be bewailed with tears of blood.[7]

The August grand jury, which indicted Phillips for murder, correctly expressed the majority opinion. The General Court enacted a law placing heavy penalties on dueling and adopting the quaintly barbarous English custom of burying both victim and murderer with a stake driven through the body; there was never an occasion to apply it.

There were many recent immigrants in Boston who took dueling more lightly, and many others who thought that the young exile had suffered enough. Accordingly a strong effort was made to obtain a royal pardon with the idea of permitting him to return. Eighty-eight of the leading inhabitants, headed by Governor Burnet, signed a statement which read, in part:

These may certify to all whom it may concern, that we, the subscribers, well knew and esteemed Mr. Henry Phillips of Boston, in New England, to be a youth of a very affable, courteous, and peaceable behavior and disposition, and never heard he was addicted to quarrelling, he being soberly brought up, in the prosecution of his studies, and living chiefly an academical life; and verily believe him slow to anger, and with difficulty moved to resentment.[8]

The original document with the full list of signers is lost, but the significant thing about the short list which has survived is the fact that it includes most of the royal officials and leading Anglicans, including the apostate Timothy Cutler (A.B. 1701) and the three other local Episcopalian clergymen. The man in the street shook his head and said, "gambling, dueling, and the Church of England, hand-in-hand." Unfortunately, too, there appeared the question of a royal pardon which would in effect nullify the new province law even though the crime, having been committed before its enactment, could not be punished under it.

Poor, exiled, homesick Henry wanted nothing less than to be alienated from the land he loved. From Rochelle, where he had

[7] Joseph Sewall, *He that Would Keep God's Commandments Must Renounce the Society of Evil Doers* (Boston, 1728), pp. i–ii.

[8] [Lucius Manlius Sargent], *Dealings with the Dead* (Boston, 1856), II, 554.

been welcomed by Jean Faneuil, uncle of Brother Peter, he wrote to his mother on March 24, 1729:

According to your desire I am come into France, but find it as all other places extreamly Chargeable, especially to me who have so small a Stock. Whether I am like to get my Pardon, only God knows, so must desire something may be done for me, not to let me Spend the last farthing. I do assure you Madam, I have not had one moments pleasure since I left you, neither do I expect any in this World, without I should be so happy to See my Dear mother and my Native Country, which I prefer to any I have Seen.[9]

Phillips had been sick when he arrived at Rochelle, and a recovery proved only temporary. On July 30, 1729, one of the Faneuils wrote to his Boston relatives:

[Henry] had ordered the nurses that if they thought he was near his end to go and call the two English Gentlemen which used to visit him daily, they came and was two Hours with him in Prayer, and they would have him sign his will, and told them that he would sign it the next day, they told him that he might be worse, and that he ought not to put it off nevertheless he did not sign it; about three a clock in the morning a cold swet took him — he then askt for a clean shirt, and told the English nurse this will be the last you will give me, half an hour after having taken some Dyet drink laid himself on his Pillow and he departed without any struggling.[10]

According to Boston tradition his death took place on May 29, 1729; according to legal papers in England it was on July 17, 1730.[11] The date of the above letter makes July 17, 1729, the more likely date. His mother had sailed to join him six weeks before.

According to the Massachusetts law covering intestate estates a good portion of Henry's property, which was valued at £4000, went to his mother and to his sister, Faith. Brother Gillam brought suit in the famous case of Phillips vs. Savage, contending that under the English common law he was entitled to the entire estate. The Province of Massachusetts, fearful lest its statute be set aside, instructed its agent in England to fight the case and, after the courts had upheld the statute, reimbursed Faith for her

[9] 2 *Proc. M. H. S.* XVIII, 241.
[10] Early Files in the Office of the Clerk of the Suffolk Supreme Court, 23,794.
[11] *Acts of the Privy Council. Colonial Series.* 1720–1745, p. 433.

legal expenses. Through his failure to sign his will, Henry Phillips had placed in jeopardy a great part of the legal structure of his beloved "Native Country."

MATHER BYLES

MATHER BYLES, poet, humorist, and Tory, was the youngest son of a Boston saddler, Josias by name, who lived on Tileston Street and served the town as constable and tithingman. Having buried two wives and raised a large family, Josias moved up the social scale by marrying Elizabeth Greenough, widow of William Greenough and daughter of their pastor, Increase Mather. Hitherto Elizabeth had been a barren vine, but on March 15, 1706/7, she put forth a son who on the next day was baptized by the name of Mather at his grandfather's church.[1] Although Josias had an Indian slave, whom he named Winchester for his own English home, he owned no real estate and was not particularly well-to-do, so when he died suddenly just after his youngest son's first birthday he left only a moderate estate. His will, which had been drawn some years before, did not mention little Mather. Accordingly the probate court ordered that one share, amounting to £52 3s 1d, be set aside for the infant.[2] Cotton Mather, who made a practice of suggesting to himself one good deed for each day, listed being a father to his little nephew as one of these. When the lad was four, he gave him a copy of *Good Lessons for Children* and promised him a piece of money for every lesson which he got by heart from it. Increase Mather died a few years later, leaving a will which provided that a fourth part of his great library should go to his

---

[1] The christening cap which little Mather wore is now in the museum of the Massachusetts Historical Society.

[2] Suffolk Probate Records, XXI, 380.

MATHER BYLES

"Fatherless Grandson Mather Byles, in case he shall be educated for, & employed in the work of the ministry, (which I much desire & pray for)."

What I give to my Daughter Elizabeth, I desire it may (if his Mother can) be improved towards the education of her only son . . . in Learning. . . . I leave it as my dying Request to his uncle my son Cotton Mather, to take care of the education of that child as of his owne. . . . To prevent his being chargeable as much as I can, I give him my wearing apparel excepting my chamlet cloak. . . .[3]

The mantle of New England's Elijah did in a very real sense fall upon this little lad. His brothers and sisters were already well established in the world of shopkeepers and artisans, but his place was in the most distinguished clerical family of New England. On his uncle's desk he saw letters from the great scholars of Protestant Europe and at his dinner table he saw governors and councillors. He must himself have been the envy of a friend of his, a sharp young printer's devil named Benjamin Franklin, who was drawn to Cotton Mather's study by the encyclopedic wisdom which was there gladly dispensed (with small change for encouragement) to any lad who showed an interest in things intellectual.

Mather Byles was ready for college at the age of fourteen, so his uncle began to levy upon his wealthy friends for funds to support him there. In the midst of this campaign the lad came down with smallpox and gave them all a bad fright. After he had recovered, Grandfather Increase, then in his last year, found an article on inoculation in the *London Mercury*, copied it out, and sent him running to James Franklin with a request that it be printed in the *New-England Courant*. That, however, was probably the extent of the boy's participation in the most important controversy of the decade.[4]

Armed with books from his uncle's study,[5] Mather presented himself at Harvard where for four years he was supported largely from the Hollis and Hulton funds. He did not live in a college

<hr>

[3] *Sibley's Harvard Graduates*, I, 437.

[4] Byles has been identified as "the young Wretch" of a "scribbling Collegian" with "just Learning enough to make a Fool of himself" who participated in this controversy, but it is more likely that it was his cousin, Samuel Mather (A.B. 1723).

[5] Several of these volumes are now in the libraries of Harvard University and the Mass. Hist. Soc. In the Harvard University Archives are Ms. copies of Charles Morton's Compendium Physicæ made by Obadiah Ayers (A.B. 1710) and Jeremy Gridley (A.B. 1725) and used by Byles.

chamber, eat in commons, or join the student societies and riots. In view of the fact that he was a social and jovial lad this would be inexplicable if it were not for the bulk and the amazingly good quality of his literary productions during these four years. Disdaining the manuscript imitations of the *Spectator* and *Tatler* which other students produced, he and a few intimate friends wrote verse and essays for the *Courant*, which continued to print their work right up until the day of its death at the hands (or on the pen point) of the Mather faction which it had so viciously attacked.[6] From this distinction Byles went on to attain heights of literary recognition beyond the dreams of mere colonials when several of his poems were printed "at London, and elsewhere, either separately, or in Miscellanies." [7] No wonder that the subject of his essay for the Master's degree was "polite literature is an ornament to a theologian."

Among the numerous poets whom the Puritan colonies had produced, Byles was second only to Anne Bradstreet in fame and perhaps in ability. True, his voluminous printed works do not compare too well with the manuscript remains of less well-known poets, but that is because he suffered from two influences which were the curse of verse in that generation. The rank and file of New Englanders, like the rank and file of Englishmen, wanted theological poetry, sermons in verse. The other handicap was the passionate worship of Alexander Pope. To the New Englander of that day poetry was not poetry unless it sounded like Pope; but unfortunately imitations of Pope have to be as good as the original or else they are bad; for technical reasons there is little middle ground. In the same way the style of the *Tatler* and *Spectator* fathered an artificiality in New England prose essays which appears in Byles's works, although he showed himself aware of the dangers of it in an essay on literary criticism.[8] Byles came nearer than did his contemporaries to mastering these forms. He had Hogarthian powers of observation and description which appear most notably in his classic account of a Harvard Commencement. In the following bit

[6] The identification of Byles's contributions must await the discovery of more annotated copies of the *Courant*. At present the last word on this subject, and Byles's poetry in general, is Mr. C. Lennart Carlson's Introduction to the 1940 Facsimile Text Society edition of *Poems on Several Occasions*.

[7] *Poems on Several Occasions* (Boston, 1744), p. (iv).

[8] Printed in the front matter prepared for the bound volumes of the *American Magazine*, I (1745); reprinted in the *Boston Magazine*, I (1783), pp. 8, 49–51.

he describes the struggle of the young blades for the conveyances waiting at the Charlestown Ferry to take the crowd to the Yard:

> Eager the sparks assault the waiting cars,
> Fops meet with fops, and clash in civil wars.
> Off fly the wigs, as mount their kicking heels,
> The rudely bouncing head with anguish swells,
> A crimson torrent gushes from the nose,
> Adown the cheeks, and wanders o'er the cloaths.
> Vaunting, the victor's strait the chariots leap,
> While the poor batter'd beau's for madness weep.[9]

We would gladly exchange all of his formal, stilted, verse for the whole of his account of Colonel Isaac Winslow's "very solemn horse" of which we have only a fragment which might well have come from the pen of Oliver Wendell Holmes. He was apparently keeping school at Marshfield when he met the horse and the Deacon, whom he described setting the psalm in church:

> The Deacon full resolved upon't,
>   To make a doleful sound,
> Twang'd thro' his Nose a murder'd grunt,
>   And all the People Groan'd.[10]

Although he deprecated "those light and idle Airs which debase the Spinetts of the Fair, in their soft and sprightly Amusements," [11] he did much to replace the twanging and groaning of congregational singing by writing hymns which are relatively singable. In them the poet occasionally crops out unexpectedly as when, after describing the horrid destruction of the earth in the last day, he dwells on the charm of the seasons in the world which has been purified by fire.

There is no sign in Byles's poetry and essays of the depression and sentimentality which one might expect from the fact that during the last half of his college years he was so wasted away by a consumption that his life was despaired of. Judge Samuel Sewall and his friends paid the expenses of the sick room, and Byles took out heavenly insurance by entering into the communion of the Mather church. So at the age of eighteen he returned from Cambridge with an established reputation as the best poet and

[9] *A Collection of Poems by Several Hands* (Boston, 1744), pp. 48–9.

[10] These bits are preserved in the Byles Letter Book (New England Hist. Gen. Soc.), Dec. 27, 1783.

[11] Mather Byles, Introduction to Andrew LeMercier, *Christian Rapture* (Boston, 1747).

essayist in a generation which was experiencing a revival of interest in literature. Of the two existing Boston papers the *News-Letter* was too conservative to be literary and the *Gazette* too devoted to the kind of news which would interest business men. So, less than a year after the demise of the brilliant but hardly respectable *Courant*, Byles and his circle interested a Boston printer in bringing out the *New-England Weekly Journal*. Byles and Samuel Danforth (A.B. 1715) appear to have been the principal editors, and their interest even extended to reading proof. The appearance of the paper began a new era in American literary history, for in it poetry, good essays, and solid articles crowded the news from the front page. Some of the essays and verses were from Byles's own pen, and as all were anonymous he took the opportunity to puff his own productions in succeeding issues.[12]

Byles became in these years the poet laureate and the official literary greeter of the province. On an occasion like the death of George I and the accession of George II he produced commemorative verse which was printed in the *Journal* and separately in handsome souvenir pamphlets. When Governor Burnet arrived, Byles greeted him in the name of "Bostonia, Mistress of the Towns, Whom the pleas'd Bay, with am'rous Arms, surrounds."

> But You, O Cambridge, how can you forbear
> In gliding Lays to charm each listning Ear?
> You, where the Youth pursue th' illustrious Toil,
> Where the Arts flourish, and the Graces smile,
> Make Burnet's Name in lasting Numbers shine,
> Ye soft Recesses of the tuneful Nine![13]

Another common occasion for commemorative verse was the death of a prominent citizen. It would appear that when Judge Daniel Oliver joined his equally distinguished ancestors, Governor Jonathan Belcher (A.B. 1699) called upon Byles to perform in his usual manner:

---

[12] The authorities for Mather Byles's participation in the publication of the *New-England Weekly Journal* are Isaiah Thomas and John Eliot (A.B. 1772), both of whom knew him. Isaiah Thomas, *History of Printing in America* (Albany, 1874), II, 41–2; I *Coll. M. H. S.* V, 211 n.; *Atlantic Monthly*, LIX (Apr. 1887), p. 525. Carlson (op. cit.) is the best modern authority, but since he wrote the American Antiquarian Society has acquired Byles's original file of the *Weekly Journal* which shows the editorial corrections which he made in his articles and poems for later printings.

[13] *Poems on Several Occasions*, pp. 70, 71.

Pensive, o'ercome, the Muse hung down her Head,
And heard the fatal News, — 'The Friend is dead.
Dumb, fixt in Sorrow, she forgot her Song,
The Tune forsook her Lyre, the Voice her Tongue:
'Till, Belcher, You command her Strains to rise,
You ask, she sings; You dictate, she replies;
That well-known Voice awakes her dying Fires,
And, instant at Your Call, the Pow'r inspires.[14]

Much of this verse is in the sickening sycophantic manner of the century, but when Byles wrote of the *Horae Lyricae* of Isaac Watts he was giving voice to a passionate admiration:

What Angel strikes the trembling strings
    And whence the golden Sound!
Or is it Watts or Gabriel sings
    From yon celestial Ground.
Tis thou seraphic Watts: thy Lyre
    Plays soft along the floods
Thy notes the answring hills inspire
    And bend the waving woods.[15]

Having admired his own poem for something over a year, Byles sent it to Watts with an effusive letter.[16] That kindly poet replied with praise and criticism, which is important evidence of Byles's place in the literary circle of his day: "Your Poetical Address to Gov. Burnet I have twice read over: The 4th page of it has admirable Verse. There is but one line in it which I could wish altered & that is 'Rough Winter feigns a youthful Tread' which I don't understand in the metaphor." [17] Byles replied in stuttering gratitude: "The Kind things which you are pleased to say . . . call aloud for me to blush and be grateful x x x x x etc etc etc." [18] As the correspondence continued, Watts advised the young poet not to be too enthusiastic in praise of politicians and warned him that too much admiration for the heathen classics might cause him to lose touch with "the sacred Institutions of Moses or Christ."

I lately read over again your Funerall Poem on Mr Oliver & it gave me fresh delight. May all those Spritely Talents which must be em-

---

[14] "An Elegy," in Thomas Prince, *The Faithful Servant* (Boston, 1732).
[15] Ms. copy in Byles's hand at the New York Public Library, photostat at the Mass. Hist. Soc.
[16] Byles Letter Book, May 3, 1728.
[17] Byles Family Papers (Gay Transcripts, Mass. Hist. Soc.), I, 1.
[18] Byles Letter Book, p. 19.

ployed in such an Elegy be consecrated to the Service of your Lord &
mine & shine & burn in all your Discourses, & under the influence of
the blessed Spirit may they enlighten & warm the souls of men with
Devotion & true Godliness.[19]

Watts sent Byles a set of his works, a kindness which the young
man repaid in an introductory ode to the New England edition of
Watts's hymns. "You surprise me," commented Watts, "when I
hear you say you have no Genius for writing Poetry, which I think
appears abundantly and Demands the Approbation of Persons of
a good Taste among you."[20] Modern taste reacts differently to
lines like this one from Byles's "Ode on the Death of Queen
Caroline": "On Royal milk the Royal infants feast."

It was one thing to establish a correspondence with Isaac Watts,
who was the epistolary friend of all of New England's intellectual
leaders, and it was another to win the attention of England's great
literary men, who knew nothing and cared less about the colonies.
But at the age of twenty Byles boldly aimed his quill at no less a
giant than Alexander Pope. In a letter to him he described how
his soul melted away before the flame of Pope's poetry, and he
sent copies of the *Journal* containing his own "Eternity" and "Pane-
gyrick on Milton."[21] The great man replied with a letter which
the proud Byles evidently wore out showing it to his friends; we
know the contents only from the somewhat amused account of John
Eliot (A.B. 1772), who saw it in its later, tattered, days.[22] The gist
of it was that Pope had long supposed that the Muses had deserted
the British Empire, but Byles's verse made it evident that they had
only emigrated to the colonies. Either the Bostonian, never a
sensitive man, failed to see the irony, or, like a true eighteenth-
century commoner, he would rather have a kick than no attention
from a great man. Pope offered to send a set of his works to some
American library, evidently believing from what he had seen that
their influence was needed. Byles replied, telling with what awe
his circle had read the letter:

[19] Byles Family Papers, I, 4–6.
[20] *Ibid.*, p. 8.
[21] Byles Letter Book, pp. 1–2. The letters to Pope are printed in the *Publications of the Modern Language Association of America*, XLVIII (Mar. 1933),
pp. 61 ff.
[22] Joseph T. Buckingham, *Specimens of Newspaper Literature* (Boston, 1850),
I, 111.

The polite and learned part of my Countrey-men agree with me, to look upon such exalted Genius's as Mr. Pope, O'tother Side the inconceivable Breadth of Ocean, in the same Light, in which You behold the admired Classicks, We read you with Transports, and talk of You with Wonder. We look upon your Letter, as You would upon the original parchment of Homer.[23]

Byles was not exaggerating his awe. He wrote again but got no answer unless a copy of the *Odyssey* which came to him was, as he hoped, a gift from the great man. Later he tried to get Pope to autograph a set of his works in the shop of some bookseller from whom he could buy them.

In the same way Byles tried to open a correspondence with Lord Barrington and with George Granville, Lord Lansdowne. The latter replied in a kindly way and so laid himself open to Byles's usual technique of obtaining presentation copies:

Let Others attend Your Lordships Levy for places of Honour and profet: If I was to give s[   ] to my Ambition, I should only Wish for Your Lordships Works from Your own Hand. — But what have I said! My Lord, I am ashamed at my Freedom, and ask pardon.[24]

Similarly he sent James Thomson some of his writings and begged for a set of Thomson's works in return.

Literature was, however, never more than an avocation with Byles. He was too much of a Mather to believe that it was the First End of Man. First and last he was always a minister of God. When Cotton Mather lay dying, he caught sight of Byles in the room. "He called him to come near, laid his trembling hands on his head, and pronounced his dying blessing. . . . 'My dear child, and my son, my son, I bless you. . . . You have been acquainted with my poor manner of living, even in the more secret strokes of it: follow what you have found in it, according to the pattern of a glorious Christ.' "[25] Joshua Gee (A.B. 1717) printed an account of this incident in the funeral sermon, of which Byles sent a copy to Thomas Bradbury, the English author, with an apology for "the Defects of the Composition" and the old-fashioned theology.[26]

In 1729 Byles was a candidate for the pulpit of Dorchester, where he received only 15 out of 66 votes. Later he was a candi-

[23] Byles Letter Book, p. 15.
[24] *Ibid.*, p. 27.
[25] Joshua Gee, *Israel's Mourning* (Boston, 1728), p. 29.
[26] Byles Letter Book, pp. 27–8.

date to succeed Cotton Mather and become Joshua Gee's colleague, but the choice fell on Samuel Mather (A.B. 1723), Cotton's son. He was no doubt beginning to worry about his inability to find a pulpit when an unexpected opening presented itself. Boston's seven Congregational churches were grouped in the northern part of the town although the population was spreading out along the lower part of the present Washington Street. In this growing district Governor Belcher had a fine country house and real estate which would rise in value if more people could be induced to build there by the establishment of a church which would make unnecessary the long Sunday walk into town. So on November 14, 1732, the Hollis Street Church was organized, and six days later it elected Byles its pastor. Having been duly dismissed from the Second Church, he was ordained on December 20.

According to Governor Belcher the new congregation consisted of "a number of sober, good Christians . . . not in the most plentifull circumstances," but he regarded himself as the chief patron of the new organization and induced Thomas Hollis, the younger, to present it with a bell which, when it arrived, was "generally thought the best in the country." The new minister became automatically a Harvard overseer but, although he attended meetings regularly, he otherwise took small part in college affairs.

Having obtained a bell for the new church, the Governor turned his mind to the next necessity, a wife for the parson. On February 14, 1733/4, Byles was married by his friend Thomas Prince (A.B. 1707) to Anna, daughter of Oliver Noyes (A.B. 1695), widow of Azor Gale, and niece of the Governor. The ceremony took place in the great state room of the Province House, the official residence of the governors of the province. Belcher was very fond of the young man's company and on one occasion invited him to be chaplain of an expedition to treat with the Eastern Indians. Byles, although he was later, if not then, a proprietor of Brunswick and Topsham, was not interested in a voyage to Maine and declined to accompany his patron. Not to be deprived of his fun, the Governor arranged to have Byles exchange pulpits with the chaplain of the Castle, off which lay the *Scarborough* man-of-war ready to sail for Maine. After the service Belcher invited Byles to take tea with him on board. While the cups were circulating, the *Scarborough* made sail with its unwilling chaplain. This excursion gave rise to a literary exchange which Bostonians for the next century treasured as excruciatingly funny. The second Sunday found the expedition at Casco Bay without a psalm book, but Byles improvised for the service a "Psalm of the Sea" which was pious and good verse although one part in particular stuck in the crop of the Boston wit, Joseph Green (A.B. 1726):

> Round thee the scaly nation roves,
>   Thy opening hands their joys bestow,
> Through all the blushing coral groves,
>   These silent gay retreats below.

Green delighted Boston with a parody, to which Byles made a doggerel answer of the crude and violent sort which appealed to their generation.[27]

At just about the same time the young minister had another and very different kind of public service to perform, one in which he showed a kind of Christianity nearer to Jesus than to Calvin. Rebekah Chamblit, who had been condemned for the murder of her illegitimate child, begged that he be the minister designated to accompany her to the gallows. As they walked along through the streets of Boston, he urged her to submit her will to God and to accept Jesus as her Mediator.

"Yes Sir, O I do it with all my heart and soul," the poor girl cried.

"Frighted and in agony as you are, why, poor Criminal, should not you hope in this free grace of God in Christ, as well as any Sinner in the World?"

"Oh! I'm afraid, I'm afraid."

"Fear not, for I bring you good tidings of great joy; God is infinitely ready to pardon you, upon your Repentance, and Faith in Christ. . . . You say you are a Sinner: Hear what the blessed Saviour says . . . 'I came not to call the righteous, but Sinners to repentance. . . . Tho' your sins be as scarlet, they shall be white as snow.' "

But at the sight of the gallows, terror swept from the girl's mind these words of comfort, and Byles turned back defeated.[28] This

---

[27] This story comes from Byles's grandnephew and friend, Jeremy Belknap (A.B. 1762). Belknap Mss. (M. H. S.), 161.C.56; 5 *Coll. M. H. S.* II, 70–3.

[28] Thomas Foxcroft, *Lessons of Caution* (Boston, 1732), pp. 64, 65.

emphasis on merciful and comforting teachings was typical of his preaching. Once he told a condemned Negro murderer:

Poor Criminal, here you stand condemned before God; within a few Moments of Death and Eternity. . . . But, O poor Malefactor, I would not drive you to Despair. I bring you good Tydings of great Joy, unto you is offer'd a Saviour. . . . Now betake yourself to God in Christ. Here, poor Criminal, here's Mercy and plentious Redemption for you. . . . O the Joys among the Blessed, at the Conversion of such a Sinner as David, and as You! More than over ninety nine just Persons who need no Repentance. With how much affection will your heavenly Father run out to meet you; and in Embrases and Kisses, proclaim, This my Son was dead and is alive, was lost and is found.[29]

The picture of God running out to welcome with a kiss a black slave who repented a particularly nasty murder is strong Christianity. To Byles repentance did not mean only a formal submission to a tyrant god but demanded a change in the individual which altered his outlook and conduct. For the person who would not submit his will to God, Byles offered no Universalist discount in the heat or eternity of hell. He could preach the horrors of the tomb as well as could Jonathan Edwards, to whose *Humble Inquiry* he signed an approving preface. His own sermon, *God Glorious in the Scenes of the Winter,* is not, despite its title, an appreciation of winter sports; it is a picture of winter as an awful example of God's way of punishing sin. Never during his preaching life did he shift far from the theological stand which he took in his youth.[30]

It was unfortunate for Byles's reputation as a preacher that he was the greatest wit of his day. The memory of his wit has obscured the fact that he was in some ways the best and most popular preacher of his generation. His sermons were short, simple, clear, and free from theological arguments. Commonly they dealt with such practical matters as the eternal rest which Heaven affords for the weary. They, theology and all, would today be heard with approval and pleasure in many of the evangelical churches. Benjamin Franklin, a most exacting connoisseur of good preaching, chose two of Byles's sermons to reprint in the *General Magazine* for January and March, 1741. Later, in a space of two years, no less than four of Byles's thirty-year-old sermons were accorded

[29] Mather Byles, *The Prayer and Plea of David* (Boston, 1751), pp. 16 ff.
[30] Ebenezer Parkman, Diary (Am. Antiq. Soc.), July 17, 1766.

the rare honor of being reprinted, and two went to three editions.

The other popular preachers of the day were either noticeably ahead of or behind the times in the matter of church policy as well as theology, but here, too, Byles held a middle ground. When Lieutenant Governor William Dummer gave the Hollis Street Church a large folio Bible, Byles saw to it that the gift was made conditional on the reading of it as a portion of the service, thus bringing the practice of the church up to that of the times.[31] According to tradition he was the first Congregationalist minister in Boston to wear a gown with bands in the pulpit, his being a gift from the Archbishop of Canterbury.[32] The story may well be true, for it would be like both Byles and the Archbishop; but such a gown would be the kind of controversy-provoking nonessential which the Hollis Street minister usually avoided. His position among the clergy of the province is shown by the fact that from 1748 to 1755 he was scribe of the Massachusetts convention.

At the ordination of his son Mather (A.B. 1751), Byles described the ideal minister in terms which very closely approximate his own pulpit personality:

Paul was a Scholar, before he commenced a Divine. . . . A Minister then, should be a Man of universal Knowledge. Especially, he should have an intimate Acquaintance with his Bible; and be a thorough Student in Divinity, in all its Branches and Connections. . . . He should understand the Controversies of the Polemical Systems; and be a ready Causist to the doubting Mind. He should have a Good Taste for Writing: And be truly Learned, without Pedantry; and truly Eloquent, without Stiffness and Affectation. The Style of the Pulpit should be Solemn and Manly: but yet it were well to be Rich and Polite. . . . A publick Speaker should be a Person of graceful Deportment, elegant Address, and fluent Utterance. He must study an easy Style, expressive Diction, and tuneful Cadences.[33]

Tradition agrees that this tall, powerful man with his deep, clear, voice and his careful diction was the great pulpit ornament of Boston. Of the three portraits of Byles, that by Pelham and the earlier of those by Copley (both at the American Antiquarian Society) show this pulpit personality, while Copley's later portrait,

[31] The Bible is at the Mass. Hist. Soc.

[32] Zachariah G. Whitman, *History of the Ancient and Honourable Artillery Company* (Boston, 1842), p. 292 *n*.

[33] Mather Byles, *The Man of God* (New London, 1758), pp. 10, 11.

here reproduced, shows his weekday manner. This last is the one to which his granddaughter Rebecca referred when she wrote from Halifax:

I had always an idea that my Grandfathers Picture was not like [him] but I never met with a likeness that struck me more forcibly than his did, when the case was first open'd. He appears to have the same kind of Countenance, he use'd to assume, when viewing Mather Brown and me, courting in the Rainwater Hogshead.[34]

Without doubt the author of the poem on the Boston ministers had this same side of Byles's personality in mind when he wrote:

> There's punning Byles invokes our smiles,
>   A man of stately parts;
> He visits folks to crack his jokes,
>   Which never mend their hearts.
> With strutting gait, and wig so great,
>   He walks along the streets,
> And throws out wit, or what's like it,
>   To every one he meets.[35]

On only one recorded occasion did the Hollis Street minister permit this other personality of his to enter the pulpit. That was when, having waited one Sabbath long and patiently for Thomas Prince to appear to preach for him, he mounted the pulpit himself and announced that he would say a few words on the text "Put not your trust in princes." When the dying Cotton Mather urged his nephew to follow in his footsteps, he did not have in mind his own unholy love of puns, but there is no doubt of the inheritance. Byles's puns and accompanying practical jokes were the chief comic relief of eighteenth-century Boston and they were retold and collected with loving care.[36] A sample which escaped the later collectors told of his entering a printing office and declaring that since the printer had "played the deuce" with him by leaving two copies of his paper, he would "raise the devil" with him, which he did by tossing the printer's apprentice into the air.[37] There were some people who

[34] Byles Family Letters, I, 166. The Pelham portrait is the original of the mezzotint. The portrait by Copley not reproduced here is in Arthur Wentworth Hamilton Eaton, *The Famous Mather Byles* (Boston, 1914), op. p. 196.

[35] *New England Hist. Gen. Reg.* XIII, 131.

[36] The two major collections of Byles stories, gathered from men who knew him, are in William Tudor, *Life of James Otis* (Boston, 1823), pp. 156–60, and [Lucius Manlius Sargent], *Dealings with the Dead* (Boston, 1856), II, 367–72.

[37] Daniel Lancaster, Notes Relating to Harvard Graduates (N. Y. Hist. Soc.).

disliked this show of wit. Jacob Bailey (A.B. 1755) declared that "the perpetual reach after puns" made "his conversation rather distasteful to persons of ordinary elegance and refinement," and his own physician, James Lloyd (A.M. 1790), said that he "was a most troublesome puppy . . . there was no peace for his punning." [38] Byles, in defense of his manner, once told Nathaniel Emmons (Yale 1767) "that the genuine Christian denied his profession if he was not continually jolly; for, his 'calling and election' being sure, he had no occasion to feel any anxiety on any subject whatever." [39]

There were few men whose conduct in the pulpit and on the street was more distasteful to the New-Lights who brought the Great Awakening to New England. Byles, on his part, reviewed in his mind the cautious way in which Cotton Mather had handled similar excitement and trances at the time of the witchcraft affair and resolved to follow the same policy. [40] As a result there was no revival in the Hollis Street Church although Whitefield on September 26, 1740, preached from a scaffold erected before it. When Eleazar Wheelock brought the revival back to Boston the next year, he took the opportunity to attack Byles before an Old South audience largely composed of Harvard students. "I believe the children of God were very much refreshed," he wrote in his diary. "They told me afterwards, they believed that Mather Byles was never so lashed in his life." [41] The next year Byles publicly opposed both Whitefield and James Davenport (Yale 1732) but interceded with the justices when they proposed to visit civil punishment on the latter for his disorderly conduct. [42] He refused to partake in the pamphlet controversy, however, or to participate in the provincial assembly of the clergy which met in July, 1743, to review the Awakening. When Whitefield returned in January, 1744/5, Byles was one of the ministers who conspicuously avoided inviting him to preach. [43] His only other excursion into public theological controversy came some years later when he upheld the action taken

---

[38] Sargent, II, 367. The minister of the French church suggested that Beelzebub was his guardian angel. William Smith, Diary (M. H. S.), Sept., 1728.

[39] Nathaniel Emmons, quoted in the *New England Magazine*, XVI, 734.

[40] *Proc. M. H. S.* XLIV, 685–6.

[41] Joseph Tracy, *The Great Awakening* (Boston, 1845), p. 203.

[42] *Christian History*, II, 407; *Boston Evening-Post*, July 5, 1742; *American Mercury*, Sept. 16, 1742.

[43] *Boston Evening-Post*, Jan. 28, 1745.

against the Universalist, John Murray.[44] Much as he disliked these itinerant preachers and the things they stood for, he believed in avoiding public controversy wherever possible.

In 1741 Byles bought a new three-story house in Tremont Street, opposite the modern Shawmut Avenue. It was a plain and simple dwelling, originally built by a bricklayer, with a pleasant green enclosure, shade trees, and a summerhouse in back. Although the busy traffic of Tremont Street now rumbles over the site, it was until long after the Revolution well out in the country:

On one Side . . . we have a fine open Prospect of the Neck, from which by a gradual removal of the Eye, you are conveyed by a pleasing succession thro' the Towns of Roxbury, Dorchester, Cambridge etc. The Water which separates them from my Chamber-Window, forming a most delightful Landscape: On the other side, the great City of Boston presents itself to your View. . . . On the Back you behold a well-regulated Garden.[45]

Standing on what was then a quiet lane ending at the water's edge, removed from the traffic of Orange (Washington) Street but elevated enough to afford a view of the towns ringing Boston, it was an ideal parsonage. Anna Byles did not long enjoy it. As she lay dying on the afternoon of April 26, 1744, the thunder of an unseasonable storm rolled along the marshes of the Charles. Although she had always been afraid of thunder, she listened quietly.

"Are you not at all afraid now?" her husband asked.

"Not at all," she replied.

As her end drew near he said, "You are now in the height of the dying agonies; are you in any pain?"

"None at all."

"Are you afraid?"

"Not in the least." To the others in the room she said, "Come learn of me to die. Death is not so terrible a thing as he seems to be to you." Looking at her husband, she quoted with serene cheer, " 'Smiling and pleas'd in death.' I bless God that ever I saw you: The Doctrines of Grace, in the comforts of which I die, have been more clearly explained and applied to my heart, under your preaching, and in your conversation, than ever they were by any one else. And I say this for your encouragement in your minis-

44 *Boston News-Letter*, May 12, 1768.
45 Byles Letter Book, Sept. 10, 1781.

try." [46] Her death was followed in less than three weeks by that of their second son, Belcher, "a pleasant Child . . . near two years old," who "fell into a Tub of water and was drownd." [47]

This double blow was probably responsible for Byles's decision to gather his poems (all of which he regarded as light) into a collected volume and so to bid "adieu to the airy Muse." The publication of his *Poems on Several Occasions*, followed quickly by the *Poems by Several Hands*, which contained perhaps sixteen other works of his, was a landmark in American literary history. Although he was certainly fully aware of the place of his poetry, he, like Cotton Mather, regarded this form of literary activity as an amusement not to be taken too seriously. With the possible exception of the "New England Hymn," he wrote no more verse.[48]

Like a true Puritan, Byles regarded widowership as an abnormal and undesirable state. Tradition has it that he turned his attention to a lady who spurned him and married a Quincy. Meeting her on the street, he remarked, "So, madam, it appears you prefer a Quincy to Byles." "Yes," she replied tartly, "for if there had been anything worse than biles, God would have afflicted Job with them." [49] The minister's puns fared hardest at the hands of women. One about whom he had talked too much set the wits laughing by sending him the curb of a bridle.[50] However, on June 11, 1747, he was married by Joseph Sewall (A.B. 1707) to Rebecca, daughter of the late Lieutenant Governor William Tailer. Within a few years their "violent squabbles" became common gossip, and even his old friends were inclined to blame him because she came to look "like a poor ruind Woman." [51] The only explanation afforded by the family papers is that Byles's humor was enough to make a wreck of any woman. Once, for example, he asked a distiller to come over and "still" his wife. On another occasion the poor woman had been at housework when guests arrived, and she hid in a closet to avoid being seen. They asked to see Byles's curiosities,

[46] Mather Byles, *Character of the Perfect and Upright Man* (Boston, 1744), pp. 34, 35.

[47] *Boston News-Letter*, May 17, 1744.

[48] Carlson, *op. cit.*

[49] It has been stated that this lady was Elizabeth Wendell, who married Edmund Quincy (A.B. 1722), but the fact that they were married before Byles graduated from college would seem to disqualify her.

[50] Wallcut Mss. (M. H. S.), I, 5.

[51] Ebenezer Parkman, Diary (M. H. S.), May 29, 1755, May 28, 1765.

and he obliged by exhibiting them, at last throwing open the closet and showing his wife as his "greatest curiosity."

Byles made a large and fantastic collection of curiosities. Besides mathematical and scientific toys (his interest in science was not very deep) it included worthless accumulations such as five or six dozen pairs of spectacles, about twenty walking sticks, a bushel of whetstones, a dozen jestbooks (worn), and several packs of cards (new and clean).[52] When the grandchildren came to the house, the parson would take them on his knee and ask, "What is the Chief End of Man?" If they rattled off the answer properly, he would play with them among the curiosities for hours.

This interest in science, slight as it was, kept warm the childhood friendship between Byles and Franklin. It was the parson who suggested that Harvard give the philosopher an honorary degree, and in return Franklin suggested that Aberdeen give one to his old friend, whom he described as "a Gentleman of Superior parts & Learning an Eloquent preacher, and on many accounts an Honour to his Country." [53] The Scotch university, eager to please the great man, awarded Byles a D.D. in 1765.[54] In return the parson sent the university as complete a set of his works as he could gather, remarking modestly that although some had passed through several editions they were now unobtainable.

Byles served on the town committee to visit the schools down to the eve of the Revolution, but the last time that he was asked to open the town meeting with prayer came in 1763. He was entering then into a difficult period both in public and private relations. That year he lost his daughter Elizabeth, wife of Gawen Brown,[55] and the next year he lost his son Samuel, who at the age of twenty-one was "a finish'd, universal Scholar" entering the practice of medicine.[56] At the same time a rift was appearing between him and his congregation although the popular doggerel on the political sentiments of the clergy affirmed the belief that "Old Mather's Race will not disgrace Their noble Pedigree." [57] In the pulpit he did his best to avoid political controversy:

[52] 5 *Coll. M. H. S.* III, 234–5
[53] Belknap Mss. 161.A.57.
[54] The diploma is printed in *Publ. Col. Soc. Mass.* XV, 321.
[55] *Boston News-Letter*, June 2, 1763.
[56] There is a long character of Samuel Byles in the *Boston News-Letter*, June 21, 1764.   [57] Ezra Stiles, *Literary Diary* (New York, 1901), I, 492.

I have thrown up four breast-works, behind which I have intrenched myself, neither of which can be forced. In the first place, I do not understand politics; in the second place, you all do, every man and mother's son of you; in the third place, you have politics all the week, pray let one day in the seven be devoted to religion; in the fourth place, I am engaged in a work of infinately greater importance: give me any subject to preach on of more consequence than the truths I bring you, and I will preach on it the next sabbath.[58]

The Doctor was certainly not so ignorant of political affairs as he made out. He subscribed for the reprint of the *Vindication* of John Wise and in his Artillery Sermon spoke learnedly of the law of nations and of the necessity of sanctions to enforce it. It was not his views but his wit which got him into trouble, for he could not suppress the inclination to jeer at what he regarded as the posturing and empty oratory of the Whig politicians. On the other hand, he felt at ease among the Loyalists with their solid attachment to the old order.

It was the bell of the Doctor's church which awoke the others at one in the morning when the news came of the repeal of the Stamp Act. To the next number of the *News-Letter* he contributed a happy account of the celebration.[59] There was nothing offensive in his classic remark on the landing of the Regulars that "our grievances" were "red-dressed," [60] but frequently he angered the mob and the agitators by boldly voicing thoughts which most men of his social position shared but kept to themselves. Nathaniel Emmons, who fifty years later defended the Doctor, recounted a typical conversation:

I stood with Parson Byles on the corner of what are now School and Washington Streets, in March, 1770, and watched the funeral procession of Crispus Attucks — that half Indian, half negro and altogether rowdy, who should have been strangled long before he was born. There were all of three thousand in the procession — the most of them drawn from the slums of Boston; and as they went by the Parson turned to me and said: "They call me a brainless Tory; but tell me, my young friend, which is better — to be ruled by one tyrant three thousand miles away, or by three thousand tyrants not a mile away." [61]

[58] Tudor, *Life of James Otis,* p. 156.
[59] Identified in the list of his works by Mary Byles in Byles Family Papers, II, 175.
[60] John Adams, *Works* (Boston, 1850–56), II, 213.
[61] *New England Magazine,* XVI, 735.

His social connections were mostly with the aristocratic Tory group, and he welcomed the appointment of Hutchinson as Governor:

Dear Sir, (For I have been used to Write to Governors in the Province House by the Title of Dear Sir: And it comes from my Heart to one who has been personally dear to me from his Infancy) — This Petition humbly shewith, That Your Petitioner has been long confined by Sickness, and his Physician advises to a short Ride. Your generous Father, were he living, would approve that Your Petitioner should apply to You to assist him with Your Charity upon this Occasion.

I congratulate Your Excellency to the Chair, and pray that every Blessing of the upper and nether Springs may be Yours. "Your Monarch's Plaudit and Your Peoples Love!" [62]

When it was a matter of having fun, he was impartial. Once he threw the Regulars into a dither by telling them "that on the 14th of June forty thousand Men would rise up in opposition to them with the clergy at their head, and left 'em to suppose it a fact, without explaining the matter, that on that day a general fast was to be observ'd through the province." [63]

During the siege Doctor Byles was one of the inner social circle of Tories and Regulars. Earl Percy serenaded his daughters with the regimental band and walked arm-in-arm with them on the Mall. This did not, however, save the Hollis Street Church from being converted into barracks. After this the Doctor officiated at the Old Brick. When the rebels seized Dorchester Heights, he moved out of his house for a few nights to avoid the cannonade, but he benefited from the fact that the commander of the artillery on the Heights owned the house next to his.

The return of the victorious rebels made not the least difference to the Doctor's wagging tongue. As he watched the portly General Knox make his entry into the delivered town, he remarked loudly, "I never saw an-ox fatter in my life." The General was angry, but the Anglo-American sense of humor protected the parson. The church in Hollis Street had had enough of him, however. The pews had been removed and the building damaged, and the members were determined to try and eject the parson whom they blamed for their troubles. The hearing was set for August 9, 1776, and on that day the members gathered in the galleries to sit as a jury.

[62] Byles Letter Book, Apr. 3, 1771.
[63] John Andrews to William Barrell, Andrews and Eliot Mss. (M. H. S.), p. 34.

In due time, the door opened slowly, and Dr. Byles entered the house with an imposing solemnity of manner. He was dressed in his ample flowing robes and bands, under a full bush wig that had been recently powdered, surmounted by a large three-cornered hat. He walked from the door to the pulpit with a long and measured tread, ascended the stairs, hung his hat upon the peg, and seated himself. After a few moments, he turned with a portentous air towards the gallery where his accusers sat, and said, — "If ye have ought to communicate, say on."

Then arose one of the deacons of the church, a man of diminutive stature and feeble voice; and, having unfolded a manuscript, commenced reading — "The church of Christ in Hollis street." — "Louder," said the Dr. in his deep toned, sonorous voice. The deacon raised his voice and began again — "The church of Christ in Hollis street" — "Louder," said the Dr. in a higher key. The little man in the gallery exerted himself to throw out his voice with more force, and read the third time the same words. "Louder," shouted the Dr., "Louder, I say." At this the deacon strained himself to the utmost; and trembling with the effort and with dread of the angry man who sat before him, proceeded to read specifications of unministerial and otherwise improper conduct alleged by the church against their paster. When the third or fourth had been read, Dr. Byles rose and shouted out upon the top of his stentorian voice, — " 'Tis false, 'tis false; and the church of Christ in Hollis street knows that 'tis false." [64]

Then the Doctor clapped his hat on his head and for the last time walked out of the meetinghouse over which he had presided for thirty years. The charges against him were too serious for him to dismiss by making fun of the little deacon:

First, His associating and spending a considerable part of his time with the officers of the British army, having them frequently on his house, and lending them his glasses, for the purpose of viewing the works erecting out of town for our defence.

Second, His neglecting to visit his people in their distress, and treating the public calamity with a great degree of lightness and indifference; likewise using his influence to prevent people from going out of town, and saying the town would be inhabited by a better sort of people than those who had left it, or words to that purpose.

Third, His being, as we think, officious to lend his aid and assistance to furnish our enemies with evidence against the country, by signing a certain paper at the request of Gen. Gage, relative to what one Hogshaw said (or did not say) respecting the battle at Lexington.

Fourth, His being unwilling to preach on a fast-day appointed by

[64] Related by the father of Samuel J. May (A.B. 1817), quoted in William B. Sprague, *Annals of the American Pulpit* (New York, 1857–69), I, 381.

Congress, when with difficulty he was prevailed with to preach one-half the day; and further, his refusing to have two services on the Lord's days. . . .

Fifth, His frequently meeting, on the Lord's days, before and after service, with a number of our inveterate enemies, at a certain place in King-street, called Tory Hall.

Sixth, His praying in public that America might submit to Great Britain, or words to the same purpose.

Seventh, His taking away the fences belonging to the society, the seats of the pews, etc.[65]

Doctor Byles refused to give these charges serious consideration, whereupon the church, without benefit of ecclesiastical council, proceeded to discharge him. Charles Chauncy (A.B. 1721) approved this action on the ground that he was "not fit for a preacher," but the Whig clergy in general protested. John Eliot refused to preach at Hollis Street:

Notwithstanding I despise Dr. Byles as much as man can hold another in contempt, yet I think the proceedings of that church with him were irregular & unwarrantable, & hath held up a precedent for a practise that will cause the ruin of our ecclesiastical constitution.[66]

As much as he disliked the Doctor, Eliot could not resist passing on his latest pun, which was calling the French allies "all-lies."

On August 23, 1776, the Boston Committee of Correspondence and Safety held a hearing to question people who had heard Doctor Byles "express himself very unfriendly to this Country." [67] Although the testimony against him was clear, no action was taken.

The Doctor struts about town in the luxuriance of his self-sufficiency, looking as if he despised all mankind. He never attends any meeting. How he doth for a maintenance, nobody knows besides him, & the only account he can give is, "That he doubles & trebles his money." He is a virulent Tory, & destitute of all prudence. Before I leave him, I will give you one more effort of his genius in the punning way. He observed Dr. Cooper to go by his house often, & one day meeting him, Dr. C., says he, you treat me just like a baby. I hardly take you, Sir, said the Doctor. Why, you go by, by, by.[68]

Instead of showing appreciation of the toleration shown him, Byles became more arrogant and his shafts more barbed. In May, 1777,

[65] Thomas S. King, *Losses and Gains of a Church* (Boston, 1852), p. 17 *n.*
[66] 6 *Coll. M. H. S.* IV, 104, 106, 107.
[67] Minutes of the Committee in *N. E. H. G. R.* XXXIV, 17.
[68] 6 *Coll. M. H. S.* IV, 106.

the selectmen placed his name on a list of public enemies which was presented to a special sessions of the peace. On June 2 he was called, the first on the list. It was an unseasonably cold day, and the judges courteously invited him to sit by the fire. "Gentlemen," he replied, "when I came among you, I expected persecution; but I could not think you would have offered me the fire so suddenly."

He appeared without counsel, and upon the nomination of the jury he objected to one Fallas, commonly called Fellows, because he said he would not be tried by fellows. The evidence was much more in favor of him than against him. All that could be proved was that he is a silly, impertinent, childish person; I should say inconsistent, if his whole conduct did not manifest him to be one consistent lump of absurdity. When he was going out of court, he observed that the ceiling was very high, & he could not discern, but asked if none of them discovered a star.[69]

He was found guilty of being an "enemy to the United States" whose continued presence was "dangerous to the public peace and safety." "You are therefore," the court instructed its officers, "directed immediately to deliver the said Mather to the board of war of the State to be by them put on board a guard ship or otherwise secured until they can transport said Mather Byles off the continent to some part of the West Indies or Europe agreeable to a late law of said State."[70] The people of Boston were amazed that he had been found guilty:

His general character has been so despicable that he seems to have no friends to pity him, tho all allow upon such evidence he o't not to be condemned. The women all proclaim a judgment from Heaven as a punishment for his ill treatment of his wives. Vengeance has at length overtaken him, they say, & his present sufferings will now bring him to reflection.[71]

Such was the public attitude that instead of being transported he was only confined to his house, ordered not to correspond with his friends abroad, and forbidden visitors. The last was not enforced.

The treatment of Mather Byles, a convicted enemy of the state and a sharp and ungrateful critic of its society, is a startling example of the sense of fair play, of the generosity and kindness of the Anglo-American. Other Europeans would have done away with

[69] *Ibid.*, p. 122.
[70] Eaton, pp. 164–5.
[71] 6 *Coll. M. H. S.* IV, 122.

him in a moment. The Bostonians remembered that he had been a great preacher and their most distinguished literary man, and they forgave the silly conduct of his later years as the inevitable result of the impact of circumstances upon his peculiar character. For ten years they supported him by delicate and unostentatious charity although some of the donors were hurt by the puns with which he received their gifts. At Thanksgiving, 1779, the church in Hollis Street took up a collection for him, but otherwise all appearance of charity, so distasteful to Yankee recipients, was avoided. One day the John R. Livingstones visited the Doctor and contrived to hide $300 on a closet shelf but were soon detected:

Dr Byles sends his Blessings to his dear Mrs. Livingstone, alias the lov'ly Peggy, & desires her to make What Discoveries she can, of certain Persons who have made Ditanidations [!] in his Closet, by misplacing an Ivory Folding stick, value Three Pence, which he finds upon an upper shelf. . . . Three Hundred Dollars Reward is proposed for the Discovery.[72]

Margaret Livingstone was a daughter of his old Tory friend William Sheaffe (A.B. 1723), but other donors, like Mrs. John Rowe, were Whigs. When the Doctor was snowed in, his friends dug him out, thawed his pump, and supplied fuel and food. Once the gift of a loin of bear furnished candles as well as food. The Byleses sold their summerhouse and other property (the lawyers refused their fees) but otherwise suffered no discomforts.

For two years the Doctor was confined to his house, part of the time under the eye of an armed guard whom he described as his "observe-a-Tory." Once he induced this soldier to go on an errand and himself shouldered the musket and marched up and down in front of his house guarding it ferociously. When the puzzled military withdrew the guard, the Doctor remarked that he had been "guarded, reguarded, and disreguarded."

Mrs. Byles died on June 23, 1779, "after eleven Days Confinement with a Dysentery." If her life had been unhappy, her husband never seemed aware of it: "That amiable & good Woman has gone a little before us, to a World for which she was well prepared & desirous to be in. Our loss here is irreparable. . . . It is not common for a Person to die so universally esteemed & lamented."[73]

[72] Byles Letter Book, May 24, 1780.
[73] Mather Byles, *ibid.*, June 23, 1779.

During these years the Doctor observed the prohibition against correspondence by dictating to his daughters politically innocent letters in the third person. To his old friend Copley he wrote to further the fortune of his grandson, the artist Mather Brown:

To Mr Copley, Europe.   To the care of M———— B————.
A certain ancient Gentleman in New-England, dictates the following Words.

Boston, N. E. Dec. 6, 1780.
My Dear Copley:

Do you forget your old connections? I am always rejoyceing to hear of your Reputation, & Felicities on your side of the Water. You will, I am certain, be pleased to see the Grey-Eyed little Boy, that you left upon the Entry Floor at New-Boston: See how Time had tanned Him. I may not write, & I need say no more to one on whose Friendship I have so firm a Reliance. Tell Mr. Pelham, he partakes in my Affection, & Prayers. Here the old Patriarch leavs off.

Mr. Copley, in the Solar System.[74]

These were years of happiness and health for the old gentleman; he said that the Twenty-third Psalm was the exact history of his life. When allowed once more to go out in public, he let his hair grow in the new style and attended the services of Samuel Parker (A.B. 1764) at Trinity Church, fully aware that he was still a handsome figure of a man. Although there are references to his having changed his religious views,[75] he cannot have gone over to the Episcopal Church for on his deathbed he impishly told Parker, the future Bishop of Massachusetts, that he was "going where no Bishops will ever enter." [76] The two men were warm friends, and the Doctor prepared a complicated arrangement of Watts's "Shepherds Rejoice" for the Trinity Christmas service of 1782.[77] He was interested in the musical work of William Billings, to whose published collection he had contributed his "New England Hymn," which had been sung at the Pilgrim celebration of 1772. His best known hymn, however, was "When Wild Confusion Wrecks the Air," which his grandnephew Jeremy Belknap (A.B. 1762) included in his *Sacred Poetry,* which was for forty years

---

[74] *Ibid.,* Dec. 6, 1780.
[75] Byles Family Papers, I, 127.
[76] *New Hampshire Spy,* July 22, 1788.
[77] Byles Letter Book, Dec. 14, 1782, *et passim.* His arrangement was not used in the service.

the most popular hymnal in New England. This particular hymn was included in other collections down to the time of the Civil War. He wrote nothing in these years but he had the pleasure of seeing his old poems and essays reprinted in the current magazines, as they were elsewhere after his death.[78]

In July, 1783, the Doctor suffered a shock:

After retiring to Bed in his usual Health, he rose about 11 oclock in order to drive off an Indian Woman who had placed herself just under his Study Window, & as I [his daughter] imagine took cold, for on his return to his room he complained of a faintness or Numbness in his right Leg & Arm. . . . At 2 o Clock we were alarmed by a Noise in the Study succeeded by the Groans of my Father, on entering the Door we found that in attempting to rise, he had fallen on the Floor, occasioned by the intire uselessness of his right side . . . & he complained of a considerable decay in his right Eye.[79]

After this he was kept close at home by the resulting paralysis and palsy. Jeremy Belknap visited him and found his wit still active, however:

He is seventy-eight years old, and usually sits in an easy chair which has a back hung on hinges. In such a chair I found him sitting, and as I approached him he held out his hand. "You must excuse my not getting up to receive you, cousin; for I am not one of the rising generation." [80]

On June 11, 1786, his house was struck by a bolt of lightning which melted the rods and knocked in the window of the study in which he was sleeping. This caused him to send a letter to his old friend Franklin, to whose lightning rods he gave credit for the preservation of his life.[81] The great man answered in his usual kindly way:

It gives me much Pleasure to understand that my Points have been of Service in the Protection of you and yours. I wish for your sake, that Electricity had really prov'd what it was at first suppos'd to be, a Cure for the Palsy. It is however happy for you, that, when Old Age and that Malady have concurr'd to infeeble you, and to disable you for writing, you have a Daughter at hand to nurse you with filial attention, and to be your Secretary. . . . I remember you had a little Collection

---

[78] Lyon N. Richardson, *A History of Early American Magazines, 1741–1789* (New York, 1931), pp. 217–8. Some of the writings formerly attributed to this part of his life are now known to be reprints and probably all are.

[79] Byles Letter Book, July, 1783.

[80] Eaton, p. 199.

[81] Byles Letter Book, May 14, 1787.

of Curiosities. Please to honour with a Place in it the inclos'd Medal, which I got struck in Paris.[82]

To a request from Ezra Stiles for historical information Doctor Byles had to reply that his mind and memory were too enfeebled to help. "I am sorry I cannot assist you: but from the borders of a World Where there is no more prayer, I send you down my tenderest wishes and blessings and drop my mantle."[83] We can well applaud the shedding of the mantle of the Mathers upon the curious and scholarly Stiles, but Byles was in the habit of shedding it indiscriminately on his correspondents. When the great fire of April 20, 1787, swept that part of Boston, Belknap kept an eye on the old gentleman:

Dr. Byles's house was in imminent danger: his hoards of books, instruments, papers, prints, etc., etc., were dislodged in an hour from a fifty years' quietness to an helter-skelter heap in an adjoining pasture. He removed for the night to a neighbour's house, and returned the next day. This morning I made him a third visit since the fire. One of his daughters observed that "her pappa was the first thing they thought of moving." Upon this he began to distinguish between persons and things, and would have brought on a long criticism, if I had not changed the discourse to some enquiries about the great fire in the year 1711, which he remembered. You know he is a curiosity.[84]

He died on July 5, 1788, and was buried by his friends in the Granary Burying Ground. They ignored the epitaph which he proposed:

> Here lies the renowned Increase Mather.
> Here lies his son Cotton, much greater.
> Here lies Mather Byles, greater than either.[85]

The two daughters, Katherine and Mary,[86] lived on in the old house for nearly fifty years, firmly maintaining that they were subjects of the English King and wearing the costume of the day when Earl Percy squired them. Although the town grew out around their house, in its last days a curiosity, black with age, Boston remembered that Mather Byles had been a clergyman and

---

[82] Benjamin Franklin, *Writings* (Smyth ed.), IX, 656.

[83] Byles Letter Book, Apr. 13, 1787.

[84] 2 *Proc. M. H. S.* XIV, 65–6 n.

[85] *N. E. H. G. R.* II, 403. There is a panegyric of Byles in the *Gentlemen and Ladies Town and Country Magazine*, I (1789), 496–7.

[86] For the children, see Eaton, pp. 205–23.

never taxed it. Without the slightest recognition of the kindness of the Bostonians (which "Kitty and Polly" ascribed to "kind Providence"), they ordered that after their deaths all the treasures of the house be sent out of this rebel land to loyal Nova Scotia, where some are yet in the possession of Lieutenant Colonel W. B. Almon of Halifax. Transcriptions of the family letters among them are in the Massachusetts Historical Society, to which (no doubt because of Belknap's interest) the daughters also gave some of their father's "curious" manuscripts. These, however, are worthless compared with his general correspondence, which appears to have been lost. His library of some 2,500 volumes was sold at auction; from it came many of the Increase and Cotton Mather items in modern libraries.

BENJAMIN KENT

BENJAMIN KENT, heretic and lawyer, was the third son of Joseph and Rebecca (Chittenden) Kent of Charlestown. They lived out on the Cambridge road and attended the First Church of Cambridge where Benjamin was baptized on June 13, 1708. He joined this church on November 21, 1725, perhaps in a fit of repentance for the offence which caused him to be degraded for a time to the Class of 1728. After graduation he kept school for a time at Framingham[1] but in May, 1729, he was awarded a Hopkins scholarship to finance a year of graduate work in preparation for the ministry. For four years he kept a study at the college although he was out of town much of the time. Indeed he missed the Commencement of 1730, when the members of his class took their second degrees, although he is recorded on the *Quaestio* sheet as being prepared to defend against all comers the statement that the doctrine of the Trinity is revealed in the Old Testament. In view of his later troubles it is tempting to believe that he stayed away from Commencement because of his *Quaestio*, but the fact is that he was shortly thereafter, if not then, serving as chaplain of Fort George and preaching to the settlers at Brunswick. The Province on June 18, 1731, paid him £20 for his services there.

Kent soon gave up the dangers and discomforts of frontier Maine for the pursuit of a more comfortable pulpit in the settled towns. Such were the complaints about his theology, however, that in June, 1733, he felt compelled to protest his orthodoxy to his old friend Parson John Swift (A.B. 1697) of Framingham:

A Report . . . has been circulated concerning me with respect to Arminianism. But I can't think but that I am much abus'd by it, for as far as I understand what Arminianism is I am Actually the farthest from it. . . . Freewill & Universal redemption, are the two corner stones of Arminianism. . . . I am entirely in the Disbelief of both those Things, And because I have preach'd & discoursed lately Agreeable hereto I have been blam'd for & Taxd with Calvinism. . . . Because I have some peculiar ways for accounting for some Religious Phenomina, which Some times in the opinion of others favor one Scheem & Some times another they have agreed to make me a Bastard on both sides. . . . I am very well Acquainted that the openness & freedom of my conversa-

[1] Ebenezer Parkman, Diary (Am. Antiq. Soc.), Sept. 3, 1728.

tion, has been a very unreasonable handle that Some have Taken hold of to conclude that I have Spoken my Sentiments when I have Hardly tho't of what I said, or at most nothing more than to see what cou'd be Said pro, or Con.[2]

In writing to a layman of Sandwich, Kent boldly defended the manner of his preaching in that town:

Dear Friend I'm heartily Sorry that I have occasion to believe, that my Successor loves a Sort of peace rather too well to make a brave Souldier. A People had better have their Teacher Turn'd into a Corner, than that they shou'd be led in the way of their own Ignorance or Prejudice, or at least Tollerated herein, by their Teacher, for the Sake of his own private interest. . . . I heartily rejoice that I have had so happy an opportunity, of setting the most material Things respecting Faith & manners in a plain Light, before the people to whom I preached at Sandwich, & I cou'd heartily wish to be with you again.[3]

It was, however, not the church at Sandwich but that at Marlborough which gave Kent a call. On the same day, August 21, 1733, the town of Marlborough ratified the choice and voted to offer the candidate a settlement of £400 and an annual salary of £180 which was to be increased in proportion as the progress of inflation debased the paper money below its then value of 20s to the ounce of silver.[4] Kent thereupon surrendered his Cambridge study and removed to Marlborough where upon September 25 his acceptance of the call was reported to a town meeting.

Usually in the New England of that generation the testing of the orthodoxy of a candidate was a formality which occupied only a few minutes of the morning of his ordination. In Kent's case the ministerial association sat for two days, October 16 and 17, 1733, at Parson Swift's in Framingham to give the candidate a thorough examination. The Congregational clergy of that generation usually tried to avoid detecting heresy among their own number lest they engender fruitless debates with the lay theologians, so it is not surprising that the Association found Kent to be an orthodox Calvinist. However, one of the members later admitted that "it was . . . intimated to Mr. Kent . . . that our Satisfaction then manifested to him must Continue upon no other Condition than

[2] Early Files in the Office of the Supreme Judicial Court, 42,172.
[3] Benjamin Kent to William Bassett, July, 1733, in Benjamin Kent's Family Papers (A. A. S.).
[4] Supreme Court Files, 35,809.

our finding by his Preaching & Conversation that those were his true and real Sentiments." [5] Another member of the council had the same doubt as to the candidate's sincerity, and "exprest his uneasiness that K. should be exposed to such a temptation to Deny his principles." He expressed his "sorrow that Mr. K. had acknowledged so much & seemed to question whether he was indeed of the opinions he appeared to be before the associations, for the Dr. said he always understood him that Original Sin was not damnable & that we were not Justified by Faith alone." [6]

When the council met to ordain Kent on October 31, 1733, it learned that there was considerable dissatisfaction in Marlborough, and that even the leading man, Colonel Woods, in whose house they were then sitting, was opposed to him.

Woods was sent for and Desired in the presence of the council to Declare what was the Cause of his uneasiness but he Declined it saying he was Very uneasy However he Would make No Disturbance. Yet the Deponent being at said Woods's House in the Evening after the ordination of Mr. Kent he heard the said Col. Woods say to Mr. Justice Whitman that he Would out Mr. Kent or Have him out in a year or some such short space and that He Never Undertook a thing but that he accomplished it. [7]

Colonel Woods could not make good his boast, perhaps because Kent succeeded in delaying the inevitable by publishing a sermon in which he savagely attacked Socinianism and Arianism. He also continued to protest that he was not an Arminian, and so confused was his theology that he may have been sincere in his claims, but he was so indiscreet that he could not be tolerated as more judicious Arminian ministers were. The ecclesiastical council which met at Marlborough during the first week of February, 1734/5, to hear the case against him, was composed of men of diverse views, but in every point in the long process they were unanimous. [8] So great was the public interest in the trial that the result, or verdict, was published. In part this reads:

As to the Matters of Scandal, such as profane and filthy Expressions that Mr. Kent has been charged with, and convicted of, which we think very

---

[5] *Ibid.*, 43,428.

[6] Nathan Stone to Israel Loring, Oct. 20, 1733, in Nathan Stone Mss. (Mass. Hist. Soc.).

[7] John Gardner, July, 1737, in Misc. Mss. Bound (M. H. S.).

[8] *New England Weekly Journal*, Feb. 10, 1735.

unbecoming any Person, much more a Gospel Minister, who should be an Example, even to Believers themselves, in Word, in Conversation, in Purity; yet he having before this Council, and the Persons offended, acknowledged his sinful Inadvertency, and asked Pardon of God therefor, we think a charitable Forgiveness should be exercised towards him.

However, his public questioning of the doctrines of the Trinity, of Absolute Election, and of Infant Damnation could not be so easily disposed of, so he was suspended until May 27 with the idea that he might conform by then.[9] On that day he met the adjourned council in Boston and presented his case, but was found to be still unsound on the Trinity and the Satisfaction of Christ, and was, therefore, advised to cease preaching pending further proceedings.[10] Four days later the church at Marlborough accepted the decision of the council. When this body met again at Marlborough on June 17, 1735, "Mr. Kent being Called in that the Council might Know what he had further to offer signified that he had thots of Requesting a Dismission from his Pastoral office & then withdrew." Thereupon the council voted unanimously that he be dismissed "according to his request." [11]

So ended one of the few heresy trials in the history of New England Congregationalism. Actually it was more the fact that Kent was offensively erratic than that he was not orthodox which brought him to trial, for he represented a tendency of the times which was recognized and feared. More orthodox pastors complained that his Arian principles were beginning to be very modish among the young gentlemen of Boston,[12] and Timothy Cutler (A.B. 1701) regarded them as proof of the collapse of the Congregational system and of the necessity of abandoning it in favor of the Church of England with its firm orthodoxy:

One Kent, a Dissenting Teacher, is now suspended by a Council for Arianism and Arminianism, though the latter is grown so venial that it would have been hushed up had it not been for the former. It is expected he will entirely be laid aside; however, that he will find friends enough to make him a new congregation and support him.[13]

[9] *At a Council of Ten Churches Convened at Marlborough on February 4, 1734,* Boston, 1734.

[10] Supreme Court Files, 39,604.

[11] *Ibid.,* 39,652.

[12] John Cotton of Halifax to John Cotton of the Isle of Wight, in Harvard University Archives, HUG 300.

[13] John Nichols, *Illustrations of the Literary History of the Eighteenth Century,* London, 1817–58, IV, 298.

Kent was less interested in organizing another church than in obtaining the money which was due him from the town of Marlborough. The town had not paid any part of his £400 settlement and had stopped his salary when he was suspended instead of continuing it while the members of the ecclesiastical council in rotation filled the pulpit, as the council intended that the town should do. Kent sued and in the Inferior Court in December, 1736, won a decision for £200. The town hired Edmund Trowbridge (A.B. 1728) to carry the case to the Superior Court, but there it was held over while the town was indicted for failing to have a settled minister. Finally, in July, 1737, the Superior Court found for Kent and awarded him costs.[14]

It may have been this success which decided Kent to enter the practice of the law in which he was active in the Boston courts as early as 1739. There were then in Boston but seven lawyers, among whom he established himself as "the Chimney sweeper of the Bar, into whose black dock entered every dirty action." [15] He handled divorces and the suit of Benjamin Faneuil's slave Pompey for his freedom,[16] cases which in those days were regarded as hardly respectable. He failed to win Pompey's freedom but in 1749 he won a case for Governor Shirley in which no less a lawyer than Benjamin Prat (A.B. 1737) was his opponent. He faithfully followed the Superior Court on its circuit and in one session entered no less than forty cases. There were, however, complaints that he sometimes failed to appear in behalf of clients who had engaged him, and that he otherwise neglected their interests. When one of these clients, Dr. Sylvanus Gardiner, took to the newspapers to prove Kent "guilty of the most wilful and corrupt perjury," the lawyer used two and a half pages of the supplement to the *Boston Evening Post* of April 23, 1770, to refute the charges. Dr. Gardiner then took up most of the supplement of the issue of May 14 to prove that Kent was, among other things, a liar. The Doctor was a Tory, but this dispute can not be dismissed as politics in view of the statement that there was "not one person in Boston" but believed that Kent's attack on the Gardiner estate was "a most in-

---

[14] Superior Court Files, *passim*; Journal of the Superior Court of Judicature, 1736–1738, p. 106.

[15] *The Herald of Freedom*, Feb. 2, 1790.

[16] Superior Court Files, 69,970; Journal of the Superior Court of Judicature, 1752–1753, p. 238.

famous transaction." [17] Indeed there were Whig leaders among his bitterest enemies. [18]

Lawyers then as now tended to leave their professional differences in the court room. Kent was an organizer of the Suffolk Law Club and, as the oldest member present, usually presided. [19] He was occasionally a guest at the Old Colony Club. [20] On these occasions Father Kent, as the younger men called him, was the life of the party. "Kent is for fun, drollery, humor, flouts, jeers, contempt," wrote young John Adams. "He has an irregular, immethodical head, but his thoughts are often good, and his expressions happy." [21] It was the delight of his life "to tease a minister or deacon with his wild conceits about religion," such as this:

We are here probationers for the next state, and in the next we shall be probationers for the next that is to follow, and so on through as many states as there are stars or sands, to all eternity. You have gone through several states already before this. [22]

Of Benjamin Kent's personal life there is not much to tell. On November 6, 1740, he was married to Elizabeth, the eighteen-year-old daughter of Samuel and Elizabeth (Shute) Watts of Chelsea. Probably they rented a house on the west side, for their children were baptized at the West Church. In 1760 Benjamin inherited the estate of his brother Jonathan (A.B. 1739), and three years later he bought four-sevenths of a brick house on the north side of King Street, a site now mostly covered by the intersection of Devonshire and State streets. Kent had years before contributed £20 toward the purchase of the bells for Christ Church, and now that he was on the same side of the hill with them he began to attend one or the other of the Episcopalian churches. His house on King Street became one of the focal points of Boston life. Although it was not until the eve of the Revolution that he received the honor of appointment to the committee of dignitaries which inspected the schools, he was, beginning in 1750, a member of more town committees than any other Bostonian. These dealt with all kinds of practical problems, such as the sweeping of chimneys, but

[17] E. Alfred Jones, *The Loyalists of Massachusetts*, London, 1930, p. 141.
[18] John Rowe, Diary (M. H. S.), Feb. 5, 1772.
[19] 1 *Proc. M. H. S.* XIX, 147.
[20] 2 *Proc. M. H. S.* III, 409.
[21] John Adams, *Works*, Boston, 1850–56, II, 75.
[22] *Ibid.*, p. 290.

the ones which are interesting, in view of the fact that he has been listed among the Loyalists, are the political ones. One of these committees, in 1750, was instructed to appeal to His Majesty against a tax laid by the General Court and to engage a town agent to act in London, but the later ones followed a more familiar pattern. Among them were committees to ask the opening of the courts in defiance of the Stamp Act, to testify the gratitude of Boston to the English Whigs for the repeal of the Act, to obtain the stay of the warrant issued against Samuel Adams as a "defective Collector," to petition the Governor in the *Liberty* affair, to ask him when the Regulars were coming, to collect patriotic pledges not to eat lamb, to defend Boston against the aspersions of the customs officers who had been driven out by the mob, to instruct the Representatives, and to consider "some suitable Method to perpetuate the memory of the horrid Massacre perpetrated on the Evening of the 5th of March 1770." On December 14, 1774, the open meeting of the Committee of Correspondence at the Old South chose him to head a committee of ten to go with Captain Francis Rotch to the Collector to apply for a clearance of one of the tea vessels.[23] He had already in town meeting publicly defended the activities of this illegal body.[24] He was a member of the North End Caucus and a Son of Liberty, and as the senior member of a committee of the latter body he entered into correspondence with John Wilkes, whom he admired greatly:

I was a dissenting minister about thirty years since; but I was so happily industrious as to finish my Great Work of the Gospel Ministry in about three years, and then was cashiered for Heresy. Of late I have attended the Church of England. . . . For my own part I have the most profound veneration of the surpreme Being, and yet I really think the Caracter of the God of the Vulgar . . . in many respects is as bad as that of the D - - - l, and in one respect much worse, for I think the Devil is never allow'd to be really omnipotent. . . . I never expect to see what you wrote on Women, but if I should find anything which is called too Luscious, I assure you I am well fortified by the revolution of sixty cold North American winters.[25]

Even Kent's business letters combined the extremes of piety and patriotism. On September 15, 1774, when a member of the Suffolk

[23] 1 *Proc. M. H. S.* xx, 15: *Acts of the Privy Council, Unbound,* p. 552.
[24] 2 *Proc. M. H. S.* x, 86.
[25] *Proc. M. H. S.* xlvii, 195–6.

County Congress, he wrote to Robert Treat Paine that he was pleased by the abrogation of the provincial charter:

Every One is well Acquainted with the present Aims of Great Britain upon the Colonies, & I am pleased that those demands are so insufferable as they Are. I dont believe she will Attempt with force & Arms to Conquer all the Colonies which Oppose her Measures. But this is now the unshaken Faith of Our Tories. . . . For my own part though a good Churchman, I believe in the Congress as much as I do in the Holy Catholick Church.[26]

Happy in his faith in bluff he urged the raising of military forces against the Regulars.[27] When events proved that he had miscalculated the handling of the explosives involved, he showed no regrets.

During the siege of Boston, Father Kent entertained with his visits his exiled friends in the neighboring towns. He returned to town after the evacuation and as early as April, 1776, was writing to John Adams at Philadelphia urging that he press for a declaration of independence from "that accursed kingdom called Great Britain," his chief reservation being that the new government must tolerate all religions.[28] To Sam Adams at Philadelphia he wrote in a less optimistic vein saying that "the Soul or Spirit of Democracy is Virtue" and that no state can long preserve its liberty "where Virtue is not supremely honord," but he admitted that this was not the universal practice in Massachusetts.[29]

From September, 1774, to April, 1776, Massachusetts got along without an attorney general, but the Suffolk Court of General Sessions then feeling the need of a prosecuting officer appointed Kent. He described himself as Attorney General of the State for Suffolk, but the General Court, feeling that it could not pay him a salary in that capacity, voted him a salary as Attorney General of the State. After the appointment of his friend Robert Treat Paine to this office, the General Court referred to Kent as Attorney General for Suffolk.[30] Boston appointed him to a committee to deal with inhabitants who were "suspected of being inimical to the American States," and one of his painful duties as attorney general was to

[26] Robert Treat Paine Mss. (M. H. S.), II, 2.
[27] Paine Mss. II, 3.
[28] John Adams, *Works*, II, 291; IX, 401.
[29] Harry A. Cushing, *Writings of Samuel Adams*, New York, 1907, III, 305.
[30] 2 *Proc. M. H. S.* X, 290; *Acts and Resolves of the Province of Massachusetts Bay*, Boston, 1869–1922, XX, 161.

issue a warning to his friend Edward Winslow (A.B. 1741), the Episcopalian minister of Braintree:

This acquaints you, that Several persons have informed me that you have disregarded a late Law of this State, relative to talking, preaching or praying disrespectfully of the present Government, the penalty is not less than 20s nor more than £50. My good friend Edward Winslow Esq. assures me, you have no disposition to oppose the Laws of the State. I rejoice at it, Therefore, I suppose you will for the future follow the good Example of my good Parson Parker — and tho' as Attorney General, I am obliged to do that duty; yet you may depend upon me, I will do every thing which you can reasonably expect from a Whig & good Christian.[31]

There were rumors that Father Kent was making a fortune as an owner of and attorney for privateers, and was not losing any money in his dealings with the estates of Loyalist exiles. One of his Tory connections was very fortunate for all concerned. In 1774 his youngest daughter, Sally, had married Samson Salter Blowers (A.B. 1763). Although Blowers was a Loyalist, Kent certainly had no objection to the match, for Sally's dowry amounted to the incredible figure of £5000 sterling cash and £1000 in slaves and goods.[32] Blowers fled at the outbreak of hostilities, but soon returned and put his father-in-law into the unpleasant situation of having to jail him. However, Blowers was soon safe with Sally behind the British lines at Newport where Elizabeth, Kent's oldest daughter, was allowed to join them. They later withdrew to New York where either Elizabeth or the middle sister, Ann, was very sick. It may have been for this reason that "Old Madam Kent, a worthy good Woman," joined them there and went with them to Halifax.[33] Kent moved into the Blowers mansion which the Massachusetts committee of sequestration turned over to him.[34]

After the Evacuation, Father Kent was even more the town father. He served on committees like that to obtain provisions and he was regularly elected moderator of town meeting. The comments of the young men who listened to him on such occasions

[31] J. M. Robbins Mss. (M. H. S.), Mar. 1777.

[32] Jones, *Loyalists*, p. 37.

[33] *Winslow Papers*, St. John, N. B., 1901, p. 116; Coffin Mss. (M. H. S.), Aug. 12, 17, 1783.

[34] *Acts and Resolves*, XXI, 476.

varied from "Old Kent blundered away" [35] to "tho 80, he is alert and gay as if but 18." [36]

In 1777 Father Kent wrote to Edward Winslow (A.B. 1765), the exiled register of probate for Suffolk, a friendly letter asking for the return of the records which he had carried away.[37] However at Halifax the records had passed into the possession of Foster Hutchinson (A.B. 1743) who maintained that he was still judge of probate for Suffolk and refused to surrender the books. After several years of experience with the resulting difficulties Governor Hancock asked Kent to go to Halifax to obtain the records. The choice of agent was a natural one for not only was Kent's family there but he had spent several months in that town before the war. He sailed on May 7, 1784, but the negotiations with Hutchinson dragged out until September and reached a successful conclusion only because of the intervention of Governor Parr. For his services Kent billed the State £6 a month but was finally allowed only £15 for both services and expenses.[38]

Perhaps it was this experience which made Kent feel more kindly about living under the jurisdiction of "that accursed kingdom called Great Britain," or perhaps it was that the two Elizabeths and Sarah insisted on remaining in Halifax and that the only son, Benjamin, had disappeared.[39] At any rate, when Father Kent returned to Boston in April, 1785, it was to wind up his affairs there in anticipation of spending the rest of his days at Halifax. One of his State Street houses he sold to Dr. Samuel Danforth (A.B. 1758). In June, 1785, he returned to Halifax where he died on October 22, 1788, and was buried in St. Paul's cemetery. Of his taking off, an old friend, Mrs. Elizabeth Partridge, wrote:

Our Friend Mr. Benjamin Kent has taken His departure but for what Land is uncertain. He thought He should be one of the Happy Few that escaped Stoping at Purgatory I wish he may not be Mistaken but have arrived safe at the Elisian Fields. . . .

To this Benjamin Franklin replied:

[35] 6 *Coll. M. H. S.* IV, 188.
[36] William Pynchon, *Diary*, Boston, 1890, p. 138.
[37] *Winslow Papers*, p. 17.
[38] 2 *Proc. M. H. S.* XVI, 118–20.
[39] See L. Vernon Briggs, *Genealogies of the Different Families Bearing the Name of Kent*, Boston, 1898, pp. 47 ff.

Our poor Friend Ben Kent is gone; I hope to the Regions of the Blessed, or at least to some Place where Souls are prepared for those Regions. I found my Hope on this, that tho' not so orthodox as you and I, he was an honest Man, and had his Virtues. If he had any Hypocrisy it was of that inverted kind, with which a Man is not so bad as he seems to be. And with regard to future Bliss I cannot help imagining, that Multitudes of the zealously Orthodox of different Sects, who at the last Day may flock together, in hopes of seeing [Kent] damn'd, will be disappointed, and oblig'd to rest content with their own Salvation.[40]

SOLOMON PRENTICE

SOLOMON PRENTICE, minister of Grafton, Easton, and Hull, was born in Cambridge on May 11, 1705, the oldest surviving son of Solomon and Lydia Prentice whose farm included the site of the Botanic Garden. The lad proved to be a quiet and frugal undergraduate, and on January 16, 1725/6, he joined the First Church of Cambridge. During his Senior year he held the post of monitor and received some aid from the Hollis funds. A Hopkins scholarship brought him back to college for another year during which he studied theology under his pastor, Nathaniel Appleton (A.B. 1712). In 1730 he was awarded his M.A. in absentia, being listed on the *Quaestio* sheet as holding the affirmative of "An Pœnæ Infernales, magis in damno, quam in sensu Consistant?" He was probably preaching at Lunenburg, where he performed an errand for the college.

Not long after this Prentice was called by the absentee proprietors and the nine white families living in the Indian settlement of Hassanamisco to be the minister there. They offered a salary of £90 but he held out for £100 and showed testimonials signed by parsons Appleton, Trowbridge (A.B. 1710) of Groton, and Parkman (A.B. 1721) of Westborough. The plantation, with the absentee proprietors voting to contribute toward his salary until a town government with the power to tax could be organized, agreed to give the candidate £100 a year with land, wood, and grazing privileges thrown in. The new church was organized on December 28, 1731, and Prentice was ordained the following day.

That Prentice was married then or very soon after is indicated by an attested statement made several years later to the effect that "he said to his Mother Clark on the Day of his Wife's Funeral. . .

[40] Benjamin Franklin, *Writings* (Smyth ed.), IX, 683.

Mrs. Clark, your Daughter is dead, and you are glad of it, are you not." [1] The wife who was known to his grandchildren was Sarah Sartell, a daughter of Captain Nathaniel and Sarah Sartell of Groton. She was a genius who knew most of the Bible by heart and could, it was said, preach as good a sermon as any man. They were married on October 26, 1732. The Captain, a wealthy Huguenot immigrant, was pleased with the match and promised to educate their children, but, thanks to another relative, his will was buried with him and his grandchildren were thus cut off. The Prentice descendants fought the case until the town of Groton hired Daniel Webster to stop them — but that is a century ahead of our story. [2]

The incorporation of the plantation of Hassanamisco as the town of Grafton in 1735 was a step in the direction of civilization, but the parish was still wild enough for the parson to shoot a bear on the way to church. His subscription for the *Chronological History* of Thomas Prince represents culture, but his subscription to the Land Bank inflationary scheme suggests frontier optimism. He was one of the few Land Bank subscribers who paid up when that scheme crashed. [3]

Prentice welcomed the Great Awakening and enthusiastically recorded its progress in Grafton:

Our Young People which were but few very much adicted to frolicking and mirth. And this I often Warned them against, and pressed upon them the Expediency & Necessity of Remembring their Creator in the Days of their Youth; it was very Ineffectual till in March 1740/1 two young Men came up from Cambridge, who had been wrot upon by the Preaching of the Rev. Mr. Whitefield and Tennant; who were invited by our youth to attend their Sabbath Evening Meeting . . . at which Meeting these Youth were desired to Relate . . . to the Society what God had done for their Souls: which they did: and the Lord was pleased to affect our youth. . . . About the Middle of March 1741/2 Mr Buell [Yale 1741] came and Preached two Sermons here: Never did I see so many Tears shed in an assembly before tho as yet no crying out in the house of God. . . . A young woman of about 14 years of Age . . . was in such wracking Horrer, & Distress about her soul,

[1] *A Result of a Council of Churches at Grafton, October 2d. 1744*, Boston, 1744, Article 21.

[2] C. J. F. Binney to J. L. Sibley (Harvard University Archives), Mar. 30, 1877.

[3] *New England Hist. Gen. Reg.* L, 190–1, 312.

which she thot was droping into Hell, that She Cryed out at that rate she might be heard far off. Now She complained of her disobedience to parents, breach of Sabbath, mispence of holy time, pride & Hardness of Heart, enough to Make the Ears of all that heard tingle: in which distress She continued about an hour. . . . While her Soul is in this distres her body Seems as if it would Shook all to peices; when all on a Suddain, her dolful Crys are turned into Surprising Joy & Rejoycing . . . "O my Dear Saviour is Coming is Coming! Behold how he Comes leaping over the Mountains & Skipping on the Hills, like a Lamb that had been Slain." [4]

By 1742 Prentice was very much at odds with Israel Loring (A.B. 1701) and other conservative ministers of the neighborhood who did not like to have uninvited exhorters whipping their congregations into a frenzy. In February, 1742/3, Prentice, Daniel Bliss (A.M. 1738), and Eliab Byram (A.B. 1740) were working the town of Mendon into a state of wild confusion by their preaching,[5] and in July of that year Prentice was one of the ministers who endorsed the revival without reservation. Even the New-Light clergy thought that he was going too far when he said that one ought not to love the unconverted, that hardly one in four was converted, and that many unconverted ministers were leading their flocks to hell. He held that the converted could tell the unconverted at a glance, and he eyed some of his colleagues when he said it. Once he said that some of his flock who stayed away from church did so because of "their cursed Wills" and that "he pray'd they might be struck dead in a Moment in their Sins." Moreover, he claimed the authority to judge his sinners down to hell.[6] On the complaint of the people of Grafton an ecclesiastical council met there on October 2, 1744, and without recorded dissent reproved the minister for his conduct. At the moment he accepted the rebuke, but as he was "very much in Raptures" one day and deep in depression the next, it is not surprising that he was soon conducting himself as of old and justifying himself by saying that "he never did come in with" the council.[7] He joyfully greeted George Whitefield, and in July, 1745, with Mrs. Prentice accompanied him on a preaching raid to Northampton.[8] The town

---

[4] Solomon Prentice, Narrative (Mass. Hist. Soc.), pp. 3, 8.
[5] Boston Evening Post, Mar. 14, 1743, p. 2.
[6] A Result of a Council of Churches at Grafton, 1744.
[7] Ebenezer Parkman, Diary (Am. Antiq. Soc.), Jan. 15, 1745.
[8] Ibid., July 18, 1745.

of Grafton continued its complaints which finally resulted in another church council which met there on February 25–7, 1747. Before this body Prentice "was convinced of his Errors & Misconduct and was ready to come into some Retractions," but some of the parish "would not consent to receive him again as a Minister." [9] Again the ministers, although unanimously against Prentice, smoothed things over temporarily. However, in March an uneducated itinerant preacher, Solomon Payne, invaded Grafton and made so much trouble that Prentice decided that he would have to resign.[10] It was a case of dog bite dog, and the conservatives enjoyed it. On March 12th the ministerial association told him that some of his people refused to accept his retraction of his heresies and advised that he ask dismission.[11]

Meantime more troubles were piling upon poor Prentice. His wife's visions and revelations had reached the point where she decided that she was immortal and, formally seceding from her husband's church, she joined the sect of "Immortals" which followed Shadrack Ireland. It was typical of this group that it tried to get even with the church of Hopkinton by marching around its meetinghouse and blowing rams' horns to bring it down after the manner of the walls of Jericho. Mrs. Prentice was as mad as any and, it was said, used to lie with "Old Ireland" as "her spiritual husband." [12] Under the circumstances it was not surprising that the minister took to beating her, tearing her clothes, and punching her fellow Immortals. In the face of pending civil prosecutions, an ecclesiastical council which met on July 7–11, 1747, could only ratify the fact of his dismission. Typically he turned against the ministers who, while unanimously reprobating his conduct, had for years held him in his pulpit.[13]

The council had voted not to recommend Prentice for another job, but he was soon preaching at Bellingham, and early in August, 1747, the church at Easton, which had fundamentalist leanings, was sounding him out. On August 28 the Easton church gave a formal call, and on September 14 the town of Easton voted to offer him a salary of £230 with the ministerial land "to plant

[9] *Ibid.*, Feb. 25, 1747.
[10] *Ibid.*, Mar. 9, 1747.
[11] *Ibid.*, Mar. 12, 1747.
[12] Ezra Stiles, *Itineraries*, New Haven, 1916, p. 418.
[13] Parkman, July 7–11, 1747.

and sow or moo or pasturing, together with cutting of fierwood
for his own fier and fencing stufe for to fenc the ministerial land
with all." He offered to exchange his right to the ministerial land
in return for more salary, so on October 2 Easton offered £400,
"Beef att twelve pence per pound to be the standard for to Easti-
mate said salery." He was still at Grafton three weeks later when
his eldest son was killed in a blasting accident.[14] On November 2
he accepted the Easton call, and on November 18 he was installed.[15] Some voices were raised against him but the installing
council was controlled by Nathaniel Leonard (A.B. 1719) who
was that anomalous creature, a theologically-liberal revivalist.[16]

Among Prentice's first acts was the reassertion of the judicial
authority of the church over the sinners of the parish. He joined
in driving Daniel Baker (A.B. 1706), an Old-Light, from the
Norton pulpit, and showed his appreciation of Jonathan Edwards
by subscribing for his *Life of Brainerd*. However, his old troubles
had followed him, for he made an entry in the church book to
the effect that his wife was a dissenter from the constitution and
doctrine of the New England churches. Two years later he indignantly added: "She is an Anabaptist. She was immersed by
a most despicable layman. . . December 5, 1750, her husband
being absent." [17]

For three years the relation of the parson and the parish were
pleasant. He turned back a portion of his salary for the use of
the poor, and when it was voted to build a new meetinghouse
he, although tax-exempt, offered to pay toward the cost a sum
equal to the payment of the fifth-largest taxpayer. However, the
meetinghouse issue was to ruin his ministry. The town meeting,
by a small majority, voted to build the new meetinghouse in a
more central location. On second thought, a large minority of the
town voters and two-thirds of the church members decided that
they liked the old site better. The minister sided with the church
majority and when asked to announce that services would be
removed to the new meetinghouse he refused to do so. On November 11, 1750, he preached the last sermon in the old meeting-

---

[14] *New York Weekly Journal*, Nov. 9, 1747, p. 3.

[15] Prentice's own account is in Mass. Archives, XIII, 222.

[16] Early Files in the Office of the Clerk of the Supreme Judicial Court, 63,590.

[17] William L. Chaffin, *History of the Town of Easton*, Cambridge, 1886, p. 136.

house but he refused to use the new one. Although he had originally approved of the new site, and no objections to it had been made until the building had begun, he accepted a vote of the majority of the church members that they should meet in private houses "until there Could be a Meeting House Built they Could attend in." [18] The town meeting tried to break the deadlock on December 27 by sending the parson a respectful invitation:

Sir we would make our Humble address unto you in behalf of the Town of Easton that You Would attend the Publick worship of God in our new meeting house . . . which we Humbly Request. [19]

To this he replied:

I must Scruple Your athority by proper Deligation from the Town Determining to Attend the Publick Worship of God in said new Meeting house to treat with me and with me to Determin & Settle Said Momentuos affair until which Scruples is Removed it must (at least) be Expected I must attend on the appointment of the Church. [20]

The town replied by stopping the minister's salary on January 15, 1750/1, and appealing to the General Court. His reply to the order of the General Court to show why the petition of the town should not be granted shows that the only issue was the supremacy of church over town: "Finaly Beging your Honours' Protection of the Churches, and their Pastors, in their Rights, & that your Honours Will not see nor suffer their Privileges to be Curtailed." [21]

Without waiting to hear the decision of the General Court, Prentice gave the land for the construction of a new meetinghouse by the church faction. A party of the inhabitants, seeing the building going on,

asced Mr Prentis wheither they would go forward with building there meeting House and he said he se nothing to hender: and they said sit may be the Cort will send a Commitey to pul it down: and Mr Prentis made this reply let them Come into my feild I will breake thear Heads . . . split theare braines out. [22]

[18] Mass. Archives, XIII, 223–4.
[19] Supreme Court Files, 67,406.
[20] *Ibid.;* Mass. Archives, XIII, 712.
[21] Mass. Archives, XIII, 222–4.
[22] *Ibid.,* XIII, 760.

On April 13, 1751, the General Court ruled that if Mr. Prentice persisted in his refusal to conduct services in the town meeting-house, the town was no longer obligated to support him as its minister. Immediately upon reading the order of the General Court, Prentice called upon the selectmen to summon a town meeting to dismiss him. This was absurd, for the town had no authority to dismiss a minister, although unquestionably it did have the right to fix the site of a meetinghouse. The town took the proper step of calling an ecclesiastical council to dismiss Prentice. This body, with the church of Easton refusing to cooperate, met on June 4 and tried gently to bring the parson to reason. Upon his obstinate refusal to give information, the council advised the town to give him a few weeks to come around and to call another council to dismiss him if he did not.[23] In defiance of this the church completed its meetinghouse and the minister began to preach in it. In a formal meeting on July 1 it demanded of Prentice's opponents "satisfaction . . . for those Scandalous & Sinfull Reflections they had cast upon him," which being refused, the critics were suspended from church membership.

An ecclesiastical council called by the town met on July 9, 1751, but the church refused to participate beyond sending a committee to defend it. The council found that Prentice was "too much given to wine, or strong drink," and had "Spoken unworthily, contemptuously, & Even A-d-sly of the great & Generall Court;" and it ruled that he should be dismissed if he did not admit his mistakes and move his services into the town meetinghouse.[24] The church committee reported that this Result was "founded upon falsehood and Lies, to the Damage and Defameing both Pastor & church." Deacon Edward Hayward, the leader of the town faction, made a move toward reconciliation by apologizing, but the church replied by voting that he "should be thrust out from all the offices he did sustain." The Hayward minority of the church then, acting as though it were the church corporation, joined with the town in voting to dismiss the minister. Now, however, he refused to accept the dismission and with his faction called another ecclesiastical council which met on September 24, 1751. This body cleared him of the charge of "Tipling and Vain

[23] Mass. Archives, XIII, 710.
[24] Ibid., 718–9.

Conversation" on training days, but held that he and his friends had been sinfully obstinate and urged that the two factions join services in the town meetinghouse.[25] In accordance with this recommendation the two parties renounced their errors and agreed to unite "to all intents as if there had niver been no discord among them." [26]

This accord lasted until the next communion day, when the Hayward faction of the church refused to partake with the Prentice group. The minister then led a movement to split off the eastern part of the town as a separate parish, which the town of course resisted. The ecclesiastical council met again on April 23, 1752, and found that the man who was making the moral charges against the minister was lying, but it also reproved Prentice for his extraordinary and unbecoming language before them and ordered him to cease meddling in the civil affairs of the town, particularly by his efforts to split it into two parishes.[27]

After this the neighboring ministers washed their hands of Prentice and refused to participate in more councils at Easton. The town faction now refused to accept reconciliation, so on October 17, 1752, the church voted "to Renounce and come off from the broken Congregational Constitution, and Declare for and come in with the Disapline and order of the Ancient and Renouwned Church of Scotland." [28] On November 5 "Mr prentes preacht the Last or fare well sermon in the Towns meeting house, and sayd that we shuld se his fase nor hear his voise nomore in that hous as menestor." The town thereupon refused to pay his last six months' salary, sued him for breach of contract, and taxed his followers for the support of the preachers which it hired for the town meetinghouse. Prentice in turn sued the town for large sums and two years later finally won a small judgment in the Superior Court.[29]

In the meantime the Prentice faction sent emissaries to Londonderry, New Hampshire, and obtained admission to what was then known as the Irish Presbytery, the churches of which claimed in-

[25] *Ibid.*, 720; Supreme Court Files, 68,528.
[26] Supreme Court Files, 72,813.
[27] *Ibid.*, 69,271.
[28] Mass. Archives, XIII, 722-4, 746.
[29] Supreme Court Files, 72,813; Records of the Superior Court, 1753-1754, p. 234.

dependence of the governments of the towns in which they were located. However, on November 12, 1754, Prentice was "Silenced pro Tempore by the Irish Presbytery . . . for adhering too much to his Wifes notions of Immortality etc." [30] This has been described as an example of Presbyterian intolerance of the Baptists, but as good a Congregationalist as Ebenezer Parkman investigated the case and recorded:

I talkd closely with him of his Wife's pretence to Immortality: he gives in to it, and thinks She is, as She declares, in the Millennium State. I also enquird strictly into their Sentiments and practices respecting their Conjugal Covenant. He utterly denys Every Thing of uncleanness, Fornication, or adultery among them.[31]

The Boston Presbytery also disavowed Prentice,[32] and to his great indignation, the subject of his suspension was never reopened. His own version of the affair is this:

Because I had received a few of my fellow creatures (and fellow christians, so far as I knew) into my house, and suffered them to pray and talk about the Scriptures, and could not make any acknowledgement therefor, to some of my Brethren who were offended thereatt, nor to the Presbytery, Voted, that he, the said S. Prentice be suspended from the discharge of the public ministry, until the Presbytery meet again next April. Because by said vote I was deprived of the small subsistence I had among my people in Easton, I thought it necessary for the Honour of God, and good of my family, to remove to Grafton, which accordingly was done, April 9th, 1755.[33]

The "Rev. & crazy" Mr. Prentice, as his Harvard contemporaries called him, still found those who were willing to listen to his preaching, particularly in Bellingham and Hull. In the latter town, which was somewhat mad in an ecclesiastical way, he was installed on March 21, 1768, on an annual salary of £26 12s 4d. There he remained for four years, after which the sea air drove him back to Grafton. Here his old friend Parkman visited him again:

Thence I went to visit old Mr. Prentice . . . he being far gone in a Dropsy. Found him pleasant, & his wife also Sociable. She Speaks of her Husband under the name of Brother Solomon: She gave me Some

[30] Josiah Cotton, Memoirs (Harvard College Library), p. 431.
[31] Ebenezer Parkman, Diary (M.H.S.), July 26, 1755.
[32] Mass. Archives, XIII, 737.
[33] Samuel H. Emery, *Ministry of Taunton*, Boston, 1853, II, 199–200.

account of the wonderful Change in her Body — her Sanctification — that God had shewn to her His mind & Will — She was taught hence-forth to know no man after the Flesh — that She had not for above 20 Years — not so much as Shook Hands with any Mann. . . . I could not know much about Mr. Prentice.[34]

The remainder of the old minister's days were quiet. The Solomon Prentice who was taken up for passing counterfeit dollars,[35] a common employment in that region, was his son, who later moved to North Carolina.[36] The elder Solomon died at Grafton on May 22, 1773, and his widow (after making people believe that there really was something in her belief in her immortality) died at Ward, now Auburn, on August 28, 1792.

JOSIAH QUINCY

COLONEL JOSIAH QUINCY was born on April 1, 1710, in the mansion which his immigrant great-grandfather, Edmund Quincy, had built in Braintree. His father was Judge Edmund (A.B. 1699) Quincy and his mother was Dorothy, sister of Tutor Henry Flynt (A.B. 1693), and daughter of Josiah Flynt (A.B. 1664), for whom he was named. In July, 1722, little Josiah was sent to live with the Reverend Benjamin Wadsworth (A.B. 1690) for six months while going to school in Boston.[1] When he went over to Harvard, a

---

[34] Parkman, Diary (M. H. S.), Feb. 23, 1773.

[35] Ibid., Mar. 16, 1773.

[36] For the children see C. J. F. Binney, *History and Genealogy of the Prentice, or Prentiss Family*, second edition, Boston, 1883, pp. 14 ff. For the estate see Worcester Probate Records, XII, 345.

[1] Benjamin Wadsworth, Commonplace Book (Mass. Hist. Soc.), p. 37.

jump ahead of President Wadsworth, his father urged Brother Flynt to find him a study with some "stay'd" person, preferably a Master of Arts. He was a disorderly lad in spite of the fact that Uncle Henry kept an eye on him, paid his minor fees and his horse bait, bought him a knife, fork, candles, and a lock for his chest, lent him a mug, and saw to it that his money was "carefully buttoned into his breeches pocket." [2] Under this oversight Josiah achieved his two degrees, although he did not appear in person to take his M.A. in 1731. Perhaps his absence was connected with a natural unwillingness to argue in public his *Quaestio*, which was a denial that there was any change in the organs of Balaam's ass when it spoke.

While riding the Superior Court circuit, Judge Quincy met and was charmed by Hannah, the daughter of his colleague John Sturgis of Yarmouth.[3] When he got back to Braintree he advised Josiah to ride down and court the young lady without delay. Josiah was dutiful and Hannah was willing; they were married on January 11, 1733. New England towns then had a practice of demonstrating democracy by electing elegant young men to disagreeable offices, and Braintree promptly elected the bridegroom a constable. He refused to serve, was reelected, and fined for not serving. His excuse was that he was about to remove to Boston where he would be nearer the seat of business of the partnership in which he and his brother Edmund (A.B. 1722) were engaged. He moved, and in 1736 was elected constable by the town of Boston. This time he was let off with an excuse, but seven years later he was reelected and fined for not serving.

The truth was that Josiah could not well hold town offices, for most of the time he was on the road, or rather on the sea, for the firm of Quincy, Quincy, and Jackson, the last being brother-in-law, Edward Jackson, of the Class of 1726. On December 20, 1737, Josiah sailed for England with his father, who was commissioned to represent the Province in its border dispute with New Hampshire. They went up to London and were enjoying the company of fellow

---

[2] Henry Flynt, Diary (Harvard University Archives), *passim*. Harvard has also one of his college texts, his copy of Adrian Heereboord, *Meletemata Philosophica*.

[3] Edmund Quincy, *Life of Josiah Quincy*, Boston, 1868, p. 6.

JOSIAH QUINCY

New Englanders when the Judge died suddenly as the result of an inoculation. It was Josiah who broke the news to the family:

How shall I begin the mournfull Story? Our Dear Father is Dead! ever since last Fryday evening . . . he has grown worse & worse till this morning about 5 a Clock he left this world. . . . Youl readily suppose this Awfull stroke will oblige me to leave this place Sooner than I intended so that you may expect me in midsummer next.[4]

Within a week the kind administrations of friends had revived Quincy's interest in Europe:

I find my self surrounded with a great many sincere good Friends especially in good Coz Phillips's Family where I now Lodge and by whom I am treated more like a Brother then a Stranger. . . . I yesterday din'd with Mr. [Slingsby] Bethel and among other things acquainted him with my W. I. Scheme which he much approves of. . . . I have a very Strong inclination to see Holland before I return to Boston and if I find I can & not be late home shall certainly go.[5]

In London he made business contacts with Bethel, who was later Lord Mayor, and in Amsterdam he made similar connections with the house of Hope. To his partners at home he wrote:

I heartily wish the goods may come to a good Market that so if possible we may give encouragement to our Employers for I do Assure you our Fate for Comission Business very much Depends on this Tryall of our Abilities: and the Harder the Times are the greater our Merits will be if we can give satisfaction especially to those who have done us justice in putting up their goods.[6]

Quincy's travels also carried him to Paris, Cadiz, and Leghorn where we have one startling glimpse of him talking Latin and passing himself off in clerical company as a priest.[7] He escaped the Inquisition and returned to Boston, but revisited Europe in 1740, 1741, and 1742.

Josiah's mansion stood on the corner of what are now Washington and Avon streets, and his gardens extended back to join those of Edmund's house on Summer Street. In his home he entertained

---

[4] Josiah Quincy, Jr., Mss. (M. H. S.), I, 29.
[5] *Ibid.*, 30.
[6] *Ibid.*, 31.
[7] John Cremer, *Ramblin' Jack*, London [1936], p. 221. As Cremer tells the story, the date and the details are obviously wrong, but under some such circumstances he met a Harvard Quincy who could have been no other than Josiah.

distinguished guests, including the four Mohawk chiefs who attended the Commencement of 1752.[8] Another guest described the hospitality which he encountered at Josiah's hands:

We Dined with Several Gentlemen and Ladies, After dinner Past the time Very Agreeable being very Merry upon Various Subjects untill about 4 a Clock and half an Hour after Six Mr. Quincy Waited on me According to Appointment to go to the Assembly, He being Steward or Master of Ceremonies a Worthy Polite Genteele Gentleman, The Assembly Consisted of 50 Gentlemen and Ladies and those the Best Fashion in Town Broke up about 12 and went home.[9]

Naturally Josiah's name is connected with all kinds of good causes. He contributed £20 toward the purchase of the bells for Christ Church,[10] and £100 toward the establishment of the Linen Manufactory, a work-relief project.[11] Like most such solid citizens, he was a hard-money man, subscribing to the Silver Bank [12] and to the agreement not to accept the inflationary Land Bank bills.[13] His name heads the petition which a group of minor Boston merchants presented to the royal Treasury to have the Louisbourg money used to retire the Massachusetts paper currency.[14]

From a warehouse which it had bought from Spencer Phips (A.B. 1703) on the town Dock, the firm of Quincy, Quincy, and Jackson carried on a very diverse commission business. Josiah put up a good political battle to have himself appointed secretary to General William Pepperrell, no doubt with an eye to the fat contracts sure to result from the Louisbourg expedition.[15] Failing at that, in 1748 he went to Paris and obtained from the French government a contract to furnish Louisbourg with supplies when it should be restored, and returned by way of London where he obtained a similar contract to supply the British garrison at Cape Sable. Another type of business in which he was interested was the agency for the sale of condemned prize vessels.

Quincy returned from Europe in 1749 to find that a single stroke of fortune had made him so wealthy that he need no longer travel.

[8] *New England Hist. Gen. Reg.* XIV, 239.
[9] Captain Francis Goelet, *ibid.*, XXIV, 56.
[10] *Ibid.*, LVIII, 69.
[11] *Ibid.*, XLIV, 103.
[12] *Ibid.*, LVII, 281.
[13] *New England Weekly Journal,* Sept. 16, 1740.
[14] Mass. Archives, XX, 446.
[15] 6 *Coll. M. H. S.* X, 101.

The *Bethel*, a little vessel owned by Quincy, Quincy, and Jackson, had by bluff captured a Spanish treasure ship from which they landed in Boston one hundred and sixty-one chests of silver and gold. The town was wild with excitement as the money was carted through the streets under an armed guard of sailors and dumped into Josiah's wine cellar. Josiah then withdrew from the firm, the three partners dividing the sum of three hundred thousand dollars, which made them perhaps the wealthiest individuals in the Province. The town of Boston in 1750 and 1751 elected Josiah to the committee to visit the schools, and also chose him tax collector, an honor which he declined. In 1752 and 1753 he was engaged in legal suits involving slander charges which he unsuccessfully carried to the General Court.

Josiah had early expressed an intention of settling on the "lower farm" of the family estate in Braintree, and his father for that reason had bequeathed him this property and a dwelling house near the river.[16] Tradition has it that in 1750 he retired from business and became a country gentleman, which is a Yankee way of saying that he shifted from trade to manufacturing, and moved to Braintree to keep an eye on the plant. In what was later known as the Germantown district of Braintree he established a glass factory in which he furnished the capital, an Englishman named Joseph Palmer provided the management, and skilled Low German immigrants, the labor. The General Court permitted him to hold lotteries to benefit this factory and a cider mill.[17] Palmer and Quincy also set up a factory to make sperm candles and applied to the General Court for a monopoly. Of the candles, we know only that it was said that they stank, but the glass factory, which was said to be the first in America, flourished until the Revolution.

When Josiah returned to Braintree as Squire Quincy he was not elected constable. On the contrary, in 1753, probably the first year of his renewed residence, he was chosen moderator, the highest honor at the town's disposal. In subsequent years he was reelected and chosen to committees to handle commons, finances, boundaries, and schools. Once he gave the town £50 to mend the highways. In similar ways he served the North Church.

In 1754 and 1757 the town of Braintree sent Squire Quincy to

[16] Suffolk Probate Records, XXXIII, 461.
[17] *Boston Gazette*, Sept. 26, 1757, p. 2.

the House of Representatives where he supported the cause of colonial union.[18] This evidence of breadth of vision caused Governor Shirley in February, 1755, to send him to Pennsylvania to enlist the aid of that colony in a Crown Point expedition. At the same time Thomas Hutchinson (A.B. 1727) was sent to New York and Thomas Pownall to New Jersey. When he arrived in Philadelphia, Quincy carried his problem to Benjamin Franklin, whom he had entertained at Boston the winter before. Franklin tells the story:

As I was in the Assembly, knew its temper, and was Mr. Quincy's Countryman, he applied to me for my influence and assistance. I dictated his address to them, which was well received. They voted an aid of ten thousand pounds, to be laid out in provisions. But the Governor refusing his assent . . . the Assembly . . . were at a loss how to accomplish it. Mr. Quincy laboured hard with the Governor to obtain his assent, but he was obstinate. I then suggested a method of doing the business without the Governor. . . . The Assembly with very little hesitation adopted the proposal. . . . Mr. Quincy returned thanks to the Assembly in a handsome memorial, went home highly pleased with the success of his embassy, and ever after bore for me the most cordial and affectionate friendship.[19]

The remaining years of the war Squire Quincy passed in inglorious service, enlisting recruits, catching deserters, and carrying draftees to Castle William from which they could not run away.[20] By 1762 he had become colonel of his father's old militia regiment, the Third Suffolk. His handling of the many civilian problems of wartime was facilitated by his appointment, in 1757, as a justice of the peace. Five years later he was added to the quorum on which he sat as a judge.

These years saw great changes in the Colonel's personal affairs. His wife, Hannah, died on August 9, 1755, and on February 19, 1756, he married Elizabeth, daughter of the Reverend William Waldron (A.B. 1717). On May 17, 1759, his home burned. "That house and furniture clung and twined round his heart, and could not be torn away without tearing to the quick," said his neighbor,

---

[18] *Journals of the House of Representatives of Massachusetts*, Boston, 1754–55, Dec. 14, 27, 1754.

[19] Josiah Quincy, *Memoir of the Life of Josiah Quincy, Junior, 1744–1775*, Boston, 1874, pp. 379, 380.

[20] Executive Records of the Province Council, *passim*.

John Adams, who added that the Colonel's grief at this loss was more eloquent than that of his brother Ned who had just lost a son.[21] Adams thought that Quincy had polished and graceful manners, was elegant in dress and appointments, and most eloquent in speech. The Colonel had something of the same estimate of himself, for he told his young neighbor:

I learned to write letters of Pope, and Swift, etc. I should not have wrote a letter with so much correctness as I can, if I had not read and imitated them. The faculty has come to me strangely, without any formed design of acquiring it.[22]

General Joseph Dwight (A.B. 1722) gives a different picture of his manner in an account of a visit which he paid to Tutor Flynt:

J. Q. has lately paid his Visit pretending to the old Gentleman That he feared he should never see him again alive fell into many Sobs and Sighs and among the few things he was able to utter in his Great Distress and anxiety for his Good Uncle being scarce able to speak. He told him he come out of pure Affections to see him etc: The Old Man apprehending his real Design and seeing his Hypocracy as he supposed — said, Aye, I doont Caare for you nor your Affections Noother.[23]

It is not impossible that the Colonel's manner on that occasion was connected with the fact that he lost his wife Elizabeth that same year, 1759. On June 26, 1762, he registered his intention of marrying Anne, daughter of the Reverend Joseph (A.B. 1705) and Ann (Fiske) Marsh. She was an intimate friend of her neighbor, Abigail Adams, and, like her, a very able and active woman. On December 6, 1769,[24] the Colonel's new house burned to the ground. He then built the handsome mansion which still stands at 20 Muirhead Street, Quincy, but which then stood in the country commanding a lovely view of Boston Harbor. Fear lest he lose this house tormented him for the rest of his life.

During the first years of Quincy's public life he saw eye to eye with Hutchinson. They served together in 1747 on a Boston town committee appointed to consider the conduct of a mob which had offered insults and abuses to the Governor and Council and had "Committed great outrages and Disorders putting the Inhabi-

[21] John Adams, *Works*, Boston, 1850–56, II, 75, 76.
[22] *Ibid.*, p. 67.
[23] Joseph to Abigail Dwight, Sedgwick Mss. II (M. H. S.), Aug. 20, 1759.
[24] *New England Hist. Gen. Reg.* LXXXIV, 261.

tants of the Town in great Terror of their Lives." The mob was rioting against impressment, but such violence had not become popular with the solid citizens of Boston as it was to be a couple of decades later. The committee reported that this disorder was the work of a "Riotous Tumultuous assembly . . . of Foreign Seamen, Servants Negroes and other Persons of mean and Vile Condition," and that the town had "the utmost Abhorence of such Illegal Criminal Proceedings" and would "to their utmost Discountenance and Suppress the same." [25] However, Quincy and Hutchinson parted when the former was appointed to a town committee to vindicate Boston from the aspersions of the Governor in regard to that same affair.

Quincy's experience as an army contractor in London made him indignant at the ignorance of American affairs shown by royal officials:

The Ministerial Writers have lul'd the Nation to Sleep about the Importance of Cape Breton as a Place of little or no Consequence to Great Britain and of none to France only as a fishing Harbour. . . . There was not One Gentleman in the Kings P - - - y C - - - - - l that was Geographer enough to find out Chebucto Harbour when advice came that the Duke D'Anvile was there tho they had a large Map of America lying before them and I heard that One of the Gentlemen ask'd if it did not lye some where about Cape Horn. [26]

During the French and Indian War he shared Hutchinson's grasp of the whole picture and took such an interest in the European aspect of the struggle, that he rejoiced extravagantly at the victories of "Our beloved Hero the King of Prussia." This statement, in view of the fact that Prussia had been up until then a traditional foe of England, shows the emotional weakness which marked the political thinking of that generation of Quincys.

Once the political parties in Massachusetts had formed, there was never any doubt of the stand of Colonel Quincy. He contributed £7 to the fund collected to square the accounts of Sam Adams with the town, [27] and it was not he but his partner Engineer Gridley who had to be attached to the Whig cause by a loan to pay off the mortgage on a furnace which they owned. [28] When Boston

[25] *Boston Town Records, 1742 to 1757* (*Record Commissioners*, XIV), p. 127.
[26] Josiah Quincy, Jr., Mss. I, 35.
[27] *New England Hist. Gen. Reg.* XIV, 262.
[28] *Ibid.*, XXX, 307.

called upon the other towns to send representatives to a convention to consider the political situation in September, 1768, Colonel Quincy was one of those chosen by Braintree, and the next year his town elected him to the committee to instruct its representative in the General Court. Apparently he was one of the Sons of Liberty, an organization of which his son Josiah (A.B. 1763) was one of the leaders.

The Colonel was horrified to hear that Josiah had agreed to act as defense attorney for the soldiers concerned in the Boston Massacre:

My dear Son,

I am under great affliction at hearing the bitterest reproaches uttered against you, for having become an advocate for those criminals who are charged with the murder of their fellow-citizens. Good God! Is it possible? I will not believe it. . . .

I must own to you, it has filled the bosom of your aged and infirm parent with anxiety and distress, lest it should not only prove true, but destructive of your reputation and interest; and I repeat, I will not believe it, unless it be confirmed by your own mouth, or under your own hand.[29]

That the mob which he had denounced in 1747 had a similar personnel and a better cause than that of 1770, he did not notice. Josiah ably vindicated himself although he was as much a leader of the Revolution in Boston as his father was in Braintree. When political trouble flared up again in 1774 the town of Braintree appointed the Colonel to committees to report on public affairs and to draw up a town covenant.

The fact that the Colonel's second son, Samuel (A.B. 1754), was a firm Loyalist, made Josiah all the more the apple of his father's eye. When Josiah was sent as the ambassador of the Massachusetts Whigs to England in October, 1774, his proud father wrote to him:

Perhaps, there never was an American, not even a [John] D[ickinson] nor a F[rankli]n, whose Abilities have raised the Expectations of their American Brethren more than yours. God Almighty grant, if your Life and Health is spared, that you may exceed them in every Respect.[30]

[29] Josiah Quincy, *Memoir*, pp. 26, 27.
[30] *Proc. M. H. S.* L, 480.

His letters of exhortation to Josiah in London take the same extreme view of the political situation.

What! have we Americans spent, so much of our Blood & Treasure in aiding Britain to conquer Canada, that Britons and Canadians in Conjunction, may now subjugate Us? forbid it Heaven! Surely the Heart of a Canadian Savage wou'd recoil at the horrid Thought.[31]

Josiah's mission to London revived the correspondence between the Colonel and Franklin on political and economic subjects. Concerning the Loyalist group to which his son Samuel belonged, Quincy wrote:

You would hardly be persuaded to believe, did not melancholy Experience evince the Truth of it, that such a number of infamous Wretches could be found upon the Continent, as are now group'd together in Boston, under Pretence of flying thither from the Rage of popular Fury; when every Body knows, and their own consciences cannot but dictate to them, that all they aim at, is, to recommend themselves to the first Offices of Trust and Power, in Case the Plan of subverting the present Constitution, and establishing a despotic Government in it's Stead, can be successfully carried into Execution.[32]

To the tragedy of the division of the Quincy family was added the death of Josiah as he returned from his mission in April, 1775.

The Massachusetts militia was a training and reserve organization rather than a group of fighting units, so Colonel Quincy and his regiment did not engage in the siege of Boston. Rather, they were assigned such tasks as gathering hay at Squantum for the army, an assignment which the Colonel protested bitterly because it left Braintree exposed to raids.[33] As the first months of the war passed his worries became an obsession:

Scarcely a Week has past without our being alarmed either from the Land or Water: great Part of our valuable Effects are removed for although we have 5 Companies stationed near us, yet the Carcases thrown from the floating Batteries, & the flat Bottom Boats, that row with 20 Oars, carry 50 Men each, and are defended with Cannon & Swivels; keep us under perpetual Apprehension of being attacked.[34]

From a shipwright who had been a slave in a Turkish galley, the Colonel obtained the idea of building similar vessels to carry on a

[31] Josiah Quincy, Jr., Mss. 1, 80.
[32] Misc. Mss. Bound (M. H. S.), Mar. 25, 1775.
[33] Thomas Mss. (M. H. S.), p. 34.
[34] Josiah Quincy, Jr., Mss. 1, 103.

dodging Indian-fight among the islands of Boston Harbor and to serve as floating batteries. He sold his idea to the Massachusetts leaders, but Washington, who had become a personal friend, gently explained that there just were not enough cannon to arm such boats.

From the windows of his mansion he observed the movements of the British fleet and reported to Washington. While watching he scratched upon a pane, "October 10, 1775, Governor Gage sailed for England with a fair wind." As the situation of the besieged became worse, the Colonel redoubled his pleas to Washington for protection against the expected reprisals by sea:

Where shall we indeed, where can we go for Relief, but to you Sir! The whole force of the Continent is under your Command, and at your Disposal. Let us not therefore plead in vain for that Help which is nowhere else to be found! We beseach you to grant us Protection before it is too late which we fear it will be if not speedily granted! [35]

His joy at the surrender of Boston was lyrical, and he wrote congratulating Washington:

Nothing less than an inveterate nervous head-ache has prevented my paying, in person, those compliments of congratulation which are due to you from every friend to liberty and the rights of mankind, upon your triumphant and almost bloodless victory, in forcing the British Army and Navy to a precipitate flight from the capital of this Colony. A grateful heart now dictates them to a trembling hand, in humble confidence of your favourable reception. . . . If my wishes must not be gratified, either in a visit to, or from your Excellency, the best I can form will constantly attend you.[36]

He was terrified by Washington's reply warning him that the retreating British might raid as they went, and asking him to hire a dozen natives to haunt the lines of communication in order to detect spies.

After the shifting of the scene of the war from Boston, the arrival of the French fleet provided the Colonel with a very pleasant experience. He dined on the flagship with the Comte d'Estaing and entertained the officers in the parlor of his mansion while the sailors sat in the kitchen and helped the maids with their knitting.

[35] Sparks Mss. LXVI (Harvard College Library), Letters to Washington, I, 30–2.
[36] Peter Force, *American Archives*, ser. 4, vol. 5, col. 455, 456.

When the Massachusetts courts reopened, Quincy was again active as a justice of the peace, for his commission was early renewed by the rebel government. In the affairs of the new State he was a conservative and was elected to the town committee to instruct the representative how to act in regard to the proposed constitution. As to interstate affairs he had lost none of his breadth of vision of twenty years before, for as early as April, 1776, he was urging Franklin's support of a strong central government:

When is the Continental Congress by general consent to be formed into a supreme legislature; alliances, defensive and offensive, formed; our ports opened; and a formidable naval force established at the public charge? [37]

Neither had his old views as to hard money undergone any change. At the outbreak of the war he had loaned the town £150 in specie, and when in later years people became reluctant to accept the depreciated paper, the town again turned to him for hard money to pay the enlistment bounty. Suits against him because he refused to accept State and Continental bills from his debtors were carried to the General Court.[38] His defence was that financing the war by inflation was a violation of Locke's axiom that the primary purpose of government was the preservation of property.[39]

Naturally his property included Negro slaves, toward whom he took the normal point of view. Early in the war, because of his "grateful sense of the faithfull services" of his "black female servant," he asked James Bowdoin to try to find her absent husband.[40] As late as 1779 he was asking General Gates to assist him in recovering this or a similar piece of "property" which had run away and joined the army.[41] If he opposed emancipation in Massachusetts, which he lived to see, there is no record of it.

On March 4, 1782, the historian, William Gordon, wrote to Washington regarding their mutual friend at Braintree: "He is breaking fast; but the powers of his mind remain strong. I wish he may live to see and enjoy a happy peace; but I much question

---

[37] Franklin, *Writings* (Smyth ed.), VI, 446.

[38] *House Journals*, June 14, 1777, and Feb. 23, 1778.

[39] Josiah Quincy, Jr., Mss. I, 104, 105. There were copies of several essays on public topics which he wrote at this period among his Mss. a few years ago. Washburn Mss. (M. H. S.), XVIII, 51.

[40] 6 *Coll. M. H. S.* IX, 392.

[41] N. Y. Hist. Soc. Mss., May 20, 1779.

it." [42] He did, however, live long enough to receive from Franklin a letter saying, "The Definitive Treaty was signed the third instant. We are now friends with England, and with all mankind! May we never see another war! for in my opinion there never was a good war or a bad peace." [43] The Colonel died, after a short illness, on March 3, 1784.

Quincy's will summed up his character and his beliefs. He ordered that his funeral be "as frugal as the Tyrany of Custom will allow," and that his debts incurred before the depreciation of the "disgraceful ruinous Paper currency" be honorably adjusted according to an act of George II. The children of his son Samuel, from whom he was separated by "the present unnatural war," received special bequests. The American Academy of Arts and Sciences was to receive £500, "the Interest arising therefrom to be annually applyed as a Premium or Reward for the best Literary Production upon the subject of Justice its Nature its Ends and inestimable Importance to the Happiness of Mankind in civil Society," but this clause was replaced in a codicil by an unrestricted bequest of £100. The mansion house went to his grandson, Josiah Quincy (A.B. 1790), the future president, as did the Copley portrait which is here reproduced through the kindness of its present owner, Mr. Edmund Quincy of Boston. The library was not large, but strong in European political works. [44]

JOHN SECCOMB

JOHN SECCOMB, [1] the rhymester, was a son of Peter Seccomb, a merchant of Medford. His mother was Hannah, a daughter of Stephen and Hannah (Eliot) Willis. John was born on April 25, 1708, and in due time proceeded to college where he waited on table, which, in that day of sharp social distinctions, suggests that his family was not then as wealthy as it later became. He lost this job in December of his Sophomore year when, "for stealing & Lying," he "was publicly admonished in the Hall, degraded to the lowest

[42] *Proc. M. H. S.* LXIII, 460.
[43] Josiah Quincy, *Memoir*, p. 425.
[44] The inventory of the library is in Suffolk Probate Records, LXXXIII, 866.
[1] So he spelled his own name; other members of the family, and the college catalogue, have always made it "Seccombe."

in the classe, and turn'd out of his waiter's place." [2] This had small effect on John, for three months later he was concerned with his classmate Phips in the demise, under suspicious circumstances, of "two fowles in the school house."

Sacomb, was examin'd concerning two fowls suspected to be stoln, under Examination he was found guilty of horrid lying (though not of theft) for this he gave in a written confession which was read in the Hall, and he admonished.[3]

After the Commencement of 1728, Seccomb remained at the college as a resident graduate. On November 22, 1728, he and Hinsdell '27 in a debate in the college Hall held against Hancock '27 the negative of the topic, "Libertas competit humano generi." Just five days later the weary Faculty voted "That Sir Lovel, Sir Seacomb, Maudsley, Stoddard & Frost Junior, be punished seven shillings apiece, for contriving to take, & for taking the third Goose lately stolen on the Common." [4] In his letters to Nicholas Gilman (A.B. 1724), Seccomb jovially described other thefts and similar student activities:

All Those that have ever played Cards at College have been found out and had an admonition which made up about 60 odd. . . . The Batchelors Play Batt & Ball mightily now adays which Stirs our bloud greatly etc.[5]

Such exploits he celebrated in mock heroic epics, like that of the goose roasted at "Yankey Hastings'," which were treasured in manuscript for many years.[6]

One of these bits of nonsense verse which celebrated the college sweeper, Matthew Abdy, was picked up by the *Weekly Rehearsal* and published under the title of "Father Abbey's Will" in its issue of January 3, 1732. This was answered shortly by a verse reply from the Yale college sweeper which was at that time attributed to Seccomb, but more recently to John Hubbard of New Haven. Governor Belcher (A.B. 1699) sent copies of these verses to London where they were seized upon with glee. Father Abbey's Will was printed in its original form in the *Gentleman's Magazine* for May,

[2] *Harvard College Records*, III (*Publ. Colonial Soc. Mass.* XXXI), p. 454.

[3] *Ibid.*, p. 457.

[4] Faculty Records (Harvard University Archives), I, 21.

[5] Nicholas Gilman Mss. (Mass. Hist. Soc.), Mar. 30, 1729.

[6] *Massachusetts Magazine*, Aug. 1795, pp. 301–2.

1732 (p. 770), the Yale reply in the issue for June, 1732 (p. 821), and an expanded form of the original in the issue for October, 1732 (p. 1024). Ignorant of the fact that Seccomb was writing nonsense verse, some of the historians of literature have quoted Father Abbey's Will to prove the low level of literary culture in colonial New England. These verses were popular in England where they were reprinted, with changes, in the periodicals as late as the issue of the *European Magazine* for May, 1781. In 1782 the Loyalist Samuel Curwin (A.B. 1735), walking the streets of London a homesick exile, bought from a vendor of broadside ballads one of these altered versions of his old friend's verses.[7] In the colonies they were many times printed in broadside form and were set to music and republished by Oliver Ditson as late as 1850.[8]

In 1729 Seccomb was awarded a Hopkins scholarship, there being more funds available for the encouragement of graduate study than there were likely candidates. In the spring of 1731 he accepted reappointment for a third term, but unexpectedly left college without performing the assigned exercises.[9] However, he was back in Cambridge by Commencement when he entertained the audience by arguing the affirmative of the question, "An Justitia originalis, si Adamus non Peccasset, Posteris suis communicaretur?"

On March 11, 1732/3, Seccomb preached for the first time at the town of Harvard, which had recently been created by detaching fragments from Lancaster, Groton, and Stow. This community, dominated by Squire John Martyn (A.B. 1724), showed every promise of supporting a strong and reasonably rich church. In April it gave Seccomb a call, and offered him a settlement of £300 and a salary of £120 "in Publick Bills of Credit or Province Bills as they are now Currant." In May he gave a rather dubious answer:

Having some time since received your Invitation to Settle with you in the Work of the Ministry, wou'd now thankfully take notice of the Respect which you have shewn me herein, and wou'd signify to you the Pleasure and Satisfaction it has been to me to see you so happily and religiously united and so hearty and Sincere in this your call and invita-

---

[7] This copy is now owned by the American Antiquarian Society.

[8] An edition with historical notes was published by John Langdon Sibley in Cambridge in 1864.

[9] Harvard College Papers (H. U. A.), I, 79 (item 164).

tion. . . . I am Perswaded that you will readily make an Allowance as need Requires so that your Liberality may appear to all men as has begun already to appear. . . . As to the Circumstances of Cutting and Sledding my Wood for me I hope that wou'd not be thought an Unreasonable Desire. I do but hint at these things which I presume is sufficient.[10]

Although cross at the indefinite nature of the candidate's hints, the town voted him an additional £40 in labor. In June, the meeting-house was raised. The caloric expenditure expected may be estimated from the fact that the town for the occasion provided 175 pounds of veal, 110 of mutton, 120 of pork, 2 barrels of beer, and 3 of cider. The church was gathered and the ordination held on October 10, 1733. John Gardner (A.B. 1715) of Stow began with prayer, John Prentice (A.B. 1700) of Lancaster gathered the church and gave the charge, Ebenezer Turell (A.B. 1721) of Medford preached from the text, "Giving no offence in anything, that the Ministry be not blamed," and Caleb Trowbridge (A.B. 1710) of Groton gave the Right Hand of Fellowship. The entertainment of the guests, including nine Seccomb relatives and as many scholars, cost the town £32.

The Seccomb family was outside of the circle of Harvard-educated ministers, judges, and generals, who ruled Massachusetts, but John was established in its center by his marriage on March 10, 1736/7, to Mercy, daughter of the Reverend William (A.B. 1705) and Hannah (Stoddard) Williams, and granddaughter of the Reverend Solomon Stoddard (A.B. 1662). Parson Seccomb built in the town of Harvard a magnificent mansion which with its gambrel roof, several chimneys, and fine landscaping was for more than a century the showplace of the region. Tradition explains its size by the story that Father Williams had made the mistake of promising to furnish as large a house as Father Seccomb would build for the young couple. From the parsonage to the meeting-house was planted what was described as the longest rows of elm trees in New England. Nor was this all, for on an island in Bare Hill Pond Parson Seccomb built a pleasure house which became famous for the lavish and jovial entertainment which he there offered to the ministerial association and other guests. He showed

---

[10] Henry S. Nourse, *History of the Town of Harvard*, Harvard, 1894, pp. 180, 181.

his wealth also in subscribing for good books such as the *Chrono-logical History* of Thomas Prince.

From almost the first Seccomb's ministry was under a cloud. Tradition has it that the trouble was caused by his wife's bringing charges that he had improper relations with a servant, and assumes that he was innocent.[11] Mrs. Seccomb was certainly queer. On one occasion she had a fit when, or possibly because, some of the neighboring ministers called.[12] On the other hand there is the evidence of the fact that Seccomb on January 1, 1738/9, "offered Christian satisfaction for his Offence." A majority of church and town accepted this evidence of reformation, but a minority demanded his removal on criminal grounds. To forestall any such action, the church sent to the neighboring ministers an envoy with this letter of introduction:

Whereas we Understand there are a few (four or five at Most) of the Church and Several of the Inhabitants of this Town about fifteen, and divers of these not Legall Voters, that remain dissatisfied with our Revd. Pastor Mr. Seccomb notwithstanding the Declaration & Acknowledgment that he has offer'd, both to the Church and Town, and have endeavoured, some of them, to raise and Carry on the Uneasiness and have applied themselves to severall of the Revd. Pastors in the Neighbouring Churches, as we hear, for their advice, and not being without apprehensions that a Partial or imperfect representation may have been made of things, We have therefore desir'd our Brother (Capt. Whitney) the Bearer of this to wait upon you with it, and give you a more full Representation.[13]

We can judge the substance of Captain Whitney's verbal message from the fact that Parson Parkman recorded in his diary that the "ill Conduct" of his friend Seccomb was a source of distress to his colleagues, particularly as "Seven Towns in that Neighborhood had had ministers guilty of Scandalous offences." [14]

Some similar disputes can be traced to the fact that the ministers were more liberal than their churches and were opposed to the excesses of the revival which was then beginning. This was not the case in the town of Harvard. Seccomb was in theology an old liberal, and his church had early shown its attitude by voting not

---

[11] W. Gilbert to J. L. Sibley (H. U. A.), Oct. 30, 1854.
[12] Ebenezer Parkman, Diary (A. A. S.), Dec. 22, 1748.
[13] Nathan Stone Mss. (M. H. S.), Feb. 27, 1738/9.
[14] Parkman, Diary, Apr. 10, 1739.

to require either written or verbal evidence of conversion as a qualification for membership. Both parson and church favored the Awakening, which was just then beginning. Seccomb thus describes its workings:

The first visible Alteration among my People for the better was some Time in the Month of September in the Year 1739, when several began to grow more thoughtful and serious. . . . After a while religious Discourse began to be introduced among Persons of Lord's-Days, between Exercises, which had been shamefully neglected, and could not before this be obtain'd. And by many it was look'd upon as a Sign of Hypocrisy, and accordingly such were much scorned by the less serious and considerate among the People. . . . Some while under the Spirit of Bondage were sensibly affected with their Danger that they dare not close their Eyes to sleep lest they should awake in Hell: And would sometimes arise in the Night and go to the Windows under alarming Fears of Christ's sudden Coming to Judgment, expecting to hear the Sounding of the Trumpet to summon all Nations to appear before him.[15]

Seccomb denied that the revival in Harvard was due to exhortation or emotional preaching on his part, or was marked by hysteria or quarrelling. When the Provincial assembly of ministers divided on this issue, he stood staunchly by the Awakening.[16] For several years his meetinghouse was packed, and to it in August, 1745, he brought the great English revivalist, George Whitefield.[17] The Marlborough association of ministers, of which he was a member, was conservative, and when he preached before it a sermon which seemed "to have been composed in A Strain which many would Term New-Light," it "did not go down well with some Gentlemen." [18]

There may well have been no connection between Seccomb's early failings and his final dismission from the Harvard pulpit on September 7, 1757; but on the other hand there is no reason for assuming, as has been done, that the fact that he was dismissed by an ecclesiastical council indicated that his reputation was vindicated. That the town was partly at fault is suggested by his remarkably accurate prophecy that it would never have a minister whose connections with it would not be prematurely severed.[19]

[15] *Christian History*, II (1744), pp. 13–21.
[16] *Ibid.*, I (1743), p. 165.
[17] Parkman, Diary, Aug. 13, 1745.
[18] *Ibid.*, June 17, 1747.
[19] Seth Chandler, *Historical Discourse Delivered Before the First Church Society in Harvard, 1882*, Boston, 1884, p. 11.

Having in his youth been settled in an established community in comfort which most of his classmates must have envied, Seccomb in his age moved to a frontier more bitterly hostile to settlement than any in New England. With the first body of settlers he went to Chester in Lunenburg County, Nova Scotia, and on their first Sabbath in the new land, August 9, 1759, he preached to them from the text, "Moreover, I will appoint a place for my people Israel, and will plant them, that they may dwell in a place of their own, and move no more." [20] His own farm was on an island now known as Seccombe's Island, west of Chester. When a committee was appointed to manage town affairs he was one of its members. [21]

Parson Seccomb was spared some of the misery which was the lot of most Nova Scotia settlers because for twenty years he spent much of his time at Halifax supplying the pulpit of Mather's, a Congregational church originally named for the great Boston family, but in later days y-clept "St. Mather's" and finally "St. Matthew's." [22] His classmate, Chief Justice Belcher, welcomed him to Halifax and with an occasional johannes eased his lot. When Belcher lost his wife, Seccomb preached a funeral sermon which was printed in Boston. Several times he returned to Massachusetts, preached for his friends, visited his old neighbors at Harvard, and gathered supplies. After one such trip early in 1770 he wrote back:

I have enjoy'd a good measure of health ever since I came into Nova Scotia, but have had it in a higher manner since I was last at New England. I took a great deal of pleasure and satisfaction in visiting my friends there, & very particularly at Harvard. I sensibly grew fatter and stronger than before, & continue so to this day. . . . The grain, butter and other things, which were given me by my friends at Harvard were very acceptable, & will be very beneficial to us. . . . You live in a country where there is a plenty of all the necessaries & comforts of life. It is far otherwise with us at Chester, for sometimes it may be said (of some at least) that they are in want of all things. [23]

A friend in Nova Scotia more frankly described the parson's condition in a letter to mutual friends in Boston:

[20] Mather Byles DesBrisay, *History of the County of Lunenburg*, Toronto, 1895, p. 278.
[21] John Bartlet Brebner, *The Neutral Yankees of Nova Scotia*, New York, 1937, p. 53.
[22] Ian F. MacKinnon, *Settlements and Churches in Nova Scotia, 1749–1776* [Montreal, 1930], p. 74.
[23] *Acadiensis*, VII, 339–40.

He has never had any Establish'd Salary, but receives about £20, per annum from his Parish, which contains a few Industrious, but poor People, He has expended all the Money he brought with him into this Country (and which we are inform'd was considerable) in Buildings & other improvements, on a new Farm, which has reduced him to very necessitous Circumstances: He has had some small relief from this Town. We cannot avoid Earnestly recommending this Gentleman, now advanc'd in years, — as an Object very worthy of a Charitable Assistance.[24]

In response to this plea the Provincial assembly of the clergy assigned some of its charitable fund to the Chester minister.

In December, 1776, Seccomb was brought before the Provincial Council of Nova Scotia charged with preaching a sermon "tending to promote sedition." In refutation of this charge he offered to produce the manuscript of the sermon, but he failed to clear himself of the charge that he had prayed for the success of the rebels. In consequence he was forced to give £500 security for his future behavior and was forbidden to preach until he had signed a formal recantation.[25] Either he did recant or the matter was soon forgotten. Although he was described as superannuated and retired in 1784, he still had several years of useful activity ahead of him.

Seccomb was a Calvinist and the warmest friend of the German Calvinist immigrants. A revival in 1788 lead to the merger of his little group of Congregationalists at Chester with the Baptists in a much stronger church. On the subject of baptism the articles of the new church allowed divergent opinions,[26] but in the matter of the Halfway Covenant and the admission of members without public description of their regeneration, it could not accept the minister's liberalism.[27] He therefore declined to join the church, but took charge of it and administered the Lord's Supper.

The kindly humanity of Parson Seccomb endeared him to the people regardless of theological differences. In 1788 an admiring friend wrote to President Stiles of Yale praising

The Revd. Mr. Seccombe . . . whose Abilities as a Poet and Divine highly recommend him as the first Character in this Province; his long

---

[24] 2 *Proc. M. H. S.* IV, 70.

[25] Brebner, p. 340.

[26] David Benedict, *General History of the Baptist Denomination*, New York, 1848, p. 525.

[27] *Christian Messenger* clipping, date line, Halifax, May 16, 1860.

Service and great Age have secured him the Name of being the Father of all the Churches in this Province.[28]

He continued to preach even when so enfeebled with age that he had to be assisted into the pulpit, but on October 29, 1792, this long ministry came to an end. The obituaries credit him with almost every Christian virtue, leavened with the understanding of one who had in youth sinned, more or less, and had never lost the humor of the lad who wrote nonsense verse.[29]

Parson Seccomb appears to have been survived by his wife. They had at least five children: (1) Hannah, b. Mar. 2, 1738; m. Eben Fitch, Nov. 18, 1770. (2) John, b. Apr. 27, 1740. (3) Willis, b. Mar. 3, 1741/2. (4) Mercy, b. Feb. 3, 1743/4. (5) Thomas, b. Mar. 16, 1757.

WILLIAM WILLIAMS

COLONEL WILLIAM WILLIAMS of Pittsfield, Massachusetts, was born on May 14, 1711, in the west parish of Watertown, now the town of Weston. His parents were the Reverend William Williams (A.B. 1705) and Hannah, the daughter of the Reverend Solomon Stoddard (A.B. 1662). "Being of a sprightly genius," he was sent to college where, according to his contemporaries, he formed "a habit of anticipating his income, which lasted as long as his life," a habit said to have arisen "either from the parsimony of his parents, or from their inability to supply him liberally." [1] There is a gendre touch in the fact that one of his term bills was paid by "his father in part Received on the backside of Stoughton College." For three years he enjoyed one of the fat Hollis scholarships which took care of the greater part of his expenses, for he lived in town much of the time and generally avoided trouble. During his Junior year he joined the First Church of Cambridge. He was graduated with his class and returned with it to take his M.A. in 1732, being then prepared to argue the negative of the question, "An Quartum Decalogi Præceptum sit Juris Ceremonialis?"

---

[28] George Gillmore to Ezra Stiles, Stiles Mss. (Yale), Nov. 12, 1788.
[29] *Halifax Royal Gazette*, Nov. 27, 1792, and *Nova Scotia Advertizer*, Nov. 27, 1792, quoted in the *Groton Landmark*, Apr. 7, 1888.
[1] *Vermont Gazette*, Nov. 30, 1789, p. 2.

Williams settled in Boston where he tried both business and medicine without much success in spite of the contacts which he enjoyed because of his social position. In 1733 he was elected to the Ancient and Honourable Artillery, in which he rose to the rank of fourth sergeant. With a number of other merchants, he agreed not to accept Rhode Island paper money.[2] In Boston town meeting he was elected scavenger and constable, but he declined the latter post. His name appears, with those of the other literati, on the list of subscribers for the *Chronological History* of Thomas Prince. On September 27, 1733, he married Miriam Tyler, "a lady of good sence," a daughter of Andrew and Meriam (Pepperrell) Tyler of Boston. The young couple first attended the Church in Brattle Square, but at the time of the organization of the West Church, Williams transferred to it his membership in the First Church of Cambridge. He lived in a house rented from Rebecca Royall and unwisely anticipated a bequest from his wife's grandfather, the elder William Pepperrell. When the bulk of this estate went to the younger William Pepperrell, the future baronet, Williams wrote indignantly accusing him of having exercised undue influence on the old gentleman. Pepperrell's reply places Williams rather well:

I believe no person who has heard of Mr. William's education could think that he wrote such a letter. . . . Such hints would be worth minding if they were written by a man of more years than yourself. I thank you for all the kind services you have done for me, and if you please to get the account for those things you bought for me ready, hope to be in Boston next week and make you full satisfaction.[3]

Fortunately Williams was not sensitive, or a man to harbor a grudge, so the family breach was soon healed.

As a physician, Williams exhibited an inventiveness which might well have brought him a profitable practice. Once when approached for treatment by a man blind from infancy, "Doctor W. fertile of invention, pulverized a small quantity of a stone jar, and placed it upon the eye of the patient, which soon ate off the film."[4] In spite of this success in making the blind see, he felt that the practice of medicine was "by no means consonant to his genius," and at the

---

[2] *New England Hist. Gen. Reg.* LVII, 279.
[3] Usher Parsons, *Life of Sir William Pepperrell*, Boston, 1856, pp. 31–2.
[4] *Vermont Gazette*, Nov. 30, 1789, p. 2.

outbreak of the war with Spain "he determined to quit the lap of ease, and try the rugged paths of war," "the sphere for which nature formed him." He enlisted as a private soldier, was sent to Carolina, and thence with Oglethorpe against St. Augustine. In view of the medical record of this army it is no wonder that he was quickly made a surgeon, being attached to the regiment of Colonel Van der Durren.[5] Although he found no profit in this, his first campaign, on June 24, 1741, he accepted a commission as ensign in Colonel William Gooch's regiment for the projected expedition against the Spanish West Indies. Under General Wentworth and Admiral Vernon he served in Cuba and at Carthagena, but when he returned he had nothing to show for his sufferings but his right to an officer's half pay.[6]

Fortunately for Ensign Williams, he was a nephew of Colonel John Stoddard (A.B. 1701) and Colonel Israel Williams (A.B. 1727), and thus a member of the aristocracy which defended the western frontiers of Massachusetts and managed most of the settlement in that region. In 1743 he moved to Deerfield, which was then one of the farthest settled towns, and signed an agreement with Uncle Stoddard and Jacob Wendell by which he promised "Spedily in person to bring forward a Settlement in a Township Lying on the Branches of Housatonock River at a place called Pontusuck" in return for a grant of two hundred acres there.[7]

The project for the settlement of Pontoosuc was abruptly halted by the outbreak of King George's War. Uncle Stoddard was in command of the defense of the western frontier, and Williams was commissioned a captain and ordered to construct a line of forts against the French and Indians. In June, 1744, he began work on forts Shirley, Pelham, and Massachusetts, in what later became the towns of Heath, Rowe, and North Adams. In general he employed the traditional type of construction with squared and dovetailed logs laid horizontally.[8] The winter of 1744–45 he spent in Fort Shirley, which stood on a bleak mountain top overlooking the frozen wilderness which extended unbroken to the north and west. The

---

[5] William Williams to Christopher Kilby, William Williams Mss. (Berkshire Athenaeum), Aug. 1749.

[6] *Ibid.*, Oct. 3, 1745.

[7] *Ibid.*, June 17, 1743.

[8] There is an account of these forts in Arthur Latham Perry, *Origins of Williamstown*, New York, 1896.

Province, appreciative of his service, promoted him to the rank of major.

That was the winter during which the Louisbourg expedition was planned. In the spring Major Williams raised a company to join Uncle Pepperrell's army, but to his great disappointment he was ordered to stay on the frontier where he was more useful. But in June, 1745, when sudden need of reenforcements at Louisbourg arose, the Province turned first to Major Williams as the man most likely to raise them quickly. Within six days of the time that he had received the express, he enlisted seventy-four men and marched them the hundred and fifty miles to Boston.[9] The Governor commissioned him lieutenant colonel in Colonel John Choat's 8th Massachusetts and ordered him to Louisbourg with his company and other reenforcements. On June 25 the transport sailed from Nantucket, taking on board Stephen Williams (A.B. 1713) as chaplain. At three in the morning of July 6 they cautiously spoke to a vessel coming out of the harbor of Louisbourg and to their "inexpressible joy" learned that the fortress had fallen. At five in the afternoon they dropped anchor in the harbor, and Colonel Williams went ashore to participate in the council of war.

During the next few months Colonel Williams was an active member of the administrative council at Louisbourg. He served on committees to issue a manuscript currency and to obtain fuel, and he drew up a plan by which Uncle Pepperrell could make £1500 a year victualling his own regiment.[10] With nearly half of his regiment down sick, he served as surgeon and billed the Province for his services.[11] While acting as provost he had a clash with Lieutenant John Shaw who, while sitting in the Louisbourg brig on Christmas Day, 1745, penned this complaint to General Pepperrell:

Last Night being Christmas Eve was making merry with my friends in My own house and in the Meantime Colonel Williams with others Enter'd my Room and used me in A manner beneath a Gentleman by Striking of Me and ordering people to force me out of my house and to goe to the Main Guard.

Wherefore presume to Entreat your honour would be pleased to have Justice done me as Soon as Possible he an officer; that Insults of this

---

[9] William Williams to Christopher Kilby, William Williams Mss., Oct. 3, 1745.

[10] Pepperrell Mss. (Mass. Hist. Soc.), II, 27.

[11] Executive Records of the Province Council, XI, 542.

Nature May not goe Unpunished, Since I can bring Sufficient proof of the Ill Treatment I have Received from the above Colonel Williams, and I understand that he has his Liberty to goe and Come as he pleas's I shall Submit to your honour whether if he continues So if my Life may not be attempt'd again and with More Success than the last.[12]

This letter brought no action from the General, who probably had his nephew to dinner; so the next day Lieutenant Shaw appealed again:

Desire a hearing this Day, in order to make plain my fidelity to his majestie's arms, my innocency in this confinement, alsoe the Trechery of Colonel Williams against my life, and Interest, which shall make appear in open Court by Credible Evidences, from the mouth of said Colonel Williams in a Court Martial, who Declared his picque against me, in these words (viz) that he did not want to hurt the men, but to quash my reputation.[13]

Against this attack on the character of Colonel Williams should be set the fact that from this time to the day of his death he was familiarly called "Captain Bill" or "Colonel Bill Williams" by almost everyone who knew him, from common soldiers in his own regiment to the Governor himself.

Soon after his arrival at Louisbourg, Colonel Williams had joined in a protest to Governor Shirley against keeping the 8th Massachusetts on garrison duty. However, it was not until May, 1746, that it was finally relieved by Regulars, and allowed to return home. Instead of quitting the service, Williams now obtained a transfer to the regiment which Colonel Joseph Dwight (A.B. 1722), who had been with him at Louisbourg, was raising for the invasion of Canada. Indian raids, and particularly the burning of Fort Massachusetts, put an end to this project, and on April 21, 1747, Williams, who was then at home in Deerfield, was ordered to take three companies and rebuild the fort. A month later he was thus engaged when "an army of the enemy came upon them, with a design to beat them off, and frustrate their purpose," but was in turn beaten off by the opportune arrival of a supply column.[14] Later his horse was shot either by the Indians or by his own soldiers, a question which perplexed the General Court for some time.

---

[12] Pepperrell Mss., II, 11.
[13] Ibid., p. 12.
[14] Samuel Niles (A.B. 1699) in 4 Coll. M. H. S. V, 375.

The regiment to which Lieutenant Colonel Williams was attached was disbanded on October 31, 1747, but he was appointed subcommissary to the army on the western frontier. Between the difficulties involved in obtaining supplies and his own financial ineptitude, he was soon in his customary state of financial disaster. On November 15, 1748, he resigned as subcommissary, and three days later he was appointed a justice of the peace for Hampshire, an expensive and time-consuming honor. The next year he signed a contract to collect the excise taxes in the county on a commission basis.[15] He now bought a house in Deerfield where he dispensed petty justice, collected taxes, and ran a store. Again he undertook to serve as subcommissary, collecting 2½% on the army supplies which flowed through his Deerfield store. As one of the leading inhabitants, he was elected selectman of the town in 1751.

For a time Colonel Bill thought of removing to Bennington, Vermont, of which he was the most prominent promoter. He was one of the original grantees of the plantation and the agent of the proprietors.[16] In October, 1749, his old friend Governor Benning Wentworth (A.B. 1715) authorized him to lay out a township on the western frontier of New Hampshire, and the following January confirmed this as the plantation of Walloomsack. Williams was also a proprietor of the New Hampshire towns of Holderness and New Holderness.

This interest in New Hampshire settlement disappeared after the death of his wife, which occurred about the middle of August, 1751.[17] On August 11, 1752, he married Sarah, daughter of Thomas Wells of Deerfield, and shortly thereafter he turned his attention from remote New Hampshire to more sheltered Pontoosuc. Colonel John Stoddard had gone to his last reward, but Aunt Stoddard was ready to renew the contract by which Bill Williams would receive a hundred acres there in return for his services as a leader of the settlement. In November, 1752, the deed was signed, and early in the next spring Colonel Williams went up to the new plantation where he installed himself temporarily in a house which had been

[15] *Acts and Resolves of the Province of Massachusetts Bay*, Boston, 1869–1922, XIV, 588.
[16] *Provincial Papers . . . of New Hampshire*, Concord, etc., 1867–1944, XXVI, *passim*.
[17] William Williams of Weston to Thomas Williams, August 25, 1751 (in the possession of Mrs. Mary Fuller of Deerfield).

fortified but abandoned on account of the last war. On June 23, 1753, the plantation of Pontoosuc was duly incorporated.

There was, however, to be yet another war before the English were secure in their hold on these rich meadows in the heart of the Berkshire Hills. In August, 1754, the Indians resumed their raids on the northern frontier. The Colonel, determined that Pontoosuc should not be abandoned again, rallied the settlers to fortify the house which he had built for himself and named Fort Anson. They labored busily and effectively, although the number of gills of rum charged in his account book would suggest that they were trying to float a seventy-four instead of build a fort. His report of his activity endeavored to establish the fact that he was conducting a public work:

Upon the [news of the Indian] mischief, protection was sent us both from this Province and Connecticut. Upon their arrival, I offered to join them with all my strength, in fortifying wherever they should choose; but none of them would undertake, either to billet or build. Upon which, rather than no stand should be made, I proposed, if they would fortify with me, I would billet them, the inhabitants and soldiers, pay the broad-axe men three shillings and narrow-axe two shillings per diem: which they accepted, and I performed, and built a handsome, strong, and very tenable fort; and if I had not thus done, the soldiers would have all returned, and no one soul would now be at P.[18]

To which Uncle Israel, who had succeeded Uncle Stoddard as commander on the western frontier, replied that Bill and the settlers at Pontoosuc might fortify if they liked, but the Province would not pay for it. Undeterred, Colonel Bill set out to recover his costs from the General Court:

I dont know but my Account will Surprize some of your Strait laced people, and Taylors account of Fortifying his house being so much Short of mine may give umbrage to some. But pray how can Taylor save his Fort from Fire & what has he done but build two mounts & picket — And what are all the Forts that are built about Houses but Scare Crows that they pull down as soon as ever the War is over. Knowing that Poontoosuck would allways be a frontier I have built a Strong fort for a place of refuge that I shall not pull down for my House is clear & open only take the plank out of the Windows so that I Shall not have the Temptation Others have nor the Province lyable to pay for a new Fort every War which commonly is the case where houses are Forted.[19]

[18] J. E. A. Smith, *History of Pittsfield*, Boston, 1869, I, 107.
[19] Israel Williams Mss. (M. H. S.), I, 92.

The General Court was surprised to receive a bill for the excessive sum of £408, which it declined to pay, although the Colonel continued to petition after this fort of his had, in spite of his arguments, been replaced by more serviceable blockhouses.[20]

In September, 1754, Uncle Israel informed Governor Shirley that the commandant at Fort Massachusetts was incompetent and unpopular, and asked that William Williams be commissioned to replace him.[21] The Governor was pleased at the prospect of obtaining the services of Bill Williams and sent Uncle Israel blank commissions for the purpose. In October Major Bill took over a command which consisted of Fort Massachusetts and the plantation of Pontoosuc, twenty miles away.

Through the winter which followed, Williams sat in his isolated fort and pondered:

My Scituation is in Such a remote Corner of the earth that what ever reaches one comes with so much uncertainty, and many times loaded with so much Absurdity, that it nauceates although my Stomach is keen sett for news; I have ever since My being Capable of Observation noted a Time or War to be a Time of lying.[22]

When he had winnowed the rumors of French aggression, he began to urge upon his correspondents an English offensive to forestall the French attack. He was delighted in January, 1755, to accept a captaincy in a regiment which Uncle Pepperrell was raising for the invasion of Canada, and at once began to recruit for it.[23] In February, Governor Shirley offered him a captain lieutenancy on a Crown Point expedition and directed him to march to the Hudson as soon as he had raised his company.[24] This commission Captain Bill diverted to his cousin, Captain Ephraim Williams, for it was far inferior in prestige to his own in Pepperrell's Fifty-First, a royal regiment clothed in scarlet and the assurance of half pay. Captain Ephraim put on the blue coat and marched into the fatal ambush at Lake George.

On June 8, 1755, Captain Bill received his orders to proceed to Albany and up the Mohawk to the Great Carrying Place, near the

[20] Executive Records of the Province Council, 1759–61, pp. 159–60.
[21] Israel Williams Mss., I, 101.
[22] William Williams Mss., Jan. 1, 1755.
[23] *Ibid.*, Jan. 26, 1755.
[24] Israel Williams Mss., I, 113.

modern city of Rome, which divided the waters which flowed into the Hudson from those which flowed into Lake Ontario. On the way up the Mohawk he stopped to consult with General Johnson about the Indians, then pressed on into the wilderness until he reached his post on June 28. His part in the Oswego campaign which followed was to hold this most important link in the line of communications.

Captain Bill found the Indians very resentful of his presence at the Great Carrying Place, and he complained bitterly of their conduct:

The Indians now increasing upon us, coming in to Carry . . . gave us a Vast Deal of Trouble and many Hours loss of Sleep and made it a Life not to be desired. At the time a report coming to us that Mischief was done by French Indians near Gen. Johnsons, I petitioned our Indians to let us Stockade or Intrench ourselves which they refused, saying that I was going to get their lands away. . . . I . . . have been obliged to bear things from them that were really unsufferable. Been many a Time calld Sun of Bitch, Damned by them, bid to go to Hell, had a gun once catched out of Capt. Jaselyins hand and cocked at me, and another time like to have lost my Life with a pick ax which If I had not dodged would have Split me Down. And for no Other reason than that I would not let them have Rum. . . .

Upon the least Mischief Done them they would make the most bitter Complaints and I have always Satisfyed Them even to the Twig of an Apple Tree. . . . After having been with them some time . . . perceiving that I had no Design to injure them, They seemed to contract a Friendship for me which was very perceivable & daily increased to the Admiration of the Traders. . . .[25]

The Latter End of September John an Indian was agoing to build an House with Bark. Told him that I would build an House with Logs as big again as he proposed and finish the Body with a partition in the midst and he should cover it with his Bark and I would have one Half & he the Other which he gladly accepted, and this is the House I have lived in till now, And He with his Family or some other Indians in His appartment till the 1st Mischief was done . . . and Scarce ever a night but more or less lay in the room with me and many times so many that I could not set my Foot out of Bed without Treading upon them, and at last they cluttered me so that I moved my Bed into the Indians apartment, he not having half the Company I had.[26]

[25] William Williams Mss. (transcripts), I, 117.
[26] Ibid., I, 118.

While Shirley was before Oswego the first question which his officers used to ask of every batteau coming down from the Carrying
Place was whether Williams and his men had as yet been cut off by
the savages. When the General returned eastward on October 29
he ordered Captain Bill to build a fort at the Carrying Place, and
was pleased to direct that it be named Fort Williams.[27] By the end
of November the fort and barracks were done. The Indians had
complained to Shirley that the rum which they obtained from the
Dutch was so watered that it froze and had asked that Honasada,
which was their name for Captain Bill, be allowed to supply them.
The General granted their request, which may in part explain the
crowded condition of the Captain's apartment.

The winter went quietly enough until March 26, 1756, when a
sleigh train bringing supplies to the Carrying Place was captured.
A party which Captain Bill sent to investigate was ambushed and
the fort at nearby Wood Creek was burned. Apparently his Indians
had met the advancing enemy and induced them to spare him and
Fort Williams while concealing from him the proximity of the
hostiles until the damage was done.[28] Rumor had it that a great
army of French and Indians was closing in on the Carrying Place,
but it failed to materialize, and on March 30 General Johnson
arrived with reinforcements.

Captain Bill was at Fort Williams as late as June, 1756, but soon
after that he was suspended from his command and ordered to
Albany to defend himself against charges which were brought
against him. One source of his trouble was the fact that he had
reached into his own pocket to pay for the transportation of supplies
to the army at Oswego.[29] Another was the charge made by Sir
William Johnson that Captain Bill's conduct toward the Indians
had so disgusted them that they were clamoring for his removal.
The royal officers sided with the Captain, saying that they could
never condescend to the savages as he had done, or bear with them;
and General Abercrombie cleared him with the remark that it was
not to be supposed that every officer could adapt himself to the
temper of an Indian.[30] However, Johnson had his way, and the
Captain remained in Albany awaiting a new assignment.

[27] *Ibid.*, I, 119.
[28] *Ibid.*, II, 164–5.
[29] William Williams to — (New York Hist. Soc.), June 28, 1756.
[30] Israel Williams Mss., I, 247.

There troubles swarmed about the poor soldier. His wife, finding their washwoman with child, charged that he had an interest in the affair, to which he replied by pointing out that he had been three hundred miles away at the critical period: "I never was able to do such a manly Feat." His letters home are a savage indictment of the petty politics and the incompetence of Shirley which had led to disaster and the abandonment of Fort Williams. He despaired of successful resistance to the French in New York and hoped that "the Great Rogers" and his rangers could be obtained to guard the Massachusetts frontier: "And if I first mentioned them, I hope you will not be Affronted, since the Colour of Red is so Offencive to some peoples Eyes." [31] The prejudice of the provincials against the Red Coats was one of the factors which made the situation seem hopeless:

I know upon many Accounts its disagreable to have such Soldiers among you as commonly compose the King's Troops — but the Highlanders are a Sett of quiet Civil fellows, I have been here this day 6 Weeks, and I never saw one of them Drunk, I never heard one of them Sware, nor ever saw any quarrells or uneasiness among them, they are no more trouble than so many Chickins in Town.[32]

He was not himself free of provinciality, for of one man he remarked, "he is a Dutchman and that is Saying every thing that can be said that implys any thing Vile." There were some provincials who would not have used to any man the language which he used in his petition to the new commander, Lord Loudoun, to be permitted to sell his captaincy and retire on half pay:

My Lord
    Permit me with the ·Deepest Humility to prostrate myself with these lines at your Lordships Feet; with awfull thoughts and a trembling hand, to make known my case.[33]

His tale of seventeen years of service as a soldier with consequent financial insolvency did not suffice to obtain a discharge, but in the spring of 1757 it was decided to disband the Fifty-First. On his way from Boston to New York to receive his discharge he paused to write to Uncle Israel:

[31] *Ibid.*, p. 263.
[32] *Ibid.*, p. 246.
[33] William Williams Mss., I, 148–9.

We are ordered (the Officers) all to New York to receive our Clearances When I design to bid Adieu to the Service if a possible Thing 17 Years is long Enough. . . . I should incline to go for England, but I hear Mr Pownel is soon to be here with his Commission this Time, and he doubtless will be able to tell me my Doom as to S - - - - - ys debt. My half pay I can get without the Expence of going for it — I mean as Ensign — And now what to Do with myself is the Question. To go to Poontoosuck to be killed and Scalped just as I have got into a Capacity to Live looks foolish.[34]

On March 27 the Fifty-First was broken up, and Williams left the service. He remained in New York trying to obtain another commission, but the only one which offered was in the Royal Americans, and he declined to serve in that foreign legion, as he called it. Was there, he inquired, to be a sufficient garrison at Fort Anson to make his scalp safe in Pontoosuc? In May he applied to Lord Loudoun for permission to go to England to collect the money due him, but in the end he left the job to an agent.

Captain Bill spent the winter at Deerfield writing letters about the money which the government owed him and blindly buying lottery tickets in an effort to recoup his fortune. On April 24, 1758, he was commissioned to raise a provincial regiment for the invasion of Canada under Abercrombie. A month later he marched his troops for the Hudson by way of Northampton, Pontoosuc, and Fort Massachusetts. In June he was at Greenbush where, as the senior of Colonel Timothy Ruggles (A.B. 1732), he was acting commander of the Massachusetts forces. Thence he marched to Lake George, went by water to Ticonderoga, suffered the shattering defeat there, and fell back with his regiment, exhausted with fatigue and bitter at the incompetence and the losses. His report to Lieutenant Governor Thomas Hutchinson (A.B. 1727) was the first official report of the disaster to reach Massachusetts and was shown in the taverns by the postriders. To his relatives he wrote with his usual whimsical twist:

We have Made A fine Kittle of Fish; And I challenge all the Cooks from Cain, who first began to Kill his fellow Mortals, to the present day, to Show such another. From an Army in health & high Spirits determined to do what they came out for, To an Army disappointed, bereft of Confidence . . . I defy Pope, Milton or Job to Draw the Contrast. . . .

The Loss of the Surprising Man, for thought, Activity, Goodness of

34 Israel Williams Mss., II, II.

Temper, vertue, Religion, and undaunted Bravery — Lord Howe was the Loss — I had almost said of New England — I cant mention his name without Tears, those remembrances of the Curse, which I hope will one day be wiped from our Eyes.[35]

It was a long time before he ceased to mourn Howe.

Williams family reunions must have been a marked feature of all such provincial military camps as this at Lake George in the summer of 1758. Now Colonel Bill was responsible for a younger generation which included his son William, a surgeon in his regiment. The misfortunes of another member of the family gave rise to this letter which the Colonel sent to "Monsieur Monsieur St Luke Le Corn":

I have a young Brother in Law named Agrippa Weeks belonging to Deerfield in New England who being crossed in Love, Enlisted himself a ranger with Major Rogers and about a Month since was most Shamefully Captivated by a Party of yours with 19 others without Firing a Gun.

What I would request of you is that you would if possible procure inlargement to him so far as to take him to your house and from Time to Time Supply him with such Things as you shall think necessary for his Comfort, and by the First Ship that may be sent to Hallifax or Boston for the Exchange of Prisoners to put him on Board, or if any other Door opens for his return you would be so good as to Embrace it. Any charge or Expense you may be at on his behalf shall directly and punctually be paid to your order.[36]

During the rest of the summer he remained at Lake George, arguing with James Montressor about building forts and complaining that he was not being consulted on matters of policy. His regiment was detached to participate in the brilliant dash which captured Fort Frontenac and Lake Ontario, but he remained at Lake George where, he complained, "We Guard & Escort, and build Hog trougs, and Hogs stys, fell Trees, and Fatiegue ourselves, Sicken, & Die." [37]

To Governor Pownall, Colonel Williams bitterly complained of the way in which Massachusetts, in contrast with Connecticut, was allowing the sick soldiers to be neglected:

[35] William Williams to Col. Thomas Williams (N. Y. H. S.), Aug. 18, 1758.
[36] Livingston Mss. (M. H. S.), 2nd ser., XII, July 22, 1758. The Frenchman to whom the letter was addressed was the famous Saint-Luc de La Corne.
[37] William Williams to Jonathan Ashley, Mass. Mss. (Library of Congress), Sept. 26, 1758.

The Sick that we have had passes for, whose Friends did not come with Horses to get them Home, got along as they could, we not being Allowed to send Well men to Take care of them, some by that means sufferd & Died in one place & some in another. Those that got as far down as Albany dare not go into the Hospital the Small pox being in it. And lay about in the Barns & Streets in need of all things. Upon hearing of this their Miserable Scituation we obtained leave to Send down an Officer & Surgeon to take care of Them, they have got them into an House below Green Bush, have Expended all their own Money for their Comfort, and now write us that they can Do no more.[38]

When the neglected sick were eaten in their blankets by rats, he asserted, they had reason to complain. When winter came he quit the army, after twenty years of service, and retired to Pontoosuc.

In spite of the military disasters of the year, the Colonel had made up his mind to risk his scalp at the new plantation, for he had obtained a fifty acre grant there from Jacob Wendell and another five hundred acres from his father, who had probably inherited it through the Stoddards. On May 30, 1759, he took the oath as proprietor's clerk of Pontoosuc plantation and settled down to his life work of guiding the development of that settlement. That summer, for the first time, it had grain to spare. Two years later he obtained the incorporation of Pontoosuc as the town of Pittsfield, and at the first town meeting he was chosen clerk, surveyor, and selectman. He was one of the founders of the iron works in nearby West Stockbridge. When the church was gathered he brought his letter from the West Church of Boston with a recommendation from Jonathan Mayhew (A.B. 1744) saying that his life and conversation were as became a professor of the Gospel.

The Colonel championed the division of Hampshire County and was not surprised when, in 1761, he was appointed a justice of the peace and quorum and a judge of the Court of Common Pleas of the new county of Berkshire. As the only justice of the peace for Pittsfield he was kept busy, and the Court of Common Pleas sometimes sat in his house. In 1765 he succeeded General Joseph Dwight (A.B. 1722) as presiding judge of the Court of Common Pleas and judge of probate. There is some suggestion that he used his place on the bench to further the family ventures in real estate, but there was nothing unusual in this practice.

[38] William Williams Mss., II, 236.

In 1762 the proud new town of Pittsfield elected Williams to the General Court, but when he presented himself he was informed that new towns did not enjoy the privilege of representation.[39] Taxation without representation was perfectly proper when practised by the colonial legislatures. However, the town was later awarded a seat which the Colonel occupied in 1765, 1766, 1771, 1772, and 1779. The few recorded roll calls of the House show that he stood with the patriots in their opposition to such acts of tyranny as a provincial tax on liquor.

For a majority of the remaining years of his life Williams was simultaneously chief justice of the Court of Common Pleas, judge of probate, colonel of militia, representative, selectman, assessor, moderator of town meeting, town clerk, and hog-reeve. As the last of a family dynasty of public servants, autocrats, and lords of the marches, he was the leading man in Berkshire north of Stockbridge down to the Revolution. As colonel of the Berkshire militia, he tried in the four compulsory training days a year to whip his cheerfully ragged companies into something resembling the royal regiments he had once known, but he urged successive governors to allow as an option for this futile compulsory training a payment of four shillings a year which would provide funds to uniform and pay volunteer regiments which might by longer training periods be developed into a really efficient force.[40]

Transportation was another problem which naturally concerned the Colonel. "With great Care and Cost" he "explor'd a Road from Hampshire to Albany," and tried in vain to get the Province to pay him for his service.[41] But when, in 1766, he urged upon the Provincial Council the dangers of the boundary dispute with New York, he was ordered to take Uncle Israel to Albany to lay the matter before Sir Henry Moore.[42] When, five years later, this dispute came to a head, Colonel Bill led the north Berkshire regiment in a dash to aid Colonel Elijah Williams' south Berkshire regiment in repelling the Yorkers' invasion of Egremont, while Colonel Israel kept the Hampshire regiment in reserve.

These hostilities, however, gave Colonel Bill much less difficulty

---

[39] *Journals of the House of Representatives of Massachusetts*, Boston, 1762–63, p. 9.

[40] William Williams to Thomas Hutchinson, Mass. Archives, xxv, pp. 463–5.

[41] *House Journals*, pp. 15, 30.

[42] Executive Records of the Province Council, xvi, 147.

than did those which resulted from his third marriage, on January 30, 1765, to Aunt Hannah Dickinson, daughter of Samuel Dickinson of Deerfield. As one of his Stoddard relatives put it succinctly, Colonel Bill had been married "to Miriam Tyler, for good sense, and got it; to Miss Wells, for love and beauty, and had it, and to Aunt Hannah Dickinson, and got horribly cheated." [43] But not even this catastrophe could dull the Colonel's enthusiasm for the town which he, more than anyone else, had created. To his brother-in-law, Nathaniel Dickinson, he wrote:

Languor, Sickness, and Excrusiating pain, were my portion, while I chose, or rather was Obliged, to Tabernacle in the narrows between the West & East Mountains of Deerfield. Since my removal to this place, I challenge any man in the Government, that has not had half the Fateague, to compare with me for Health or freedom from Pain. . . . Another indisputable proof of the goodness of the Country is the prolifick behaviour of the Female sex among. Us. . . . Your Sister has got me twice so fat as ever you saw me; I can scarce Buckle my Shoes. . . . I am Serious in what I have said as to the Women, More Especially, as I, or some one else, has gotten your Sister with child.[44]

The Colonel lived elegantly, served by a Negro girl named Pendar and a Negro man named Hartford. Hospitable, generous, prodigal, and fond of display, he found himself harassed by suits in his own court, some of which were carried to the Superior Court. By 1767 he found himself so heavily in debt that he had to turn to John Hancock, whose purse was always open to those to whom favors might have political value:

These Wait on You by my Spouse who will Let you know my Miserable Circumstances A Case truly pittyable even by the unrelenting; and cannot fail of making an impression on a Mind like yours; the Sterling Excellency of which reaches the remote corners of New England, filld with a sence thereof my pen had like to have led me into a recital of them but when I came to review them in my own thoughts the enumeration of Them I welld to such advise tho all expressd in varity; Seemd to Savour of Adulation; than which nothing can be more disgustfull to a noble and pure mind.[45]

The fact that he was a debtor like nearly everyone else on the frontier made him all the more popular.

---

[43] Stephen W. Williams, *Genealogy and History of the Family of Williams,* Greenfield, 1847, p. 188.

[44] William Williams Mss., II, 260, 262.

[45] *Ibid.,* II, 265.

At the time of the Stamp Act, Judge Williams suddenly became too ill to hold the probate court, but he recovered suddenly when repeal made the hated stamps unnecessary. Like most of his class he was disgusted by the lying vilification of the conservative leaders by the radical politicians:

The Greenwilliam Spirit against America rather increases and the late wretched doings at Boston about the beginning of this month will increase that Spirit in England I mean the Scurilous Libel Against the Governor not much if any short of a Blasphemy & the disregard with which the House treated it. I am setled in my opinion the late Conduct of the House will bring on a demolition of our Charter Unless we are treated by King & parliament as a people Insane & so not to be punished until we come to our Wits.[46]

The town shared his aversion to the conduct of the Boston mob and made him chairman of a committee which brought in a resolution condemning the Tea Party. However, later in that same year, 1774, he lost the political leadership of northern Berkshire to a young radical, Israel Dickinson (Yale 1758). The conservatives did not go down without fighting, for the Colonel, Israel Stoddard (Yale 1758), and Woodbridge Little (Yale 1760), publicly accused the Reverend Thomas Allen (A.B. 1762) of "rebellion, treason, and sedition," which charges the town voted to be "groundless, false, and scandalous." Two days later, on August 17, 1774, Judge Williams bowed to rebellion and closed the probate court.

The line between Loyalist and rebel in Berkshire was a thin one, but it is surprising to find Colonel Bill chosen in March, 1775, to head a committee "to take care of disorderly persons," meaning Tories. Probably he was chosen because a majority thought that he would manage this business better than would the radicals who believed that the proper way to deal with Tories was to hang them by their thumbs. At any rate, he served for two years or more and induced most of the Loyalists of the region to take the oath of allegiance to the United States. In 1777 he was restored to the Board of Selectmen, and two years later he was sent again to the General Court. There he was appointed to a committee to auction off the confiscated Tory estates in Berkshire.

The invasion of 1777 rekindled the military fires in Colonel Williams, and brought from him a denunciation of the "infamous, ignominious, and cowardly" abandonment of Ticonderoga. Pitts-

---

[46] *Ibid.*, to Oliver Partridge, Mar. 21, 1768.

field was threatened, and in its behalf he appealed to the State Commissary:

I inclose you the last account we have from our northern army, in the close of which you will perceive the scarcity of powder. All that we had in store which was upwards of 100wt went forward last night for their supply. The dishes, plates, and spoons at Bennington and many towns this side are and have been melted into ball, and I immagine that the powder left in Pittsfield is not sufficient to make an alarm and all the lead that we had in the county went up last night. . . . If you cannot think of a more proper person I will be accountable if you send 100 stands, also powder, lead, and flints as much as you please which I will indeavour to forward to whom you shall direct to at Bennington or elsewhere, taking out so much as has been taken from us. I immagine it is imposable but that you will answer this my request, as there is no select or committee man but what is gone forward, nay a few squaws would take the whole town unless we have arms and ammunition.[47]

In spite of this proof of the dependence of Berkshire upon the State, Colonel Williams that same summer served as chairman of a county convention of some eight hundred delegates who voted not to permit the State to open courts in the county until it had adopted a democratic constitution. In accordance with this sentiment he resigned from the bench of Common Pleas in 1778, and in October of that year accepted the chairmanship of a town committee appointed to take over the administrative and judicial functions of the old court.[48]

There were practical limits to this spirit of independence, and in 1779 Colonel Williams accepted a State commission as a justice of the peace. In September of that year he attended the constitutional convention which met at Cambridge and was amazed that such a respectable body of men could be collected in Massachusetts. In 1781 he surrendered the probate office because of his advancing years, but at the same time he became a proprietor of Barre, Vermont. He died at Pittsfield on April 20, 1785,[49] and was buried in the cemetery on the corner of North Street and Park Square. Since then the growth of his plantation of Pontoosuc into a great industrial city has forced him to make two removals to less valuable land.

Colonel Bill was an object of respect and affectionate amusement

[47] 7 *Coll. M. H. S.* IV, 142–3.
[48] William Williams Mss., III, 372–5.
[49] *Sic.* The Harvard catalogue is wrong.

throughout western New England, and the newspapers gave considerable space to obituary notices:

He was of a public spirit and loved works of munificence — Took no pleasure in amassing wealth, but in using it for his own comfort, the hospitable entertainment of his friends and others and in relieveing the distresses of the miserable. He was of a pacific spirit, moderate in counsel, an healer of divisions, a promoter of union in civil and religious matters, moderate, not violent or loose in his religious principles. Generous, humane, hospitable, easy of access, and condescending to the meanest — benevolent, and compassionate — A friend to the clergy and public order — a professor of christianity and a constant seasonable attendant on all divine institutions.[50]

> His generous spirit no cold medium knew;
> His mind expanded, look'd creation through;
> In him concentred virtue, well refin'd;
> Religion's prop, a wise and lib'ral mind.
> Fond to excess of jollity and mirth,
> His friends he chose for their intrinsic worth;
> Nor did he spurn the converse of the poor,
> But always kept a free and open door,
> To sooth the remnant of declining age,
> Was his delight; the pleasure of a sage;
> To wipe the tear from the sad orphan's eye,
> Cheer'd his good soul, & learnt him how to die.
> These are some laurels, these his high renown,
> For which he was, and is, & will be known.[51]

Aunt Hannah married one Shearer of Pittsfield and, according to the family, turned out the younger Williams children, who were forced to take refuge with the Shakers of New Lebanon.[52] The Colonel had at least seven children: (1) Miriam, b. Aug. 23, 1735; d. young. (2) William, b. Jan. 31, 1736/7; a physician of Salisbury, Conn., died of smallpox in the army, 1760. (3) Marjory, bap. June 3, 1739. (4) Miriam, b. Feb. 6, 1756; m. Capt. James D. Colt of Pittsfield. (5) Sally, b. Oct. 31, 1758; m. Leonard Chester (Yale 1769), Sept. 19, 1776. (6) Sylvia, m. one Easton. (7) William Pepperrell, m. Katie Blanchard. The Colonel's manuscripts are preserved in the Berkshire Athenaeum at Pittsfield, the officers of which have most generously helped in the preparation of this sketch.

---

[50] *Connecticut Courant*, May 23, 1785, p. 3.
[51] *Vermont Gazette*, Nov. 30, 1789, p. 2.   [52] Stephen W. Williams, p. 189.

PETER OLIVER

CHIEF JUSTICE PETER OLIVER was born on March 17, 1713, the second son of that name and the youngest of the Honourable Daniel and Elizabeth (Belcher) Oliver of Boston. His place at the head of the Class of 1730 was due to the distinction of his family and not, as some have said, to his academic ability. He boarded with President Wadsworth for one year,[1] but when freed from this supervision he opened a long career of disorder, his most serious offence being that which brought this Faculty vote upon him:

That Oliver & Gardner Senior, for being concerned in stealing the Goose lately taken on the Common; and also for stealing a Turkey, be degraded, viz. Oliver below Steel, & Gardner be degraded below five.[2]

He was restored a year later after confession, petition, and proof of reformation. When he took his M.A. in 1733 his *Quaestio* was the negative of "An Tautologia unquam Oratori sit Ornamentum," which he had the opportunity to demonstrate by delivering the valedictory oration at the Commencement exercises.

The death of Daniel Oliver in 1732 was a major public event, and the Reverend Thomas Prince (A.B. 1707) devoted the next Boston Public Lecture to an obituary sermon which he dedicated to the widow and the two surviving sons. Andrew (A.B. 1724), the elder son, inherited the new house which their father was building, and Peter received the old mansion on Purchase Street and the pew in the South Meetinghouse.[3] Such Boston houses had lovely gardens, and Peter improved his by importing fruit trees of all sorts from London. On July 5, 1733, he obtained a mistress for this mansion by marrying Mary, daughter of William and Hannah (Appleton) Clark, and sister of Richard (A.B. 1729). The young couple joined the Old South shortly before the birth of their first child. Peter also joined St. John's Lodge,[4] but he was always too much of an aristocrat to enjoy the society of Masons.

The partnership of Andrew and Peter Oliver, sometimes in conjunction with Thomas Hubbard (A.B. 1721), carried on a normal

---

[1] Benjamin Wadsworth, Commonplace Book (Mass. Hist. Soc.), p. 86.
[2] Faculty Records (Harvard University Archives), I, 22.
[3] Hutchinson-Oliver Mss. (M. H. S.), I, 58.
[4] *History of St. John's Lodge*, Boston, 1917, p. 221.

business in foreign trade, but extraneous matters had a way of creeping into the younger brother's business letters:

I now inclose you a bill of Lading for one bundle of Bone . . . which I desire you to dispose of & when You are in Cash to send me a very good Microscope, of the Large Sort, with its proper apparatus; I also have delivered the Master part of a Perspective glass, which I would beg you to get fitted; it wants one Slider & a glass, there is also one glass pretty much scratchd & I should chuse a new one in its room, unless by grinding out the Scratches, it will do full as well.[5]

Books and London magazines interested Peter more than trade, so it is not surprising that he dropped out of the partnership about 1737. Three years later the two brothers asked to lease the un-fenced town lands on Fort Hill for a pasture, but were sharply repulsed by a committee headed by the elder Samuel Adams who reported that their project would interfere with the health of the inhabitants by depriving them of the "many fair Prospects" to be had by "Walking and Disporting" on the hill.[6] The Oliver brothers had been associated with the several efforts to check the inflation which Adams strongly favored.

The only town office to which Peter Oliver was elected was that of constable, which he avoided by paying a fine. He showed a pub-lic spirit, however, by contributing no less than £80 to the work-relief project of 1735. He was a proprietor of Long Wharf, and in 1738 he bought a warehouse on it. He soon mortgaged this, and several other parcels of Boston real estate, perhaps to raise the £100 which was his original investment in the Middleborough project which was to be the chief business venture of his life.

It is probable that Peter Oliver's attention was first drawn to Middleborough by some property which his father had owned there. He was attracted by the industrial possibilities of a site on the Nemasket River, the location of the first White settlement, burned in King Philip's War, and since then an Indian reservation. The natives had recently asked to be allowed to sell these lands in order to move to a wilder part of the town where there would be more game and fresh fields. On January 15, 1742, Oliver began buying land and mills from the men who had acquired the site from

---

[5] Misc. Mss. (Am. Antiq. Soc.), May 4, 1734.
[6] *Boston Records from 1729 to 1742* (*Boston Record Commissioners*, XII), p. 254.

PETER OLIVER

the Indians, and within a few years he had acquired twenty parcels. After 1744 Jeremy Gridley (A.B. 1725) was a partner in the venture. Tradition has it that the only slitting mill in the colonies was at Milton, and that Oliver sent one of his workmen disguised as a fool to steal its closely-guarded secret processes. The truth is that on December 26, 1744, he and Gridley bought half of a slitting mill then standing on the Nemasket.

It is unfortunate that industrial history is beyond the scope of this sketch, for the Middleborough iron works were an experiment of great importance. There were oddities about it: for instance, the title of the general manager was "the Skipper." In 1750 the plant included a "A Mill or Engine for Rolling & Slitting of Iron . . . as also a Plating Forge to work with a Tilt Hammer."[7] Eight waterwheels on a single dam provided power, and five ponds in chain provided a reliable source of water. Household ware was cast from bog iron and artillery from "mineral ore." To obtain the latter Oliver enlisted the aid of his uncle, Governor Jonathan Belcher (A.B. 1699) of New Jersey, who sent him pig iron from that province.[8] Some of his largest orders were for ten-inch "hawlitzers" and shells for the Province of Massachusetts.[9] The business made a net profit of four or five hundred pounds a year, a great sum for those days. On September 1, 1758, Oliver bought out Gridley's interest in the works.

Gridley had elected to remain in Boston, but on July 1, 1744, Oliver moved out to Middleborough,[10] which was henceforth his home although his children were born in Boston. On a hill overlooking the river he built Oliver Hall, which was reputed to be the most elegant mansion in the province. It was modeled after an English manor house, and its framework, wainscotting, and hangings were ordered from England. The library occupied a separate wing and "pleasure houses" studded the lovely gardens. For the out-of-door dinners which were so popular, the wine, in bottles blown with the Oliver name, was cooled in a springhouse. The most brilliant of the social occasions for which Oliver Hall was famous was the four-day reception which followed the marriage of

[7] Report to the Board of Trade in Mass. Archives, xx, 653.
[8] Jonathan Belcher, Letter Book (M. H. S.), Nov. 27, 1755.
[9] Executive Records of the Province Council, 1755–1758, p. 248, *et passim*.
[10] Peter Oliver, Jr., Diary (British Museum).

Peter Oliver, Jr., (A.B. 1761) to Sarah, the daughter of Governor Thomas Hutchinson (A.B. 1727). Oliver Hall was a favorite vacation spot for the Hutchinsons and other leading Massachusetts families, and the usual place to entertain such distinguished guests of the Province as Benjamin Franklin.

Life at Oliver Hall was quite as brilliant as any on the Southern plantations, but in some ways it was very different. There were only two black slaves among the servants, and the function of one of these was chiefly to entertain the guests with his wit. The Olivers took an active part in the life of Middleborough and were popular because of their hearty cooperation. True, the more conservative villagers disliked some of the improvements, such as singing by note, which the Olivers fostered. One man who disliked the new method of congregational singing complained that Squire Oliver "was bawling in the gallery with the boys." [11] The Squire joined happily in church controversies, and it was he who carried the case of Parson Thomas Weld (A.B. 1723) to the General Court.[12] He was the regional authority on the latest scientific farming, and it was he who published the complete edition of Jared Eliot's (Class of 1703) classic *Essays Upon Field Husbandry*, to which he contributed the appendix.[13] Eliot declared that it was to Oliver that he was leaving the task of carrying on the reporting of American scientific agriculture.[14] The British Society for Promoting Agriculture recognized the Squire's work by electing him to membership.

Like so many men of his century, Oliver was interested in all kinds of science. In such matters he was sometimes credulous to the point of being foolish, and much of his experimentation was on the exotic fringes of science. In a skeptical society he was courageous enough to announce that he had enjoyed remarkable success in the use of a divining rod to locate minerals in the earth.[15] Such was his local reputation for universal knowledge that, he once complained, there was only one man in the town of Middleborough who would express a contrary opinion if he had previously stated his views on a subject.

[11] *New England Hist. Gen. Reg.* XL, 248.
[12] Mass. Archives, XIII, 116, 118.
[13] Peter Oliver to Jared Eliot (Yale University Library), July 19, 1761.
[14] *Ibid.*, Dec. 14, 1761.
[15] Oliver to Ezra Stiles (*ibid.*), Mar. 31, 1756.

Oliver was one of the largest subscribers for the *Chronological History* of Thomas Prince (A.B. 1707). According to John Eliot (A.B. 1772), "He had the true spirit of an old colony man. Every relick, or document, which related to the settlement of the country, or was curious, had a value stamped upon it." [16] He collected many papers and records, and with his own hand copied a three or four hundred page manuscript of Hubbard's History. His library at Oliver Hall was extensive, and its books were marked by an engraved plate. [17]

Without doubt Oliver's wit and his interest in poetry had much to do with his friendship for Jeremy Gridley. His gravestone verse was much above the ordinary jingle. [18] His longer published poems show a maturity and a boldness of style unusual among the provincials of that generation. His verse on the death of Province Secretary Willard (A.B. 1698) shows at once these qualities and his political leanings:

> His Heart nor aw'd by Power, nor yet his Ear
> Tickled with Popularity's false Chimes:
> In Græcian and in Roman Fate well read,
> He saw their Grandeur and their Ruin too;
> The fatal Rocks, on which they split, he mark'd,
> And, tho' conceal'd, he pointed where they lay.
> Licentious Liberty's wild frantick Dreams
> He fled abhorrent, for he knew no Curse
> The social Bonds would sooner disunite. [19]

With views like these he would have been most happy to contribute to the volume of verse dedicated by Harvard College to George III, and indeed contribution No. 29 in this *Pietas et Gratulatio* has all the earmarks of his work.

Oliver's public addresses were laden with those literary flowers which shed their sickening perfume over the work of the next generation. Commemorating Judge Isaac Lothrop (A.B. 1726) he said:

Little did I think that the first Time I was to discharge my Duty from this Bench would be on this solemn Occasion, of Celebrating the Virtues of the deceased; but the Task, although irksome, is what Duty strictly

---

[16] John Eliot, *Biographical Dictionary*, Salem and Boston, 1809, p. 350.

[17] There is an example of the book plate at the American Antiquarian Society.

[18] Ezra Stiles, *Itineraries*, New Haven, 1916, p. 166.

[19] Peter Oliver, *Poem Sacred to the Memory of . . . Willard*, Boston, 1757, p. 10.

demands, and I obey. The Memory of a departed Friend makes Nature recoil, revives my Passion, and tells me I am too unequal to the attempt. But shall publick Merit be passed by unregarded? No! publick Justice calls too loud and claims the Tribute of a Tear for a publick Loss.[20]

This style was a new English literary importation which was to stifle the magnificent prose of the Revolutionary generation.

In 1744 Oliver was appointed a justice of the peace for Plymouth County, and three years later he was promoted to the bench of Common Pleas. At this time the Massachusetts courts were beginning to take over the full wigs, scarlet robes, and some of the barbarous precedents of the courts of England. The oath which Oliver took when mounting the bench in 1748 is perhaps the earliest example of the use in Massachusetts of certain clauses prescribed by English law:

I do believe that in the Sacrament of the Lord's-Supper, there is not any Transubstantiation of the Elements of Bread and Wine into the Body and Blood of Christ . . . And that the Invocation or Adoration of the Virgin Mary, or any other Saint, and the Sacrifice of the Mass, as they are now used in the Church of Rome, are Superstitious and Idolatrous . . . without any Evasion, Equivocation or mental Reservation whatsoever; and without and Dispensation already granted me for this Purpose by the Pope.[21]

Such needless affectation was so dear to his heart as to suggest it might have been he who introduced here the use of this oath.

In spite of the lack of any formal legal training, Judge Oliver served Plymouth County well. He planned and supervised the building of the courthouse and served as one of the guardians of the Titticut and Mattakeeset Indians. It must be admitted that he showed small interest in mission work and had to be poked by his brother Andrew into doing such small chores as distributing blankets.[22] He was universally respected and admired, and his promotion to the Superior Court in 1756 was a very popular move. Here was an even better background for the ostentation which he loved, for the Court met in each county in rotation. He travelled from one county seat to the next in a coach emblazoned with his

[20] Peter Oliver, *Speech*, Boston, 1750, pp. 1–2.

[21] Mass. Archives, XLII, 667–8.

[22] Mass. Archives, XXXII, 352; Peter to Andrew Oliver, Misc. Mss. Bound (M. H. S.), Jan. 12, 1762.

arms and accompanied by postillions and outriders in scarlet livery.
Everyone enjoyed the show, and the sheriff, the bar, and the leading inhabitants of the county seats habitually rode out to greet the
judges of the Superior Court and escort them in glory to the court
house.

Judge Oliver had sat for Middleborough in the House of Representatives in 1749 and 1751, and in 1759 he was elected to the
Council. This made him an Overseer of Harvard College, but he
never showed much interest in his duties as such. On the Council
he was for the first time exposed to the rising wind of revolution,
and he showed no interest in remaining there to suffer for king and
cause. He blamed the political troubles on the corruption of the
Yankee moral sense which resulted from the practice of smuggling:

The Inhabitants of the Massachusetts Bay were notorious in the smuggling Business, from the capital Merchants down to the meanest Mechanick; & whereas in England it is dishonorable, to a Merchant of Honor,
to be guilty of such base Subterfuges to increase their Estates, it is in New
England so far from being reproachfull, that some of the greatest Fortunes there were acquired in this disgracefull Trade; & the Proprietors
of them boast of their Method of Acquisition.[23]

He had small use for the vaunted democracy of the New England
system for, he said, "the People in general . . . were like the
Mobility of all Countries, perfect Machines, wound up by any
Hand who might first take the Winch." [24] Naturally he supported
the policies of Parliament, holding that the purpose of the Stamp
Act was only to reimburse the Crown in part for the expense it had
suffered in the war to rescue the colonists "upon their own repeated,
earnest & humble Supplications, from being subjugated by the
French of Canada." [25] The Stamp Act riot he blamed largely on
Charles Chauncy (A.B. 1721) who had, he said, urged the people
to fight up to their knees in blood.

Such was the Frenzy of Anarchy, that every Man was jealous of his
Neighbour, & seemed to wait for his Turn of Destruction; & such was
the political Enthusiasm, that the Minds of the most pious Men seemed
to be wholly absorbed in the Temper of riot.[26]

[23] Peter Oliver, The Origin and Progress of the American Revolution (M. H.
S.), p. 63.
[24] Ibid., p. 90.
[25] Ibid., p. 69.
[26] Ibid., p. 73.

He actually believed the statement of one of the rioters who had plundered the Hutchinson house "that the first Places which they looked into were the Beds, in Order to Murder the Children." [27] On the other hand, he had a kind word for Mackintosh, the leader of the mob, who, he said, "was sensible & manly, and performed their dirty Jobs for them with great Eclat." [28] He was as quick to blame Pitt for his open approval of the use of terror to effect political ends in America.

At this stage Oliver thought that he could sit out the political struggle. Much to the irritation of Lieutenant Governor Hutchinson, he refused to go up to Boston to do his duty in the Superior Court and the Council. In letters full of classical quotations and merry tales, he presented such excuses as having to stay in Middleborough to pick over his potatoes. The attacks of the Whig press he accepted as an honor: "Write on Oh! Grubstreet Patriots. Triumph in your Efrontery, for glorious are your Laurels." [29] Jokingly he demanded compensation for his services in sheltering political refugees from the Boston mob:

When you Petition his Majesty again, be pleased to let me write a Postscript, & ask for my Reward; for my Claim is just & high. Last Summer I protected his first Officers in my Pigeon House; about 3 Months ago his Secretary [Andrew Oliver] was protected by me here from the Mob: & about a fortnight since, the Lieut. Governor fled to my House for the same Shelter. This is resented . . . in the Country, & yet I am still unhurt. If I am not a Man of the first Consequence then there is no man of any Consequence at all. I expect a very great Reward for my Importance.[30]

He was entirely serious, however, in his opposition to the opening of the courts without the stamps required by law. It was only, John Adams noted, because of his "fear of violence of the Sons of Liberty" that he took his place at the March, 1766, session of the Superior Court.[31]

At this meeting, the chief Justice not attending, one of the Judges Mr Peter Oliver, said that he attended the Court according to his duty; that he understood that it would be expected that he and his brethren should

[27] *Ibid.*, p. 72.
[28] *Ibid.*, p. 75.
[29] Oliver to Thomas Hutchinson (New York Public Library), Feb. 1, 1766.
[30] *Ibid.*, Feb. 25, 1766.
[31] John Adams, *Works*, Boston, 1856–60, II, 189.

proceed in business in defiance of the late Act of Parliament; that such proceeding was contrary to his Judgement and opinion; and that if he submitted to it, it would be only for self preservation, as he knew he was in the hands of the Populace: and therefore he previously protested that all such Acts of his if they should happen, would be acts done under Duress.[32]

He asked to be excused from attending the next session of the court because he wished to be at home to face a mob which was gathering in one of the neighboring towns to prevent the Middleborough people from taking river fish, and had announced its intention to "demolish" any justice of the peace who attempted to interfere.[33]

Judge Oliver was now marked as a prerogative man, so in the election of May, 1766, he was dropped from the Council. The next year he "made no show at all" in the election, and it is not recorded that anyone was brave enough to vote for him after that. He took his defeat philosophically and remarked that he would rather be left out with Hutchinson than to sit without him. As for the new General Court, he quoted with approval the remark of a country-man that "if the D — — l don't take them Fellows, we had as good have no D — — l at all." He approved of the repeal of the Stamp Act on the ground that it is better to be without laws than "to leave those that are made unexecuted & trampled upon by the dirty Foot of Rebellion or Faction." [34] Of the popular celebration at the repeal he remarked that it exhibited not gratitude but triumph; that America had found out a way of redressing her own grievances without applying to a superior power. Observing the popular re-action to the Declarative Act, he remarked, "the Daughters of the uncircumsised Americans cried, 'Pish! Words are but Wind.' " [35] The argument based on a distinction between internal and external taxation he dismissed as "a Distinction without a Difference . . . an Affront to the Understanding of those who advance it," and among these he definitely included Pitt. The Townshend Acts he held to be no more unreasonable than others which had been ac-cepted without question, but to be peculiar in that they pressed on the smugglers of tea, who were a majority of the merchants. He said that the non-importation agreements were generally violated

[32] Sparks Mss. IV (Harvard College Library), Bernard, IV, 216.
[33] Oliver to Thomas Hutchinson (N. Y. P.), Mar. 24, 1766.
[34] Origin, p. 71.
[35] Ibid., p. 77.

by the merchants who made them, but he was troubled by the fact that the resulting riots demonstrated that "the civil Power of the Country was not sufficient to protect any one who was obnoxious to the Leaders of the Faction." [36] He jeered at the patriotic tealess tea parties of the ladies and "the Enthusiasm of the Spinning Wheel . . . the good old cause." The idea of American self-sufficiency in manufactures he held to be absurd; he paid his iron workers 2*s* a day while his English competitors had to pay but 6 or 8*d*.[37]

Still Judge Oliver refused to take the political situation seriously. On the news of the modification of the Townshend Acts he wrote to Hutchinson: "We hear the acts are repealed, and that the mob killed 80 soldiers at New-York; killing time was formerly in November but tempora mutantur, et &c." [38] He defended the inactivity of the Tories, holding that men of sense would as soon combat a hurricane as a government which was in the hands of the mob, and offered to sell auger holes to the Boston Loyalists to stick their heads into when the Sons of Liberty got after them. Gradually the political jokes in his letters became more forced as he became sensitive to the signal whistle of the Boston mob which, he said, like the Iroquois yell, forever echoed unpleasantly in the ears of everyone who had ever heard it. Naturally he welcomed the arrival of the Regulars, remarking that for the first time in years one could walk the streets of Boston at night without fear of insults. He had other troubles, however, for in 1768 the Sons of Liberty devised a new way of getting at him. Like all business men in the colonies he had to operate chiefly on credit, for there was by no means enough cash to lubricate the movement of goods; so when the Whigs put pressure on his creditors to demand cash, he had to offer to mortgage his entire holdings to raise £800.[39] This may throw some light on the accusation of the Whigs that he was heavily in debt.[40]

The time finally came when Judge Oliver had to take the political situation seriously, and then he faced his duty with conscientiousness and courage. In the Richardson Case of 1770, when the Superior Court was ready to brave the mob to the extent of letting

[36] *Ibid.*, p. 85.

[37] *Ibid.*, p. 89.

[38] *Boston Gazette*, Oct. 2, 1775.

[39] Oliver to Thomas Hutchinson (N. Y. P.), Apr. 27, 1768.

[40] Benjamin Lynde to James Bowdoin, Bowdoin-Temple Mss. (M. H. S.), June 5, 1776.

cal harangues to the juries.[45] His appointment did not come, however, until January 24, 1772, when time had made him the only
possible candidate. Hutchinson recommended him to the home
government as the man most likely to resist popular pressure. His
first important task was to serve on the Gaspee commission of 1773
where his work was of less interest to us than his long conversations
which were recorded by Ezra Stiles (A.M. 1754).[46] In one of these
he said that he thought that the Ashfield Baptists had been somewhat oppressed, but that he would rather not discuss the matter
lest it prejudice him before the case came before him on the bench.
When that happened, he upheld the law, and was condemned therefor by the Baptists.[47]

As a result of the political reign of terror maintained by the
Boston mob, and the use of the public purse for political purposes
by the House of Representatives, it had become amply apparent
that it was necessary that something be done to prevent the courts
from becoming rubber stamps for the Whig party. One such
proposal was the payment of the salaries of the judges of the
Superior Court by the Crown. Oliver was not averse either to
the change in source or to the increase in the amount which accompanied it, for at the moment he was being dunned by the town of
Boston for cash to meet certain obligations. When it was discovered,
early in 1773, that he was drawing only half of his salary from the
Province on the expectation that the rest of it would come from
the Crown, the Whigs resolved to make a political issue of the matter. The storm needed considerable exorcising before it would rise.
The Judge had discussed his situation with many members of the
House, offering, if the General Court would repay him half of
the £2000 that he was out of pocket for his expenses in seventeen
years of service on the bench, either to decline the Crown grant or to
resign. Privately these members advised him to accept the King's
grant, "but there was so great Virtue in the Boards & plaistering of
the Assembly Room, that upon setting their Feet over its Thresh-
old, they at once changed Opinions."[48] As public indignation
built up, the other judges, much his junior in service, hastily re-

[45] *Boston Gazette*, Apr. 25, 1774, p. 1.

[46] Ezra Stiles, *Literary Diary*, New York, 1901, I, 329–33, 345–51, 376–84.

[47] Isaac Backus to Samuel Adams, Jan. 19, 1774, quoted in Alvah Hovey,
Memoir of Isaac Backus, Boston, 1859, p. 197.

[48] Origin, p. 156.

off the accused with a finding of manslaughter, Oliver refused to go that far "and with great skill charged the death of the Boy upon the promoters of the effegies & their Exhibition which had drawn the people together & caused unlawful & tumultuous assemblies and he did not excuse such as had neglected suppressing those Assemblies." [41] In spite of hisses and shouts of "guilty" from the crowd he admitted the testimony of the Richardson children, and, according to the Whigs, in summing for the jury he accepted every fact presented by the defence and threw out everything testified by the witnesses for the prosecution.[42]

With the Boston Massacre trial impending, the newspapers began to threaten the judges, particularly Oliver. Hutchinson had the greatest difficulty in prevailing upon three of the five judges to sit at all. When the case opened Trowbridge (A.B. 1728) contented himself with a formal statement of the law. It was Oliver who really shook out the case. In his charge to the jury he denounced the untruthful and malignant newspaper attacks on the court and on the defendants, and he summed the evidence so as to expose the deliberate intention of the mob to provoke the soldiers to violence.[43] The public reaction to the decision was favorable, and for a little while Judge Oliver was almost popular. Hutchinson was enthusiastic over his courage and his success, and he proposed to General Gage that the government reward the Judge by taking off of his hands several tons of eighteen-pound shot which the Province had turned back to him as being in excess of its requirement for that size.[44]

Judge Oliver frequently threatened to quit the bench. His salary was less than that of the doorkeeper of the House of Representatives, and his duties required that he travel twelve hundred miles a year and support himself away from home for nine months of the twelve. The Whigs did not believe that he was sincere, but accused him of scheming to obtain the chief justiceship from which he could better deliver what they claimed were his habitual politi-

[41] Mass. Archives, XXVI, 463.

[42] Acts of the Privy Council, Colonial, 1766–1783, p. 251; Boston News-Letter, Apr. 23, 1772, p. 1.

[43] The Trial of . . . Soldiers for the Murder of . . . the 5th of March, 1770, Boston, 1770, pp. 197–207.

[44] Gage Mss. (William L. Clements Library), Mar. 31, 1771.

nounced the Crown grant. At the session of the Superior Court held in Boston in August, William Molineaux refused to serve on the jury as long as Oliver was on the bench. One of the newspapers took the Judge to task for his manner on this occasion:

I was by no means surprized at the very indecent manner in which you treated a gentleman, in many respects your superior. You interrupted Mr. Molineux in a very harsh, and I thought, ungentleman-like manner, notwithstanding you had engaged him audience, saying, "then you refuse to serve," "do you then refuse to serve?" Mr. Molineux answered, he came to request your Honors to excuse him, and begged you would hear the reasons for his request, and was again proceeding to deliver them: You again interrupted him, saying, "the court will consider of it."[49]

It is certainly understandable that the Chief Justice was unwilling to listen to a political indictment of himself under those circumstances.

By September, 1773, when the Superior Court was due to meet at Worcester, public opinion had become more inflamed. He had not appeared by the time when the Court usually opened, and the jurymen refused to be sworn lest he appear. They held to their determination until late in the day when it became obvious that he was not going to sit at that session.

During the first week in February, 1774, the House of Representatives took up the question of Judge Oliver's salary. In reply to the questions of the House he argued that his salary from the Province had never been sufficient, that he had several times offered to resign, and that he could not decline the Crown grant lest he "incur a Censure from the best of Sovereigns."[50] On February 11 the House passed a resolution saying that he should be removed from his office because he had "contrary to the Usage and Custom of the Justices of the said Superior Court . . . contrary to the plain Sence and Meaning of the said Charter and against the known Constitution of this Province . . . received the said Salary and Reward out of the Revenue unjustly and unconstitutionally levied and extorted from the Inhabitants of the American Colonies," and had "by his Conduct . . . perversely and corruptly done that which hath an obvious and direct Tendency to the Perversion of Law and

[49] Boston Evening-Post, Jan. 24, 1774, p. 2.
[50] Journals of the House of Representatives of Massachusetts, Boston, 1774, pp. 134-5.

Justice in the said Superior Court." [51] The House sent this resolution to the Governor, and, unconstitutionally, ordered the Superior Court adjourned pending the removal of the Chief Justice. Hutchinson replied that there were no grounds for action. According to Oliver,

Adams now addressed his Gallery Men, to attack the Chief Justice when he came to Court; & they perfectly understood his Meaning — even one of the Assembly, a Col. Gardiner, who was afterwards killed at the Battle of Bunker's Hill, declared in the general Assembly, that he himself would drag the chief Justice from the Bench if he should sit upon it.[52]

The House itself suggested rioting as the next step,[53] and John Adams, one of the leaders in the effort to remove the Judge, "shuddered at the expectation that the mob might put on him a coat of tar and feathers, if not put him to death." Adams would have regretted this, for he regarded him as an amiable man, "very respectable and virtuous." [54] Although well aware of the danger, Oliver set out from Middleborough for the next Superior Court session, but was turned back by a snow storm. When the storm let up a messenger arrived from the House ordering him not to attend the Court. The bearer of the message was an old friend who wept as he warned the Chief Justice that he might be murdered in Boston, and turned back dinnerless lest he incriminate himself by breaking bread with the Olivers.[55] The Boston Loyalists whom Oliver had once chided for resisting the popular party now unanimously urged him to keep away from the Court lest he provoke a tumult.[56] He was, said Hutchinson, the only man who had the courage to stand up against the popular party then, but it was not to be expected that he would dare to attend the Court in defiance of the order of the House.

The Governor, on his part, faced a visitation of the whole House and flatly refused to remove the Chief Justice. Thereupon the House took steps to impeach Oliver for "High Crimes and Mis-

---

[51] *Ibid.*, pp. 146–7, 150; *Boston Gazette*, Feb. 14, 21, 1774.

[52] Origin, pp. 157–8.

[53] *House Journals*, p. 167.

[54] Adams, *Works*, II, 328.

[55] Origin, pp. 157–8.

[56] Hutchinson to Dartmouth, Feb. 14, 1774, in Sparks Mss. X, New England Papers, IV, p. 64.

demeanors." As adopted by a vote of 92 to 8 on February 24, 1774, the articles began:

Peter Oliver . . . hath ungratefully, falsely and maliciously laboured to lay Imputation and Scandal upon this his Majesty's Government, insolently and contemptuously insinuating, that by the Parsimony, Injustice and Ingratitude of the said Government, in withholding from him an adiquate and due Reward for his Services as a Justice of the said Superior Court, he hath been greatly impoverished, and that therefore he was obliged to take his Majesty's Grant from a Principle of Justice due to his Family and others. Whereas in Fact, the Rewards granted to him by this Government, were always fully equal to the Merit of his Service as a Justice of the said Court; as it is well known that the said Peter Oliver, Esq; before his Advancement to a Seat in the Superior Court, had been usually employed in the Business of Trade, Husbandry and Manufactures, to which he had applied his Mind; and that he was appointed to said Office without previous Education and regular Study in the Law.[57]

Hutchinson refused to entertain the impeachment because it carried the implication that the King was offering a bribe, and because the Governor and Council had no such jurisdiction under the Charter. The Council was anxious to proceed, but could not legally do so because Hutchinson refused to preside.

On March 1, 1774, the House withdrew the impeachment and offered the same articles as a criminal indictment.[58] Again the Council tried without success to make the Governor cooperate in the action.[59] There was nothing more that the House could do but to instruct Franklin to take what action he could against Oliver in England.[60]

At this moment Andrew Oliver died. Hutchinson tried to obtain from the Whig leaders a safe-conduct by which the Chief Justice might attend his brother's funeral without fear of being mobbed, but it was refused. Instead there was displayed in the streets of Boston a large transparency showing Peter Oliver "in the horrors." There was now undertaken a campaign to discredit the Chief Justice

---

[57] *House Journals*, pp. 198–9; *Boston Gazette*, Mar. 7, 1774, p. 1. There are contemporary copies of most of these documents in Mass. Mss., Elwyn Gift (M. H. S.).

[58] *House Journals*, pp. 211–7.

[59] 6 *Coll. M. H. S.* IX, 343–53.

[60] The statement of the House as addressed to Franklin is in Mass. Mss., Elwyn Gift, II, 108.

as "the lack-learning Judge," or "Sir Peter Lack-Learning." [61] He was lampooned in Mercy Warren's *Adulateur* as Lord Chief Justice Hazelrod, a name which has puzzled historians, but is explained by Oliver's report to Stiles, an ardent Whig, of his experiments with divining. In *M'Fingal*, John Trumbull (Yale 1767) described him as an ignoramus. The Chief Justice made no attempt to answer. His only political writings were some contributions to the *Censor,* a sober Loyalist sheet.[62]

For years Oliver had found Middleborough a quiet haven, for the country people disliked the activity of the Boston mobs, and his influence over his fellow townsmen was such that an indignant Whig once told them: "If Judge Oliver told you that that rock had been moved during the night, you are all d — — fools enough to believe it." In January, 1774, he could aid Elisha Hutchinson (A.B. 1762) after the mob had stoned him from Plymouth. Had Oliver agreed to cease his opposition to the popular movement and remain among his friends at Middleborough, he might well have out-ridden the storm. As it was, his conscience drove him to his duty on the bench. When he took his place on the Court at Charlestown in April, 1774, the Grand Jury protested:

We . . . think it our incumbent duty to remonstrate & protest Against the Hon. Peter Oliver Esq., his Setting as Chiefe Justice on the trial of any Offences by us presented untill he shall be acquitted of the Crimes he is Charged with. The impropriety of the Chief Justice setting to Judge of the offences of Others while he himself lyes under an Impeachment for high Crimes & Misdemeanors, will we flatter ourselves sufficiently apologize to the Hon. Court for this remonstrance.[63]

At Worcester the jury presented a similar complaint and served only because he did not appear. Oliver was furious that his colleagues had listened to such charges against him, and complained that if they had been similarly insulted he would have quitted the bench rather than permitted such an indignity pass unnoticed.[64] He dared not attend the Plymouth session of the court, but on May 8

---

[61] For example, *Boston Gazette*, Apr. 25, 1774, p. 1.

[62] Eliot, p. 351. His reputation as a judge quickly recovered in Massachusetts histories, and Emory Washburn praised him: *Sketches of the Judicial History of Massachusetts*, Boston, 1840, pp. 303–4.

[63] Early Files in the Office of the Clerk of the Supreme Judicial Court, 92,139.

[64] 1 *Proc. M. H. S.* xv, 68.

he wrote to Judge Trowbridge about attending the Barnstable session:

One Dr. [Nathaniel] Freeman of Sandwich was at a Tavern in this Town, and before Several of my neighbours asked whether I intended for Barnstable Court. He was told Yes. He then said that there was an high Insult designed me at Barnstable if I went there, and sent his Compliments to me to call at his house if I went. . . . This last I Took as the beginning of an Insult, for . . . his Character I am sufficiently acquainted with, for I was Informed that he left money at a Tavern at Plymouth to treat a set of infamous Villians who destroyed some of my property in Plymouth last Winter. . . . I apprehend that his mind is so callous to every sentiment of Virtue that I should not dare without defensive weapons to guard me . . . to pass within the reach of that Malevolence.[65]

Specifically he wanted Trowbridge's opinion as to the legality of the sessions of the Court without its full complement of judges, and he wanted specific assurance that anyone who refused to serve on the jury would be fined as the law directed:

If the Court will not support me I will not go down to Plymouth. I am more anxious to try the force of the Law, and support the Dignity of the Court, as I can have enough in this Town and in Plymouth to guard my Person and I am Chiefly concerned for the Honour of the Court.[66]

Instead of committing himself on a paper which might fall into the wrong hands, Trowbridge went down to Middleborough to inform Oliver that his Boston friends thought that he ought to keep away from the Plymouth court. His Plymouth County friends promised him an armed escort and urged him to attend, but Trowbridge's refusal to promise the punishment of the recalcitrant jurymen kept him away.[67]

In spite of this Oliver determined to preside over the session of the Superior Court at Boston in August, 1774, and so finally burned his bridges behind him. The situation was becoming uncomfortable even at Oliver Hall, where on August 24 a deputation of "Middleboro Brutes" paid a visit. It may have been on this occasion that he was compelled by threats to sign a promise not to exercise his office.

---

[65] *Mass. Law Quarterly*, XIII, No. 4, p. 40.
[66] *Ibid.*, pp. 40–1.
[67] *Boston Gazette*, Sept. 18, 1775, p. 3.

He was brought to this because a more recalcitrant neighbor, Silas Wood, was saved from drowning only by the pleas of his children who clung to him as he was dragged to the pond by the mob.[68] Oliver appears to have made no bones of the fact that he regarded a promise made under duress as invalid, or of the fact that he intended to attend the Boston session of the Court. He "was threatened to be stop'd on the high way in going to Boston court, but his firmness and known resolution . . . intimidated the mob from laying hands on him." On August 30, the date set for the opening of the Court, a threatening mob of over a thousand gathered in Boston. General Gage offered military protection, but the Chief Justice, with an obstinacy which must have dismayed his colleagues, insisted on relying on what he called the force of civil power. According to the newspapers, it was only the fear of the troops which kept the mob in hand. The judges took their place on the bench, but the jury, one and all, refused to take the oath as long as Peter Oliver was there. By foresight the jury was provided with a neatly printed indictment of the Chief Justice and statement of their own position, drawn in the form of a court document. So the judges adjourned the court and filed out to the hisses of the mob.[69]

From this day Oliver was besieged in Boston. His carriages were stopped and turned back on the roads, and no goods and effects were allowed to reach him from Middleborough. In November, he was dragged from the bench by a mob, but the next day he was back doing his duty under the protection of an armed guard.[70] When the February term of the Superior Court began the Chief Justice was on the bench as usual, but the record stops abruptly after a single case.

Fortunately Oliver had other work to do. As far back as 1769 he had been nominated by Governor Bernard, whom he had defended from the "falsehoods" of the Whigs, for the proposed Mandamus Council. On June 1, 1774, he was appointed, and on

[68] Origin, p. 226.

[69] *Boston News-Letter*, Feb. 23, 1775, p. 2; E. Alfred Jones, *The Loyalists of Massachusetts*, London, 1930, p. 223; John Andrews to William Barrell, Andrews and Eliot Mss. (M. H. S.), p. 41. There is a copy of the jury's broadside in the library of the American Antiquarian Society.

[70] *Boston News-Letter*, Dec. 1, 1774, p. 3.

August 16, he was sworn in. From that day until its death he was an active member of that body.[71]

The Whigs believed that Peter Oliver was an active candidate to succeed his brother Andrew as lieutenant governor, and that he had missed the honor only because the London authorities had by error inserted the name of Thomas Oliver (a.b. 1753) in the commission. There is no evidence that Peter wanted the job, and there is ample proof that the reason why he, the obvious candidate, was not appointed was that Hutchinson did not think that it would look well to have the offices so closely held in their family.

It was not unnatural that the Chief Justice argued that Gage ought to send the Whig leaders to England to be tried for treason. He defended the Boston Port Bill on the ground that the people had forfeited their charter by their invasion of royal prerogative. The social revolution which was going on merely amused him: "Never did the World exhibit a greater Raree Show, of Beggars riding on Horse back & in Coaches, & Princes walking on Foot." [72] His world was further changed by the death of his wife on March 25, 1775.[73] Six years later he had painted a picture of himself standing beside her tomb, with a glowing tribute to her which clearly shows the nature of their relationship.[74]

The most unexpected of Oliver's friendships is that for General Putnam. He states baldly that Putnam offered his military services to General Gage for 10s a day and regretfully went over to the rebels when Gage could not meet that figure. He held that Putnam was humane and would have been guilty of none of the "Barbarities" which were "committed by Washington & his Savages." [75] High on the list of such barbarities he placed the fighting from behind walls and fences on the Nineteenth of April.

Finding himself shut up in Boston, Oliver sat down to renew his correspondence with the Hutchinsons and the other refugees in London:

---

[71] Letters and Doings of the Council, 1774 (Mass. Archives), *passim. Publ. Colonial Soc. Mass.* XXXII, 472.

[72] Origin, p. 173.

[73] *Essex Gazette*, Mar. 28, 1775, p. 3.

[74] The picture and inscription are reproduced in *Mass. Law Quarterly*, XIII, No. 2, p. 7.

[75] Origin, pp. 175, 178.

This intercourse hath been interrupted by the Sons of Anarchy. . . . I feel the miseries which impend over my country: may Heaven avert them by the people's being convinced of the horrid crime of Rebellion, before it is too late. The God of Order may punish a community for a time with their own disorders: but it is incompatible with the rectitude of the Divine Nature, to suffer anarchy to prevail.[76]

He noticed, hopefully, the work of Divine Nature in the poor health of the republican leaders.

Even the hardships of the besieged could not dampen his spirit:

You who riot in pleasure in London, know nothing of the distress in Boston: you can regale upon delicacies, whilst we are in the rotations of salt beef and salt pork one day, and the next, chewing upon salt pork and salt beef. The very rats are grown so familiar that they ask you to eat them, for they say that they have ate up the sills already. . . .

We have little else to do now but to take snuff; we snuff in the air for want of food: we take snuff at the rebels for their barbarities: and we enjoy the snuff of candles, when we can get them to burn; and that sort of snuff often recovers expiring life.[77]

His writings add little to our knowledge of the war, but his account of meeting with one of the wounded returning from the Battle of Bunker Hill is startling:

I was walking in one of the Streets of Boston, & saw a Man advancing towards me: his white Waistcoat, Breeches & Stockings being very much dyed of a Scarlet Hue: I thus spake to him; "my Friend! are you wounded?" He replied, "Yes Sir! I have 3 Bullets through me." He then told me the places where; one of them being a mortal Wound: he then with a philosophical Calmness began to relate the History of the Battle; & in all Probability would have talked 'till he died, had I not begged him to walk off to the Hospital; which he did, in as sedate a Manner as if he had been walking for his Pleasure.[78]

During the siege Oliver was politically active. He regularly attended the meetings of the Mandamus Council, and was appointed by Howe to be one of the managers of the Loyalist Association.[79] For his services he drew a comfortable salary.[80]

---

[76] Peter Orlando Hutchinson, *The Diary and Letters of Thomas Hutchinson*, London, 1883, I, 457–8.

[77] *Ibid.*, I, 469, 573.

[78] Origin, p. 185.

[79] Letters and Doings of the (Mandamus) Council, p. 69; *New England Chronicle: or Essex Gazette*, Nov. 2, 1775, p. 3.

[80] Benj. F. Stevens, *Facsimiles of Manuscripts in European Archives, 1773-1783*, London, 1889, No. 2024, p. 18.

The Judge believed that during the siege he had frequent narrow escapes from assassination,[81] but he would have remained at his post had not Howe assured him that he could be of no further service in Boston. Moreover his health had failed during the siege, he had a long-pending suit over some Irish property — and the British were evacuating, anyway.[82] So on March 10, 1776, Howe having requisitioned quarters for him, he embarked with his niece Jenny Clarke in the *Pacific*, Indiaman. For two weeks they lay at anchor, the monotony being broken for the Judge by a dinner with Admiral Shuldham. On the night of the 20th he stood at the rail with other Loyalists and watched the terrible sight of the destruction of Castle William, symbolic to them of the destruction of their world:

> The blowing up of the Castle Walls continued: and at night all the combustible part of the Castle was fired. The conflagration was the most pleasingly dreadful that I ever beheld: sometimes it appeared like an eruption of Mount Etna; and then a deluge of fire opened to the view; that nothing could reconcile the horror to the mind, but the prevention of such a Fortress falling into the hands of rebels, who had already spread such a conflagration of diabolical fury throughout America, which scarce anything can quench.[83]

According to his housekeeper at Oliver Hall, the Judge took advantage of the confusion of his last hours in New England to slip ashore and ride to his old home, where, without daring to stop to eat or rest, he gathered a few of his valuables into his saddlebags.

Back on shipboard the next morning, March 27, Oliver watched the sandy shores of Nantasket slip by and recorded in his diary thoughts more sober than his public pose:

> Here I took my leave of that once happy country, where peace and plenty reigned uncontrouled, till that infernal Hydra Rebellion, with its hundred Heads, had devoured its happiness, spread desolation over its fertile fields, and ravaged the peacefull mansions of its inhabitants, to whom late, very late if ever, will return that security and repose which once surrounded them; and if in part restored, will be attended with the disagreeable recollection of the savage barbarities, and diabolical cruelties which had been perpetrated to support rebellion, and which were in-

---

[81] Jones, p. 223.
[82] Oliver to ——, Apr. 29, 1776, C. O. 5:175, Library of Congress transcripts.
[83] Peter Oliver, Diary, in Thomas Hutchinson, *Diary and Letters*, II, 47.

stigated by Leaders who were desperate in their fortunes, unbounded in their ambition and malice, and infernal in their dictates. Here I drop the filial tear into the Urn of my Country. . . . And here I bid A Dieu to that shore, which I never wish to tread again till that greatest of social blessings, a firm established British Government, precedes or accompanies me thither.[84]

Many of the Loyalists found that they could not endure the homesickness which their exile entailed, but so great was Judge Oliver's attachment to the greater whole of the British Empire, that the only pangs of separation which he experienced were literary. He went home happily and never regretted the change.

On April 3, 1776, Oliver landed at Halifax where he was taken in by Edward Lyde. He soon discovered that he detested Nova Scotia, and on May 12 he and Jenny embarked happily, as guests of Governor Legge, on the *Harriot* packet.[85] The Judge was seasick during the entire voyage, a fact which probably colored his reflections on the sight of Falmouth Harbor:

We harbored near that Island of Peace and Plenty where Government is, and can be supported, and where Rebellion hath formerly been check'd in its wanton career, and whose authority it is hoped, will suppress that American one, which exceeds in cruelty, malice, and infernal ingratitude, the united Rebellions recorded in history.[86]

As he rode the three hundred miles to London his eager mind took in "the Agriculture brot to Perfection, Churches, Gentlemens Seats, publick Works," and the like, until he was exhausted by the impact of the new and the curious.[87] On June 13 he reached the refuge of London:

Thanks be to Heaven, I am now in a Place where I can be protected from the Harpy Claws of that Rebellion which is now tearing out its own Bowels in America, as well as destroying all, who in any Degree oppose its Progress.[88]

The Judge and Jenny lodged first in Jermyn Street, where they found three furnished rooms for a guinea and a half a week. Thence they moved in with Governor Hutchinson for a time, and finally

[84] *Ibid.*, II, 48.
[85] *Connecticut Courant and Hartford Weekly Intelligencer*, July 1, 1776, p. 2.
[86] Hutchinson, *Diary and Letters*, II, 52.
[87] Oliver to Byfield Lyde, June 24, 1776 (New York Public Library).
[88] Peter Oliver, Diary (British Museum), June 13, 1776.

settled in a house in High Street, Marybone Gardens, with Peter, Jr., who had preceded them.[89]

Hutchinson welcomed Oliver, took him on the round of the government offices, and in July carried him to Oxford where he was awarded the "Degree of Doctor, in Jure Civili, honoris causa," with scarlet ceremony which enchanted the heart of the Judge.[90] Looking about him he praised God that "Cromwell's Rabble" had spared Oxford. With Hutchinson he toured England, recording carefully in his diary his impressions of such things as a horse race at Guilford:

The Rabble with their Trulls, threw into my Mind the Recollection of a New England Cambridge Commencement on a showery Day. Such an Expense of Time and Money is perhaps necessary in a Nation of so much Liberty, where otherwise the People would run mad with Politicks, instead of imploying their Time & Thoughts about Diversions; & if so, then farewell to the Liberty of old England forever.[91]

There were some aspects of social England, such as the employment of children as chimney sweeps, which he rationalized with difficulty:

A Person is apt to pity these Boys, but they are always so merry in the Chimny as well as out of it, that it is a Pity to disturb them of their Mirth even by a contracted Muscle but if you incline to it, a single half penny plunges them into as much Happiness as they can grapple with.[92]

Chain gangs he approved as a better deterrent of crime than executions, and he found nothing objectionable in the swift justice of the Old Bailey:

There were 8 Criminals tried for their Lives in the Space of two Hours, & I never as yet have seen fairer Trials. The Prisoners had no Counsel allowed them; it being always taken for granted that the Judge is Council for the Prisoner.[93]

The only thing which he admitted finding disagreeable in this barbarous judicial system was the ever-present town gibbet.

Somewhat to Hutchinson's irritation, the Judge took a great

[89] *Continental Journal*, Dec. 19, 1776, p. 3; Nov. 28, 1776, p. 2.
[90] Oliver's long account of the ceremony is in Hutchinson, *Diary and Letters*, II, 77.
[91] Oliver, Diary (B. M.), May 23, 1777.
[92] *Ibid.*, Aug. 6, 1776.
[93] *Ibid.*, July 11, 1776.

interest in magicians and similar popular shows. He even took the trouble to make a picture of a merry-go-round with wooden horses for children.[94] Public affairs interested him not at all, although he complained that he had no work to do. If he had any inclination to polish his old practice of polite literature as a means of taking up his time, he found himself discouraged by its state in England:

As in New England, every Person you meet with is an Adept in Laws, Physick & Divinity, so in England the Rage of Poetry is such that You see the Fragments of it upon every Sign Post & hear the Bawlings of it from the Mouth of every Billingsgate & Cinder Wench Who infests the Streets.[95]

As an old iron-master he was amazed by the development of British heavy industry with its steam engines, railways, and other labor-saving devices.

Judge Oliver kissed the King's hand on August 7, 1776, and petitioned that his salary of £400 as chief justice be continued. He received a pension of half that sum, and he was compensated for only about half of his property loss of £5000. His estate in Massachusetts was confiscated on April 30, 1779, and he was proscribed as an absentee.[96] On November 4, 1778, Oliver Hall, which was standing empty, took fire, as his friends expected that it would. The townspeople made no effort to save the house but took the opportunity to plunder it effectively. For many years their woman-folk on gala occasions wore in their hair bits of the gilt stripped from the woodwork. The orchards and the ornamental trees were maliciously cut down, and in a few years the site had reverted to the wild. The housekeeper's account of the process sounds like something out of Gibbon, and indeed it marked the passing of a culture.[97]

Safe in London, Judge Oliver was not at all concerned in what was passing in America. Like Hutchinson, he had his hands full supporting a large family in England and scattered relatives in

[94] *Ibid.*, July 7, 1777.

[95] *Ibid.*, Oct. 23, 1776.

[96] There is a detailed description of the property in the *Independent Chronicle*, Mar. 16, 1780, p. 2. It was occupied as early as September, 1775, by James Bowdoin (A.B. 1745) who entertained Washington and Franklin there. — Samuel Cooper, "Diary" in *Am. Hist. Rev.* VI, 320, 323.

[97] Thomas Weston, *History of the Town of Middleboro*, Boston, 1906, pp. 371–3.

America. Sometimes he borrowed money to help other refugees.[98] As little as he wanted to go back to America, he appears to have accepted appointment in 1780 as governor of New Ireland, a province to be erected in northern Maine, on a salary of £1200 a year.[99] When the course of the war put an end to that dream he philosophically put himself to writing his "Origin and Progress of the American Revolution," the thesis of which is that the American rebellion was historically peculiar in that it did not result from tyranny, but from the over-indulgence of the mother country. When peace was signed, he warned that America would attack England again as soon as she had restocked with English manufactures.[100]

Much to the bewilderment of their friends, Judge Oliver and Jenny in May, 1778, moved to Birmingham, intending to find a country place near that industrial beehive. He preferred the smoke and noise of manufacturing to the high society of London, which he never wanted to see again. In his new home he enjoyed the personal acquaintance of Dr. Priestly and James Watt of the steam engine,[101] and set himself to the compilation of a Biblical lexicon which ran through several editions and is said to have been a textbook at Oxford. His correspondence was as gay and amusing as ever, but during the winter of 1785–86 his health failed. In 1788 he became a helpless invalid, faithfully served by Jenny.[102] From this sickbed he proudly paid off the last of his pre-war indebtedness to the Hutchinsons. He died at Birmingham on October 12, 1791, and was buried under St. Philip's Church, where his son Peter erected a monument to him.[103] His sons Daniel (A.B. 1758) and Andrew (A.B. 1765) had died before him.[104]

Some of the references to the diary of Judge Oliver are misleading. The manuscript at the British Museum, which is the earliest of which there is any evidence, does not begin until March 24, 1776. Portions are quoted in the *Diary and Letters* of Thomas

[98] Gardiner Mss. (M. H. S.), II, 89.
[99] 3 *Coll. Maine Hist. Soc.* I, 151; Oliver to William Knox, Aug. 19, 1780, C. O. 5:175, Library of Congress transcripts.
[100] Hutchinson Mss. (B. M.), Jan. 18, 1784.
[101] Elkanah Watson, *Men and Times of the Revolution*, New York, 1856, p. 155.
[102] *Winslow Papers*, St. John, 1901, p. 548.
[103] Peter Oliver, Jr., Diary, Oct. 19, 1791.
[104] For the children see the *New England Hist. Gen. Reg.* XIX, 104–5.

Hutchinson and there is a microfilm of the whole at the Massachusetts Historical Society (Hutchinson Mss. No. 4). Later volumes of the diary, said to be purely travel journals, were not turned over to the British Museum by the family. The family carried to England a number of manuscripts relating to American history, the most important being a manuscript of Hubbard's History of New England. The efforts of the Massachusetts Historical Society in 1814 to borrow this for purposes of publication led to an unhappy exchange with Peter Oliver, Jr., who proposed that the Society return Hutchinson manuscripts which had been "stolen by the mob." [105]

The Olivers apparently recovered and carried to England the eight family portraits which were taken from Oliver Hall and stored at Plymouth by the custodians of confiscated goods. There are probably three original portraits of Judge Oliver surviving. The youngest is of the group of Oliver boys and is now at the Museum of Fine Arts, Boston. It is reproduced in *One Hundred Colonial Portraits*, Boston, 1930, p. 59. A portrait by Smibert belonging to the daughters of the late Peter Oliver (A.B. 1922) is reproduced in this book. Mr. William H. P. Oliver of Morristown, New Jersey, has the original of the miniature portrait of the Judge by Copley which is reproduced in the *Massachusetts Law Quarterly*, XIII, No. 2, p. 1. There is a copy in miniature in the possession of the Boston Olivers, and a large portrait based upon it in the Social Law Library of Boston. Mr. William H. P. Oliver also has the original Copley of the Judge standing by his wife's tomb, reproduced in the *Massachusetts Law Quarterly*, XIII, No. 2, p. 7. There is a copy of this attributed to an English artist in the possession of Miss Susan Oliver of Boston. The Morristown Olivers also have a photograph of a portrait of Judge Oliver as an old man wearing his judicial robes; the location of the original is unknown. The Harvard Medical School has a handsome tall clock which he presented in 1790 to Benjamin Waterhouse, husband of Elizabeth Oliver, daughter of his son Andrew.

---

[105] 2 *Coll. M. H. S.* III, 286–90.

JOSEPH SECCOMBE

Oliver Wendell Holmes expressed the popular conception of the policy of the New England settlers toward the Indians when he said that the Puritans on landing "fell upon their knees and then upon the aborigines." Actually, the English settlers made a great effort to integrate the natives into their civilization. As a result, there were, at one time or another, one hundred and one Christian Indian villages, some with native ministers and town officers. For the reasons for the failure of these efforts see the sketches of the missionaries, such as Gideon Hawley and Stephen Badger in Volume XII of the Biographical Sketches. *Seccombe lacked the devotion of his more successful fellows.*

JOSEPH SECCOMBE, missionary to the Indians and third minister of Kingston, New Hampshire, was a grandson of the immigrant, Richard Seccombe of Casco, and a son of John and Mehitable (Simmons) Seccombe of Boston, where he was born on June 14, 1706. John was a very obscure innholder who rented at the lower end of Wing's Lane and attended the North Church, where Joseph was baptized. Mehitable was a member of the Old South, which Joseph joined on October 6, 1723. That church had generous charity funds from which young Seccombe was educated. In June, 1726, it sent him to Ipswich Hamlet to study under the Reverend Samuel Wigglesworth (A.B. 1707), and with its bounty followed him through college.[1] He entered Harvard with the Class of 1732, in which he was placed surprisingly high, considering his age and poverty. In October, 1729, he was "advanced a year, and placed

---

[1] *New England Hist. Gen. Reg.*, XV, 309–13.

lowest in the Junior Sophisters Class," not because of his scholastic ability, but because of his age, financial standing, and purpose to enter the ministry. The aid which he received from the Old South was supplemented by a Hollis scholarship, and these between them met most of his expenses. After graduation he kept school at Ipswich Hamlet for a time, but returned to Cambridge as a Hopkins scholar. However, he forfeited the scholarship [2] early in 1732 by leaving Cambridge to take part in a missionary venture which excited the pious in both New and old England.

The Edinburgh Society for Propagating Christian Knowledge among the Indians had instructed Governor Jonathan Belcher (A.B. 1699) and the Reverend Benjamin Colman (A.B. 1692) to choose three missionaries to work at its expense. Seccombe was the first who offered, and in the depth of the Winter of 1731–32 he sailed for Fort St. George's (in the present town of St. George, Maine) carrying this letter from Governor Belcher to Captain John Giles:

This I intend you by the hands of Mr. Joseph Seccomb, who comes as a missionary (from the Society in Scotland for propagating Christian Knowledge) in order to Christianize the Indians in your parts. I therefore desire you to receive him into your fort, and treat him kindly and respectfully; and as Mr. Pierpoint [A.B. 1721] is come hither, not intending to return to you again, Mr. Seccomb may for the present preach in the garrison, and you must perswade as many Indians as you can to come and hear him, and you must interpret to them, and assist him what you can to gain the language. He must also pray and read the Scriptures to the garrison morning and evening, and so supply Mr. Pierpoint's place.[3]

Colman's first report on this mission to the SPCK was made on November 14, 1732:

I come now Sir to a Particular thing, which nearly concerns me, having much encouraged your Missioner Mr. Secomb at St. Georges River in the East, early to enter Himself in that difficult Post. He has been serving there already nine or ten Months before we coul'd find the other two to be his companions. . . . I humbly move therefore that the Honourable and Reverend Society would please to allow and remit for Mr. Secombe, what they shall think just and reasonable for his three-quarters of a Year Advance before his Brethren. . . . Your first

---

[2] College Papers (Harvard University Archives), I, 79, item 164.

[3] Mass. Hist. Soc., *Collections*, series 6, VI, 103.

Missioner has already highly approved himself to us in his zeal and Labours, and poor as he is has already, to ingratiate himself with the Salvages, scattered among them many Pounds in little things which are acceptable to them: And he has done this without Knowing that his Allowance from the Society could not begin till the other two were entered after him; which when he heard of and wrote to me, I have ventured to encourage Him that the Society would in their Goodness and Justice please to advance to Him something in proportion to the time he has been before the rest in their Service.[4]

Seccombe's reports to Colman are personally revealing:

Reveiwing my Minutes I find I have writ of my design to present several curious Furrs, with a description of the Creatures, of their Manner of living, etc, which I had received from my Captain and the Indians I have conversed with: having these Intimations and with something of my own observations by dissections etc. and finding little of the Nature but what we derive from the French (to whom, to our Shame it may be mention'd, we are beholden for the knowlege of our own Country) this fired my Soul with the Emulation of the present mentioned. But upon deliberate Consideration, I have thot it may be rather offensive than pleasing . . . that by making these observations I obstruct the design of the Mission, tho' I hereby take the Indians in their own talk, discourse with them in the most pleasing manner.[5]

The three young missionaries, Seccombe, Stephen Parker (A.B. 1727) and Ebenezer Hinsdell (A.B. 1727) were recalled to Boston to be ordained on December 12, 1733, before a vast multitude which regarded the young men as heroes, no doubt because they were an expression of the uneasy social conscience of New England. The ceremony took place

at our South Brick Meeting House, before a great and reverend asembly of Ministers and People. The Rev. Mr. Cooper [A.B. 1712] began the Exercise with Prayer, Dr. Sewal [A.B. 1707] preach'd from Acts xxvi, 18. To open their Eyes and turn them from Darkness to Light, and from the Power of Satan, unto God, etc. Then Mr. Webb [A.B. 1708] Pray'd, Dr. Coleman gave the Charge, and Mr. Prince [A.B. 1707] the Right Hand of Fellowship: Several other Reverend Ministers of the Town Joyn'd in the laying on of Hands.[6]

The temperature in the meetinghouse that day was so bitter that even those cold-hardened ministers cut short the service and ac-

[4] Colman Mss. (Mass. Hist. Soc.), Nov. 14, 1732.

[5] Ibid., July 17, 1733.

[6] Boston News-Letter, Dec. 13, 1733.

cepted with alacrity when "His Excellency invited the Commissioners, Ministers and all the Messengers of the Churches to a Noble Supper at His own Seat." [7]

Even the newspapers caught the missionary spirit and said that the young men were "returning (by the Will of God) accompanied with the Prayers of all the People of God, for their Success in a Mission of so great Difficulty and Importance." [8] Indeed the instructions which the young men received called upon them to accomplish a much more difficult work than their Jesuit rivals attempted:

You are to instruct the Heathen People . . . in the Principles of the Christian, Reformed, Protestant Religion: And in order thereunto, not only to preach and catechize, but also to school and teach them to read the Holy Scriptures of the Old and New Testament, and other good and pious Books. You are also to teach them Writing and Arithmetick, and to understand and speak the English Language. . . . And when you shall judge them fit to receive the Seals of the Covenant of Grace, you are to administer the same unto them. [9]

This program to its contemporaries seemed by no means too ambitious, for there were in that generation in the settled parts of New England many Christian Indian villages, some with their own churches and native preachers.

When the Commencement of 1734 came around, Seccombe sent in a particularly suitable argument to the effect that in Apostolic times, evangelists were distinct from pastors and presbyters. The Corporation, "considering his distance, and his being Imploy'd in public Service," voted him his M.A. in absentia.

The news of the triple ordination of the missionaries excited great interest in England. Captain Thomas Coram, the philanthropist, wrote enthusiastically to Colman, "Mr. Seccombe Stands high in the opinion of the Earl of Egment and some others here I hope he will deserve always to be so and that it may be for his hon'r and advantage." [10] The Bray Associates poured upon the young men shipments of books, a commodity which naturally attracted the notice of the Boston newspapers. Seccombe alone received 220 volumes in one shipment from the Bray Associates, and Coram sent

[7] *New-England Weekly Journal*, Dec. 17, 1733.

[8] *Ibid.*

[9] Joseph Sewall, *Christ Victorious*, Boston, 1733, p. 30.

[10] Mass. Hist. Soc., *Proceedings*, LVI, 26.

him more with the stipulation that they belonged to him personally and not to the mission.

In his letters of acknowledgment to the Bray Associates, Seccombe told of his "solitary hazardous Life," of lodging with the Indians in the woods in January, and of spending most of his salary of £40 on necessary presents.[11] He was discouraged by his lack of progress, as well he might be. He was bound by his instructions to teach the savages English, reading, writing, and arithmetic, and then, if they seemed likely subjects, to admit them to the church, where by reading and reasoning each might fight his own way to God. The Jesuits, on the other hand, could instantly and painlessly bring any Indian to the same end if only he consented to be baptized. In one of his letters Seccombe described this hopeless contest for the souls of the Indians:

As to Religion they are extremly Frenchify'd, having for above Thirty years had Jesuits among them, who have bro't them to be strenuous Biggots to the Church of Rome. I have seen Two of these Jesuits, and one of them is a subtle old Gentleman, very warm and zealous in his Disputations, which are frequent, either Verbal or in Writing; when he keeps to the Bible he never gets the better of Me; but when he comes to Fathers and councils, I am at a Loss, for he will not acknowledge our Translations, and the Originals are not to be purchased in this Country. I have been at considerable Pains and Charge, to get some of their Children to live with me till they have learn'd to read; but the Jesuit, threatens them so much with Excommunication, that they dare not tarry long with me.[12]

The "subtle old Gentleman" was Father Stephen Lauverjat, or Lauveriat, who sent Seccombe incredibly long theological documents addressed to "Domine et Fili charissime Pax Christi." [13] He passed these on to Colman who addressed his replies to "Domino Stephano Lauveriat, E Societate Jesu, Paunabscot. Domine, et Frater Charissime, Pax Christi." [14] Seccombe's covering letters to Colman were not so courtly:

Mr. Lauverjat follows me with continual essays to make a Pervert [sic] of me. I have sent the most plausible of them, in the same foul Condition I receiv'd them; not knowing but that some Gentlemen might

[11] *Boston News-Letter*, Apr. 17, 1735.
[12] *Ibid.*
[13] Colman Mss., Apr. 13, 1734.
[14] Ebenezer Turell, *Life and Character of . . . Benjamin Colman*, Boston, 1749, pp. 64–9.

have the Curiosity to peruse it, if its Sordidness should not render it nauseous at first View.[15]

The picture of Seccombe "Grapling with the Jesuit" excited wide interest in England, and Captain Coram showed his letters to the principal men of Liverpool who promised their aid. In both England and America the Captain urged "the Purchasing proper books for the better Enabling those young Davids on the Borders to Beat down the old Goliahs French Jesuits." To Colman he wrote:

I am fully Satisfyed Mr. Seccombe is as you say not so learned as one or both of the other Missioners, w'ch cannot be wondered at if as I have heard his Parents are poor, And the others Rich who could be at greater Expence to Edefy them but as he was sent to the place of the Greatest Difficulty and at a vast Distance from the emidiat Inspecktn of the Commissioners . . . and behaved well in it and was thankfull for what he had received and gave some notable perticuler accounts of the Indians etc. I thought he deserved some perticuler Notice.[16]

When the news reached Coram that the New England commissioners for the SPCK were about to abandon the St. George's mission, he tried to enlist the aid of Isaac Watts who, although willing to help, thought that they ought to defer to the men nearer the scene: "I must own I know not what to say about his continuance, since he must contend at once with heathenism and popery." [17]

Early in 1737 Seccombe laid down his mission. He petitioned the General Court for the £100 which he said was due him "for his Essays as Missionary to the Indians, for above five years, at the Garrison on St. George's River." The Council granted him only £33 8s.[18]

On August 10, 1737, a committee from the town of Kingston, New Hampshire, invited Seccombe to preach. Evidently he gave satisfaction, for on October 17 the church called him with "utmost Unanimity," and a few days later the town ratified the call "with the greatest Unanimity." This is his own record of his installation:

Nov. 23d. I was Installed, The Reverend Mr Fogg [A.B. 1730] being ordained the same Day, who was Born and had ever lived in the

---

[15] Colman Mss., June 27, 1734.
[16] Mass. Hist. Soc., *Proceedings*, LVI, 34.
[17] Mass. Hist. Soc., *Proceedings*, series 2, IX, 354.
[18] Mass. Archives, XI, 501; Executive Records of the Province Council, X, 141.

Neighbourhood, I being a Stranger in these Parts, had few of the Neighbouring Ministers. My Master, the Reverend Mr Wigglesworth, being sick, the Reverend Mr Hale [A.B. 1699] somewhat indisposed and being late in the Year the Boston Ministers did not come. These excused themselves by very handsome Epistles, signifying their ardent Wishes for our Prosperity, etc. Therefore, the Rev. Mr. Flagg [A.B. 1725] of Chester introduced the Solemnity with Prayer, I preached from 7 Mark 37. The Reverend Mr. Odlin [A.B. 1702] of Exeter gave me the Care of the Church, the Reverend Mr Bacheller [A.B. 1731] appointed some Part of the 132 Psalm. The Rev. Mr. Seccomb [A.B. 1728] made the last Prayer and I pronounced the Blessing.[19]

Seccombe married Mary Thuriel on January 17, 1737/8, and busied himself building what was described as a mansion house. Almost simultaneously he petitioned New Hampshire asking for a grant of land in Kingston, and signed a petition for the annexation of the town by Massachusetts. Later he did receive a grant in Salisbury, New Hampshire, from the Masonian proprietors.[20]

It may well be that life in the Maine woods had injured the Parson's health, for in dedicating his first printed sermon to Colman, he wrote:

Several of my very good Friends . . . desired it for the Press; and the signal of their Affection I gratefully receiv'd, tho' I could not willingly then part with it; Modesty restrain'd me; but since my last dangerous Sickness, I have been less afraid.

If others have whereof they may boast, I have those Things, in which, I might glory also: sure none will envy me the Honour of glorying in my Infirmities. This, I humbly hope, may redound to the Glory of God, who suffered me to fall into very low Circumstances; low by Sickness, and (thro' the frequent Misfortunes of my Parents) low by Poverty.[21]

This last sentence was typical of Seccombe, who had the worst social inferiority complex among the Harvard men of his generation. His townspeople long remembered that he was a poor man's son, was good, able, eccentric, and extremely fond of fishing. He certainly had none of his cousin John Seccomb's poetic ability if, as his contemporaries thought, the verse dialogue *On the Death of*

[19] Sibley's Letters Received (Harvard University Archives), IV, 133.

[20] *Provincial Papers . . . of New Hampshire*, Concord, etc., 1867–1944, IX, 350; XII, 336, 338; XXVIII, 217.

[21] *Plain and Brief Rehearsal*, Boston, 1740, pp. i–ii.

the Reverend Benjamin Colman came from his pen. All that can be said for the author is that he had read English poetry.

A passion for fishing drew the Parson, a black, humble, diffident figure, to the edges of the circle of gay and gorgeous aristocrats who yearly gathered to sport at Amoskeag. They respected his education and profession as only New Englanders would, and asked him to preach the sermon without which even gay parties were not complete. To their delight he complied with what is still famous as the first American work on sport, in which he lashed at the "Popish superstition" which led men to withdraw from business and social pleasures. For several years the chief men of New Hampshire importuned him to give them the manuscript to print. When he finally complied, it had this typical preface:

To the Honourable Theodore Atkinson [A.B. 1718], Esq; and Other The Worthy Patrons of the Fishing at Ammauskeeg. Gentlemen, It's not to signify to others that I pretend to an Intimacy with you, or that I ever had a Share in those pleasant Diversions, which you have innocently indulged your selves in, at the Place where I have taken an annual Tour for some Years past. Yet I doubt not but you'l Patronize my Intention, which is to fence against Bigottry and Superstition.

The defence of pleasure against "Bigottry and Superstition" was a favorite theme all through Seccombe's life, and he preached it to the humble as well as to the great:

Recreation and Exercise are necessary for the better Support of the Body; and the Soul, during her Residence in the Flesh, must have Suspensions from her more abstracted and serious Pursuits. It has pleased the great Author of our Being, that these Things should be so, and it will never displease him that we indulge either, under such Limits as Religion and Reason appoint.[22]

Indeed he so belabored this subject that one wonders why. The educated New Englanders had never questioned the desirability of innocent pleasures. Perhaps he was protesting the souring of religion which came with the Awakening.

Seccombe had found the Kingston church under the Halfway Covenant and had conceded the election of deacons in order to obtain the use of written music for congregational singing. He was believed to be the author of a very weighty and learned argument

[22] [Joseph Seccombe], *The Ways of Pleasure and the Paths of Peace*, Boston, [1762], p. 8.

for the new music, *An Essay to Excite a Further Inquiry into the Ancient Matter and Manner of Sacred Singing*. He failed, however, in his efforts to get his congregation to abandon the horrors of the New England Psalm Book for a more singable translation. His church had a local revival without serious disruption, and he conspicuously abstained from joining in the opposition to Whitefield. His own preaching partook of the enthusiastic air:

O! the inexpressible Amazement! the intollerable Confusion of the disappointed Soul: who under Hypocritical Delusions, dies like a Lamb easy and calm, with conceit of its Innocence and Safety. . . . And in a Moment is seiz'd by ghastly Fiends, condemn'd of God and hurl'd away to gloomy Caverns of unutterable Anguish and eternal Despair! O! O! O! Heavens, Earth, Hell, Men, Devils! who, where, what am I! Confusion, Wrecks, Torture . . . Oh no! my deceiving devilish damning Hypocrisy come to this! [23]

In view of this sample of his preaching it is surprising to find him opposing the excesses of the revivalists and denouncing the "Tricks which all your former Haranguers, Enthusiasts, Buffoons, and others, whose Trade it was merely to move the Passions, knew and practic'd in Perfection." [24] He had no use for the hair-splitting theology of the New-Lights and their "Subtilities" of Grace vs. Works, but he was keenly aware that they were largely the newer immigrants, the poor, and the poorly educated, with whom he always identified himself, whereas the Old-Lightism prevailing among Harvard men was in some degree snobbery and blind conservativism: "The Prejudices of Custom and Education are evident to every one. Who is not apt to plead for the good old Way of their Forefathers." [25] In the end he threw in his lot with the Harvard Old-Lights, for he subscribed for two copies of the *Seasonable Thoughts* of Charles Chauncy (A.B. 1721).

Parson Seccombe's social inferiority complex kept him within the circle of his choosing. He had no children, but a nephew, Simmons Seccombe, and the town physician, Josiah Bartlett, lived with him. When he died, on September 15, 1760, the town laid out £510 on his funeral. His widow advertised for the return of books

[23] Joseph Seccombe, *Reflections on Hypocrisy*, Boston, 1741, p. 16.
[24] [Joseph Seccombe], *A Specimen of the Harmony of Wisdom and Felicity*, Boston, 1743.
[25] [Joseph Seccombe], *Some Occasional Thoughts*, Boston, 1742.

which had been borrowed from his library, which contained something over five hundred volumes. This was sold by the executor, who prepared a catalogue and advertised the sale.[26]

JOHN WINTHROP

PROFESSOR JOHN WINTHROP was one of the sixteen children of Judge Adam Winthrop (A.B. 1694) of Boston, who thus recorded his arrival: "1714 December 8th Wednesday about half an hour before one a Clock in the morning my Wife was de[livered] of a Son who was the next Sabbath Baptized John at the North Church by Dr Cotton Mather."[1] There must have been witchcraft in the drops of water which fell from the fingers of the first great American man of science onto the head of the infant who was to become, next to Franklin (a friend of both of them), the greatest American scientist of the colonial period. When John entered Harvard, at the age of thirteen, he was placed at the head of his class, a recognition of the public services of his father and his ancestors, but an honor to which he was entitled on his own account. The commonplace book which he began as a Freshman is one of the most interesting and mature documents of its kind to survive. It shows that his interest in science was fed by the description of recent European discoveries in Cotton Mather's *Christian Philosopher*. Poetry, however, occupies as many pages as philosophy, and includes verse from the *Spectator* and the *Tatler*, and by Pope, Mather Byles (A.B. 1725),[2] and Swift. The selections from the Dean are very vulgar and irreverent. In this, and in his interest in a defence of the art of dancing, John showed how much he differed from his father, who was reactionary in everything which he thought pertained to religion. But the detailed notes which John as a Freshman took on the sermons of Nathaniel Appleton (A.B. 1712) show

[26] *New-Hampshire Gazette*, Apr. 24, 1761, p. 2. The Ms. catalogue is in the New Hampshire Historical Society; there is a photostat of it in the Harvard College Library.

[1] Winthrop Mss. (Mass. Hist. Soc.), XX, 81.

[2] John Winthrop, Commonplace Book (Harvard University Archives), pp. 113–6.

an amazing maturity and a thorough grounding in theology.[3] Unfortunately he did not have the instincts of a good diarist, but in his diary recorded nothing more interesting than the weather when one would expect him to report such exciting Cambridge events as the theft of the body of Thomas Stokes, who had been executed for burglary, to be dissected in Boston.[4]

On other occasions when the youngest member of a Harvard class was, for reasons of family, placed at its head, he usually set a lamentable example. Not so John, whose conduct and cooperation with the college were flawless, even to the paying of his term bills with casks of flour instead of with troublesome old cows or silver spoons of doubtful character. He regarded the solid classical core of the college curriculum as delightful literature, and showed such proficiency in every branch that he was awarded a Hopkins Prize, his deturs being Bishop Beveridge's *Thoughts on Religion* and Cotton Mather's *Ratio* and *Manuductio*. He delivered the Oratio Salutatoria at his first Commencement, and then returned to continue his studies. In 1733 he joined the First Church of Cambridge with a group of students. His duties as a resident graduate were performed with the same regularity until October, 1734, when President Wadsworth with amazement recorded the first resistance which he had received from his exemplary student:

The College Law requiring, that Batchelours should dispute once a fortnight from September 10 to March 10 I spake seasonably to Sir Winthrop (Senior of the Bachelours) to prepare for disputing and also analysing; he desir'd to be excus'd from both, because he thought of leaving the College in about a fortnight or 3 weeks time. . . . A while after, I sent for Sir Winthrop again, he insisted as before, that he was about leaving the College speedily, though as yet not absolutely determin'd when.[5]

In November he finally gave up his chamber and cellar, and left the college. He was back at Commencement, 1735, when he was prepared to prove his fitness for the Master's Degree by arguing that it is not permissible for the magistrates to impose hardships on anyone for maintaining his own religious views.

In 1738, when Isaac Greenwood (A.B. 1721) drank himself out of the only professorship of science in America, the college was very

---

[3] In the Harvard University Archives.
[4] Boardman Diaries (Harvard College Library), Mar. 21, 1734.
[5] Colonial Soc. Mass., *Publications*, XXXI, 498.

JOHN WINTHROP

much at a loss as to a successor. Tutor Nathan Prince (A.B. 1718), the obvious candidate, had the same vice, among others. So, upon the urging of Parson Appleton, the Corporation on August 30 elected the twenty-three-year-old Winthrop to the Hollis professorship. When the Overseers met three weeks later they debated "whether a proper presentation agreable to Mr. Hollis's rules of the Corporations Choice of a Mathematical Professor had been now made" but finally voted that "the above question should not be put." [6] On October 3 the Overseers appointed a committee to examine the candidate "as to his knowledge in the Mathematicks," but rejected a motion that a second committee be appointed "to Examine Mr. Winthrope about his principles of religion." [7] Two weeks later the committee reported back that the candidate had made "very great proficiency" in mathematics and was "wel qualifyed to Sustain the office." Again the question of examining him as to his religious views was raised, and this time it was voted to put the question over until the next session of the General Court, when most of the lay members of the Board of Overseers would be in Boston, and to send special notice to the six clerical members. A very full meeting of the Board on December 7 voted down the motion that he be examined as to the principles of religion and ratified his election as "Hollisian Professor of the Mathematicks and of natural and Experimental Philosophy." [8]

The installation was set for January 2, 1738/9, and the House of Representatives was invited to attend. When the Overseers met for the ceremony on the day appointed, the religious conservatives made a last effort to have Winthrop's principles examined, but were overruled after a "great debate." [9]

The solemnity began near twelve of the clock. The Corporation and Overseers went down from the Library into the Hall and took their places at several Tables. The President began with prayer which was followed by an Introductory Latin Oration. Mr. Professor Wigglesworth was desired to read Mr. Hollis's rules and Statutes respecting the Professor of Mathematicks and Then the Oaths to the civil Government were read by Mr. Flynt and repeated verbatim by Mr. Winthrope. Then the printed Copys of them were signed by Mr. Winthrope who

---

[6] Overseers' Records (Harvard University Archives), I, 176.

[7] *Ibid.*, p. 177.

[8] *Ibid.*, p. 180.

[9] *Diaries of Benjamin Lynde and Benjamin Lynde, Jr.*, Boston, 1880, pp. 154–5.

was Thus Sworn before his Excellency Jonathan Belcher. . . . After this Mr. Winthrope was desired by the President to Express his declarations and Promises according to the 12 and 13th articles of the rules which being done The President Called for Mr. Winthropes Inaugural Oration which oration in Latin being finished The President asked Leave of the Overseers and Corporation to declare Mr. Winthrope Hollisian Professor. . . . After this Dr. Sewal made the Last Prayer and some stanzas of Psalm 148 were sung and the Overseers and Corporation went up to the Library whilst the Tables were spread and returned with other gentlemen to dinner in the College Hall.[10]

Newspapers as far away as Philadelphia carried even fuller accounts supplying such supplementary information as that the dinner was "very plentiful" and the whole ceremony conducted "with great Decency." [11]

The only protest came from the disappointed Nathan Prince, who declared that "that Boy" Winthrop "knew no more of Philosophy than a Fowl"; "I could teach him his A. B. C. in the Mathematicks." [12] As a matter of fact, the young professor had a nearly universal knowledge, including even an unusual command of the oriental languages then available. Even more important, he had the curiosity, the critical spirit, and the orderliness essential in a scientist. Walking on Boston Common on April 19, 1739, he observed sunspots and began the first of his important studies, in which he seems to have guessed the connection between the spots and the aurora.[13] In 1740 he reported to the secretary of the Royal Society his observations on a transit of Mercury and of an eclipse of the moon. These were published in the *Philosophical Transactions*,[14] the first of his eleven articles to appear in that important serial, and were later reprinted in the *Memoirs* of the Royal Academy of Sciences at Paris.[15] The reply of the secretary of the Royal Society was most flattering:

[10] Overseers' Records, I, 181–2.

[11] *Pennsylvania Gazette*, Feb. 1, 1739, 3/2.

[12] Nathan Prince Case Mss. (Harvard University Archives).

[13] John Winthrop, Observations on Sun Spots (Harvard University Archives); Frederick G. Kilgour, "Professor John Winthrop's Notes on Sun Spot Observations," *Isis*, XXIX, No. 2. In view of the forthcoming biography of Winthrop by Frederick E. Brasch, this sketch will not attempt a thorough coverage of his scientific career.

[14] *Philosophical Transactions*, XLII, 572; LIX, 505.

[15] Stephen Sewall, *An Oration . . . at the Funeral of . . . John Winthrop*, Boston, 1779, p. 5.

Sir, The Royal Society were greatly pleased with your observation of the transit of Mercury over the Sun . . . and with your other communications, for which I am desired to return you their thanks, and to beg of you the continuance of your observations. I am glad to find the Hollisian Chair so well filled, particularly as I knew Mr. Hollis the founder of the Professorship; and as you enjoy it, being one of that illustrious family so eminent in the history of New-England.[16]

Over the years Winthrop won more distinction for Harvard than any predecessor had done; but he had been hired to teach. His lecture hall and laboratory was the western room of the second story of Old Harvard, and he made it a place of significance in the history of science in America. There, on May 10, 1746, he gave the first practical demonstration and experiment in electricity and magnetism in a laboratory in this country.[17] Among the scientific instruments which he had Franklin obtain in London was an electric battery, lovingly assembled by the philosopher himself, with which the Professor performed demonstrations for generations of college students.[18] Jeremy Belknap (A.B. 1762) years later described his experiments in physics:

The glass bubbles . . . I remember to have seen used in Dr. Winthrop's course of Experimental Philosophy, to evince the elasticity of the air. One of them was put on a lighted candle, and exploded with a report equal to a pocket pistol. There is another sort made by dropping melted glass into water, which I think he told us was a "Nodus philosophorum," and could not be explained satisfactorily. On breaking the point, the whole mass falls into dust.[19]

He was far, however, from being satisfied with the tricks of the medieval philosophers. Once when a collector of Mathematical Theses for the Commencement exercises handed him several based on Cartesian hypotheses, the Professor told him that the design of the publication of the Theses sheet was to show the progress of knowledge, and that the ones proposed would not answer that end. When the student refused to undertake the preparation of others, he did it himself.[20]

[16] Edward Wigglesworth, *The Hope of Immortality*, Boston, 1779, p. 21 *n.*
[17] Deduced by Mr. Brasch from Winthrop's notes in the Harvard Archives: *Scientific Monthly*, XXXIII, 451. For the teaching of science under Winthrop see I. Bernard Cohen, *Some Early Tools of American Science*, Cambridge, 1950.
[18] Mass. Hist. Soc., *Proceedings*, series 1, IV, 117.
[19] Mass. Hist. Soc., *Collections*, series 5, II, 88.
[20] William Bentley, *Diary*, Salem, 1905–14, I, 158.

In the teaching of mathematics, Winthrop introduced, in 1751, the study of fluxions, out of which have grown the differential and integral calculus of modern times. The average student, like Timothy Pickering (A.B. 1763), found his lectures far too rarefied:

Mr. Winthrop also attended the class a few times, when they were learning arithmetic. . . . He touched on a few matters rapidly; the subjects of course very familiar to him — but to the novitiates, "it was all Greek." We derived no benefit from his remarks.[21]

On the other hand, brilliant and eager students like Benjamin Thompson found in him a "happy teacher."[22] According to Samuel Langdon (A.B. 1740):

He had the happy talent of communicating his ideas in the easiest and most elegant manner, and making the most difficult matters plain to the youths which he instructed. Though his temper had sufficient sensibility, yet it was perfectly under command, and with the mildest words he preserved the strictest authority; so that all were ready to obey, even a word or a look, with profound respect.[23]

Stephen Sewall (A.B. 1761) testified to the same effect and called upon his fellows for confirmation:

His diction was refinedly pure, and exquisitely elegant; his method perfectly easy, and in the highest degree perspicuous. So compleat a master was he of every subject he handled, that each new lecture seemed a new revelation. The dignity and mildness of his aspect, at once commanded both respect and love. For the truth of these things, I need only appeal to those who have enjoyed the privilege of being under his tuition.[24]

True, in his later years his methods came to seem out-moded, and before he was in his grave the Overseers voted that there was "reason to think an emendment necessary in the usual mode of teaching the Mathematicks and natural philosophy," and appointed a committee to look into the matter.[25] But this was a long generation ahead of our story.

As a young bachelor professor, he received a very small salary. The Corporation rejected his petition for an increase on the ground

[21] College Papers (Harvard University Archives), I, p. 96, item 200.
[22] George E. Ellis, *Memoir of Sir Benjamin Thompson*, Boston, 1871, p. 7.
[23] Samuel Langdon, *The High Value of a Great and Good Name*, Boston, 1779, p. 18.
[24] Stephen Sewall, *An Oration*, p. 4.
[25] Overseers' Records, III, 178.

of his lack of marital status, and when the General Court also rejected his pleas, he looked about for a wife. His choice fell upon Rebecca, daughter of James and Elizabeth (Phillips) Townsend and stepdaughter of Charles Chauncy (A.B. 1721). They were published on July 1, 1746, and on November 3, they came to Cambridge and moved into Wadsworth House with President Holyoke.[26] Later they lived in a small and lovely house on what is now the northwest corner of Mount Auburn and Boylston streets, facing south with a view of the river and the wooded hills beyond. Behind it was a garden and an orchard. The Professor had to go to the General Court to obtain the exemption from the real-estate tax of the town of Cambridge to which he was entitled. More generous than the town, the legislature began making grants of from fifty to eighty pounds a year to supplement his salary.

In 1747 Winthrop was appointed a Justice of the Peace, an honor unusual for one of his years, but one in which he took no particular interest. Law was one branch of learning which did not appeal to him. If one may judge by his diary, the most shocking experience in his life was the "terrible spectacle" of the burning of Phyllis for the murder of her master.[27] When his own slave boy, George, died of the measles, he was mourned as one of the family, not as an unfortunate investment. His successor, Scipio, was watched over like the white children of the family.

Winthrop was of little use to the Province as a Justice, but very helpful in other ways. In 1757 he undertook to survey Castle William, and during the war scare of 1759, at the request of Governor Pownall, he made a chart of Boston harbor to replace the old Admiralty chart of 1705. In a special message to the General Court the Governor urged that Winthrop be paid for this service, but the House put the matter over to the next session, and then allowed him but £16 for his labors. In 1764 the General Court, in a more generous mood, paid him £30 for his observations on the variation of the needle, and had his tables printed at the end of the *House Journal* for that year.

The death of Rebecca Winthrop on August 22 or 23, 1753, shook the Professor loose from the meticulous routine of his observations and his teaching. Lonely, he suddenly resolved to cultivate the

[26] *Holyoke Diaries*, Salem, 1911, p. 8.
[27] John Winthrop, Diary (Harvard University Archives), Sept. 18, 1755.

friendships which had sprung from his entertainment of visitors to Cambridge and from his correspondence. William Samuel Johnson (Yale 1744) was typical of the visitors whom he had charmed:

Mr. Winthrop is a gentleman of the utmost candor, benevolence and humanity as well as the best sence and good breeding. As soon as he had read my letters . . . he seemed almost to receive me at once into the number of his intimate friends and treats me as his equal without the least distance or reserve. He . . . offers to show me any experiment I want to see, and will he says repeat the whole lecture upon electricity on my account.[28]

It was partly as a result of this kindness to Johnson that Winthrop was considered when a head was being sought for Franklin's Philadelphia academy.[29]

Gathering up letters of introduction from various distinguished Boston men, Winthrop set out in April, 1754, to make his way from friend to friend through Connecticut and to the southward. Among his letters was one from Governor Belcher introducing him to Benjamin Franklin. Unfortunately neither man left an account of their meeting, but there is ample evidence that both men received pleasure and intellectual stimulation from their contacts. Winthrop was not awed by Franklin, and once told him that he "was good at starting game for philosophers." It is evidence of the greatness of the latter that he took this as a compliment.[30] At Newark the Professor visited President Burr, who was much pleased with him.[31]

Back at home, Winthrop was married on April 8, 1756, to Hannah, sister of Samuel Fayerweather (A.B. 1743), and widow of Farr Tolman of Boston. She was younger than he, a woman of prodigious energy with an omnivorous, if undiscriminating, mind. The soaring prose of her letters reminds one of the figures for her weight in her husband's careful family records. He himself remained at a steady 150 pounds, which meant that, being 5'8", he was only comfortably plump. In spite of Mrs. Winthrop's quickness of tongue, mind, and fancy, the marriage seems to have been a thoroughly happy one, as far as the inevitable ills of the generation would permit. She and the children took the smallpox in 1764, and

---

[28] Samuel Johnson, *Works*, New York, 1929, I, 120.
[29] *Ibid.*, p. 155.
[30] Mass. Hist. Soc., *Collections*, series 7, VI, 190.
[31] Jonathan Belcher, Letter Book (Mass. Hist. Soc.), May 3, 1754.

suffered from what even their contemporaries recognized as medical mispractice.[32]

If the Professor's home life was tranquil, it was not because he was in the habit of avoiding disputes or of making concessions for the sake of peace. On November 18, 1755, New England was shaken by an earthquake which he studied with eager interest. This was the first time that such phenomena in the colonies had been subjected to the scrutiny of science, and the lecture which he gave to his students on November 26 attracted very general interest. Over and above an interesting discussion of the earthquake, it was a devastating attack on the prevailing concepts of the relationship of God to the universe. In it he ascribed earthquakes to purely physical causes and said that, considering the structure of the earth, it was remarkable that there are not more of them.

> Though these explosions . . . have indeed occasioned most terrible desolations, and in this light may justly be regarded as the tokens of an incensed Deity; yet it can by no means be concluded from hence, that they are not of real and standing advantage to the globe in general. Multitudes, it is true, have at different times suffered by them . . . but much greater multitudes may have been every day benefited by them. The all-wise Creator . . . has established such laws for the government of the world, as tend to promote the good of the whole.

Earthquakes, like wind and gravity, were forces arising from nature, and not "scourges in the hand of the Almighty." Yet he maintained that his thesis did not detract from the majesty or justice of God.[33]

As radical as such doctrine might have seemed in some circles, it would not have created much stir in a society which two generations before had seen Increase Mather (A.B. 1656) reverse himself to accept the natural origins of fossils and comets, had not Winthrop gone out of his way to attack Thomas Prince (A.B. 1707), the brother of his old critic, Tutor Prince, and one of the ministers of the old school who liked to pontificate on all subjects. In 1727 Prince had printed a sermon, *Earthquakes the Works of God*, in which he said that the natural causes were secondary. Early in

[32] *Boston Gazette*, Apr. 30, (3/1), May 7, (1/1), May 28, supplement (2/1), 1764.

[33] John Winthrop, *A Lecture on Earthquakes; Read in the Chapel of Harvard College*, Boston, 1755, p. 27. Reprinted in *The Bulletin of the Seismological Society of America*, for March, 1916. See also *Philosophical Transactions*, L, 1.

December, 1755, he let a printer bring out a new edition to which he added an appendix in which he suggested that lightning rods might contribute to earthquakes by drawing electricity from the heavens. To this the Professor made a sarcastic answer in an appendix to his lecture, which was then going to press.[34] The minister replied on the front page of the *Boston Gazette* for January 26, 1756, shifting his ground somewhat from lightning rods to the Deity. It was a gentle letter, but one which rather talked down to the young scientist. In a rage Winthrop wrote a long *Letter to the Publishers of the Boston Gazette* which was printed in pamphlet form rather than in the paper. In it he exposes Prince's inconsistency, sarcastically couples him with Newton and Boyle, and generally shows a side of his own mind which his friends do not mention in their eulogies:

I would humbly propose it to the consideration of this Rev. Author, who, I doubt not, is very willing that posterity, as well as the present age, should enjoy the advantage of his learned labors; whether it may not be advisable for him to deposite a copy of this Sermon in the public library of Harvard-College, for the benefit of my Successors and of that whole Society, in all succeeding generations. . . . I cannot but esteem it an high felicity to have rescued this worthy Divine from the panic which had seized him, when he wrote his Postscript about the iron-points; and by him, consequently, a great number of others, especially of the more timorous Sex (so extensive is his influence!) who have been thrown into unreasonable terrors, by means of a too slender acquaintance with the laws of electricity.

Prince demonstrated that he was a Christian gentleman as well as a famous author by publicly heaping coals of fire on Winthrop's head, praising his scholarship and urging that the General Court give him a fat pension and appoint him supervisor of all of the surveyors in the Province.[35] With some grumbling the Professor accepted the praise and the implied recantation.[36]

To Ezra Stiles (Yale 1746), scientist, minister, and future president of Yale, Winthrop explained at length his quarrel with the popular religion:

[34] For the debate see Eleanor M. Tilton, "Lightning-Rods and the Earthquake of 1755," in the *New England Quarterly*, XIII, 85–97.

[35] *Boston Gazette*, Feb. 23, 1756, 2/1.

[36] *Ibid.*, Mar. 1, 1756, 1/3.

Tis a great pleasure to me to find my self supported in my sentiments concerning earthquakes, which were a little singular, by a Gentleman of your ingenuity and judgment. Since the earthquake, our pulpits have generally rung with terror and earthquakes have been represented only as "indications of the particular displeasure of the Almighty." One rev. Gentleman in Boston was pleased lately to observe to his audience, that not one Protestant place had suffered in the late European earthquake. What a narrow spirit of party do such observations discover . . . and how inconsistent with the express declaration of our Savior in the case of the 18 who perish'd by the fall of the tower of Siloam. . . . Were the inhabitants of Lisbon greater sinners than the inhabitants of London. . . . I freely confess, my notions of the design of earthquakes are very different. I doubt not they are intended to answer very valuable purposes in the natural world; some of which that most readily occurr'd to my tho'ts I threw together in my Lecture. But the grand moral purpose of these, as well as of other terrifying phaenomena, I fully believe to be . . . to keep up in mankind a reverent sense of the Deity.[37]

In part, his dislike of ministers like Prince was due to the fact that his command of languages and literature made him a much better Biblical critic than they, and scornful of their pretensions of certainty in matters of religion. In part, it was due to a shrewdly critical appraisal of ideas no less than of the facts of the world around him. Stiles wanted to believe that the inscription on Dighton Rock was the work of ancient wanderers from the eastern hemisphere, but Winthrop remarked that it in no way resembled the written languages of the eastern world but did look like the kind of drawings Indians made in idle diversion.[38] Never did he permit the Bible to interfere with observation and reason. He argued at length that the physical characteristics of the different races were due to climatic environment, and did not mention the Bible stories of their origins.[39]

It is significant that the student most frequently mentioned in Winthrop's diaries is Jonathan Mayhew (A.B. 1744). When death cut short that brilliant career, the glowing tributes which the Professor contributed to the press made the liberality of his own religious opinions abundantly clear.[40] Charles Chauncy (A.B. 1721), the other outstanding religious liberal in New England, thus praised him to their common friend, Stiles:

[37] Stiles Mss. (Yale University Library), Apr. 17, 1756.

[38] Ibid., July 21, 1767.

[39] Ibid., July 19, 1759.

[40] Printed in pamphlet form as From the Public News-Papers. Boston, July 14 & 17, 1766.

He is by far the greatest man they have among the officers of the College in Cambridge. Had he been of a pushing genius and disposition to make a figure in the world, he might have done it to his own honor, as well as to the honor of the college. I suppose none will dispute his being the greatest Mathematician and Philosopher in this Country; and was the world acquainted with his accomplishments, he would be rankt among the chief for his learning with reference to the other sciences. He is, in short, a very critical thinker and writer; Knows a vast deal in every part of literature, and is well able to manage his Knowledge in a Way of strong reasoning, as any man I know. He went along with me in a particular study for nearly two years. I had many written communications from him, and he from me, not so much by way of dispute, as by joining our forces in order to the investigation of some certain truths.[41]

It was to be many years before Stiles realized that this joint investigation which Chauncy mentioned was in preparation of a blast intended to breach the walls of Calvinism. Only after Winthrop's death did Chauncy think the world intellectually mature enough to stand the shock of his *Salvation for All Men* (Boston, 1782, London, 1784), and his *The Mystery Hid from Ages* (London, 1784). Winthrop's participation in the preparation of the argument was not publicly avowed, but was known to their Boston contemporaries.[42] Stiles, less well informed, guessed the extent of Winthrop's theological failings and said, after praising his friend's morals, virtues, and piety: "I only wish the evangelical Doctrines of Grace had made a greater figure in his ideal System of Divinity." [43]

In essence, Winthrop found repulsive the primitive type of religion which held that the Deity interfered directly and immediately in the affairs of men, like a bad-tempered and ill-informed human father. But, he was awed by the vast complexity of the universe into a humble and religious frame of mind, and after investigating all relevant human knowledge, found it contained no adequate natural explanation. So he accepted "revealed religion" as necessary to explain the universe and God. Once he told Simeon Howard (A.B. 1758):

I wish . . . the present generation would remember the design upon which their forefathers came into this country, and the regard they

[41] Stiles Mss.

[42] Mass. Hist. Soc., *Collections*, series 5, II, 409–10; John Eliot, *Biographical Dictionary*, Salem and Boston, 1809, p. 508.

[43] Stiles, Diary (Yale University Library), May 11, 1779.

shewed to religion in all their transactions; and consider, that they meant to hand down their religion as well as their worldly possessions to their posterity.[44]

He showed his religious spirit in practical ways. When Cambridge built a new meetinghouse, the contribution which he made out of his meager salary was exceeded only by those of General Brattle and Lieutenant Governor Phips. Despite the marks of his teeth in Parson Prince, organized Congregationalism made a practice of exhibiting him as the model Christian, New-England style.

Mr. Prince had a brief laugh at the Professor's expense not long after their lightning-rod encounter. Winthrop was interested in the Philadelphia project to publish *The American Magazine* and agreed to collect subscriptions in Massachusetts.[45] However, when the number for October, 1757, arrived, he found in it an abstract of a work by a member of the Royal Academy of Berlin saying that earthquakes were caused by the accumulation of electricity in the earth. He hastened to print a reply in the December number. For years he strove to increase the use of his friend Franklin's lightning rods, and when Hollis Hall, which, despite his arguments, had no rods, was struck, he was provided with the text for an article which occupied most of the front page of the *Boston News-Letter* for July 7, 1768, and was copied by papers in other colonies, and even by the *London Magazine*.[46]

The Mathers had long before removed comets from the arsenal of weapons of the New England Deity, so Winthrop was free to study them without being distracted by the remarks of the clergy. He did this with such enthusiasm that it was a part of college lore that Professor Winthrop's lectures were suspended for the duration of each comet. In 1744 he had in mind the publication of a paper on comets, but he seems to have withheld it because President Clap (A.B. 1722) of Yale was publishing a pamphlet on meteors.[47] Like all men of science the world over, he awaited with interest the reappearance of Halley's comet, which would for the first time give astronomers proof that they had correctly determined the nature of such bodies. He first saw the bright spot in the sky on April 3,

[44] Simeon Howard, *Christians no Cause to be Ashamed of their Religion*, Boston, 1779, pp. 28–9.

[45] *Pennsylvania Gazette*, Nov. 17, 1757, 1/2.

[46] *London Magazine*, Sept. 1768, pp. 473–5.

[47] John Winthrop to Thomas Clap (Yale University Library), Apr. 23, 1744.

1759, and in the *Boston News-Letter* of April 12 he announced
that this was indeed the expected visitor. In the same paper on
May 3 he explained the movements of the comet to the public.
The two lectures which he read in the college were printed at the
time and reprinted in 1811 to combat the "crude and fantastic
ideas" still prevalent. Franklin communicated to the Royal Society
the Hollis Professor's attempt to demonstrate Sir Isaac Newton's
theory as to the ascent of the tails of comets, which was published
both in the *Transactions* [48] and separately in pamphlet form. Winthrop was a profound admirer of Newton, and through Franklin
presented to the Royal Society a paper correcting an unfortunate
misunderstanding in Giovanni Castiglioni's *Life of Newton*.[49]

Comparable in importance for astronomers with the return of
Halley's comet was the long-anticipated transit of the planet
Venus over the disk of the sun, which would afford measurements
to solve several fundamental points as to the nature of the solar
system. On April 18, 1761, Governor Bernard went before the
House of Representatives with a special message:

I have received a Letter from Professor Winthrop, wherein he offers
his Service to go to Newfoundland to observe the Transit of Venus.
. . . You must know that this Phenomenon (which has been observed
but once before since the Creation of the World) will, in all Probability,
settle some Questions in Astronomy which may ultimately be very
serviceable to Navigation. . . . It happens that this Phenomenon is
not to be seen in North America except the most Northern Parts, of
which Newfoundland exhibits the best view; I cannot, therefore, excuse
myself recommending to you to enable Mr. Winthrop to take a view
from thence. The best and least expensive Method I can think of is this:
that the Province Sloop, which a little before that Time will be obliged
to be sent to Penobscot, may proceed from thence with the Professor to
Newfoundland.[50]

The General Court promptly requested the Governor to place the
sloop at the Professor's disposal. Bernard wrote helpful letters to
General Amherst and others, and the Harvard Corporation authorized Winthrop to take from the college instruments "the time
piece, a small refracting Telescope given by C. Kilby, Esq.; the
reflecting Telescope lately given by the Honourable Thomas Han-

---

[48] *Philosophical Transactions*, LVII, 132.
[49] Printed *ibid.*, LXIV, 153-7.
[50] Justin Winsor, *Memorial History of Boston*, Boston, [1881], IV, 495.

cock, Esq., and Hadley's Octant given by Ezekiel Goldthwait, Esq." [51]

With these and a little party of friends and assistants, the Professor sailed on May 9. They arrived safely and, with fair success, made the observation on June 6. The Boston newspapers for the last week in July carried the story of the expedition and its return. The Royal Society welcomed the report and printed it in the *Philosophical Transactions* [52] for the use of other astronomers the world over. Stiles was as proud as if he had made the observations himself and, ignoring his friend's becoming modesty, began to push Franklin to have him elected to the Royal Society. This would, he told Franklin, embellish even his memory. The project was delayed by Franklin's return to America, but encouraged by his visit to Winthrop in the Summer of 1763. Back in London, he presented the Professor's name to the Royal Society on June 27, 1765. Due to a Summer recess and the lack of a quorum, the election was delayed until February 20, 1766, when John Winthrop became the third of the name to receive that honor. Franklin signed a bond for his contributions and paid his admission fee. [53] The Harvard Corporation, gratified, voted to pay his dues if he would put in the college library the volumes of the *Philosophical Transactions* which he received as a member. Stiles rejoiced in the honor America had received from the election.

For several years Stiles had been urging the formation of an American "Philosophic Society," but Winthrop, who at this time was less fervently patriotic, had discouraged him:

I must doubt whether, all things considered, it will answer at present to do any thing of this sort, at least in a public way. Two or three attempts of this nature that have been made within my memory, quickly failed. It seems to me our Country has hardly arrived yet to a proper state of maturity for such an undertaking. [54]

Instead, he increased his flow of articles on comets, meteors, and whirlwinds to the *Philosophical Transactions*. However, he and Stiles were both elected to the rejuvenated American Philosophical Society of Philadelphia in February, 1768.

[51] Corporation Records (Harvard University Archives), II, 142.
[52] *Philosophical Transactions*, LIV, 279.
[53] *William and Mary Quarterly*, series 3, III, 249–50.
[54] Stiles Mss., Jan. 18, 1763.

Three years before, Winthrop had been elected a Fellow of the Corporation. This was a ripe honor, for he had long been the chief servant as well as the chief distinction of the college. Periodically for years he had climbed to the eaves of Stoughton College to measure the ominous bulge of the walls. In 1746 he had planted an elm on either side of the Chapel, one of which survived until modern times as the Class Day Elm.[55] He was credited with having contributed one of the Latin poems to the volume with which Harvard welcomed the accession of George III.[56] His reputation was largely responsible for the magnificent gifts which made good the losses resulting from the burning of Old Harvard. He accepted election as a trustee of the Hopkins Charity, but declined when the General Court proposed to make him manager of the college lottery. Because of the illness of President Holyoke, he presided at the Commencement of 1765.

In the eighteenth century both Harvard and Yale tended to elect men of science to the presidency, so Winthrop's name was often mentioned in that connection as the years grew upon Holyoke. He disliked the administrative duties which he was obliged to perform, particularly when he had to tell his old friend Stiles that a Yale Senior who came up to Cambridge to transfer, bringing a warm letter of recommendation, had, upon examination, turned out to be hopelessly far behind the Senior class at Harvard. He could only say that he was "very much a stranger to the method of education now practiced at Newhaven," and imply that its shortcomings were due to the student disorders which had crippled instruction there.[57]

Within the year the situation at Harvard was nearly as bad. In spite of the assertions that Professor Winthrop could quell students at a glance, the fact remains that when Holyoke was sick and the administration was in his hands, there were serious disorders. On April 4, 1768, the three junior classes proposed to leave the college.[58] A year later Holyoke died and was put to rest with the greatest funeral Cambridge had ever seen. Winthrop unhappily took up the reins while the Corporation tried to find a younger and stronger man than he who would accept the presidency. After

[55] John Winthrop, Diary, Apr. 7, 1746.
[56] But the poem is also attributed to John Lovell. See *Library of Harvard University, Bibliographical Contributions*, No. 4, p. 6.
[57] Stiles Mss., May 5, 1767.
[58] Edward Marrett, Annotated Almanac (Harvard College Library).

every other possible candidate closely connected with the college had joined him in firmly refusing the honor, it was finally accepted by Samuel Locke, to whom he surrendered the keys on March 21, 1770. When that administration ended miserably, Winthrop was older and more racked by his asthma than when this had been regarded as disqualifying him as Holyoke's successor; but now the need for clear moral leadership was so great that the Corporation, without stopping to consult him, elected him president on January 31, 1774. A week later he flatly refused the honor, and in October he had the pleasure of surrendering the administration to Samuel Langdon.

In view of his administrative duties, it is surprising that Winthrop in these years contributed as much as he did to science. The center of his interest was the transit of Venus which was to occur in 1769, for the inaccuracy of the observations of the last had left several problems unsolved. Franklin and Astronomer-Royal Maskelyne urged him to go in person or to send one of his students to Lake Superior, the nearest point at which the entire transit would be visible. Although it was out of the question for the ailing Hollis Professor to make the difficult expedition himself, he beat the bushes for funds and sent James Bowdoin (A.B. 1745) to General Gage with Franklin's and Maskelyne's recommendation. The General promised "all the assistance in his power," but did not offer financial help. The Province Council voted that Winthrop should be asked to head the expedition, but referred the question of funds to the General Court. The political activities with which the politicians were then busy made any such constructive action impossible. Failing action from the General Court, Governor Bernard finally proposed that a public subscription be taken to finance the expedition. It was too late, however, so Winthrop resigned himself to observing the part of the transit which was visible from Cambridge. His report went to the Royal Society in a series of letters on related subjects.[59]

Although George III had never heard of Professor Winthrop,[60] his name was familiar in all of the British universities. However, Oxford was reputed to have replied to the suggestion that he be

[59] *Philosophical Transactions*, LIX, 351; LX, 358; LXI, 51.
[60] Peter Orlando Hutchinson, *Diary and Letters of Thomas Hutchinson*, Boston, 1881, I, 161.

given an honorary degree with the statement that "Mr. Winthrop most undoubtedly was every way deserving of that Honor; but as it was supposed he was a dissenter from the Church of England the Favour could not be granted." [61] This was quoted with anger when Oxford so honored Thomas Hutchinson (A.B. 1727) and Peter Oliver (A.B. 1730), who were Congregationalists, but Loyalists. At Franklin's suggestion, William Robertson, the historian, induced the University of Edinburgh to award him an LL.D. in 1771; and two years later Harvard, granting this degree for the first time, conferred upon him the same honor.

Dr. Winthrop was innocent of politics. He was indignant at the Stamp Act and the Massacre, but he had no interest in or comprehension of the underlying political problems. He first became familiar with the Whig leaders when the General Court was sitting at the college, when he was much in the company of Otis, Hancock, and Sam Adams. Hancock exercised his usual means of influencing impecunious gentlemen by giving him an order on his own tailor for a suit of clothes of superfine broadcloth.[62] Governor Hutchinson believed that he was "a high son of Liberty" and responsible for the Corporation's protest against the sitting of the General Court at Cambridge.[63] As a matter of fact, it was Mrs. Winthrop who was the violent patriot. She had, coupled with the normal emotional irrationalism of some women, a flow and felicity of words which made her a geyser of patriotism. Mercy Warren was her congenial friend.

Asthma and the prospect of pleasant companionship sent Dr. Winthrop to the eastward in the Province packet in the Summer of 1772, in company with John Hancock, General Brattle, and like friends.[64] Business was probably not a factor, for such wild lands as he owned were in the interior of New Hampshire, in the towns of Dummer, Ellsworth, and Stockbridge. His new friendship with the Whig politicians and his old friendship with Franklin made it natural that he was one of the five to whom the latter sent the stolen Hutchinson letters. Not unnaturally the Governor reported that the Professor was engaged in "treasonable" correspondence

---

[61] Ezra Stiles, Diary, June 10, 1777.
[62] Chamberlain Mss. (Boston Public Library), A 4.2.
[63] Mass. Archives, XXVI, 478.
[64] John Rowe, Diary (Mass. Hist. Soc.), Aug. 6, 1772.

with Franklin.[65] And Winthrop had reason to think that a package containing a volume of the *Philosophical Transactions* had been opened in the mail to remove a letter from Franklin.[66]

Except for his limited service as a Justice of the Peace and a term on the Cambridge school committee, Dr. Winthrop had held no public office until his election to the Province Council in 1773. To Franklin's letter of congratulation and praise, he replied:

No considerate Person, I should think, can approve of desperate Remedies, except in desperate Cases. The People of America are extreamly agitated by the repeated Efforts of Administration to subject them to absolute Power. They have been Amused with Accounts of the pacific Disposition of the Ministry, and flattered with Assurances, that upon their humble Petitions, all their Grievances should be redressed. They have petitioned from time to time; but their Petitions have had no other Effect than to make them feel more sensibly their own Slavery. . . . If the Ministry are determin'd to inforce these Measures, I dread the Consequences: I verily fear they will turn America into a field of Blood.[67]

As a member of the Council he served on the committees to consider the tea petition of Richard Clarke (A.B. 1729) and to urge upon Gage the removal of Chief Justice Peter Oliver (A.B. 1730). In April, 1774, Cambridge elected him to the Provincial Congress; but in May, Gage disallowed his reelection to the Council. This did not reduce his participation in public life, however, for in the negotiations with Gage he was mentioned with Hancock, Bowdoin, and Joseph Warren as one of the Whig spokesmen. In October the town of Cambridge instructed him to return to the Provincial Congress and there support action "most proper to deliver ourselves and all America from the iron jaws of slavery." This sounds like the exaggerations in his private correspondence in which he declares that the Americans "see themselves treated like a parcel of slaves on a plantation, who are to work just as they are ordered by their masters." [68] Totally blind to the problems of imperial administration, he described the regulatory acts of Parliament as nothing but an effort to support "an imaginary dignity of government." He talked of violence much more openly than did the

[65] Sparks Mss. (Harvard College Library), 10, IV, 33.
[66] Mass. Hist. Soc., *Proceedings*, series 1, VII, 123.
[67] Benjamin Franklin, *Writings* (Smyth ed.), VI, 274.
[68] Mass. Hist. Soc., *Proceedings*, series 2, XVII, 285.

politicians, and bluntly declared that Massachusetts ought not to submit to the alteration of its charter.

This bold public stand made Dr. Winthrop quite as popular a figure in politics as he had been as an expounder of the causes of earthquakes and the nature of comets. Henry Pelham proposed to the mezzotinters of Newport that they make a print of him as a companion piece to their print of Franklin, and offered to prepare "an exact drawing in Black and White taken from a Painting of Mr. Copley's, which is a elegant Picture and a very striking likeness." [69] The more cautious mezzotinters raised the question of how many people would pay 5s for a picture of the Doctor.

Before this question had been decided, the war about which the Professor had talked too glibly was upon him. His wife's account of the events of the Nineteenth of April is much fuller than his:

Not knowing what the event would be at Cambridge, at the return of these bloody ruffians [from Concord], and seeing another brigade dispatched to the assistance of the former, looking with the ferocity of barbarians, it seemed necessary to retire to some place of safety till the calamity was passed. My partner had been confined a fortnight by sickness. After dinner we set out, not knowing whither we went. We were directed to a place called Fresh-pond, about a mile from the town; but what a distressed house did we find it, filled with women whose husbands had gone forth to meet the assailants, seventy or eighty of these (with numberless infant children,) weeping and agonizing for the fate of their husbands. In addition to this scene of distress, we were for sometime in sight of the battle; the glittering instruments of death proclaiming by an incessant [fire] that much blood must be shed. . . . Another uncomfortable night we passed; some nodding in their chairs, some resting their weary limbs on the floor.[70]

On the morning of April 20 the Winthrops joined the crowd of frightened women and children making their way westward. Their path was the bloody, body-strewn, road through Menotomy. That night they found refuge with the Reverend Jonathan French (A.B. 1771) at Andover. Ten days later they went to board with a neighbor, Nehemiah Abbot.

After one exploratory trip to Cambridge, Dr. Winthrop returned to Andover where he remained until the middle of June, when he

[69] Mass. Hist. Soc., *Collections*, LXXI, 294.
[70] Elizabeth F. Ellet, *The Women of the American Revolution*, New York, 1853, I, 94–5.

went back to the college to pack the apparatus and library for removal. During the stay of the college in Concord the Winthrops
occupied the house later made famous by Hawthorne as "Wayside." In the meantime, son James (A.B. 1769) set up the provincial post office at their house in Cambridge.[71]

The Professor's accounts of the Battle of Lexington and Concord
are certainly no credit to his powers of judgment and observation,[72]
but on the other hand he saw more of the canvas of the war than
did most of his contemporaries. As early as June he was urging
upon John Adams in Congress the necessity of Congressional support for the army around Boston.[73] On July 21 he was reelected
to the Council, where one of his first duties was to visit Washington
to enquire the extent of the power conferred upon him by Congress. In his diary for August 12 he remarks laconically that he
spent the day with Washington. The new government appointed
him a Justice of the Peace and Quorum for Middlesex, and later
a Justice "thro' this colony." [74] In September, he was appointed
Judge of Probate for Middlesex. In the Council, he served on
committees to list the Boston Tories and their offences, and to revise
the Test Act. John Adams urged that he be elected lieutenant
governor. There is some reason to think that he was responsible
for the choice of the motto of the Commonwealth.[75] All these
public duties were too much for a sick man who had to teach for a
living, so on May 28, 1777, he declined reelection to the Council.[76]
That Fall he served on the Cambridge committee to find quarters
for the British officers captured at Saratoga.

Dr. Winthrop was one of the very small salaried class which
faced starvation as war inflation progressed. The college increased
his salary to £1238 a year, but this fell hopelessly short of such
social obligations as his having Lafayette and the Count d'Estaing
to dinner. They must have been charmed by the republican simplicity of the great philosopher in his little wooden house, but unfortunately their accounts of the occasion have not survived.

[71] *Boston Gazette*, June 12, 1775, 4/3.

[72] Mass. Hist. Soc., *Proceedings*, series 2, XVII, 289–90.

[73] Edmund C. Burnett, *Letters of Members of the Continental Congress*, Washington, 1921, I, 130 *n*.

[74] Executive Records of the Province Council, XVII, 161.

[75] Mass. Hist. Soc., *Proceedings*, LI, 259–82.

[76] *Journals of the House of Representatives of Massachusetts*, Boston, 1777.

The outbreak of fighting did not check the leftward progress of Professor Winthrop's ideas. He was charmed by Tom Paine's *Common Sense* and became one of the early advocates of independence. In April, 1776, he wrote to John Adams:

Our people are impatiently waiting for the Congress to declare off from Great Britain. If they should not do it pretty soon, I am not sure but this colony will do it for themselves. Pray, how would such a step be relished by the Congress. . . ? I hope "Common Sense" is in as high estimation at the southward as with us. It is universally admired here. If the Congress should adopt the sentiments of it, it would give the greatest satisfaction to our people.[77]

True, the Massachusetts radicals were going further left than he thought safe. He was distressed by the mobs which would not permit the new courts to sit in some counties, and by some of the wild proposals: "There is such a spirit of innovation gone forth as I am afraid will throw us into confusion. It seems as if every thing was to be altered. Scarce a newspaper but teems with new projects." [78] Among those which he disliked were a legislature for every county and a registry of deeds and probate for every town. Some men were driven into the ranks of the Loyalists by such radical programs, but Winthrop never wavered. He wrote to his congressmen protesting the Howe peace negotiations: "This grand affair must be decided by war, not by treaty. Our inveterate enemies will never give it up, till they find themselves compelled to it." [79] Many thousand colonists could never bring themselves to believe that in a few years the Englishmen on the other side of the Atlantic had become their "inveterate enemies" and the long-dreaded French, their friends; but Winthrop in writing Franklin about the French alliance showed complete detachment from the ancient loyalties of his family:

Certainly they [the French] never had so fair an opportunity of depressing and weakening their great rival. This seems to be the critical moment for them to step in; and if they act from national views, and with their usual policy, I should think they would not let it slip.[80]

[77] Mass. Hist. Soc., *Collections*, series 5, IV, 298.
[78] *Ibid.*, p. 308.
[79] *Ibid.*, p. 312.
[80] Sparks Mss. XVI, 209.

The extent of his detachment from England would not have been surprising in a politician whose views were bounded by the Province, but in a member of the Royal Society with friends in every center of learning in the British Isles, it is remarkable.

On November 9, 1778, the Corporation adjourned to meet at the house of the Hollis Professor because of his "great bodily weakness." During the Winter he was unwell. His diary for April records the progress of a "tremendous cough" which, with his old asthma, sometimes compelled him to sit up all night. He made his last entry in his diary on May 1, and died two days later. Suitable preparations delayed his funeral until the Eighth:

The Corps, preceeded by the Governors and Students of the College, was borne into the Meeting-House; where Professor Sewall delivered an elegant English Oration on the melancholy occasion. The reverend President prayed, and the Students sung a well adapted Anthem. A solemnity appeared in every countenance, expressive of the esteem and respect borne to the deceased, and of the most unfeigned sorrow for the loss of so eminent and worthy a person.[81]

He was placed to rest in the family lot in the burying ground by King's Chapel.

The newspapers carried long obituaries and advertisements for the three funeral sermons which were printed. A poem by Mercy Warren occupied two-thirds of the first page of the *Independent Chronicle* for October 21, 1779. A sample will suffice:

> A Seraph shot across the plain,
>   The lucid form display'd,
> The starry round, which Winthrop trod,
>   And cry'd Great Winthrop's dead.
> Down, through the planetary fields,
>   Where thousand systems roll,
> A Newton's glorious kindred shade,
>   Descends to meet his soul.[82]

Andrew Oliver (a.b. 1749) made another essay in the same direction:

> Ye Sons of Harvard! Who, by Winthrop taught,
> Can travel round each planetary sphere;

[81] Wigglesworth, p. 21 *n.*
[82] Also printed in her *Poems*, Boston, 1790, pp. 235-9.

And, wing'd with his rapidity of thought,
Trace all the movements of the rolling year;
Drop on his urn the tribute of a tear.[83]

More moving than the published letters was the simple comment of Franklin to Samuel Cooper (A.B. 1743):

Our excellent Mr. Winthrop, I see, is gone. He was one of those old Friends, for the sake of whose Society I wish'd to return and spend the small Remnant of my Days in New England. A few more such Deaths will make me a Stranger in my own Country.[84]

Professor Winthrop left most of his property to his widow for her lifetime, after her death (which occurred in May, 1790) to be divided equally among his three sons, John (A.B. 1765), James (A.B. 1769), and William (A.B. 1770). Adam (A.B. 1767) had died before his father. William inherited the library and left it to Allegheny College.[85] In 1787 there was a movement to erect in Lincoln County a Winthrop College, named to honor the Professor and the two governors.[86] Nothing came of that suggestion, but a century and a half later Harvard named one of the divisions of the college Winthrop House in their memory. In it hangs the Copley portrait which is reproduced near the beginning of this sketch.

John Winthrop was an inveterate annotater of interleaved almanacs, sometimes keeping three for different purposes in one year. The notes are very brief and the hand varies remarkably, making some of them difficult to identify. As a whole they are extremely disappointing, some of the later ones being devoted largely to a list of those who came to "T." His meteorological journal, which runs from December 11, 1742, to April 29, 1779, belongs to the American Academy of Arts and Sciences, but is deposited with the diaries in the Harvard University Archives. It contains very full records of the weather for nearly every day of that period.

[83] *Independent Chronicle*, June 17, 1779, 4/2.
[84] *Writings*, VII, 407.
[85] For the children see Lawrence Shaw Mayo, *The Winthrop Family in America*, Boston, 1948, pp. 191 ff. Anyone seriously interested in Professor Winthrop should read Mr. Mayo's sketch as well as this, for they are deliberately supplementary.
[86] *Independent Chronicle*, June 14, 1787, 2/3.

THOMAS BELL

THOMAS BELL was one of the best-known Americans of his century, and is still a prominent figure in folklore for reasons thus succinctly stated long ago:

Tom Bell . . . greatly excelled in low art and cunning. His mind was totally debased, and his whole conduct betrayed a soul capable of descending to every species of iniquity. In all the arts of theft, robbery, fraud, deception, and defamation, he was so deeply skilled, and so thoroughly practised, that it is believed he never had his equal in this country.[1]

For years any exceptionally successful criminal was called "almost a second Tom Bell," [2] and Judge Richard S. Field of New Jersey in reviewing his career declared that one so accomplished in villainy must have been fitted for his profession in some foreign school.[3] Although folklore has forgotten it, he had a Harvard education which was a considerable element in his success.

Tom was born on February 18, 1712/3, the eldest son of Captain Thomas and Joanna (Adams) Bell of Boston. The Captain was a mariner and shipwright who owned two household slaves and considerable real estate, including a house on what is the east side of Hanover Street, near Commercial. He died in Virginia on July 29, 1729, and lies under a New England gravestone in the burial ground of the Upper Parish of Nansemond County.[4] His will ordered that Tom be sent to Harvard, and provided £200 for that purpose. His estate was less than he had expected it to be, and the Boston Selectmen refused his widow a liquor license.

Hardly had Tom settled at college when Jabez Richardson, a recent graduate who happened to be in Cambridge, beat him so badly that he needed repairs costing 10s. The Faculty fined Jabez but privately admonished Bell "for his Saucy behaviour to Sir Richardson." [5] In his Junior year they punished him for stealing two bottles of wine, and on February 8, 1732/3, they finally lost patience:

[1] *The General Assembly's Missionary Magazine; or Evangelical Intelligencer,* II (1806), p. 156.
[2] As in the *New York Gazette Revived in the Weekly Post Boy,* Oct. 9, 1752.
[3] New Jersey Hist. Soc., *Proceedings,* VI, 32.
[4] *William and Mary College Quarterly,* series 2, XXI, 47.
[5] Faculty Records (Harvard University Archives), I, 31.

Whereas Bell has been heretofore convict of theift . . . and whereas he has been Since found guilty of other acts of theft, particularly of stealing private Letters out of Gray's study, and lies under the strongest suspicion of having stolen a cake of chocolate this week from Hunt, and has been guilty of the most notorious complicated lying both formerly and also more lately, and particularly in the affair of the Chocolate . . . whereas to all the rest he has added, a scandalous neglect of his college Exercises, and is now become a disgrace to the society, and unworthy to be continued a member of it, therefore agreed and ordered, that Bell be forthwith expelled the College.

Memor. This was read in the Hall after evening prayers Feb. 19, 1732/3 and Bell's name (he being absent) rased off from the Buttery Tables, in the Hall, and he was pronounced expelled, and the scholars forbidden to entertain him at their chambers.[6]

His bills were written off as a bad debt, but one credit, "to Burgis," may, in view of one of his later acts, have some significance which is not apparent to us now.

At the time of his separation, Tom was but ten days short of his twenty-first birthday, an event anxiously awaited by his tailor, Henry Laughton of Boston, who was eager to collect a bill of some £30 for such items as a silk jacket and hose. The day that Bell passed from the protection of his minority, Laughton sued. Tom lost in the Inferior Court, appealed to the Superior Court, and lost again.[7]

Thus inauspiciously began the career of the famous Tom Bell. Although we cannot precisely trace his activities during the next few years, they were a busy period, for the *Boston Evening Post* of September 10, 1739, declared:

This young Blade . . . for some Adventures of a high Nature, to avoid Prosecution, quitted his Country, and has been Fortune hunting for several Years past; and if he has play'd half the Pranks that are imputed to him we may venture to pronounce, that the English Rogue was a meer Idiot, compar'd to him.

In 1736 he was in Virginia and Maryland where he impersonated a member of the Fairfax family and drew bills on the head of the clan. Two years later he was touring Virginia as Francis Partridge Hutchinson when he was detected and ordered committed to the

---

[6] *Ibid.*, pp. 46–7.
[7] Early Files in the Office of the Clerk of the Supreme Judicial Court, 34,755; Records of the Superior Court of Judicature, 1730–1733, p. 246.

Isle of Wight County Gaol. On the way thence he escaped from the constable, who advertised for him as "a middle-sized Man, with his own Hair," wearing "a Brown colour'd Broadcloth Coat, lin'd with Ash colour'd Silk, which has had a Rent in the Back of it." [8]

After a visit to New York, where he was arrested and tried for robbery,[9] Bell made his way to Barbados. There he presented himself as Gilbert Burnet, son of the Governor, displaying an imitation of His Excellency's seal which was copied from the engraving on the Burgis view of Boston. He was welcomed with open arms by the highest society of the Island, who found the fact that he was temporarily out of funds a welcome means of winning favor with the great. And who could doubt the honesty of a descendant of Bishop Burnet?

Late in July, 1739, this visit came to an exciting climax which was thus reported on the mainland:

A Son of the late Governour Burnet, happening to be at a young Jew's Wedding, whose Father was nam'd Lopus, this Lopus made very much of him, and being at Dinner, Burnet, as his Name goes, complain'd of the Head-Ache; whereupon Lopus advis'd him to step to his House hard by and lay down, accordingly he accepted and went: About half an Hour after, Lopus with five or six of his Companions came and took him by the Collar, strip'd him and gave him several Blows on the Face, and charg'd him with robbing him of some Money, which so surpriz'd him that he could hardly speak: Upon this the People of Speights Town, in which they were, were all in an Uproar; and tho' the Jew was not able to fasten the Theft upon the young Man, yet a Justice bound him over to the Grand Sessions; and the Jew is sued by the said Burnet in Two Actions of Ten Thousand Pounds.[10]

An indignant mob drove all of the Jews out of town and burned the synagogue, the gentile merchants revived the ancient slanders and painful truths, and the Island seemed on the verge of a nasty religious war. The *New York Gazette* shared the indignation of the Barbadians at the treatment of "Burnet," but Boston recognized the fine forging hand of her son. At first the Islanders refused to listen to the warnings from Massachusetts, but by October, 1739, even Tom Bell's bondsman was disillusioned:

[8] *Virginia Gazette*, July 21, 1738, 3/2.
[9] *Boston Evening Post*, Dec. 10, 1739, 2/1.
[10] *Boston Evening Post*, Sept. 10, 1739.

Your Advice and Judgment of Mr. Burnet has at last proved true; he is now in Goal, and the greatest Villain that was ever born. I went great Lengths to serve this Fellow, was his Security for his Appearance at the Grand Sessions . . . and never doubted but he really was Mr. Gilbert Burnet. . . .

He absconded, and was to have run off the Island in a Boat; I discoverer'd him, and had a strict Watch round the Island, by which Means I stopt him, after searching a Week for him. I found him last Wednesday in Disguise, about 3 Miles out of Town, then I took him and brought him to Town and delivered him up, upon which he was committed, I say, to Goal, where I hope he will always lie, except he comes out to be hanged. . . . He owes about Two hundred and fifty Pounds to several People; if he had got off, his Intentions were to Jamaica, there he would have got larger Sums than here, being well equipped with several costly Suits of Cloaths, one of black Velvet. . . .[11]

At the December session he "was Whipt and Pillor'd . . . and . . . order'd to be branded with the Letter (R) on each Cheek . . . but the Goodness and Clemency of his Excellency Governour Byng, released him from that Punishment."[12] The Governor's mistaken kindness proved a curse for the mainland colonies.

One Spring day in 1741 Tom Bell walked into a tavern in Princeton, New Jersey, wearing someone's gray frock coat, and was startled to be mistaken for the Reverend John (Hell Fire) Rowland, the associate of Whitefield and the Tennents. Some Harvard men would have taken this as an insult, but Bell too well remembered the injunctions of his Puritan ancestors regarding the improvement of opportunities not to make the most of his resemblance to the revivalist. Going into Hunterdon County, where his classmate Guild had been driven from the pulpit of the church at Pennington by the warmer preaching of the Reverend Mr. Rowland, he introduced himself as the latter, enjoyed the fat of the land, and accepted an invitation to preach the following Sunday. Riding to meeting in a wagon with the ladies while the farmer who was his host rode beside on an unusually fine horse, "Mr. Rowland" remembered that he had forgotten his sermon notes, borrowed the horse, and galloped back after them. When the puzzled farmer got back home later in the day he was enlightened by the discovery that his house had been rifled and his valuables had gone with the horse and the parson.

[11] *Ibid.*, Dec. 10, 1739, 2/1.
[12] *New-York Gazette*, Mar. 31, 1740, 4/1.

Bell must have really enjoyed the fact that his visit to Hunterdon split the Presbyterian Church of the Middle Colonies into factious disputes almost as violent, and far longer lasting, than the quarrel between the Jews and gentiles of Barbados. When Parson Rowland next visited that part of New Jersey, he was jailed for stealing the farmer's horse. The Reverend William Tennent, Jr., provided an alibi by swearing that the two of them had been preaching in another colony when the crime occurred. Thereupon Tennent was arrested for perjury. The Presbytery was split wide. Tennent appears to have courted conviction in order to have the opportunity to preach dramatically from the gallows. Whether or not he was saved from jail by a supernatural summoning of witnesses is a matter to be determined by faith.[13]

During the rest of the year 1741 Bell was very busy. After a visit to Barbados he landed at Newport and toured New England. Thence he worked from mansion to mansion down the Hudson as Winthrop, De Lancey, Wendell, and Francis Hutchinson; and then took a turn in New Jersey and Pennsylvania. This not proving profitable, in February, 1741/2, he announced "an entire and universal Reformation," and passed the hat as a suitable subject for charity.[14] His reformation was less than entire, for later in the year he was practicing his old trade in Maryland. Since the story of his return to Philadelphia on February 5, 1742/3, was told throughout the colonies, it is perhaps worth retelling as a sample of his methods:

Last Night came to Town young Mr. Livingston, Son of Robert Livingston of Albany; formerly by the Name of Gilbert Burnet; since by that of Rip Van Dam, jun. . . . He imposed the Name of Livingston on one Mathers of Chester . . . and . . . appearing like a Gentleman, persuaded Mathers to let him have two Horses and a Man to bring him up here: He's now taking his second Degree in our College; [15] from which Place he will soon have his Trial, and if he be found fit for the Pillory, doubtless he will receive his Ordination. He imposed himself on the Lower Counties for a Minister. I was present when he was before the Mayor; he there said he was well acquainted with Books, skill'd in Law and Divinity, and believ'd no one could sell Him any Thing he

---

[13] The best critical account of this affair is in the Trenton Historical Society's *History of Trenton*, Princeton, 1929, II, 626–9. See also Presbyterian Hist. Soc., *Journal*, XII, 94–5.

[14] *New-York Weekly Journal*, Feb. 15, 1742.

[15] A poke at Harvard where Bell nearly took his first degree.

was not acquainted with before; he rejoic'd in his Sufferings, and hoped to clear himself from all Aspersions.[16]

Instead he cleared himself from jail on June 11 by dexterous manipulation of the lock. According to the Sheriff, he left Philadelphia with an elegant wardrobe,[17] but was quickly separated from it in Long Island where he failed to move fast and far enough after obtaining money as "a son of Col. Floyds." His captor conducted him as far as the ferry to Manhattan before he escaped.[18]

On this occasion Bell's entry into New York lacked its usual éclat:

Yesterday was sevennight the notorious Tom Bell . . . was at the Ferry opposite this City: He cross'd over . . . early in Morning in a Pettyauger, without either Coat or Waistcoat, and went to a Public House there, where having breakfasted, he called for Pen, Ink and Paper, and pretended he was going to write to some of the most considerable Gentlemen of this City, with whom he was well acquainted, but could not see them until he was provided with Cloaths to appear in, and said he had lately come from Sea, where he had met with Misfortunes and had lost upwards of £600 and insinuated he was the Son of a Gentleman of Fortune at the East End of Long Island; but being soon known and challenged, he denied that he ever heard of such a Man as Tom Bell; yet he said that whoever Tom Bell was, he was a Man that deserved Compassion; and taking his Hat, he thought fit to make the best of his Way out at the Back Door.[19]

Finding his welcome exhausted in his old haunts, he went up to Portsmouth, New Hampshire, where he practiced successfully for a month before this public notice straitened his activities:

August 19. Notice hereby is given to the Publick to be upon their Guard, for in all probability, the famous, or rather infamous Tom Bell is upon the Line: A Person exactly answering his Character . . . has this Week broke open a Chest, and carried off considerable Moneys and Goods of value from several Persons. He says he has been Prisoner at St. Augustine 9 or 10 Months, and that he came from Eustatia: He is very shy in telling his Name; tho' to some he said it was Winslow. He went hence Yesterday at Noon, had on a blue riding Coat, strip'd Holland Shirt, a white Cap and Handkerchief. The Catchpole is after

[16] *Boston News-Letter*, Mar. 10, 1743, 2/1; *Pennsylvania Gazette*, Feb. 10, 1743; *New-York Weekly Journal*, Feb. 21, 1743.
[17] *Pennsylvania Gazette*, July 14, 1743, 4/2.
[18] *Ibid.*, July 21, 1743; *Boston Evening Post*, July 25, 1743.
[19] *Pennsylvania Gazette*, July 14, 1743.

him, and if he can haply lay his ample Hand upon his Shoulder, his Body obsequious to the touch, will be convey'd to our inchanted Castle. . . . The above Chap is very grave and serious, has a ready Invention, with a good Elocution; and has ('tis tho't) already deceived many here.[20]

Being thus billed in Massachusetts as a thief, Bell reverted to his role of clergyman, and preached and exhorted around Plymouth County to the edification of the New-Lights.[21] Finding the ministry then ill-paid, he went over to Martha's Vineyard where a touch of burglary yielded him £90 or so. Thence he went to Stonington and Newport, but in the latter town he was recognized and arrested for an old offence.[22] Paying his fine from the Vineyard proceeds, he went again into the Bay Province, where he had the misfortune to be taken up at Woburn and committed to the Charlestown jail "on Suspicion of Male-Practices." [23] As a matter of fact this was one of the few crimes which could not be proved against him.

The Charlestown jail was bitter cold in December, 1743, so Tom let himself out in his usual way, to the rage of the Sheriff whose notice provides a few additional lines to the portrait of his temporary guest:

He is about 30 Years of Age, of a midling Stature, of a ruddy Complexion, of a pleasant Countenance, and has a handsome Set of Teeth, and shows his upper Teeth when he speaks. He has a variety of Cloaths, and appears sometimes in one Dress, and sometimes in another. . . . His Discourse is polite, and he is of a spritely Look and Gesture.[24]

He was, however, back in Charlestown in time to enjoy the Christmas season there:

One Day last Week the very famous Mr. Thomas Bell, who lately left his Lodgings in Charlestown in a pretty abrupt Manner, in order (as he says) to pay a few Visits, and settle his private Affairs, was conducted back to his old Cell, where 'tis thought he will be obliged to continue at least for three Months, the Court being Just over at which he had a case depending, but was unhappily absent.[25]

[20] *Boston Post-Boy*, Aug. 22, 1743.
[21] *Boston Evening-Post*, Sept. 12, 1743; *American Weekly Mercury*, Sept. 22, 1743, 4/1.
[22] *Boston Evening-Post*, Oct. 3, 1743, 1/2.
[23] *Ibid.*, Nov. 14, 1743, 2/1.
[24] *Ibid.*, Dec. 19, 1743, 3/2.
[25] *Ibid.*, Dec. 26, 1743, 4/1.

There he remained until the Court of General Sessions of Middle-sex met in March, 1744, and ordered that he receive twenty lashes at Charlestown market place and be fined £75. If he could not pay, he was to be sold for three years.[26] How he beat the rap on that occasion is unknown. At any event, in July, 1744, "the famous sharper, TOM BELL," was again in the common jail of the City of Brotherly Love.[27] On this occasion he obtained his release by turning in a gang of counterfeiters who had applied to him to assist in signing their bills.[28] In September, "being on the Pad" in the Jerseys, he came to rest in the Monmouth County jail.[29] In Novem-ber the solid burghers of Manhattan were warned that "the noted Tom Bell was last week seen by several who knew him walking about this city with a large Patch on his face and wrapt up in a Great Coat, and is supposed to be still lurking." [30]

Finding such publicity embarrassing, Bell made his way to South Carolina, where he introduced himself to the Honorable Edmond Atken of Wands Neck as Nathaniel More. Recognizing his guest, Atken accompanied him into Charleston under pretext of seeing him on his way, but had him committed to the work-house as a vagabond. "A great Concourse of People" daily flocked to see him, but after several weeks were deprived of their pleasure when the judges decided that he could not be held because he had com-mitted no crime in Charleston.[31]

From South Carolina, Bell traveled overland to Maryland as a deserving shipwrecked mariner. Thence he journeyed to Jamaica, but in October, 1745, was in North Carolina equipped with a forged letter of credit and the nom de plume of Robert Middleton. Working his way north, he successfully evaded the Edenton and Williamsburg jails, which were hopefully awaiting him.[32] At New Castle he was befriended by a trusting sea captain who carried him to Philadelphia, only to lose his scarlet breeches for his kindness.[33] In April, 1746, Bell was arrested in the town of New York for selling a horse which he was alleged to have stolen on Long

[26] Ibid., Mar. 19, 1744, 2/1.
[27] Pennsylvania Gazette, July 12, 1744; Boston Post-Boy, July 13, 1744.
[28] Pennsylvania Gazette, Aug. 9, 1744, 3/2; New York Post-Boy, Aug. 6, 1744.
[29] New York Weekly Post Boy, Sept. 17, 1744.
[30] Ibid., Nov. 5, 1744.
[31] South Carolina Gazette, Feb. 18, Mar. 11, 1745.
[32] Virginia Gazette, Oct. 31, 1745, 4/1.
[33] Pennsylvania Gazette, Aug. 14, 1746.

Island.[34] Claiming to be innocent, he took some exception to being called a horse thief and was very much irritated to have the rough and uncouth offender described as Tom Bell.[35]

It may have been as a means of getting out of jail that Bell enlisted as a private in Captain Campbel Stevens' company which on September 1, 1746, embarked at Newark for Albany and Canada.[36] No chronicle of Bell against the wilderness has been found, but during the next two years the newspapers report him in and out of New York and New Jersey jails for activities ranging from vagabondage to coining money. He took exception to these accounts and filled nearly half of the *New York Evening Post* of September 4, 1749, with this vindication:

I have bought and paid for some score Pounds Worth of Good, and Merchandize for the Space of 12 Years, have sail'd as a Fore-mastman out of this Port, enlisted as a Soldier in the neighbouring Province, taught School in almost every Government in America, survey'd Land, etc etc. Notwithstanding I am treated here as if I was justly deem'd either *Bell* or the *Dragon*, the Destroyer or the Devouerer of the Lives and Properties of Mankind . . . no sooner than I make my Appearance in this City, but the alarm's given, *Tom Bell's in Town*: Great is the Company of them that publish it; Misers, Informers and busy Bodies, trudge apace, and they that tarry at home protect and defend their Mammon from falling a Prey into the Jaws of so voracious an Animal: To Arms, to Arms, to Arms . . . for the troubler of our Israel is again come amongst us: What shall be done unto this incorrigible Villain? Forthwith all their Artillery is level'd and employ'd against me. . . .

Thus I am harshly and closely pursued and chas'd from Street, to Street, and from Ward, to Ward, With the eyes of Argos, light as Air, swift as an Eagle, and innocent as a Dove: — Yet perhaps it may be objected why so much Vigilancy and Veloaty, where there's so much Innocency: The wicked flee, when only the Sting of corporal Punishment . . . pursues . . . Yet, I rely upon't, it very often happens, that the Innocent and Righteous are oblig'd to flee, when every Body pursues.

As to the generality of the Inhabitants of this City, how they are inclin'd to treat me, I wont say: yet I cannot be perswaded but that I appear to the Gentry and Clergy, as an object of pity, and subject of prayer. — However, in the main I am strangely oppress'd on every side,

[34] *New York Weekly Post Boy*, Apr. 14, 1746; *Virginia Gazette*, May 15, 1746, 3/2.
[35] *New York Evening Post*, Sept. 4, 1749, 2/1.
[36] *New York Weekly Post Boy*, Sept. 15, 1746.

when ever I reside a Day or an Hour in this City. . . . For the space of twelve Years, it has not been in the power of any of my Enemies, to charge me justly with actually transgressing any act or Law of this Government, except the intentional Design, etc. towards Robert Livingston, Esq, Junr, transacted near a 11 Years ago: And for which . . . that Gentleman and the then commander in Chief, afforded me their generous Pardons: Notwithstanding, 'tis insinuated by some, that, for that only Offence, I ought to be debar'd the liberty of the City. . . . I am not wholly lost to all Sence of Honour or Honesty, nor so far incorporated into Vice or Immorality; but what with a little Indulgence, less Assistance, a favourable Censure, and a generous Pardon, I may be easily reclaim'd: But my Enemies cares for none of these Things . . . I promise them if they'll prevail upon any Gentleman of Ability and Education to undertake to set my Enormities and Foibles, in a fair public View, I'll furnish him with Facts and materials . . . And I do solemnly promise them, that from this Day foreward I'll use my utmost Endeavours to introduce myself into some lawful, visible, Vocation and Calling. . . .

N. B. If any Gentleman, being Merchant Owner, or commander of any Ship . . . bound to any Port in North America, wants an able and expierenced Pilot or Sailor they may be supplied by applying to,

The slandered and unworthy Person of
T. Bell.[37]

The rival newspaper, far from convinced, the next week reported that the notorious Tom Bell was still lurking in the town:

Had it not been for some of Gery's Monkeys, who flock round him in every Place, grinning Applause to his redundant Chattering, and thereby supporting him in his unparallel'd Impudence; 'tis more than probable he had long ago took the Swing his Merits deserve.[38]

Saddened by this answer, he shook the mud of New York from his shoes and withdrew to keep school in Virginia.

It would have been quite impossible for Bell to hide his light for long under a ferule, so it is not surprising to find this in the *Virginia Gazette* of July 17, 1752:

This Day the ingenious THOMAS BELL, the famous American Traveller, made his public Appearance in this City. As his former Character, and romantick Life, have made a great Noise in every American Colony, 'twill doubtless be a Satisfaction to all who have any Knowlege of him, to hear in what Manner he has lived, during his

---

[37] *New York Evening Post*, Sept. 4, 1749, 2–3.
[38] *New York Weekly Post Boy*, Sept. 11, 1749.

Retirement from the Public, — He has resided in Hanover County, in this Colony, near two Years past, in the private Station of a Schoolmaster, and has, during that Time, behaved himself with Justice, Sobriety, and good Manners, of which he has produc'd a Certificate, sign'd by the principal Gentlemen of that County. By this his Behaviour, and his future Conduct, he hopes to wipe off the Odium that his former Manner of Life had fix'd on him, and thereby to approve himself a useful Member of Society.[39]

When, four days later, he left Williamsburg for Hanover, the newspaper remarked that if he was "really sincere in his Professions of Reformation, and his Intentions of living an honest and industrious Life," it would "perhaps be a Surprize to many." [40] Having thus revived interest in himself, he advertised in the *Virginia Gazette* of August 14 that if he received enough advance subscriptions he would have his autobiography printed at Williamsburg in a large and handsome volume, hoping thus to obtain the means to settle down as an honest man. He claimed to have received considerable advance subscription.

For the next three years Bell made his home in Charleston where he kept school when not traveling to other colonies to sell subscriptions to his memoirs.[41] Now he advertised his presence in the newspapers, and was apparently kept so busy gathering in the proffered money that he never got around to writing the book. The last certain record of him was at St. John's in Antigua late in 1754:

Some time last Month arrived here that renowned American Traveller, the tragic-comic, the reformed TOM BELL, and, after . . . a few Days, set Sail for some other Port. He was in a poor state of Health, and imagined that the Air of this Climate would recover it so as to enable him to pursue the Design of publishing his Life and Adventures. . . . As a reputed Liar is seldom believed when he speaks the Truth, so a Man who has once acquired the Name of a Villain, is liable to the Imputation of the Crimes and Enormities which he never committed; that this is Bell's Case, admits not more Doubt than the Falsity of common Report, and may be as easily proved as that he has been a Rogue and an Imposter. Who can believe that he never swore a prophane Oath, or was drunk? Yet, this he affirms: Who can credit, that he never took Advantage of, or debauched, Virgin Innocence? he declares it for the Truth: Who can imagine that He never stole a Horse? he clears himself in that Particular, and informs us, that some of his Pranks

[39] Also reprinted in the *Boston News-Letter*, Aug. 27, 1752, 1/2.
[40] *Virginia Gazette*, July 24, 1752, 3/2.
[41] *South Carolina Gazette*, July 18, Aug. 8, 1754.

. . . had. . . Humour in their Connection, and a comical Conclusion. In or about 1750 he was in London, near eighteen Months a Store Keeper and Clerk in Jamaica and had an Academy near the Ohio, besides officiating as a School-Master in every Province from Nova-Scotia towards Georgia. He has practiced Physic, pleaded Law, and acted Tar by being many Voyages before the Mast loading and unloading Vessels. Since, therefore, it appears, that his Proceedings do not partake of Tragedy, he seems he deserved the Appellation of an Universal Comedian.[42]

He may have been the Tom Bell who was hanged for piracy at Kingston, Jamaica, in 1771,[43] but it is more likely that he simply turned honest, as he had sometimes threatened to do, and so ceased to be of any public interest.

No copy of the memoirs of Tom Bell has ever been found, but Professor Carl Bridenbaugh, who assisted in the preparation of this sketch, promises a full biography. He is not taking subscriptions.

[42] *Boston News-Letter*, Apr. 10, 1755, 2/2.
[43] Purdie and Dixon's *Virginia Gazette*, July 4, 1771, 1/3.

JOHN PHILLIPS

The founders of New England placed so much value on a college education that they required every town above a certain size to maintain a grammar school, or as we would say, a Latin school, capable of furnishing a supply of students for the colleges. After a century, an expanding economy made a "liberal education" no longer an essential of a satisfying life, so a smaller percentage of boys went to college and fewer towns troubled to keep a Latin master to prepare those who did. The counter-action of the advocates of higher education was the founding of the Phillips Exeter Academy, Phillips Academy, Andover, and similar classical schools.

JOHN PHILLIPS of Exeter was born on December 27, 1719, a son of the Reverend Samuel (A. B. 1708) and Hannah (White) Phillips of Andover, Massachusetts. He was only eleven when he entered college, but although by two years the youngest member of his class he was precocious enough as a Freshman to write on the flyleaf of a textbook a note asking Tutor Davenport to step up to his chamber for a glass of ale.[1] He received Hollis and William Browne scholarships, was awarded the Hopkins Prize for excellence in his studies, and was chosen to deliver the Salutatory Oration at Commencement.

Although the voluminous tradition about John Phillips contains no mention of his having read medicine between his two degrees, it is likely that he did so, for when he returned for his M.A. in 1738 he was prepared to hold the affirmative of "An Glandularum Meatuumq; Cuticularium distentio, a Variolis effecta, istius morbi reditum impediat?" Tradition says that he read theology with his father during these years. He did join the Andover church and keep the town school. About the time that he took his second degree, he moved to Exeter, where tradition has it that he kept a private Latin school for a couple of years before being called in 1741 to keep the town school. After the manner of schoolmasters of that day he served as pulpit supply and clerk in legal matters. Among the wills which he was called to witness was

[1] Colonial Soc. Mass., *Publications*, XXI, 186.

JOHN PHILLIPS

that of "Gentleman Nat" Gilman, who died in 1741 leaving an estate of £8300 to his daughter, Tabitha, and his widow, Sarah, daughter of the Reverend Samuel Emery (A. B. 1691). Phillips entered the employ of the Widow Gilman as a clerk, and in time proposed to Tabitha who, however, preferred one of his cousins. He then turned his attentions to the Widow, who was sixteen years his senior, and on August 4, 1743, they were married.

With the proceeds of this venture, Phillips built a two-story house of the conventional type and furnished it richly. Being a typical Yankee, he saw nothing incongruous between opening a store in his new house and displaying the Phillips coat of arms. He attended the First Church, and sometimes supplied its pulpit; but when the Old-Light minister, John Odlin (A.B. 1702), was successful in having his equally Old-Light son, Woodbridge (A.B. 1738), ordained as colleague, Phillips joined the seceding New-Lights who organized the Second Church on June 7, 1744. This body first called the graceless Sam Buell (Yale 1741), who fortunately realized that he needed a wider field for his ranting. The Second Church then, on May 25, 1747, formally called Phillips, who declined "because of the delicacy of his lungs"; New-Light preachers did need leather ones. According to family tradition he thereafter preached at Portsmouth and other towns which called him in vain. He retained his residence in Exeter where he was the chief pillar of the Second Church, which he served for years as clerk, moderator, and ruling elder. Naturally he was in the front of the battle to have its members excused from paying taxes for the support of the Odlins. Being a conservative in spite of his New-Light connections, he was so troubled by his position as a member of a church which had not obtained legal recognition, that in October, 1753, he appeared before the convention of the ministers of the province to ask their advice relating to his situation. They did not think it expedient to give him direct advice, but said that if the Second Church would hold regular and Christian communion with its neighbors, they would be disposed to reciprocate.[2] Two years later the church was recognized by the General Court.

By this time Phillips was the leading inhabitant of Exeter. It

[2] New Hampshire Hist. Soc., *Collections*, IX, 23.

was a lumber town, so preoccupied with its sawmills that it did not raise enough food to feed itself. As a result, a merchant with his business genius could turn over his money quickly. He bought lumber by the hundred-thousand feet, sawed it, and loaded the boards at his own wharf, to which his returning vessels brought Virginia corn and pork. Except for the Portsmouth aristocrats who profited from land grants made by the Province, he became the largest holder of wilderness property in New Hampshire. Soon he was the leading banker and money lender of the region, with a reputation for ruthlessness in the pursuit of his debtors through the courts. By 1765 he was the richest man in Exeter, and rapidly opening the financial gap between himself and his neighbors. His instinct to save was notable even among Yankees. Tradition tells that he used to soak the backlog in water over night, and that at family prayers he would extinguish the candle as unnecessary when he closed his eyes to pray. Wendell Phillips (A. B. 1831) on the authority of his college goody used to tell that when Squire Phillips entertained a group of visiting clergy, as was his wont, all of them received ordinary table wine except the one who was to preach the sermon of the day, and he received an especially potent draught.[3] When he walked the streets of Exeter he insisted that every young man whom he met should touch his cap, and that every young woman should curtsy.

One aspect of the Squire's life about which there were no jokes was his marriage. Despite the disparity in age they were an affectionate couple, and he deeply mourned her death, which occurred on October 9, 1765. Writing to her granddaughter a week later, he said:

I have been ardently wishing for your dear grandmother's picture, and you make me happy in presenting me therewith. Methinks the lovely person lives in you, and that the old mansion house is once more enlivened and ornamented with the living image of its late inhabitant. Your fondness for my return minds me of her anxiety for me when absent, and the sweet welcome which her faithful heart discovered in her countenance, as well as with her lips, when she received me.[4]

On November 3, 1767, the Squire married Elizabeth, daughter of the Honorable Ephraim Dennett of Portsmouth and widow of

---

[3] Frank H. Cunningham, *Familiar Sketches of the Phillips Exeter Academy*, Boston, 1883, p. 94.

[4] Lawrence M. Crosbie, *The Phillips Exeter Academy*, Norwood, 1923, p. 18.

Dr. Eliphalet Hale of Exeter. Governor Wentworth (A. B. 1755) sent a note of congratulation with the marriage license.

In that generation public office was a duty which no good citizen shirked. Beginning in 1752, Phillips held arduous and often disagreeable town offices almost continuously. Three years later Exeter sent him to the General Assembly where, however, he was denied his seat because of election irregularities. In 1765 he joined in the conservative protest against public conduct which was based on the idea that because of the Stamp Act all law and order were at an end. He was appointed a Justice of the Peace in 1768, and promoted to the Court of Common Pleas four years later.

In the years preceding the War, Squire Phillips enjoyed the confidence of both factions. In 1770 he was on the town committee which instructed the Representatives that it favored the non-importation agreements, and was chairman of another which denounced the "unconstitutional acts" of Parliament. The Whigs sent him to the General Assembly of 1771 without any doubts. At the suggestion of Governor Wentworth a cadet corps was formed in Exeter and handsomely uniformed in scarlet and buff. The cadets chose Phillips to be their commander, and the Governor, pleased with the way his friend handled the company, commissioned him colonel. On the Bench, Squire Phillips supported the Governor in land cases,[5] and was rewarded by appointment to the Province Council in 1774. At the same time Exeter chose him to the Committee of Correspondence and sent him to the First Provincial Congress.

Colonel Phillips had neither the spirit nor, thanks to his recent support of the Governor, the public confidence requisite for active military service in the War. In a letter to President Wheelock he showed his attitude toward the political situation:

I can feel for you, under your late circumstances, from my own experience in time of the stamp act, when I could not avail my self of an appeal from popular Clamor, rais'd by one or two unreasonable and wicked men — and am not without disagreable apprehensions of male treatment at this unhappy time, for not acting contrary to the light of my own mind — to what I take to be the mind of Christ, and the most prudent measures for our political safety.[6]

[5] *Acts of the Privy Council, Unbound Papers*, p. 534.
[6] John Phillips to Eleazar Wheelock (Dartmouth College Archives), Dec. 25, 1775.

By retiring from business he avoided offending the rebels and attracting the attention of the mobs. His correspondence showed his doubts:

I have just now seen the Declaration of Independency, and perhaps it will not be long before we shall experience whether we are able to support it — or whether the measures taken by the Government on both sides will not be ruinous. The Lord in mercy prevent it, and make us mutual blessings to and [not] destroyers of one another.[7]

Although Exeter was favored over Portsmouth by the new government, the Squire's attitude of neutrality was not unpopular there. In 1778 the town meetings began again to choose him Moderator. He was a leader of the hard-money faction in 1786 when rioting debtors tried to intimidate the General Assembly, which was sitting in the First Church of Exeter.

When he retired from business, Squire Phillips for the first time could really apply himself to the task of distributing wisely the tithe of his income which he, according to the custom of his family, devoted to good works. He had no children of his own, but he sought out likely youths who needed financial help toward their education. Such a lad would be surprised by a letter like this:

I now send fifty pounds, hoping if [after] the frugal expenditure thereof there should be occasion for more you will be pleas'd to give yourself the trouble — no! the pleasure of letting me know what further sum wou'd be serviceable.[8]

Another £50 went to help the school which his classmate Parsons was keeping in Gilmanton. Missionary work among the Indians had appealed to him ever since he had subscribed for Jonathan Edwards' *Life of Brainerd,* and he was disappointed when the jealous Episcopalians killed the Society for Propagating Christian Knowledge among the Indians of North America. He was not interested in helping Harvard College, perhaps because it was not heathen. When Henry Sherburne (A. B. 1728) passed the hat for the Indian school which was to become Dartmouth College, the Squire pledged £100. The gift was appropriately made in the form of blankets rather than books.[9] At the Dartmouth Com-

---

[7] *Ibid.,* June 16, 1776.

[8] Mass. Hist. Soc., *Collections,* series 6, IV, 95.

[9] The Phillips-Wheelock correspondence in the Dartmouth College Archives is supplemented by a series at the New Hampshire Historical Society, largely printed in New Hampshire Hist. Soc., *Collections,* IX, 68 ff.

mencement of 1772 it was announced that he had made "a Gift of One Hundred and Seventy Five Pounds Lawful Money, towards providing a Philosophical Apparatus for the Use of the College." [10] He entrusted the money to Governor Wentworth, in whose hands it was when war broke out. The Squire sued and eventually recovered from the Wentworth estate, but his money went toward the debts of the college rather than for scientific instruments.

In 1773 Phillips was chosen to the Board of Trustees of Dartmouth College, and for twenty years he was its most useful member. However, he had to refuse the first favor asked of him, that he purchase an ill-run tavern which was debauching the college boys. He begged Wheelock to keep him informed of college affairs, particularly in time of such crises as the Quebec Act, which he feared would discourage Lord Dartmouth by exposing the college to Romanism: "Oh Liberty! Oh my Country! and Oh Dartmouth particularly! may'st thou be preserved in this Day of the Lords anger." [11] In a typical reply to Wheelock's acknowledgment of a gift, he said:

The over flowing of your pious and grateful heart upon this occasion merits my most thoughtful Notice. If our services respecting this Seminary are acceptable to God (of whose own we have given him) we are involv'd in a debt of thanks which we shall never be able fully to discharge. The assistance afforded in the Indian service by such an unworthy wretch, as I feel and know myself to be, I could wish to be known to God only if that renderd not the benefit less extensive to my poor fellow worms. [12]

No other worm emulated his generosity, so he was without rival as the first great benefactor of the college. The LL.D. which he received in 1777 was richly deserved.

The War created at once greater problems and greater opportunities of service. Phillips asked Wheelock's permission to finance Joseph Johnson's mission to the Six Nations. Being a conservative, he became suspicious of the paper money before it collapsed, and by investing in land all of the currency which came his way, he acquired vast areas, most of which he later deeded to Dartmouth. When the college permitted some of this real estate to be sold for

[10] Boston News-Letter, Sept. 10, 1772, 1/2.
[11] New Hampshire Hist. Soc., Collections, IX, 77.
[12] Ibid., p. 83.

taxes, he bought it back at the auctions. Such mismanagement of college affairs, college politics, Wheelock's death, and the movement to split the Connecticut-River towns from New Hampshire, combined to discourage him from further donations:

What gives us concern is that College affairs are in so perplexed a situation. The obstacles to it's prosperity appear so many and of such a nature, that without a wonderful interposition of Providence what hopes can we entertain of their removal? You doubt not my disposition to serve the College interest; but I assure you, dear Sir, I have little prospect of being able to render it any considerable service at present, and the difficulty of journeying so far in hot weather appears so great that [I] must hope to be excused, and therefore, beg the favor of you to request the Board at the next meeting to accept my Resignation.[13]

It was not accepted, and his donations increased again, although the situation at Hanover became steadily worse. In 1782 Tutor John Smith reported to him that he was resigning from the college after nine years because boys were being admitted too poorly prepared to educate. He thought that he could do more good keeping a grammar school.[14]

After this Squire Phillips turned more and more of his money and attention to the academies, but he continued to participate in Dartmouth business. When he urged the drawing of the college lottery even though many of the tickets had not been sold, the managers assigned three hundred to him and made the drawing. A fortunate payment of interest on his Continental securities was all that enabled him to pay for the tickets. The next year, 1789, the Dartmouth trustees sought to revive his interest by voting that when they obtained the funds to establish the first professorship, it should be named the Phillips Professorship of Divinity and have the income from his gifts allotted to it. Busy with other good deeds at that time, he could spare only 285 bushels of wheat, which were invested in a woodlot for the professor. A more gentle touch than the professorship was the request that he select the books which were to be purchased for the college library from his donation. When he finally resigned from the Board of Trustees in 1793, it sent Joseph Steward (Dart. 1786) to Exeter to paint his portrait

[13] John Phillips to Bezaleel Woodward (Dartmouth College Archives), June 4, 1779.
[14] John Smith to John Phillips (*ibid.*), Apr. 2, 1782.

for the college. Steward was not a distinguished artist, and after he had found a profession to which he was better suited he was accustomed to refer to this picture as one of his early indiscretions. The college has been ashamed that it did not employ a better artist to record the appearance of its first great benefactor, but the picture is here reproduced as a more speaking likeness than the popular one at Phillips Andover.[15]

Dartmouth was not the only college which benefited from Phillips benevolence. In 1776 the Squire and his brothers sent the College of New Jersey about $1000 to be expended in any way which the authorities saw fit. This was the largest gift which Princeton had as yet received from individuals. But from this time on, it was the academies at Andover and Exeter which absorbed most of the family thought and money. Squire John in 1777 pledged £1666 toward the purchase of a site in Andover for the school which his nephew Samuel (A. B. 1771) proposed to establish. In all he contributed about $31,000 and a third interest in his estate to that academy. He was a member of the Board of Trustees, and its president after the death of his brother Samuel (A. B. 1734). But the institution nearest to his heart was the academy at Exeter, which he established with the largest endowment ever made in America up to that time. It was enthusiastically incorporated by the legislature on April 3, 1781. The original constitution, written in the Squire's own hand, follows that of Andover except that it reserves to him authority usually delegated to trustees, such as the making of rules for the government of the academy and appointing his successor as president of the Board of Trustees.

On May 1, 1783, the academy at Exeter was formally opened with a brilliant ceremony at which the speaker of the day turned the attention of the attending multitude to the benefactor:

I must ask your indulgence, while on this occasion, I make an address, or two: — and were I permitted, I would honor my discourse, with a public acknowledgment of the generous deed, which has convened us together, by laying the deserved applause, of a grateful public, at the feet of the honorable founder of the Exeter Academy. Her future sons

[15] The latter is reproduced in Crosbie, op. p. 8. It is said to be a copy made in 1829 by Folsom after an original attributed to Gilbert Stuart but burned in 1914. However, a very similar likeness in *Polyanthos*, IV, 113, is inscribed "H. Williams pinx." The Connecticut Historical Society has what appears to be the study for the head of the Steward portrait; it is reproduced in Conn. Hist. Soc., *Bulletin*, XVIII, 1.

will pronounce the name, with affectionate and grateful veneration. May the blessing of thousands, who by means of this generosity, may be rescued from ignorance and qualified for usefulness, rest on the honorable Founder! and the rewards of a future life be the glorious recompence of his extensive charities, in this![16]

The fact that Phillips pointedly ignored Old-Light Harvard while bestowing his bounty on New-Light Dartmouth and Princeton raises the question whether he intended the academies primarily as nurseries of evangelical preachers. He had given his blessing to a revival at Dartmouth, but had quoted to the college authorities Jonathan Edwards' warning against the excesses common on such occasions. In writing to his nephew Samuel about teachers for Andover, he said:

I am convinced of the need of Scholars being under the Tuition of Instructors who are of what we call Calvinistical Principles. I would not employ any that neglected teaching the Assembly's Catechism — or if any part was objected to, should expect to know what part.[17]

The latitude suggested in the last phrase is more important than the proposed restriction. The charter and curriculum of Phillips Exeter, which can safely be assumed to express his ideal, provide for the teaching of the full scope of the liberal arts without any special stress on theology. He must have known that Benjamin Abbot (A. B. 1788), whom he chose as principal, was besmirched with Harvard theological liberalism. Two of the trustees whom he chose were latent Unitarians. Frequently he said that he expected that the time would come when it would not be necessary for Andover and Exeter graduates to go on to college to complete their educations. In view of his own high educational standards, it is evident that he expected the academies to devote themselves to providing a liberal education. A clause in his will left a third of his estate for theological education beyond the academy level. The money was to be used

for the benefit more especially of charity scholars, such as may be of excelling genius and good moral character, preferring the hopefully pious, and for the assistance of youths liberally educated, designed for the ministry, while studying Divinity under the direction of some

[16] David M'Clure, *An Oration on the Advantages of an Early Education*, Exeter, 1783, pp. 19–20.
[17] Claude M. Fuess, *An Old New England School*, Boston, 1917, p. 66.

eminent Calvinistic Minister of the Gospel, until a professor of Divinity, able, pious, and Orthodox, should be supported by this Academy, or at Exeter, in New Hampshire, or in both.[18]

This clause ultimately led to the founding of the Andover Theological Seminary. During his last years he did his best to reunite the First and Second churches of Exeter, and when that failed, he quitted the Second and rejoined the First.

Preoccupation with the business of the academies kept Phillips out of public life. He was scheduled to escort President Washington into Exeter at the head of the cadet company, but the unexpectedly-early arrival of the guest prevented that appearance. He worked at his academic offices until he felt the breath of death upon him, and then he methodically transferred them to younger hands. His death came suddenly on April 21, 1795:

He was seized with a kind of fainting fit on Monday morning, from which he in part recovered, so as to walk about the house, and was perfectly sensible, and apprised of his approaching dissolution, and spake of it to his friends with calmness, and serenity, and with apparent pleasure. . . . "My work is done. I have settled all my affairs, and have now nothing to do but to die! It is no matter how soon!" And, retaining his reason to the last, the next morning he died.[19]

By his will he left the bulk of his estate of about $100,000 to the academies, leaving to his widow only a thousand silver dollars, the household goods which she brought with her at their marriage, and the produce of his farms to the value of £50 a year. When she announced that she was going to contest the will, the trustees of the academies made a much more generous settlement, which turned out well, for she lived only two years longer. They were less fortunate with their responsibility for the Squire's last slave, an old man named Corydon. Although Phillips was assessed for the Negro as late as 1785, in his will he provided for his "man-servant," adding parenthetically, "Slave I have none." Corydon lived to be about a hundred, consuming quantities of rum and tobacco that must have irked the academies which had to pay for them. When his funeral was finally celebrated, they gamely provided more rum.

[18] Albert M. Phillips, *Phillips Genealogies*, Auburn, 1885, p. 19.
[19] Crosbie, p. 30; Cunningham, p. 101.

The old mansion of which Squire Phillips was so fond passed eventually into the hands of countrymen of Corydon in which it fell into such disrepute that when it burned in 1860 this was held to be an act of Providence rather than arson. His manuscripts may well have gone with it. His diaries, which were in existence a century ago, have not been reported since that time.

ANDREW ELIOT

ANDREW ELIOT, the third minister of the New North Church of Boston, was born in that town on December 21, 1718. The family came from Beverly, where his paternal grandfather, Andrew, had served on the witchcraft jury, and had ever after been tortured by regrets.[1] The minister never forgot his grandfather's sorrow, and because of it he was loath to follow the pack which did the hunting in his generation. The second Andrew was a pious and industrious cordwainer who moved from Beverly to Boston and bought a house on the present Cambridge Street. In his adopted town he achieved the social distinction of being appointed to the committee of visitation and of being addressed as "Mr." His wife, our Andrew's mother, was Ruth Symonds.[2]

The one thing which the Eliot family vividly remembered about Andrew's childhood was the time that he was fished from

---

[1] Mass. Hist. Soc., *Collections*, 2nd Ser., I, 229.

[2] The genealogies, following the later of the two conflicting statements in the "Memorandum of the Eliot Family," pp. 83 ff., in Ephraim Eliot's Commonplace Book (in the possession of Samuel Eliot Morison), say that our Andrew was the son of a second wife, Mary Herrick. Contemporary records all call the mother Ruth.

a tub of water apparently dead, and restored only after long labor. He attended the South Grammar School where he studied under Nathaniel Williams (A.B. 1693) and John Lovell (A.B. 1728), and used a pony of Terence.[3] He entered Harvard in the middle of the Freshman year of his class and was placed fifteenth, twelve below John Eliot, whose father had a commission of the peace. One fine and one private admonition for playing at cards taught him his lesson. An early evidence of his interest in history is the fact that he subscribed for the *Chronological History* of Thomas Prince (A.B. 1707), as did his father. At the Commencement of 1737 he spoke as one of the respondents.[4] With the aid of a Hopkins Fellowship he remained in residence, living in Massachusetts 27 with Pemberton '42. Faithfully he took part in the speaking exercises in the College Hall which most resident graduates tried to avoid. In his diary he once remarks, "The Bachelours began to dispute, I made the sis." [5] At the Commencement of 1740, when he took his M. A., he delivered an address in which he maintained that absolute and arbitrary monarchy is contrary to reason. Obviously he had found in the college library the works of Locke and Sidney which for the rest of his life he regarded with religious veneration.

Still Eliot remained in residence at the college, where in 1741 he testified before the Overseers as to the bad habits of Tutor Nathan Prince (A.B. 1718). Beginning in 1739 he had ridden out with increasing frequency to preach in the pulpit of the First Baptist Church of Boston, and in the Congregational churches in the neighboring towns. He was serving as the regular supply at Roxbury when on August 18, 1741, he was invited to preach as a candidate at the New North of Boston. On January 11, 1742, this church voted by a majority of sixty-three to nineteen, to proceed to call this candidate, and appointed a committee, headed by the pastor, John Webb (A.B. 1708), to examine his sentiments and beliefs. The committee reported back favorably on the same afternoon, and the church voted overwhelmingly to accept the report. Later Eliot submitted a confession of faith which was satisfactory

---

[3] Mass. Hist. Soc., *Proceedings*, XLIII, 40.
[4] John Ballantine, Diary (Am. Antiq. Soc. transcript), July 6, 1737.
[5] Andrew Eliot, Diary (Harvard University Archives), Nov. 2, 1739.

to the church and on February 28 he formally accepted the call.[6] Then he moved from his college room to Boston.

The New North was one of the great churches of New England, and "a vast assembly" gathered on April 14, 1742, to witness the ordination. Dr. Joseph Sewall (A.B. 1707) began with prayer, Eliot preached, Mr. Webb gave the Charge, and Nathaniel Appleton (A.B. 1712) of Cambridge gave the Right Hand of Fellowship. Compelled against his will to preach his own ordination sermon, he chose a topic which suggests that he was aware that in college circles his colleague Webb was regarded as somewhat hypocritical in his orthodoxy:

It is too evident that there are many Errors and Heresies which prevail, and threaten the Church of God with the total Extirpation of true Religion; it's high Time for the Ministers to ascend the Watch-Tower, to cry aloud, and spare not, and vigorously oppose those Doctrines and Tenets which are so contrary to the Religion of Jesus. If they are dangerous, it's no Bigotry, but an incumbent Duty, to do all that in them lies to prevent their Propagation, so far as it can be done without external Force, or the least Shew of Persecution.[7]

This idea that "the total Extirpation of true Religion" was a lesser evil than "the least Shew of Persecution" illustrates the inability to reconcile theory and fact which troubled him all of his days.

In New England the minister's marriage was hardly less essential than his ordination. Fortunately one of the deacons of the New North was in a position to help. On the evening of October 5, 1742, Eliot was married to Elizabeth, daughter of Deacon Josiah and Elizabeth (Sexton) Langdon. All surviving evidence supports the statement of William Bentley (A.B. 1777) that she was "exactly the prudent wife for a minister."[8] Friends took up a collection to buy the young couple a black boy, but the minister, always a zealous opponent of slavery, insisted that it be agreed that he could apprentice the lad to a good trade and free him at a suitable age. The friends lost interest.[9]

[6] [Ephraim Eliot], *Historical Notices of the New North Religious Society* (Boston, 1822), p. 20. The confession of faith and the numerous votes relating to his call and settlement are printed in Colonial Soc. Mass., *Publications*, XIII, 234–241.

[7] Andrew Eliot, *The Faithful Steward* (Boston, 1742), p. 21.

[8] William Bentley, *Diary* (Salem, 1905–14), IV, 151.

[9] [Ephraim Eliot], *New North*, p. 30.

Eliot's beginning salary was only £50 Sterling a year, and even after it had been doubled it was hardly adequate to care for the eleven children who arrived in quick succession. In his diaries the list of shoes purchased in a single year sometimes takes up an entire page. Of course there were the usual gifts to supplement his salary. Two unusual sources of income were the rings which he received at funerals and the gloves which were given him at funerals, weddings, and baptisms. At one point in his diary he totaled his take and found that he had received nearly three thousand pairs of gloves, which he had sold for £1441 18s 1d. In the same way he carefully recorded rings received and sold.

The Eliots rented until 1756, when they bought a house on the northeast corner of Hanover and North Bennet streets. The minister never owned any other property in Boston, and a little land which he inherited in Maine was a burden rather than an asset. Neither this comparative poverty nor the fact that his father was a shoemaker produced in the minister any of the social inferiority complex which made life miserable for some of his fellows in like situations. He accepted as normal the friendship of members of the old and wealthy families, but never in his contacts with them showed that social deference which they expected:

Thus, its strange to see, how some empty Wretches expect a Deference to their Judgment because they are rich; as if Riches procured them Brains and they were Men of Sense because they had large Possessions: So likewise as to Honour and other Things, they oftentimes put Men above themselves, so that they are in a Rage at the least Slight, yea, if they are not preferred before others.[10]

Because of his place in the upper social circle, distinguished visitors to Boston were almost always brought to see him as one of the more agreeable members of the clergy, "always cheerful and entertaining in conversation, abounding in interesting anecdotes, yet never descending to levity." [11] But, far from neglecting his swarm of parishioners, he made a point of reserving the time to visit them for social entertainment and instruction. It was notable that his mind was slow to set — as he grew older his more intimate friends were younger men.

In spite of Eliot's warm companionship with his contemporaries,

[10] Andrew Eliot, *An Inordinate Love* (Boston, 1744), p. 19.
[11] [Ephraim Eliot], *New North*, p. 30.

he was so reserved in recording his thoughts about his personal affairs that he is somewhat baffling to a modern. He carefully recorded the date on which he "began tobacco," but ignored the births of his children. He went on fishing and roast-beef parties, but did not say that he had a good time. Very definitely he did not like the theater:

I am no enemy to innocent amusements, but I have long thought our modern theatre, "the bane of virtue." Some years ago theatrical entertainments were introduced among us. I had such an opinion of their pernicious tendency, especially in a young country, that I exerted myself to procure an Act to prohibit them, and by the help of friends, succeeded. This does not wholly prevent them; but so many are engaged to suppress them, that they will not soon be publicly tolerated.[12]

On the other hand, he approved and attended Kinnersley's course of lectures on electricity.[13] Never an ardent student of science like some of his colleagues, he observed and accepted it. When two-thirds of his parishioners were down with the smallpox in 1752,[14] he had his six children inoculated.

In this and like matters Parson Eliot soon became regarded as a leader in the Boston community. Particularly he was faithful in his attendance on the committee to visit the schools. Curiously enough, he rarely served on the annual committee of visitation which interviewed and exhorted every poor or unruly inhabitant. His fellow Bostonians had, he thought, fallen away from the virtues of their ancestors:

'Tis surprising to observe the Degree of Profanity to which some have arrived; with what Ease and Impudence they can toss about the most shocking Oaths. Even our Children in the Streets are Adepts in this Sin; they can Swear and Curse without Remorse, and seem to vie with each other in this aggravated Crime.[15]

This religious decline made itself felt also in the support of the clergy. The Boston ministers had always been maintained by voluntary contributions, but even in the flourishing New North these were no longer large enough to support two clergymen. Consequently, when Webb died in 1750, the parish asked Eliot

---

[12] Mass. Hist. Soc., *Collections*, 4th Ser., IV, 403–404.
[13] Andrew Eliot, Diary (Mass. Hist. Soc.), Nov. 28, 1751.
[14] Ezra Stiles, *Itineraries* (New Haven, 1916), pp. 25–26.
[15] Andrew Eliot, *An Evil and Adulterous Generation* (Boston, 1753), p. 12.

to carry the entire burden, and sweetened the request with a substantial raise. Fortunately he was a powerful man as well as an
industrious one, quite able to do the work alone.

For several years Webb had been partially incapacitated by a
stroke, but out of respect for his feelings Eliot had accepted certain
primitive practices of the New North. Changes came quickly after
the death of the old pastor. Under Eliot's administration there
was a distinct falling off in church action to punish drunkenness,
fornication, and disorderly conduct in the meetinghouse, but an
increase in votes to expel members of the congregation who had
been converted to the views of the Episcopalians and the Baptists.
This change was not, of course, an indication of the Parson's ideas
as to the relative seriousness of the offences. The public confession
of sin was less frequently exacted, and two non-church members
were added to the Standing Committee to represent the congregation. It was voted that the Bible should be read as a part of the
service, and the minister was presented with a magnificent folio
copy for use in the pulpit.[16] It was only after long debate that he
succeeded in getting the church to abolish lining-out, to substitute
Tate and Brady for the New England psalm book, and to add a
selection of hymns. In 1763 a steeple was added to the meetinghouse, and ten years later the requirement of the public narration
of the experience of conversion as a requisite for church membership was abolished. About the same time the rules for baptism
were liberalized.

So successful was Andrew Eliot in maintaining his policy of
keeping out of public theological controversy that, if it were not
for these changes in New North practice under his administration,
we should be doubtful of his stand on some of the lively issues of
the day. However, his New-Light contemporaries always disliked
him. On the evening before his ordination, Andrew Croswell (A.B.
1728), at a public lecture in Charlestown, prayed fervently for the
New North which, he said, was about to settle an unconverted
man. Eliot publicly protested against the conduct of such revivalists as William Hobby (A.B. 1725) and James Davenport (Yale
1732), and when signing the declaration of the clergy of the
Province on July 7, 1743, in regard to the Great Awakening, he
made the reservation that the document did not sufficiently testify

[16] This Bible is now at the Massachusetts Historical Society.

against the evils of itinerant preaching.[17] In later years he frequently remarked with the satisfaction of vindication "that the
zealous upholders of these fanatics had turned out vagabonds." [18]

When a majority of his fellows savagely attacked and boy-
cotted the English revivalist, Whitefield, Eliot attended his services, but made no public comment. To a friend in another colony
he wrote his opinions:

As to Mr. Whitefield's being the ringleader of those things of bad and
dangerous tendency which have prevailed among us, I am really at a
loss what to say. . . . Has not a vein of enthusiasm run through his
writings, his preaching, and his conduct? I must needs say there has
been too much in all these which has appeared to me to border at least
upon enthusiasm, and which I always thought had a very dangerous
tendency, and I fear has had very unhappy effects. . . . I have been
ready to think that the defects in the account of his own conversion,
(if he was not mistaken in the time,) proceeded from his own ignorance
of the doctrines of the Gospel, and unacquaintedness with experimental
writers, so that he did not know how to express the real experience
which he had.[19]

Although Eliot disliked the revivals, he subscribed for Jonathan
Edwards' *Life of Brainerd* and *Original Sin* while pointedly
avoiding open support of the views of Charles Chauncy (A.B.
1721); but by 1773 he had joined James Dana (A.B. 1753) in
his resistance to Edwards' doctrine of the Will.[20]

Although never drawn into personal theological controversy,
Eliot did not hesitate to make clear his stand on a few major issues,
such as Antinomianism:

We may not now expect immediate Revelations from him, Christ com-
municates Light in a more ordinary Way: while Men attend those
Means which are in themselves suited to answer the End, he blesses
their Industry and grants them that Light and Influence which is
necessary for them. If any look for Light from Christ in any other
Way . . . what they take to be Faith is only Presumption, and is
like to lead them into Mistakes.[21]

---

[17] *Christian History*, I (1743), p. 164.
[18] [Ephraim Eliot], *New North*, p. 28.
[19] William B. Sprague, *Annals of the American Pulpit* (New York, 1857–69),
I, 417.
[20] Andrew Eliot to James Dana, July 9, 1773, Andrew-Eliot Mss. (Mass. Hist.
Soc.), p. 111.
[21] Andrew Eliot, *A Burning and Shining Light* (Boston, [1750]), p. 14.

As to the then burning issue whether a man should undertake the ministry without a clear conviction of his conversion, Eliot replied that such knowledge was a good thing, but that for a minister to lay down his work every time he had doubts and fears about his state would open the door to great inconveniences. "But, no one ought to engage in this sacred work, who has not some reason to hope, that he is a good man, and a sincere Christian." [22] Much more important than confidence in one's conversion was education:

A man who is sensible of his want of knowlege and necessary learning, and yet intrudes into the sacred ministry, is guilty of such arrogance and presumption as can scarce be consistent with grace. The only excuse which the most extensive charity can suggest in favor of an ignorant minister, is that which our Lord made for his murderers, "they know not what they do. . . ." It is at least of as much importance to enquire whether they understand any principles at all, as whether they are willing to subscribe the particular creed to which we are attached.[23]

It was on these issues of personal revelation, conviction of conversion, and formal education that Eliot came into conflict, if it could be called that, with the Baptists. He was a warm personal friend of the Reverend Jeremiah Condy (A.B. 1726) of the First Baptist Church and was much disappointed when in the choice of a president for the new College of Rhode Island, his friend was passed over and the less liberal James Manning (Princeton 1762) chosen.

The choice of another man from the central colonies, Samuel Stillman (A.M. 1761), to succeed Condy in the First Baptist Church, seemed to Eliot to be another threat to the interdenominational peace which had flourished in Boston for several generations. Like the Mathers, he had been accustomed to welcome members of other sects to the communion table in his church, and before he would accept the invitation to participate in Stillman's ordination he asked the candidate to assure him that he did not think that only those who had been baptized by immersion were entitled to take communion. Stillman agreed, but when Manning came up for the ordination he took the opposite position, and demanded that the candidate take him to see Eliot in order that there might be no misunderstanding. The minister of the New

[22] Andrew Eliot, *Sermon Preached at the Ordination of Joseph Roberts* (Boston, 1754), p. 17.

[23] Andrew Eliot, *Sermon Preached September 17. 1766* (Boston, 1766), p. 13.

North refused to discuss the matter until John Lathrop (A.M. 1768) had been brought in as a witness. Stillman tried to argue that he had qualified his acceptance of Eliot's proposition, but the latter flatly said that he had not.[24] The result was an unfortunate coldness between the two men which lasted for some years, but for the moment it seemed to Eliot that he could do more good by taking part in the ordination and extending the Right Hand of Fellowship to the candidate. On this occasion he told the gathering:

The solemnities of this day shew that christians of different sentiments can unite in offices of love. You, our beloved brethren, have invited the churches in the neighbourhood to join in one of the most sacred acts of religion. We have attended to your call, and shewed our readiness to all acts of communion and christian fellowship. And what is there to break our union, or to keep us at a distance? If our religious opinions be not just the same, we agree in owning Christ to be our Master and Lord, and in calling one another brethren in Christ. You have in your letters missive, acknowledged us to be the churches of Christ. We cheerfully return the honourable title. We own you to be a church of our Lord Jesus Christ.[25]

When Whitefield returned to Boston in 1770, Eliot permitted him to preach a series of sermons in the New North, and for this he was sharply attacked by the Old-Lights.[26] The next year the New-Lights just as bitterly attacked him for opening his pulpit to William Emerson (A.B. 1761).[27] He drew the line at John Murray (Edinburgh 1761),[28] a Presbyterian who had obtained a pulpit by misrepresentation; and he urged the Congregationalists not to omit entirely the examination of candidates because in some instances intemperately orthodox members of councils had been too strict.[29]

Every mistake in doctrine, or what we think so, is not to err from the truth. . . . We are not warranted to pronounce every one a sinner and in an unconverted state, who differs from us in sentiments on points of religion. It proceeds from the pride of our hearts, that we set up our interpretation of the sacred oracles as the standard of truth, and look on those who do not come up to this, as in a fatal error and enemies

---

[24] [Ephraim Eliot], *New North*, pp. 23–24.
[25] *Ibid.*, p. 47.
[26] *Boston Evening-Post*, Sept. 17, 1770, 1/1.
[27] *Boston Gazette*, supplement, May 6, 1771, 1/1–2.
[28] *Boston News-Letter*, May 12, 1768, 2/3.
[29] Andrew Eliot, *Sermon Preached at the Ordination of . . . Joseph Willard* (Boston, 1773), pp. 11, 12.

of the cross of Christ. We forget that we are weak fallible men, and liable to err as well as others. There are opinions which are not true, and yet may be held by men of upright hearts, by those whom Christ will own as his disciples.[30]

Eliot early entered with enthusiasm into the plans for the Christianizing of the Indians, was active in the work of the London Society for Promoting the Gospel in New England and Parts Adjacent, and was one of the Boston commissioners of the Edinburgh Society for Propagating Christian Knowledge among the Indians. Had friends not dissuaded him, he would have sent his son Samuel into the Indian country to learn the language preparatory to becoming a missionary. He was one of the most active organizers of the Massachusetts Society for Propagating Christian Knowledge among the Indians of North America, and his experience in this case was the first step in his disaffection from England. Amazed to find his "Episcopal Brethren" inclined to block this good cause, he wrote to Dr. Samuel Chandler in England imploring his aid and urging the cooperation of the two denominations in the mission work.[31] When he learned of the defeat of the charter he told Jasper Mauduit, "It is strange that Gentlemen who profess Christianity will not send the Gospel to the Heathen themselves nor permit it to be sent by others." [32]

By 1769 Eliot's zeal for the missions had completely evaporated because of their miserable failure.[33] For this reason he regarded Wheelock's efforts with a jaundiced eye:

The Commissioners for Indian Affairs, appointed by a corporation in London, (. . . of which Board I have . . . the honor to be a member,) at first encouraged this school which Mr. Wheelock set up; not because we had any great opinion of him, but because we were willing to try every method to serve the poor Indians. We soon found that he had great, and as we feared, romantic designs; and besides, he was, as we thought, very unreasonable in his charges. We therefore withdrew ourselves.[34]

However, the degree of Eliot's withdrawal from the mission work has been exaggerated. After the death of Andrew Oliver (A.B.

[30] Andrew Eliot, *Sermon Preached at the Ordination of Andrew Eliot* (Boston, 1774), p. 10.
[31] Andrew Eliot to Samuel Chandler (Harvard College Library), c. 1762.
[32] Mass. Hist. Soc., *Collections*, LXXIV, 119.
[33] Andrew Eliot to William Harris (Harvard College Library), July 3, 1769.
[34] Mass. Hist. Soc., *Collections*, 4th Ser., IV, 415–416.

1724) he carried on much of the correspondence of the Commissioners, and the missionaries regularly reported to him.[35] He was most active in this work on the very eve of the Revolution.

This was typical of the good work with which Eliot busied himself. He was scribe of the provincial convention of the clergy from 1758 to 1761, he preached the Convention Sermon of 1767, and he represented the interests of the Congregational churches of Nova Scotia. But of all these good works, the nearest to his heart was Harvard College. As a young man he had hailed the election of Holyoke as the victory of catholicism over bigotry,[36] and as an Overseer by virtue of his office, he staunchly upheld the policies of the group which controlled the college. In 1758 he was chosen clerk of the Board of Overseers, and so made its mouthpiece. Loyally he opposed the foundation of a college in Hampshire on the ground that New England could not support another collegiate institution. He opposed the Governor's claim to have the right to issue such a charter, protesting that this was dangerous because presumably all future governors would be Episcopalians.

The burning of Harvard Hall launched Eliot on activities which were to take up much of the rest of his life. The Corporation appointed him to committees to raise money to replace the scientific instruments and to prepare a catalogue of books wanted for the new library, and although he was afraid that the damage could never be made good, he made every exertion to obtain the necessary gifts.[37] The unexpected response to his efforts opened his eyes to the possibility of building up the college by such means, and led him to compile a record of all benefactions received by Harvard since its founding. This his son Josiah engrossed into the handsome Donation Book of which the present tattered state is proof of its utility over the years. Two years after the beginning of the drive, he reported its success to the London agent:

The General Court has built us an house much superior to that which was consumed. The room which contains [the library] is perhaps the most elegant in America. And when we have received all our benefactions it will be the best furnished. It is divided into ten Alcoves,

[35] The reports of the missionaries to Eliot are in Misc. Mss. (Mass. Hist. Soc.), 1773–75, *passim*.

[36] Mass. Hist. Soc., *Collections*, 4th Ser., IV, 409.

[37] The drafts of Eliot's letters as clerk and secretary of the Board of Overseers are in the Andrews-Eliot Mss.

five on each Side. . . . Care is taken when the Book's are put up in
the Library to have an alphabetic account of each Book with it's form
and edition. A Scientific catalogue you are sensible must be a work of
time, it will be undertaken as soon as we have received the donations
of which we have notice.[38]

In 1765 Eliot was elected to the Corporation, and thereafter he
became more than ever, particularly because of Holyoke's advancing years, the agent, secretary, and spokesman of the college. His
influence was fully in accord with the best in Harvard tradition.
When Dr. William Harris, the historian, tentatively sent him
certain heretical books for the college with the remark that some
things were to be learned even from enemies,[39] he replied that he
was not afraid of reading "obnoxious Books," and that he had
learned from infidels that some doctrines he had once had a high
opinion of could not be maintained.[40] So far as dogma was concerned, he set the students a high standard of intellectual honesty.
For the vain display of learning he had no use:

In the Course of my Preaching, I have not meddled with abstruse
Speculations: And, as far as Ministerial Fidelity would allow, have
avoided Subjects of Controversy. I have rather desired to impress upon
your Hearts and my own a deeper Sense of those great and important
Truths, in which good Men are agreed, and which are at the Foundation of all Religion. Being fully persuaded, that the Pulpit was not
designed to be a School of Disputation, or to display a Minister's Acquaintance with Science falsly so called.[41]

When preaching the Dudleian Lecture at the College he defined
"Natural Religion" as that revealed in nature, and "Revealed
Religion" as that revealed in the Bible. He advised the students
to keep an humble faith in the goodness of God rather than to
beat their brains out on insoluble theological problems.[42] Not,
however, that he avoided real problems. His method was to begin
with a simple question, such as "is there a God," or "is there a
life after death," and to state his case clearly in fundamental terms,
without obscuring quotations from Scripture. Although he regarded himself as opposed to Arminianism, in his openness to logic

[38] Andrew Eliot to Israel Mauduit, Oct. 30, 1766, Andrews-Eliot Mss., p. 100.
[39] Andrew Eliot to William Harris, July 27, 1767, *ibid.*, p. 102.
[40] Andrew Eliot to William Harris, Dec. 1767, Harvard College Library.
[41] Andrew Eliot, *Twenty Sermons* (Boston, 1774), p. iv.
[42] Andrew Eliot, *Discourse on Natural Religion* (Boston, 1771), *passim.*

and Biblical criticism he was sailing full in the course which in another generation was to carry the college into Unitarianism. In one matter he was trying to hold back the tide, and that was the use of flowery diction in the pulpit. His own discourses were always plain and practical, delivered carefully from notes without any physical action.

After the death of Holyoke, Eliot was the obvious candidate for the presidency, but he was so firm in rejecting the suggestion that the Corporation did not act on his name. The fact that he was "quite sensible of the state of anarchy and confusion in which the College" then was may have had something to do with his decision.[43] His son Andrew (A.B. 1762) was elected to the Corporation in 1772, affording the unique situation of father and son as members at the same time. When the presidency fell vacant in 1774, the Corporation offered it successively to three of its members, John Winthrop, Samuel Cooper, and the elder Eliot. The last refused it because he thought that the Corporation ought not to choose one of its own members, and because he felt that he should not leave his church. It was he who went to Portsmouth to persuade Langdon to accept the presidency.

There is truth behind the idea that the Revolution in Massachusetts was made by the black legion of preachers thundering politics from their pulpits, but Eliot was not one of them. He impressed upon his sons and their friends who entered the ministry that they must never as clergymen take sides in politics, but this did not mean that he was without interest in public affairs. In his sermon on the taking of Quebec he traced the glorious rise of God-favored Protestantism from Luther through Queen Elizabeth and William of Orange, and step by step through the wars with the French in America. He gave an excellent résumé of the colonial wars, written in the assurance that the British arms, like those of the ancient Jews, were God's arms.[44]

It was difficult in those busy days to keep quiet on political matters. Eliot used his influence, unsuccessfully, to prevent the discharge of William Bollan from his post as Province agent for mere political reasons. In 1764 he was appointed chaplain to the

[43] Andrew Eliot to John Winthrop, College Papers (Harvard University Archives), II, 18.

[44] Andrew Eliot, *Sermon Preached October 25th, 1759* (Boston, 1759).

General Court and invited to preach the Election Sermon. The legislature was then sitting at Concord because of the smallpox at Boston, and when Eliot pointed out that he might bring the disease from his parishioners whom he was daily visiting, he was excused. The year following, on the day before Patrick Henry made his famous speech on the Virginia Resolves, he electrified his hearers by preaching an Election Sermon which was pure political science. Quoting Grotius, Burlamaqui, and Montesquieu, he urged that

All power has it's foundation in compact and mutual consent, or else it proceeds from fraud or violence. Where the latter take place, the dominion which men claim is no better than usurpation; and they, who by these methods raise themselves above their brethren, are so far from having a right to govern, that they ought to be punished as public disturbers and the enemies of mankind.[45]

Although he praised the British system, even to the appointment of colonial governors, it was clear that his discussion of the virtues of the good ruler was a criticism of Bernard, before whom he was speaking. The English Whigs happily seized upon this sermon, had it reprinted in England, and sought the acquaintance of the author. In the resulting correspondence he developed a theory of the history of the colonies which fitted his political beliefs. The founders of New England, he said, had regarded their connection with Great Britain purely as a matter of expediency. Were the descendants of the French among them subjects of France? To assert Parliamentary authority over the English settlers after seven generations was absurd, and would be still more so when the colonies came to exceed the mother country in numbers, power, and wealth, as soon they must. But he would bitterly oppose the breaking of the compact by the colonists; independence would bring anarchy.[46]

Eliot was sincerely interested in history. He rescued old coins from silversmiths and sent them to the collectors among his English friends. He remarked that pine-tree shillings of several dies were still common, but that he had only seen one New England

[45] Andrew Eliot, *Sermon Preached before his Excellency Francis Bernard* (Boston, 1765), pp. 17–18.

[46] Andrew Eliot to Francis Blackburne (Harvard College Library), Dec. 15, 1767.

sixpence in his life. At the auction of the effects of Timothy Cutler (A.B. 1701), he bought a chair which had been made for Dean Berkeley in Rome.[47] He was interested in Harvard biography and made notes which, passing through the hands of William Winthrop (A.B. 1770), survived to serve this series. Recognizing the value of newspapers as source material, he preserved some of the files now in the collection of the Massachusetts Historical Society. Other documents he preserved and copied for his friend Thomas Hutchinson (A.B. 1727), who used them in his *History*. When in August, 1765, the mob scattered the latter's manuscripts through the streets of Boston, he posted a notice asking that they, and any of the rest of his plundered goods which might be recovered, be returned to Parson Eliot. The minister himself advertised and, to Hutchinson's surprise and joy, recovered enough manuscripts to fill several trunks. John Eliot (A.B. 1772) in the next century used to remember the picture of his father sorting and arranging the recovered material.[48]

The fact that Hutchinson attended the New North and shared historical interests with Eliot was enough to damn the pastor with the Whigs. John Adams said contemptuously that Eliot "was Hutchinson's parish priest, and his devoted idolater." [49] The minister ignored this attitude and always spoke of the young lawyer in the most complimentary terms. Actually it was Hutchinson rather than Eliot who furnished the warmth in their friendship; but any association with the chief Tory was enough to make the Parson unpopular with the Whigs, who sang of him as "Andrew sly, who oft draws nigh to Tommy skin and bones." [50]

The reserve which existed between Eliot and Hutchinson was in curious contrast with the effusive correspondence between the pastor of the New North and Thomas Hollis, which began with this letter from that famous liberal and philanthropist:

The last year I returned you my hearty thanks, by Dr. Mayhew, for the obliging present of a curious sermon, preached on an important occasion. A few books, which I had intended for that excellent man, why, seemingly, no more! with an addition to them, (and such political fry as the times have produced,) I now request place for in your study.

[47] This chair is now at the Massachusetts Historical Society.
[48] John Eliot, *Biographical Dictionary* (Salem and Boston, 1809), p. 192.
[49] John Adams, *Works* (Boston, 1850–56), X, 243.
[50] Ezra Stiles, *Literary Diary* (New York, 1901), I, 491.

They are sent in that kind of way always used by me toward my friends. . . . The similarity of turn, as appeareth by your sermon, to my late honored friend [Mayhew], the regularity of your education, the fullness of your character, your age, station, power, will to render public service, all have concurred with me . . . to take this measure. . . .[51]

The flood of books from Hollis included Harrington, Sidney, and Locke, with which Eliot was familiar, and the first set of Milton's prose works which he had ever seen. This last the pastor of the New North received with rapture and veneration. Without comment he passed Rousseau on Education along to the Harvard library. In view of his enthusiastic acceptance of these English writers it is hard to see why President Stiles called him a "Boerhaavian" in contrast to his "Newtonian" and "Lockean" fellows.[52]

With these works which passed through Eliot's hands to the several college libraries of the country, there came from the English Whigs some radical political essays which he undertook, apparently with success, to get into the American newspapers. To these he added controversial pieces of his own authorship which he sent to the papers anonymously as not suited to the cloth.[53] He was brought now to engage thus in politics because of the Stamp Act excitement. There was rather a note of approval in his report of the mob's attack on Secretary Oliver, but he thought that the sacking of the Hutchinson house, which he watched, "was a scene of riot, drunkenness, profaneness and robbery." He was unwilling to admit that Mayhew's sermon had raised the riot.[54]

Eliot repeatedly assured Hollis and his English friends that the opposition to the Stamp Act did not mean that the colonists wished to leave the Empire:

The people here have no notion of aiming at independence. They highly value their connection with their mother country. They glory in the name of Englishmen, and only desire to enjoy the liberties of Englishmen. . . . We have imprudent men among us, but the community ought not to suffer for the faults of a few. . . . We are not ripe for a disunion; but our growth is so great, that in a few years Great B————n will not be able to compel our submission. Whereas, if they treat us as brethren and friends, it will be the interest, and the

[51] Mass. Hist. Soc., *Collections*, 4th Ser., IV, 398.
[52] Ezra Stiles, *Itineraries*, p. 27.
[53] John Eliot, *Biographical Dictionary*, p. 193*n*.
[54] Mass. Hist. Soc., *Collections*, 4th Ser., IV, 406–407.

inclination of the Colonists, to be united with their parent country. . . .
The Colonies, if disunited from Great Britain, must undergo great
convulsions before they would be settled on a firm basis. Colony would
be against Colony, and there would be in every one furious internal
contests for power. . . . The latter we have seen something of in
Rhode Island, where they elect their own Governor. They are divided
into furious parties; they bribe, they quarrel, they hardly keep from
blows. The parties are so nearly equal, that they change Governors
and Magistrates almost every year. If things are so bad in that little
government, what would they be in greater. I hope not to live to see
the American British Colonies disconnected from Great Britain.[55]

Unfortunately there were other causes of tension besides the
Stamp Act. At the time of the burning of Harvard Hall, Archbishop Thomas Secker had presented the college with a set of his
own sermons for which Eliot thanked him heartily before he discovered among them a discourse which the prelate had preached
to the S. P. G. on its mission work in New England. In this he
had, with incredible ignorance, described as prevailing in New
England the barbarism in manners and religion which then existed
in the backwoods of Carolina, had urged the sending of missionaries to the benighted northern colonies, and had entirely mis-
represented the legal situation of the Church of England in Massachusetts and Connecticut. Eliot composed a long and hot rebuttal,
pointing out that the maligned New England laws were more
liberal than those which prevailed where the Episcopal church was
established, as in New York, where the Presbyterian minister,
Ebenezer Pemberton (A.B. 1721), was taxed for the support of
the Anglican, William Vesey (A.B. 1693). He bitterly denounced
the flagrant misrepresentation of the treatment of Episcopal missionaries in New England, and suggested that the preachers before
the S. P. G. were stopping at nothing to raise money.[56]

Hollis passed this manuscript along to Archdeacon Francis
Blackburne, who in his efforts to reform the Church of England
from within was quite as rabid as the New England Congregationalists. He had a portion of this essay printed in England, and he fed
the flames of Boston anger with his own denunciations of the

[55] *Ibid.*, pp. 404, 420–1.
[56] "Remarks on the Bishop of Oxford's Sermon Preached before the Incorporated Society for the Propagation of the Gospel in Foreign Parts," pp. 190–215 in
Mass. Hist. Soc., *Collections*, 2nd Ser., II.

Church.[57] Hollis reported that as a result of these attacks the S. P. G. was getting so sensitive that it would not give out copies of its latest missionary sermon. Perhaps it was because he thought that further controversy was unnecessary that he returned Eliot's manuscript instead of printing it *in extenso*, which was what the author had hoped for.

The pastor of the New North was irritated at being regarded as a fit subject for missionary activity, but he was really worried lest a bishop be appointed for the American colonies. He was not alone in this; Pennsylvanians enlisted his aid in the struggle against the establishment of the Church of England in their colony,[58] and Dr. Blackburne warned him of the "Hierarchial Yoke" which was being prepared for the colonies in "the dregs of the Stuartine and Laudean Ecclesiastical Politics."[59] Eliot became a focal point for American resistance to the project for a bishop, and the funnel through which Hollis and his friends poured anti-Episcopal and the more absurd pro-Episcopal tracts into the college libraries and newspaper offices.

The appointment of a Roman Catholic bishop for Canada came as a stab in the back. After reading the S. P. G. sermon announcing this, Eliot wrote to Hollis:

I do not wonder they endeavor to conceal these annual productions. They are such miserable performances in themselves, so full of falsehood, scurrility and pious fraud, that they are not fit to see the light. . . . I particularly observed the passage you marked, relative to a Bishop of Canada. These gentlemen seem to have lost all modesty. They first, contrary to all law, policy and religion, send a bishop to encourage the inhabitants of this newly conquered country in their fatal superstitions . . . and then argue from thence, that the hierarchy must be established in the other Colonies. Was not this the main thing they had in view in sending this popish bishop . . .? If I hear of a bishop sent to America, I shall . . . expect soon to hear that popery is tolerated in Ireland, then in England. . . . They who plead so strongly for an American bishop, have other ends in view; to make a more pompous show, by which they hope to increase their faction; to add to the number of Lord Bishops; to extend their episcopal influence;

---

[57] Francis Blackburne to Andrew Eliot, Aug. 18, 1767, Eliot transcripts (Harvard College Library), pp. 136–142.

[58] Jonathan Ewing to Andrew Eliot, Aug. 6, 1766, Andrews-Eliot Mss., p. 101.

[59] Francis Blackburne to Andrew Eliot, Jan. 23, 1767, Eliot transcripts, pp. 127–129.

to subject the American dissenters to their yoke; to tyrannize over those who yet stand fast in the liberty wherewith Christ hath made them free.[60]

When the General Court met again, Eliot encouraged some of the members to present a resolution expressing the general fear of the setting up of a hierarchy in America. He hoped, he told Hollis, that the excitement here would not "break out into a flame and produce a general conflagration. There are men with you, and men with us, who regard no consequences, if they can but gratify their passions." [61] Particularly he was irritated that now American Episcopalians had only to make the short journey to the Roman bishop in Quebec to obtain an ordination which the Church of England would hold valid while denying the validity of congregational ordinations. With amazement he watched the campaign for the toleration of Roman Catholics in England:

I am surprised at the impudence with which the papists plead for toleration. He must have lost all principles of self-preservation, who will take a serpent into his bosom, especially when he has felt his sting, and but just escaped with his life.[62]

When Parliament moved to limit Catholic liberties he wrote that he wished "success to every attempt to curb the insolence of those enemies of truth, of liberty, of mankind." [63] When these measures were defeated he lamented, and warned his English friends that people returning from Quebec reported that the Papists there, now that they had a bishop, were becoming insolent, and were turning the Protestant congregations out of the churches which they had been using jointly.[64]

Eliot's fears of the church of Rome and the Church of England were purely political. He never mentioned his theological differences with Rome, and he accepted the doctrine of the Episcopalians on all points except the descent of Christ into hell, which he regarded as an "absurd opinion, which some of the ancients entertained." [65] Although his parishioners regarded even the observance of Easter as a popish superstition, he himself attended a Good

[60] Mass. His. Soc., *Collections*, 4th Ser., IV, 410–411.
[61] *Ibid.*, p. 422.
[62] *Ibid.*, p. 406.
[63] *Ibid.*, pp. 414–415.
[64] *Ibid.*, pp. 448–449.
[65] Andrew Eliot, *Christ's Promise* (Boston, 1773), p. 7.

Friday service at Christ Church.[66] The ceremony and vestments of the older churches he regarded as "a mechanical kind of devotion, entirely opposite to the rational, manly religion which Jesus Christ inculcates on his followers." [67] Bitterly he opposed all religious establishments, because if you defend one, you have to defend them among "Papists, Turks and Heathen." He regretted the fact that Massachusetts town churches had to be supported by taxation, and was grateful that the Boston churches had always been supported by voluntary contributions. To his friend Blackburne he explained his ideas of toleration:

The fathers of New England were a set of worthy men, but they did not understand religious liberty. There was too much of an intolerant spirit among them. It was not a fault peculiar to them; it was the error of the day. But however contracted they were in their religious sentiments, *they never imposed subscriptions to any human forms.* Possibly this was because there was no suspicion of erroneous principles. But I would rather think, that their good sense taught them that it could answer no valuable end, and could lead only to prevarication and falsehood.[68]

If we judged Eliot by the prejudices aroused in him by his political fears of Rome and Lambeth, we should be unfair to him. Peter Thacher (A.B. 1769) was substantially right when he said of him that "he embraced within the arms of his christian affection all those who appeared to have an honest regard to religious truth, let their tenets or modes of worship be what they would; and possessed the happy talent of securing men's affection and esteem while he opposed their favourite notions." [69]

Many shreds of evidence show that this last sentence needs some qualification, and it is best provided by his son Ephraim (A.B. 1780):

His tone of voice was bold and positive, as though he would not be contradicted. Nor indeed did he bear contradiction tamely out of the pulpit. Over an highly irascible temper he had acquired a remarkable command. When he felt his passions rising, he would retire by himself, till he had controlled them. His influence over his parishioners was

[66] Andrew Eliot, Diary (Mass. Hist. Soc.), Mar. 28, 1766.

[67] Mass. Hist. Soc., *Collections*, 4th Ser., IV, 408.

[68] Andrew Eliot to Francis Blackburne, May 13, 1767, quoted in Francis Parkman, *Discourses Preached in the New North Church . . . December 9* (Boston, 1839), p. 40.

[69] Peter Thacher, *Rest which Remaineth* (Boston, 1778), p. 31.

great; so that, although there were a number very inimical to him, yet he never was openly opposed by them. They, out of derision, used to style him "Pope." [70]

But although Whigs and a large minority of his parishioners distrusted him, better and less biased judges regarded him as one of the ornaments of his age. Benjamin Franklin in 1767 was moving distinguished Europeans to help him to obtain an S. T. D. for Eliot at Edinburgh or Glasgow when Deacon John Barrett of the New North obtained the degree from the former for his pastor by the simpler process of paying cash. [71]

When Eliot took his diploma over to Cambridge to have it copied into the College Records, he was shown a copy of the citation on Timothy Cutler's Oxford D. D. sheepskin which declared that the latter had betaken himself to the bosom of the Anglican Church, and for that very reason had been annoyed by his own people with manifold insults and injuries. [72] Eliot declared that nothing could be more false and abusive; and this reflection on the New England Congregationalists was not copied into the records. His own degree was generally accepted as deserved despite its origin. One version of the poem on the Boston ministers declared:

> Eliot the great whose doctorate
>   Was surely well applied,
> To sermonize is wonderous wise:
>   He is the peoples pride.
>
> New North would sink they rightly think
>   If he should them foresake
> If he were sent as President
>   Their hearts would sadly quake. [73]

At the moment it was Eliot's heart which was quaking because of the gathering political clouds. He had expected the colonists to accept the Townshend Acts, and was horrified by the violence of the reaction to them. The landing of the Regulars incited him to irrational statements about standing armies and the enslavement of the colonists. On a journey into the country he was impressed

---

[70] [Ephraim Eliot], *New North*, p. 29.

[71] William B. Sprague, *Annals of the American Pulpit*, I, 418.

[72] Mass. Hist. Soc., *Collections*, 2nd Ser. II, 214; translation by Richard M. Gummere (Ph.D. 1907).

[73] Thaddeus Mason Harris Mss. (Am. Antiq. Soc.)

by the days of fasting and prayer appointed by the towns, and was put in mind of the events of the year 1641. Sometimes he gave ear to the wild charges made against Governor Hutchinson, but he always came back to the realization that his old friend truly loved his country. At first he believed that the stolen Hutchinson letters were forgeries because he did not think that they represented his views; it was a severe blow when the Governor admitted the authorship. After Hutchinson moved into Province House, he and the Parson saw little of each other.

The Whig leaders had now decided that Dr. Eliot was safe, and they took him into their confidence as to their plans. Writing to Hollis the minister justified his change:

I have sometimes given offence by opposing some measures which I tho't rash among us — but I begin to think I have been mistaken. Every Step the Ministry takes serves to Justify our warmest measures — and it is now plain that if they had not had their hands full at home they would have crushed the Colonies — and that if we had not been vigorous in our opposition we had lost all — I fear nothing is to be expected from the justice or equity of P———t. The treatment of the Colonies on your side of the water tends to alienate them from the Parent Country and to Hasten that Independency which at present the warmest among us deprecate.[74]

Pleased at this conversion, Hollis increased the flow of English Whig tracts through Eliot's hands to the Whig leaders in the colonies.

The New North pastor was not ready, however, to use the college as a pawn in the political game, for he appears to have supported Hutchinson when Chauncy and Cooper tried to use it as a means of forcing the General Court back to Boston.[75] The Doctor was not at all happy to see the legislature sitting in Harvard Hall:

I shall be sorry if the Court should be at Cambridge. It hinders the scholars in their studies. The young gentlemen are already enough taken up with politics. They have catched the spirit of the times. Their declamations and forensic disputes breathe the spirit of liberty. This has always been encouraged; but they have sometimes wrought themselves up to such a pitch of enthusiasm that it has been difficult to keep them within due bounds. But their tutors are fearful of giving too great a check to a

[74] Andrew Eliot to Thomas Hollis (Harvard College Library), July 3, 1769.
[75] Notes on the Overseers' Meeting of May 1, 1770 (Harvard University Archives).

disposition which may hereafter fill the country with patriots, and choose to leave it to age and experience to correct their ardor.[76]

He did not accept Hollis' suggestion that the holding of the General Court in Cambridge was a Tory plot to check the growth of learning in America by burning Harvard Hall a second time.[77]

The blow fell in Boston instead. The Doctor thus described the Massacre to Hollis:

It was an awful scene. There had been such an animosity between the inhabitants and the soldiery some time before this tragedy, that I greatly feared the event. The people seemed determined to be rid of such troublesome inmates, as soon as possible, but were generally careful not to be the aggressors. Capt. Preston, who commanded the party that fired on the unarmed inhabitants, had the character of a benevolent, humane man; he insists on his innocence, and that his men fired without his order. The evidence will be perplexed, if not contradictory.[78]

Eliot was more perplexed one morning to read in his newspaper that the Baptists were complaining of grievous persecutions in Massachusetts. This was the first that he had heard that they were dissatisfied with the laws regarding church support, and he at once suspected that they had formed a coalition with the Episcopalians on the issue of the American bishop. Hastening over to Mr. Stillman, he complained of the injury which the Baptists were doing to the country, and told him that if there was anything wrong with the church laws the Congregationalists would join with him in having it redressed. Then the Doctor spoke to the political leaders of the Province who promised him that the law would be altered so as to give all reasonable satisfaction to the Baptists.[79] At the moment he was driven less by a sense of justice than by a realization of the need of political unity: "We daily look for war. We are at a loss which is best — peace or war. Peace is in itself desirable — but war hath sometimes happy effects." [80]

If the Doctor sounded for the moment like a jingo, it was not his better self. His friends sometimes remarked that his knowledge was sound but not extensive; and that was particularly true in political matters. His Whiggism was soundly based on a reason-

[76] Mass. Hist. Soc., *Collections*, 4th Ser., IV, 447.
[77] Thomas Hollis to Andrew Eliot, June 2, 1770, Eliot transcripts, p. 92.
[78] Mass. Hist. Soc., *Collections*, 4th Ser., IV, 451.
[79] *Ibid.*, pp. 455, 456.
[80] Andrew Eliot to Thomas Hollis, *ibid.*, pp. 458–459.

able interpretation of history, but of the imperial political problems
of his own day he had not an idea. He never comprehended the
purposes of Parliamentary legislation for the colonies, and he
instinctively opposed every exercise of it. He deplored the violence
of the mobs and the lies heaped upon Hutchinson and other Loyalists, but he thought these a lesser evil than any restraint of "the
people." Naturally he rejected as impractical all such suggested
solutions as colonial representation in Parliament. Yet he was by
no means a last-ditch theorist, as this letter on the right of Parliamentary taxation shows:

Perhaps it might be as well not to dispute in such strong terms, the legal
right of Parliament. This is a point that cannot easily be settled, and had
therefore best be touched very gently. It cannot be supposed that the
Parliament will give up their right of taxation in express terms, it will
be prudence for them never again to exercise it. If the Colonies dispute
this right of legislation, which hath been always submitted to, particularly with respect to the regulation of trade, it may raise a new ferment,
and may create suspicions that nothing will satisfy but absolute independence.[81]

Neither the English nor the American Whigs were pleased by
the Doctor's moderate stand, and after the early part of the year
1771 he was dropped from their councils. His political correspondence with the English Whigs dwindled, but Hollis in his
will left him a legacy of £100.

The Boston Port Act reduced Eliot to the depth of despair.
England could easily destroy the seaports, but the effort to reduce
the inland country would ruin both. He put his discouragement
into a letter to Thomas Brand Hollis in February, 1775:

We are preparing for war. . . . To fight with whom? not with
France and Spain whom we have been used to think our natural enemies
— But with Great Britain — Our parent Country. . . . My heart re-
coils — My flesh trembles at the thought. . . . I have ever wished for
moderate counsels and temperate measures on both sides — But I can
have very little influence on men or measures any where. Pride and
passion, avarice and a lust of domination have an uncontrolled sway.
For a good measure sequestered from the great world — unconnected
with parties, I endeavour To attend the duties of my station and enjoy
myself never more than when I can find time for reading and contem-
plation in my own study.[82]

[81] *Ibid.*, p. 459.
[82] Andrew Eliot to T. B. Hollis (Harvard College Library), Feb. 16, 1775.

This period of sequestration from the great world came to a violent end on the Nineteenth of April. Eliot was bewildered and terrified. Such "melancholy and Darkness" he had never seen. As soon as he could, he sent his wife to their son Andrew, now the minister of Fairfield, Connecticut, and the younger children to Salem. Josiah remained in the town with him, where he felt constrained to stay for a while to care for his congregation, the largest in Boston. He worried lest his scattered children starve, and he feared that if he left his house the "licentious soldiery" would plunder it. To Hollis on May 31 he described the situation:

I have remained in this Town till this day, much against my inclination. . . . My situation is uncomfortable to the last degree. — Friends perpetually coming to bid me adieu; — Much the greater part of the Inhabitants gone out of the Town. . . . Grass growing in the public walks and Streets of this once populous and flourishing place; — Shops and warehouses shut up; — business at an end. . . . The advantage hath hitherto been on the side of the Provincials. — And it is not improbable to me, that if they attempt the Town, they will carry it. . . . These things you will easily believe keep us in perpetual alarm, and make this a very unquiet habitation. — I cannot stand it long.[83]

Although he was very timid and uncertain, the demands of his position kept him in Boston. Most of the other churches were closed, so in the New North meetinghouse in the places of those who had fled were strange faces, worried, frightened, sick, looking up to him for a comfort and assurance which he did not feel. The Battle of Bunker Hill made things worse:

It was a new and awful spectacle to us to have men carried through the streets groaning, bleeding, and dying. . . . Amidst the carnage of Saturday, the town of Charlestown was set on fire. . . . You may easily judge what distress we were in to see and hear Englishmen destroying one another, and a town with which we have been so intimately connected all in flames. . . . You will be anxious to hear of the fate of the College. I can only say that last week I received a letter from the President, informing me that there was to be a meeting of as many Overseers and Fellows as could be got together, in order to consult what was proper to be done. What was done at this meeting I do not know. I wish they may come to no sudden resolutions: it is no time to give offence. I have heard it said there was talk of moving the College to Haverhill or to Worcester. I wonder who will send their children in this time of confusion. I should think it were better to leave

[83] Andrew Eliot to T. B. Hollis, Eliot transcripts, pp. 122–123.

matters at present, and to wait the issue of things, but you know how little influence I have had of late. Perhaps it will one day be seen that it had been as well if more moderate counsels had been pursued.[84]

After the battle the Doctor was constantly with the sick and wounded prisoners until his attentions to that arch-rebel, Master James Lovell (A.B. 1756), caused him to be excluded from the prison. Outside of the town, Whigs who had not thought kindly of him for some time heard that he had been arrested and imprisoned on a man-of-war, and were indignant. Actually he was working night and day among the civilians who by the hundred were falling victim to heat, malnutrition, and disease. To a parishioner who sent some meat into Boston he described the situation:

I received the two quarters of mutton, and have divided one between Dr. Rand and Mr. Welsh. . . . Part of the other I shall send to make broth for the prisoners, who have really suffered for want of fresh meat. I shall this day make a quantity of broth for the sick around me, who are very numerous. You cannot conceive the relief you will give to great number of persons by this kind office. Perhaps your broth has been dispensed to thirty or forty sick people. I thank you for the ability of helping them. I have invited a number of friends to partake of the rest. . . . I eat very little of it myself, and yet never had so much pleasure in any provision in my life. If I could only get a little at times, I would engage not to taste it myself, and to give it only to the sick. Provision for myself is my least concern.[85]

In August he came down with the "disorders of the times," but he was so busy visiting the sick and dying that he had no time for his own troubles.

Among the well, tempers were short. Samuel Danforth (A.B. 1715) threatened to break down the minister's door to obtain money owed to him for lottery tickets. Jonathan Sewall (A.B. 1748), an old friend who had remained loyal to the king, said bitterly that the Doctor had been opposed to the policy of the radicals until the Nineteenth of April, but since then he had shunned his Loyalist friends. It was the Tories who now called Andrew sly; but it was he who boldly told General Robertson and Crean Brush that their instructions to confiscate linens and woolens for the wounded did not entitle them to take silks and

84 Mass. Hist. Soc., *Proceedings*, 1st Ser., XVI, 288.
85 *Ibid.*, pp. 292, 293.

glass.[86] By the beginning of September he had made up his mind to leave Boston and join his family:

I had prepared my things for a speedy departure from this devoted town, but heard yesterday that it was determined in a conclave of our new-fangled councillors that I should not have a pass. However, I was determined to apply. This day I waited on the Town Major, who peremptorily refused to give me a pass. . . . I shall soon wait on the General, but fear it is already determined that I should not go. It is very hard treatment. I have no fuel, and very little provision. Some of those gentlemen who have inserted themselves in this affair insinuated that I made money by tarrying. So far from this, I do not receive one-half of what I received from my people, and if I must tarry, should be willing to preach only for my wood, which would cost more than I am like to receive, if it is to be got at all. . . . I am at length allowed again to visit the prisoners. They were overjoyed to see me.[87]

When Dr. Eliot finally decided that he would not be permitted to leave Boston, he cheerfully tightened his belt and prepared for a bad winter. His health was good and he was no longer confused. As the population of the town dwindled, he and Samuel Mather (A.B. 1723), who had between them carried the Thursday Lecture into its 140th year, reluctantly and sadly abandoned that ancient institution. The minister of the New North discreetly kept out of the way of "our despots," and although he worried about the manuscripts in the Old South steeple did not risk approaching the Red Coats who frequented that meetinghouse. He was reasonably calm about the shells which the Provincials threw into the town in the last days of the siege. When he watched the Regulars leave he said that he never expected to see them or any other British soldiers in Boston again.

We have been afraid to speak, to write, almost to think. We are now relieved, wonderfully delivered. The town hath been evacuated by the British troops, so suddenly that they have left amazing stores behind them, vast quantities of Coal, which the inhabitants have been cruelly denied through the winter, cannon and warlike stores in abundance, porter, horsebeans, hay, casks, bran, etc. . . . This inglorious retreat hath raised the spirits of the colonists to the highest pitch. They look upon it as a compleat victory. I dare now to say, what I did not dare to say before this — I have long thought it — that Great Britain cannot subjugate the Colonies. Independence, a year ago, could not have been

[86] *Connecticut Courant*, Apr. 22, 1776, 2/3.
[87] Mass. Hist. Soc., *Proceedings*, 1st. Ser., XVI, 297–298.

publicly mentioned with impunity. Nothing else is now talked of.[88]

It was typical of Eliot that when the Evacuation was in progress, he went to the Reverend Samuel Parker (A.B. 1764) of Trinity Church and convinced him that it was his duty, as the youngest and least unpopular of the Church of England ministers, to remain and administer to the religious needs of the Episcopalians of New England. So the future bishop unpacked again, and was ever grateful to the ancient enemy of the American episcopate.[89]

General Washington asked the Doctor to make the renewal of the Thursday Lecture, on March 28, 1776, the official thanksgiving sermon. On this occasion the military dignitaries met in the State House, whence they marched, preceded by the Sheriff with his wand, and followed by those members of the Council who had had smallpox, to the Old Brick meeting house, where Eliot entertained them with a sermon which was a history of events in the town of Boston during the siege. Then they proceeded to the Bunch of Grapes Tavern where an elegant dinner was provided at public expense.[90]

The Doctor immediately returned to the service of the college:

I attended last week a meeting of the Overseers and Corporation at Watertown for the first time since our enlargement. We voted General Washington a degree of LL.D. He is a fine Gentleman, and hath charmed everybody since he hath had the command. I find a committee of Overseers appointed, at the motion of the General Court, to examine the political principles of those who govern the College. I hope no evil will come to several worthy men there. . . . The President is in haste to move the Students to Cambridge. The Buildings are in a shocking state, having been improved for barracks; the library and apparatus are safe at Andover.[91]

He interested himself in preparing a new edition of the Harvard catalogue, and accepted appointment as a Hopkins Trustee.

For the time being Dr. Eliot seemed to be the center of a circle of good feeling. Miles Whitworth (A.B. 1772), jailed as a Tory,

---

[88] Andrew Eliot to Isaac Smith, Apr. 9, 1776, Smith-Carter Mss. (Mass. Hist. Soc.)

[89] [Ephraim Eliot], *Old North*, p. 31.

[90] James Thacher, *Military Journal* (Boston, 1827), p. 45; *New-England Chronicle*, Apr. 4, 1776, 3/2; John Rowe, Diary (Mass. Hist. Soc.), p. 2129. Unfortunately no copy of the sermon can be found.

[91] Andrew Eliot to Isaac Smith, Apr. 9, 1776, Smith-Carter Mss.

begged the parson to get him out.[92] The Adamses and Hancocks were kind and friendly. For the first time he shared their dream of the America to come:

I have no doubt but great things are designed in Providence for America, and that Boston will emerge from its present unhappy Situation. . . . Guarded by troops and ships — an asylum for those who dare not appear elsewhere — it's trade destroyed. . . . This is certain, that there never will be a Submission to the late unrighteous acts — Our people in the Country towns are universally armed and ready to appear at a moment's warning. . . . All are in anxious expectation of the Result of the Congress. What is determined there will be Law to America. . . . We are thankful for the munificence of our Friends in the Southern Colonies — May God Reward you! Such a surprising union *will* — *must* have great effect — and make the haughtiest minister shake.[93]

In letters to American leaders of state he spoke repeatedly of the day when this would be "made a quiet habitation — a land of liberty — a land of knowledge. . . . Every thing leads us to conclude that Providence hath some great design to accomplish in this part of the world. There is a fine scope for imagination but these prospects are more than imaginary." [94]

In the spring of 1777 he lost his faith in the design of Providence, and in anticipation of the return of Howe to Boston he bought a house in Concord and sent thither all the furniture he could spare. About the time this scare blew over, his volume of collected sermons won a satisfactory acclaim in the press. He lost his old friend Pemberton, but succeeded him as treasurer of the Massachusetts convention of the Congregational clergy. In spite of the war in New York, he resumed his activity in the Indian mission work. One of his last acts was to defend Harvard against a charge of Deism made by a Yale man. His last major public appearance was on July 8, 1778, when John Clarke (A.B. 1774) was ordained at the First Church. Illness confined him to his house for only a short time. His last days were described by his son John:

Soon after he was taken ill, he deliberated against a recovery from the disease. . . . Above a week before he died, he told me that he never

[92] Miles Whitworth to Andrew Eliot, May 25, 1776, Andrews-Eliot Mss., p. 138.

[93] Andrew Eliot to ———, Sept. 14, 1776 (Harvard University Archives).

[94] Andrew Eliot to John Hancock, Oct. 14, 1776, Chamberlain Mss. (Boston Public Library), A 2.3.

[should] go out of his chamber. . . . When the physicians told him that they were destitute of hope, and that he had not long to live, his answer was, the sooner the better, "I have finished my course with joy." Saturday night, he told me he should begin an everlasting Sabbath the next morning, and with great affection wished me a good night. About 5 o'clock, I was called up and went to his bedside. . . . Turning to me, he said, "It is a question with me whether I o't to wish to die, or to wait quietly till it is the will of my Father to call me hence. His will be done." Not a minute after this elapsed before he breathed his last.[95]

The Doctor died on Sunday, September 13, 1778. With one of his last breaths he sent word to Peter Thacher, who was preaching at the New North, to tell the congregation that he still believed the doctrines which he had taught them, and that these doctrines were the greatest comfort and satisfaction in his dying hour. This message was delivered in the afternoon sermon; "Had there been an unweeping eye, the bosom of such a person must have been steeled with adamant." [96] The funeral would have done justice to royalty:

Dr. Gordon prayed with the family, Dr. Langdon with the men at New North and Dr. Mather with the women at his own meeting. The congregation and church preceded the corps: it was followed by the family, relations, women, overseers, corporation and government of the college, clergy, officers, gentlemen, carriages. It is said that the procession consisted of 600 couple beside about 30 carriages and that it extended about half a mile. He was laid in a tomb upon Cops hill, which he had purchased not long before.[97]

"Even the Church clergy, composed very affectionate and charitable collects upon this occasion." [98]

There was, however, a surprising lack of affection and charity in certain Congregational quarters. The entire absence of the usual newspaper eulogies aroused first suspicion and then indignation:

A Correspondent expresses his Grief and Astonishment, that the immemorial Custom of embalming the Memories of those illustrious Persons who have been the Ornaments and Blessings of the Community

[95] Mass. Hist. Soc., *Collections*, 6th Ser., IV, 130–131.

[96] Benjamin Guild, Diary (in private possession), Sept. 13, 1778; Mass. Hist. Soc., *Collections*, 6th Ser., IV, 132.

[97] Guild, Diary; Caleb Gannett, Diary (Harvard College Library); Eliot transcripts, end.

[98] Harrison Gray Otis, quoted in Ephraim Eliot, Commonplace Book, p. 233.

should, for the first Time, be violated in the Neglect shown to the Character of that great and good Man, the Rev. Dr. Eliot. — That those whose Lives have been a direct Contradiction to the Encomiums pass'd on them at their Decease, should be held up to the World as Patterns of every Virtue, while so striking, so amiable an Example of Piety, Purity and Charity, should pass off the Stage of Action entirely unnoticed.[99]

The good obituary which followed this protest did not put an end to the indignant letters to the papers. William Pynchon (A.B. 1743), a Loyalist, put his finger on the matter when he said that the neglect of the memory of "that polite, affable, and most agreeable gentleman, that truly good man," was due to the fact that the modern test of character was not honor or integrity, but politics.[100]

In England, Governor Hutchinson, who did not know what a change the course of events had made in the views of his old friend, also suspected politics:

Dr. Eliot was long my friend. One of my last letters from Dr. Pemberton, said his sentiments were the same they used to be. After Howe left the town, he [Eliot] wrote two letters to England which were intercepted, and carried to Halifax, and copies given. They were very strong in favour of American proceedings. Some thought he expected they would be intercepted, and that he desired to have it known in Boston that he publickly owned the cause. He said to my son at Boston he was afraid, or had reason to think his continuing in Boston had made him obnoxious to the people without the town. Great allowance must be made for the difficulty of his circumstances, but after all, as no man is without infirmity, perhaps his might be a disposition to temporize, always, I trust, having satisfied himself he was to be justified: — but this must be left. Some of the Americans speak lightly upon the news of his death. I heard the news with grief, and wished to see him again in this world. Dr. Pemberton and he for many years, were the best neighbours I had.[101]

Back in Boston the congregation of the New North generously put the Eliot family into handsome mourning and continued the Doctor's salary to his widow for a long time; but the church officers, who had been hostile to him, showed only the minimum of decency in printing one of the several memorial sermons, that

[99] *Boston Gazette*, supplement, Sept. 28, 1778, 1/3, Oct. 5, 1778, 2/3; *Independent Ledger*, Sept. 28, 1778, 3/2–3. Oct. 5, 1778, 3/1, 3/3.

[100] William Pynchon, *Diary* (Boston, 1890), p. 59.

[101] Peter Orlando Hutchinson, *Diary and Letters of Thomas Hutchinson* (Boston, 1884–86), II, 223–224.

by Peter Thacher on *The Rest which Remaineth*; and they pointedly neglected to give a copy to any member of the family. Naturally they bitterly resisted the calling of his son John, but the congregation had its way. When he was ordained, on November 3, 1779, his brother Andrew preached a sermon which deals largely with the impression which their father made upon them. Although one could not tell it from the Doctor's diary, he had eleven children: (1) Andrew, b. Jan. 11, 1743/4; A.B. 1762; m. Mary Pynchon, July, 1774. (2) Josiah, b. Jan. 31, 1745/6; d. unm. in Georgia, 1794. (3) Elizabeth, b. May 4, 1747; d. unm. Dec. 31, 1780. (4) Samuel, b. June 17, 1748; m. Elizabeth Greenleaf, May 7, 1771; d. Mar. 1784. (5) Ruth, b. Oct. 2, 1749; m. Capt. Thomas Knox, Oct. 9, 1792; d. 1803. (6) Mary, b. Jan. 24, 1750/1; m. Capt. Nathaniel Goodwin, Mar. 27, 1782; d. 1811. (7) John, b. May 31, 1754; A.B. 1772; m. Ann Treadwell, Aug. 19, 1784. (8) Sarah, b. Nov. 3, 1755; m. Joseph Squire of Fairfield, Oct. 5, 1778; d. 1779. (9) Susannah, b. Feb. 25, 1759; m. David Hull (Yale 1785), Nov. 10, 1789; d. 1832. (10) Ephraim, b. Dec. 29, 1761; A.B. 1780; m. 1st Elizabeth Fleet, Dec. 6, 1789, 2nd Mary Fleet, May 14, 1793. (11) Anna, b. Apr. 27, 1765; m. Capt. Melzar Joy of Boston, 1795; d. Mar. 28, 1799. Their mother died on June 14, 1795.

It is plain from Dr. Eliot's will that the only part of his estate to concern him particularly was his books.[102] His manuscripts were given away piecemeal by the family during the next century, but the bulk of them have come to rest in the Harvard College Library and the Massachusetts Historical Society. Most of the letters which are quoted in this sketch survive in at least two of four forms: the rough draft, the fair copy, an early nineteenth-century copy, and print. Many of the manuscripts are now too badly worn to be read with assurance. The Doctor kept a thin and disappointing line-a-day diary in annotated almanacs, of which the years 1734 and 1739 are in the Harvard University Archives, and the years 1740–52, 1754, 1756, 1758–77, in the Massachusetts Historical Society. The Hutchinson papers which he rescued are in the Massachusetts State Archives.

---

[102] Suffolk Probate Records, LXXVII, 544.

BENJAMIN PRAT

John Read (Class of 1697) represents the first generation of educated lawyers in New England, but in his day they were still regarded as unfit for the Bench. As late as 1739, Stephen Sewall (Class of 1721) was called directly from a tutorship in Harvard College to a place on the Superior Court. Prat was the first New England lawyer to obtain such an appointment, and the distinction of his service as Chief Justice of New York convinced colonial governors that the amateur did not make the best judge.

BENJAMIN PRAT, Chief Justice of the Province of New York, was born on March 13, 1710/1, the eighth son of Aaron and Sarah (Wright, Cummings) Pratt, who lived on what is now South Main Street in Cohasset. Aaron was a farmer and a part owner of a local ironworks. His highest office seems to have been that of constable. There was no thought that Benjamin should do other than follow his father until that day in his eighteenth year when he fell from an apple tree and injured his leg so badly that it had to be amputated at the hip. Never again was he long without pain, or free from his crutches; but he knew that his mind was unusually keen, and he determined to obtain a college education to offset his physical handicap:

He was without resources, without friends, and somewhat advanced in years, yet he knew human life enough to believe that every thing may be done by perseverance. He had also that opinion of himself, that he believed he should not only gain a subsistence by his learning, but make a shining figure among his contemporaries.[1]

Fortunately for his purpose, Nehemiah Hobart (A.B. 1714), the local minister, was his brother-in-law and willing to fit him for college. In five years this was done, and Benjamin was admitted as a Freshman with the Class of 1738. Because of his age and poverty he was placed at the foot, but after two quarters he was promoted to the Sophomore class.

[1] John Eliot, *Biographical Dictionary* (Salem and Boston, 1809), .p. 388.

Aaron Pratt made no effort to help Benjamin, and in his will left him only a small piece of land. Ben sold this to his brothers for £400 two days before their father died. The college helped further by giving him Hollis and William Browne scholarships, and the Faculty remitted a fine which Tutor Prince had laid upon him. It was not lack of money, but suffering which prevented him from living at the college, and finally compelled him to withdraw at the end of his Junior year. He had no difficulty in getting a job keeping the Hingham school, but he had no intention of spending the rest of his life subjected to that mixture of poverty, respect, and amused tolerance which New England had for its country schoolmasters. On June 9, 1737, he wrote this piteous letter to Tutor Prince:

I presume on your Goodness That You will not forget one of your Pupils or deny a Protection to his Cause on the Account of his Misfortune. I beseech that my deplorable Circumstances may not debar me from any academic Priviliges And that I maynt be excluded because I am unfortunate. I am not a Judge of my Qualifications — But this I think I may venture to say That my Defficiencies are not my Fault. But the Effects of a ruinous Constitution and Want of Advantages — For I presume there's no Man that loves Learning better or that more freely could spend his Life in the pursuit of it than I. But I must study how to live and how to bear the Miseries of a Wretched Life. I have here (pursuent to your Order) sent my imperfect Thoughts on a Subject the Importance of which will excuse me (as I hope) in deviating from the usual Method of young Students Vizt Commonplacing on the obvious Parts of Learning. . . . Tho my Tenet may for ought I know be new yet I am far from dreaming that I shall afford any new Speculation in the Matter my only Intentions are to present my Reasons to your Examination and tho I dare not be so presumptuous as to desire the Honor of being informed of my Sophisms — Yet I would beg your Judicious and candid examination of them.[2]

The "imperfect Thoughts" were the thesis required of candidates for the degree of B. A., and with it he sent a petition to the Corporation asking for his degree in spite of the fact that his lameness would prevent his appearing in person at Commencement. The Tutors gave "Testimony of his Superior Genius and great proficiency in Learning and also of his sober Conversation during his abode at the College," and the Corporation took into consideration the fact that he had "been Exercised with grevious bodily afflic-

_____

[2] Misc. Mss. Bound (Mass. Hist. Soc.), June 9, 1737.

tions and yet under his paines and difficulties" had "pursued his Studys with uncommon application," so suspended the college laws in order to grant his degree.[3]

In May, 1738, Prat reappeared at college and was immediately awarded a Hopkins Fellowship for the year preceding. It was renewed for two more years, with the understanding that the funds were being used to prepare him for the ministry. According to family tradition he tried preaching to the Indians on the islands in Boston Harbor; probably he preached to the soldiers in Castle William. At the Commencement of 1740, when he took his M. A., he was awarded a speaking part, and chose to hold the negative of "An Legis moralis Observationem, Christi satisfactio, quacunque Ratione solvat?" Still he remained in residence, not eating in commons but living in Massachusetts 20 and then in Massachusetts 22 with Phillips '45. In June, 1742, he was appointed college librarian, and given a salary of £30 a year; but in January, 1743, he left Cambridge for good.

Prat went to Boston and studied law, first under Jeremy Gridley (A.B. 1725), and then under Robert Auchmuty, Judge of Admiralty. Socially as well as physically he was a pitiful figure, but he attracted the attention of the Judge's daughter, Isabella, who must often have seen him, as others described him, so absorbed in his law book that he seemed unaware of the pain which made the sweat run down his face. They were married on December 30, 1749. Isabella's love was repaid with gentle kindness; whatever her husband's manners may have been in public, at home they were amiable.[4]

This was the easier for Prat because he soon made his mark in the practice of the law. There were occasional complaints as to his methods and insinuations that he did not always tell the truth,[5] but there was no question but his learning and ability were the cause of his success. Soon there were coming to his office on the north side of King Street, near the town pump and the Old State House, young men like Abel Willard (A.B. 1752), who wanted to read under one of the most learned lawyers in the Province, and the most important men of Boston, who had discovered that

[3] Colonial Soc. Mass., *Publications*, XVI, 662–663.

[4] *Monthly Anthology*, VIII (1810), p. 329.

[5] As in Mass. Archives, XLII, 849–850.

he was perhaps the most successful of the seven lawyers then active in the town. After ten years of practice he was earning £750 a year,[6] a princely income. Young John Adams (A.B. 1755) when first visiting the courtroom "looked with wonder to see such a little body, hung upon two sticks, send forth such eloquence and displays of mind."[7] James Otis (A.B. 1743), on the other hand, laughed "to see Pratt lug a cartload of books into court, to prove a point as clear as the sun" in an action already "dead as a hub."[8] There were other lawyers who jeered that Prat's learning could muddy the clearest law, but these were individuals who had reason to resent his effect on practice.

Unquestionably Prat was one of the great lawyers of the colonial period, but he was no politician, and never had a popular following. John Eliot (A.B. 1772) explained this by saying that the inhabitants of Boston "could never love a man who had no complacency in his disposition, nor urbanity in his manners; a man who emerged from low life to a high station, and despised those who formerly knew him, even those from whom he had received favours."[9] This remark brought the quick retort that Prat was "not deficient in urbanity; and in conversation and manners [was] attractive and pleasing. . . . The pride of the low will always dictate suspicions of the pride of the eminent."[10] A lawyer who had weighed all of the evidence said:

Many of the people of Boston thought him morose, distant and haughty; but they did not fully understand him. To the few for whom he felt a high respect for their worth and intelligence, he was communicative and courteous.[11]

Most illuminating is John Adams' account of the coolly correct treatment which he received from Prat when he appeared to be sworn in at the Boston bar; one can understand why he went away "full of wrath" at the great man's "ill-nature."[12]

There was indeed another side to Prat's character. With Gridley, whose manners warmed Adams' heart, he was a segment of a

---

[6] *Herald of Freedom*, Feb. 2, 1790, 1/1.

[7] William Sullivan, *Address to the Members of the Bar of Suffolk* (Boston, 1825), p. 33.

[8] John Adams, *Works* (Boston, 1850–56), II, 71.

[9] Eliot, p. 389.

[10] *Monthly Anthology*, VIII, 329.

[11] Samuel L. Knapp, *Biographical Sketches* (Boston, 1821), p. 164.

[12] Adams, *Works*, II, 47–48.

small literary circle which contributed essays to the newspapers and wrote poetry. Somewhere in print, but not now to be found, is a poem which he wrote on leaving college. In one surviving piece he laid bare the misery that was the ever-present background of his professional success:

> 'Twas adverse fortune prescious of my guilt,
> That doom'd me being, and frowning bid me live.
> Then Clotho with an angry hand began
> To form the carcase where I suffer life.
> For deep distress she drew the fatal plan,
> And ting'd each fibre in Pandora's box,
> And thus diffus'd the cause of future pain,
> And every nerve assign'd for pungent ill.
> No kind abortion rescued from her pow'r.
> Ah me! from silent nothing I must come,
> To meet the woes that fate forbids to shun.
> On painful terms then by Lucina led,
> I gasp'd in air, and join'd a hapless throng.[13]

In his most famous and commonly reprinted poem he describes man as "the ape-kind" who "eates, and drinks, and propagates and dies." [14] Once when Benjamin Kent (A.B. 1727), himself unorthodox, asked him the question from the catechism, "What is the chief end of man?" he received the startling answer, "To provide food, &c. for other animals." [15]

Prat was a proprietor of King's Chapel, where he occupied pew 35 and helped to buy the organ. At times he held that "a variety of religions has the same beauty in the moral world, that a variety of flowers has in a garden," but again he thought that all sectaries should be made Churchmen under a uniform establishment. He certainly did not increase his popularity among the New England sectaries when he said:

It is a very happy thing to have people superstitious. They should believe exactly as their minister believes; they should have no creeds and confessions; they should not so much as know what they believe. The people ought to be ignorant; and our free schools are the very bane of society; they make the lowest of the people infinitely conceited.

---

[13] *Massachusetts Magazine*, V (1793), p. 327.

[14] *Royal American Magazine*, I (1774), p. 106; *Boston Gazette*, Sept. 22, 1777, 4/1; *Massachusetts Magazine*, I (1789), p. 56; *Monthly Anthology*, IV (1806), pp. 316–317, etc.

[15] Adams, *Works*, II, 223.

Which, said John Adams, were "French-worse principles" than even Frenchmen held.[16]

The fact that Prat held such views makes it most unfortunate that he did not live long enough to execute his plan to write the history of Massachusetts for which he had made a large collection of documents. Perhaps these papers passed into the hands of Thomas Hutchinson (A.B. 1727), whose summer home was only three houses from the fine Milton estate which the Prats purchased in 1757.[17] Although never warm friends, these two men were much together. After buying in Milton they continued to be active in Boston affairs, serving on routine committees together. Prat, in addition, served on the committee of the town appointed to appeal to London to disallow certain taxes laid by the General Court, and to instruct the town's diplomatic agent in England how to fight the case. In 1757 Prat was elected Moderator of town meeting, and in 1761 he was a member of the committee to inspect the schools. One would have expected him to be a regular member of this particular committee; perhaps his lameness or his views about "the very bane of society" prevented.

In 1757 and the two following years Boston sent Prat to the House of Representatives, where he "was a constant, fearless, and independent lover of freedom, and never hesitated to support what he thought to be just, wise and expedient, without crouching to prerogative, or bending to the people." [18] He was appointed to prosecute those who bought land from the Indians illegally, and he served on the committee to care for the Acadians, and on the commission to carry on intercolonial negotiations about the prosecution of the war. Governor Pownall chose to lean on Prat and John Tyng (A.B. 1725), rather than on Hutchinson and Peter Oliver (A.B. 1730), who had been Shirley's chief advisors. There is no evidence of political differences between Prat and Shirley, who had formerly employed him as counsel. The first break with the Hutchinson-Oliver faction was over the motion, which he supported, for the erection of a monument to Lord Howe. "Pratt exerted all his eloquence," said John Adams, "and I never heard eloquence more impressive, except from James Otis, Junior, in

[16] *Ibid.*, pp. 97–98.
[17] This property is described in the *New England Chronicle; or Essex Gazette*, Mar. 14, 21, 1776, 1/3.
[18] Knapp, p. 163.

support of that vote"; but the Hutchinson party cried "extravi-
gance" and defeated him on those grounds.[19] This is probably the
period of a Tory farce which attacks "the Political Tool Pratt.
. . . a little deformed Lawyer, a pretended Despiser of Court
favours to lately, but of great Ability and knowledge." [20]

The story of Prat's political fall as given by those who had it by
word of mouth is much fuller than any version which can be re-
constructed from documents:

When Pownall left the province, Pratt lost entirely the regard of the
people. The merchants and mechanicks in the town were very indignant at his conduct in the general court in supporting a motion to send
away the province ship. This ship, though owned by the government,
was designed to protect the trade, and the merchants had subscribed
liberally towards building her. Yet, in the midst of the war, it was
proposed by Pownall's friends, that this ship should leave the station,
and the trade suffer merely for his personal honour or safety. The
clamour was so great, that the governour found it necessary to take
his passage in a private vessel. But the spirit of the people was not suddenly calmed. A larger town meeting than ever had assembled at
Faneuil hall, discovered their displeasure by leaving out Pratt and Tyng
from the list of their representatives.[21]

This version is supported by various brief contemporary references,[22] but no sooner had John Eliot printed it than he was roundly
and publicly taken to task by some of his friends who from their
own stores of verbal sources drew another account, much more
favorable to Prat, and not at all contradictory to surviving documents:

The ship was to leave the coast in time of war, it is true; but the period
of her absence was to be from fall to spring, when she was commonly
in harbour; and she would return to her station before the usual time
of her being at sea in the opening of the year; meanwhile she was to
have new sails and repairs in England, for which the governour was
willing to be in advance to the province; and she was also to bring
out the reimbursement money granted by Parliament. But the measure
did not happen to take with the merchants or the people in general;
and two persons, who wanted the places of Pratt and Tyng, finding
the populace fermentable upon the subject, managed it so adroitly as to

[19] Adams, *Works*, X, 243.

[20] In Curwen Mss. (Essex Institute), III, n. p.

[21] Eliot, p. 389.

[22] As in Letters from the Rev. Samuel Mather to his Son (Mass. Hist. Soc.),
May 8, 1760.

carry their point.[23] Several days after the general court, which had voted the ship to the governour, had adjourned, four or five hundred heroes assembled in mob, and dismantled the frigate which was preparing for the voyage; and to vindicate their conduct, when they came to vote at the next election of representatives, of course passed over these gentlemen, whose proceedings had made it necessary for them to interfere and save the country.[24]

Certainly Prat's reputation was not badly damaged, for the next year both the merchants and the government asked him to take the case, respectively for and against, the Writs of Assistance. He declined, but John Adams years later remembered his presence "In a corner of the room . . . as a spectator and an auditor, wit, sense, imagination, genius, pathos, reason, prudence, eloquence, learning, and immense reading, hanging by the shoulders on two crutches, covered with a great cloth coat." [25] Perhaps he refused this case because he knew that Pownall intended to recommend him for appointive office. It was said to have been purely an accident that he had not been appointed to the bench of Massachusetts. New York documents refer to him as "His Majesty's Advocate-General for Massachusetts," which he was not; but he may have served as Advocate General in the Court of Admiralty.

At all events, Prat was appointed in March, 1761, to the offices of chief justice and Councillor for the Province of New York. At the same time General Robert Monckton was appointed governor, and Cadwallader Colden, lieutenant governor. Prat received the news on June 4, but no copy of the commission came with it. His brother-in-law, the Reverend Samuel Auchmuty (A.B. 1742) of Trinity Church, New York, warned him that the politicians of that Province were determined that the chief justice should be appointed "during good behavior," so that he would be subject to the legislature and the governor, rather than "during His Majesty's pleasure." In August, Prat wrote to Colden asking how his commission was worded, and expressing an unwillingness to get into a position in which he was "liable to be broke by the Governour if he don't please him. And to be Starved by the Assembly if he dont please them!" As he pointed out, George III had recently

[23] The men who captured these places at the next election were Deacon John Phillips and Prat's future son-in-law, Samuel Welles (A.B. 1744).

[24] *Monthly Anthology*, VIII, 328.

[25] Adams, *Works*, X, 245.

informed Parliament that colonial judges ought to be financially independent of the legislatures and free from removal for any cause but misconduct.[26]

Colden replied that he had not heard of Prat's appointment but was happy to do so, urged him to accept, and promised to help have his commission revised if it turned out to be unsatisfactory.[27] In September Prat wrote again:

I know what I am to relinquish here, but know not what I am to expect, at New York — You know the State Advantages and Disadvantages of that Office better than I can . . . Your Advice Sir must therefore have great Weight with me. . . . All that I aspire at is, that I may, in that Post, be independent enough to be able to do My Duty with Safety; And that the Salary and legal Perquisites should afford me and Family a Decent Support. The Center of all my Wishes, in Life, has always been to be in Circumstances that would permit me to Devote a great Part of my Time, to Speculation, Literary Ingagements, Correspondence with Friends of that Taste, and to the Doing Some thing for public Emolument.[28]

Colden answered that he was resisting the demand of the legislature that all judges be appointed for good behavior, and urged him to come at once in order to get on the pay roll for the year. Early in October Prat received his mandamus and found that he was appointed "During Our Pleasure." His decision to accept and leave for New York was hurried by news, probably from Auchmuty, that the jails there were full of people who had been long confined "by reasons of some scruples by the judges," and were likely to remain there indefinitely because the other judges were determined to resign if Colden would not renew their commissions "during good behavior," which he was forbidden by his instructions to do.[29] On the point of leaving Prat wrote to Colden:

What Little I have in the World consists in a real Estate for which I can neither find Purchasers nor Tenants without Sinking one half the Value . . . and in a personal Estate all out in many Debts, chiefly Small, and many Doubtful and under such Circumstances by Reason of the Length of Time they have been due, that they cannot be collected in my Absence without great Loss. Upon the whole if I hurry

[26] New York Hist. Soc., *Collections*, 1922, p. 68.

[27] *Ibid.*, 1876, pp. 113–114.

[28] *Ibid.*, 1922, p. 77.

[29] *Documents Relating to the Colonial History of New York* (Albany, 1856–83), VII, 500.

away from this Place, I must be a very great Loser besides Leaving my Business here worth double the Profits of the Office of the Chief Justice as it has been in my Predecessors Time.[30]

All of the members of the Boston bar accompanied him on his way as far as Dedham, where, after a farewell dinner, they exchanged an address and a reply which were famous in their day, but are now lost.

Prat arrived in New York on November 5, 1761,[31] and was soon sworn in.

> When he took his seat on the bench . . . he was treated with great coldness, and even disrespect, by the side judges and the bar; but . . . he had been in his chair but a few days in the first term, when a very intricate cause, which had been hung up for years, was brought before the court. Judge Pratt entered into it with quick and keen perception, caught its difficulties with wonderful success; and gave a statement of the case so luminous, profound and eloquent, that he became immediately the object of admiration to those, who were disposed, but not able, to withhold their applause.[32]

Even a bitter political and professional foe had to make the grudging admission

> that he knew Mr. Pratt to try eight criminals in a forenoon upon different indictments, and with the same jury; that he took no notes, but summed the evidence with great exactness, remembered every circumstance of every testimony, and the names of all the witnesses, although the witnesses were Dutch people, and their names such as Mr. Pratt never could have heard.[33]

He was amazed at the excessive cost of legal proceedings in New York, and was determined to remedy this evil, and to break the power of the ring of lawyers who, by control of the courts and the legislature, had been fattening on the exploitation of the Province.

Colden was overjoyed to find in Prat a man "of abilities sufficient to restrain the licentiousness of the Lawyers," and he backed him to the hilt. Of the opposition of the popular leaders, William Livingston and William Smith, Colden said:

> These men had formerly gained so great an influence over the Judges of the Supreme Court, from their want of a sufficient foundation of

[30] New York Hist. Soc., *Collections*, 1922, pp. 81–82.
[31] *Pennsylvania Journal*, Nov. 12, 1761, 2/3.
[32] *Monthly Anthology*, VIII, 327.
[33] Adams, *Works*, II, 354.

knowledge and of that Resolution and firmness necessary to curb the insolence and petulance of a popular Lawyer, that the Lawyers obtained so great an influence in the Courts of Justice, as to become the object of dread to many and of complaint to others while they got the applause of the Mob by their licentious harangues and by propagating the Doctrine that all authority is derived from the People. Now when these Lawyers see a chief Justice on the Bench capable to restrain them, their resentments are greatly provoked.[34]

Defying Colden, the Assembly refused to vote Prat any salary so long as he held office "during His Majesty's pleasure." They were plainly determined to starve him out and get a complaisant member of their oligarchy in his place. To make matters worse, all of the other judges sided with the Assembly and refused to serve unless commissioned "during good behavior." Prat continued to sit alone and to struggle as best he could with unfamiliar local law and customs.

Colden, who knew that a chief justice who was the tool of the Assembly would be much more powerful than the governor, was determined to keep Prat; but on March 15, 1762, the latter informed Speaker William Nicholl that urgent private business compelled him to return to Boston for a considerable time, leaving New York without a single judge on the bench. The Speaker replied that it was not for him to enquire why the other judges refused to accept their commissions and the "handsome Sallaries" offered by the Assembly; "The essential Rights and security of the People" determined the terms offered, and it was not their fault if the Province encountered "the great mischief and dangerous consequences of being without a Supreme Court." He made it clear that he would not be sorry to see Prat go for good.[35] If he cared to remain and submit to the authority of the Assembly, he could have the "handsome Sallary" of £300 New York currency.

Late in April, 1762, Prat obtained leave until the following September, and set out for Milton. He was not ready to give up the struggle, for he wrote to the Lords of Trade describing his own situation as an illustration of the necessity of maintaining an independent judiciary in the colonies. Untrained or complacent judges would, he said, permit the colonies to drift away from the mother country, and would not enforce the laws of trade or protect

---

[34] New York Hist. Soc., *Collections*, 1876, p. 187.
[35] *Ibid.*, pp. 175–176.

His Majesty's lands. Independent judges would permit the governors to rule without attaching themselves to political parties.[36]

Reunited with his family and friends in Milton, and invited again to serve on the committee to visit the Boston schools, he felt less and less like sacrificing himself to the good of the Empire. On July 23, 1762, he wrote to Governor Monckton from Milton:

I hope you will not deem it unreasonable in me to desire to be excused from attending an Office in which I cannot be supported. My leave of Absence expires next September. I beg your Excellency would please to consider how extremely hard it would be for me then, either, again to Leave my Family, or be at the Trouble and Expence to remove it, and return, without any Allowance for my past Attendance Time and Expence, Or the Least Assurance that the office will, in future, be provided for. I have . . . given up my Practice, and broke all my Measures to devote myself to this Office, *this*, together with my Absence from Home for Seven Months to do my Duty in it, and wait in vain for Justice, is a Sacrifice adequate to every Demand of Honor and Fidelity. . . .[37]

Fortunately he asked for leave of absence instead of resigning, for Colden had found a solution — to pay him a salary of £500 sterling out of His Majesty's quit rents. The Privy Council heartily approved, and expressed its support so warmly that the other judges fell into line and joined Prat on the bench when he returned to New York.

According to Colden, Prat now entered into a campaign to reform the legal practice in the Province. According to some of his friends, he was tired of New York and determined to return to Boston. According to family tradition, he had been appointed to a governorship in the West Indies and had bought his uniform when his final sickness overtook him. Samuel Auchmuty thus reported the unhappy event to Isabella Prat:

The seventh day of last month he came home from council at one o'Clock we immediately sat down to dinner, but I observed that he eat very little, and seemed to be in pain. I ask'd him if he was unwell, he said, not very well. After Diner he retired to his own room. In the Evening I still found, tho' he endeavoured to hide it, that he was in pain. I propo'd him to go to Bed and take something hot to Sweat him. He took my advice and appeared next morning to be better . . . His ease however was but of a short duration, for by noon his pain returned

---

[36] *Docs. Rel. Col. Hist. N. Y.*, VII, 501.
[37] Mass. Hist. Soc., *Collections*, 4th Ser., IX, 465, 466.

with great violence, fixed as he said in his well thigh. . . . About Eleven o'clock at night I went into his Room and to my surprize found him up, sitting by the Fire in a great Agony. . . . He then told me that he was apprehensive from the extreme pain that the Limb would swell, he must have a good Surgeon. . . . But Alas! The fatal Disorder baffled all art, and every day he grew weaker.[38]

When it became evident that it would be necessary that his remaining leg be operated upon, he dictated a will to Auchmuty, directing that Isabella receive his farm at Milton, that his son Benjamin receive the wild lands in Lynesburgh, New Hampshire, and that his remaining property be sold. Auchmuty and Dr. Ezekiel Hersey (A.B. 1728) were named executors. Then the operation was performed, but the Chief Justice died on January 6, 1763. The next night he was buried under the chancel of Trinity Church, the Governor and Council serving as bearers.

The newspaper notices and the correspondence of the period show that Prat was sincerely mourned in New York. In Boston some member of his old literary group published a memorial poem:

> With ardent love for ancient wisdom fir'd,
> And with a genius Heav'n alone inspir'd,
> He rifled Rome of all its mighty Store;
> And still athirst to Athens went for more.
> Both now exhausted — from the modern page,
> Fraught with the sense of each preceding age,
> He seiz'd it's treasures; made them all his own;
> And 'midst the sons of science greatly shone.
>   In him tho' science did it's ray unite,
> And shed around him a distinguish'd light,
> 'Twas but a second merit. — Virtue more
> Adorn'd the man than all his learning's store.
>   To heaven now sped — beyond all mortal ken —
> He rivals angels as he rival'd men.[39]

When the business of the Revolution carried Massachusetts political leaders to New York, they visited the burial place of Chief Justice Prat, and exchanged reminiscences of him with the Yorkers. The publication of John Eliot's uncomplimentary biographical sketch of him aroused a storm of protest which by its violence shows the respect in which he was generally held. President Kirk-

[38] Annette Townsend, *The Auchmuty Family* (New York, 1932), pp. 26, 27.
[39] *Boston News-Letter*, Jan. 27, 1763, 2/2.

land, with the approval of the other members of the Anthology Club, wrote the rebuttal printed in the *Monthly Anthology* and quoted above. When President Adams read Eliot's article he was "deeply afflicted with a mixture of pity, grief, and indignation." He blamed Eliot's prejudice on his father, Prat's classmate, a friend of Hutchinson who, he thought, was influenced by the Governor's Tory views.[40] There is no evidence of this in either the Eliot or the Hutchinson papers. More likely the Eliots, both Congregational ministers, were influenced by Prat's high Episcopalianism and his unorthodox views as to the place of man in creation.

The Chief Justice had three sons and a daughter. Benjamin, who was baptized on January 20, 1757 or 1758, studied law with Francis Dana (A.B. 1762), and went to Carolina, where he died about 1783. The younger boys, Frederick and George, were "weak of body and mind," and "of little account." Frederick for a time kept a school for very young children, but both died early.[41] Isabella married Samuel Welles on December 17, 1772, and died on May 30, 1788. Through her the Milton mansion and the Chief Justice's manuscripts passed to the Welles family. The papers seem to have been intact as late as 1825; in 1932 some few pieces remained in the hands of Miss Georgianna Welles Sargent of New York and Lenox.

In recent years a handsome picture alleged to be a Smibert portrait of the Chief Justice was given to the Harvard Law School.[42] Unfortunately it was one of several discovered by a dealer who specialized in such finds after it became known that there was a good market for portraits of Harvard lawyers.

---

[40] Adams, *Works*, X, 243.

[41] Eleazer Franklin Pratt, "Slight Biographical Sketch of Benjamin Pratt," in Sibley's Letters Received (Harvard University Archives), IX, 15.

[42] It is reproduced in the *Massachusetts Law Quarterly*, XIV, 7 (Nov. 1928), XV, 11 (Sept. 1930), Frank W. Bayley, *Five Colonial Artists* (Boston, 1929), p. 419, and *The Auchmuty Family*, p. 26.

OXENBRIDGE THACHER

OXENBRIDGE THACHER, the Boston lawyer, was born in that town on December 29, 1719, a son of Oxenbridge (A.B. 1698) and Elizabeth, sister of Sir Charles Hobby and widow of Thomas Lillie. His father had tried the ministry but, reluctantly, had gone into business as a brazier, and had established himself on the west side of Tremont Street between Court and Beacon. It was because of his reputation for piety rather than any interest in politics on his part that the popular party of Elisha Cooke (A.B. 1697) sent him to the General Court in 1731. The younger Oxenbridge became the first Freshman to win the Hopkins Prize; later he was fined for using prohibited liquors. He remained in residence after taking his first degree, read for the ministry, testified before the Overseers as to the misconduct of Tutor Prince (A.B. 1718), and joined the First Church of Cambridge. At the Commencement of 1741, when he took his M. A., he delivered the Valedictory.[1] For his *Quaestio* he prepared the negative of "An Bruta, ab omni morali obligatione esse immunia, possit probari."

In 1740 the elder Oxenbridge had taken as his second wife Bathsheba Doggett, widow of John Kent of Boston, who brought a sixteen-year-old daughter, Sarah, into the household on Tremont Street. On July 27, 1741, Sarah and the younger Oxenbridge were married. The family fondly hoped that he would follow his distinguished ancestors into the ministry, and his own inclinations were in that direction. He was an admirer of Jonathan Edwards, for whose *Life of Brainerd* he subscribed. He never lost his relish for theology, and he became one of the would-be incorporators of the Society for Propagating Christian Knowledge among the Indians of North America.[2] In the words of a younger contemporary,

He . . . was a preacher, but with a small voice, and slender state of health, did not meet with success equal to some who have only the sounding brass to given them a reputation. Mr. Thacher was sensible, learned, pious, a Calvinist, beloved by his friends, and respected . . . yet with all these advantages, found it necessary to leave his profession, and go into a line of life, which required no abilities but a vast deal of

[1] Andrew Eliot, Diary (Mass. Hist. Soc.), July 1, 1741.
[2] Mass. Archives, XIV, 289.

drudgery to transact. He soon failed, and was persuaded to study law; for which he had no great inclination at first.[3]

Tradition says that he did not serve the usual apprenticeship in the practice of law, but it is also said that he read with Jeremy Gridley (A.B. 1725).[4]

At first Thacher took divorce cases, which were regarded as refuse below the attention of any lawyer who could do better; but in a few years he worked up a practice as good as any in Boston. Able young men like Josiah Quincy (A.B. 1763) and John Lowell (A.B. 1760) were glad to read law in his office opposite the south door of the Old State House. In a happy moment for his reputation he befriended a young lawyer named John Adams (A.B. 1755). This sharp young man observed his patron and many times described him in his letters and diary. Comparing him with James Otis (A.B. 1743), he said, "Thacher has not the same strength and elasticity; he is sensible but slow of conception and communication; he is queer and affected — he is not easy." [5] He was given to voicing "wild, extravagant, loose opinions and expressions" without considering that "these crude thoughts and wild expressions are catched and treasured as proofs of his character;" his passions were "easily touched, his shame, his compassion, his fear, his anger, etc." [6] But his "amiable manners and pure principles united to a very easy and musical eloquence made him very popular;" [7] "there was not a citizen of Boston more universally beloved for his learning, ingenuity, every domestic and social virtue, and conscientious conduct in every relation of life." [8] This respect was shared even by those who disagreed with him in political matters and therefore felt the lash of his passions and his extravagant language.

Thacher's qualities were precisely those to make him popular with his fellow townsmen. In 1745 they elected him constable, but, no doubt because of his health, he paid a fine rather than serve. He did, however, accept the office of clerk of the market, and for

---

[3] John Eliot, *Biographical Dictionary* (Boston and Salem, 1809), p. 454.

[4] Emory Washburn, *Sketches of the Judicial History of Massachusetts* (Boston, 1840), p. 223.

[5] John Adams, *Works* (Boston, 1850–56), II, 67.

[6] *Ibid.*, pp. 74–75.

[7] *Ibid.*, 124 *n.*

[8] *Ibid.*, p. 47 *n.*

many years served on the school committee. Over the years he labored on special committees to revise the bylaws, instruct the Representatives, regulate the sale of the proceeds of street sweeping, appeal to the General Court for relief from taxation, choose a Province agent, install turnpikes in Hog Alley, prosecute the parkers of carts in Dock Square, draw a bill for the better regulation of the smallpox, and many others. His most spectacular service came in February and August, 1761, when with James Otis he presented the case of the merchants against the Writs of Assistance before the Superior Court. According to Adams, he "argued with the softness of manners, the ingenuity and cool reasoning, which were remarkable in his amiable character." [9] He held that the Superior Court of Massachusetts did not share the power of the Exchequer Court of England to issue such writs. [10]

Later in the same year Thacher took issue with Lieutenant Governor Thomas Hutchinson (A.B. 1727) in another matter freighted with political consequences. For some time silver had been over-valued on the English market, with the result that it had been the chief medium for remittance and so had been drained from the Province. The House of Representatives, under the control of the debtor group which would profit from a little inflation, passed a bill making gold legal tender at a rate which would accomplish its purpose. The Council rejected the bill, and the General Court prorogued in a bad temper. Hutchinson, who on the subject of the currency was an oracle to whom even Adams listened respectfully, then published an article in which he held that there could be no fixed ratio between silver and gold, and that the attempt to make the latter legal tender at a high rate would drive out the rest of the silver. Otis, without regard to the economics involved, used the opportunity to whip up popular prejudice against Hutchinson. [11] Thacher entered the fray by publishing a tract, *Considerations on Lowering the Value of Gold Coins, within the Province of the Massachusetts Bay*, in which, although he did not show the Lieutenant Governor the deference

[9] *Ibid.*, X, 247.

[10] Josiah Quincy, *Reports of Cases Argued and Adjudged in the Superior Court of Judicature of the Province of Massachusetts Bay Between 1761 and 1772* (Boston, 1865), pp. 52–55, 469–471, 482.

[11] George Richards Minot, *Continuation of the History of Massachusetts* (Boston, 1798–1803), II, 102–106.

to which he was accustomed, he ably presented the other side of the argument.

Thacher was not an economic dreamer with rosy visions of the future of America, for he thought that the country was full.[12] Rather, his politics were dictated by his lively emotions and prejudices. Unfortunately he never moved in the circle of the Hutchinsons and Olivers so his festering suspicions of their motives were never exposed to the healing knowledge that they were honorable men devoted to what they, perhaps mistakenly, believed to be the good of the country. In a letter to his old friend, Chief Justice Benjamin Prat (A.B. 1737), written in 1762, he voiced his fears and forebodings:

We seem to be in that deep sleep or stupor that Cicero describes his country to be in a year or two before the civil wars broke out. The sea is perfectly calm and unagitated. Whether this profound quiet be the forerunner of a storm I leave to your judgment. . . . I even hear that the press now is under the dominion of our great men, and that those printers who owe their first subsistence and present greatness to the freedom of their press refuse to admit any thing they suspect is not pleasing to our sovereign lords. I will lay a guinea that they are bound to that in good behaviour, and that our sovereign lord the kings attorney [Edmund Trowbridge (A.B. 1728)] hath threatened them with a prosecution for some past freedoms.[13]

Swiftly Thacher's indignation grew as he became convinced that the Hutchinson tribe was engaged "in a deep and treasonable conspiracy to betray the liberties of their country, for their own private, personal, and family aggrandizement."[14] His unbridled public and private philippics amazed the Lieutenant Governor who had hitherto regarded him as being on the side of the government.[15] According to Adams, Hutchinson remarked that Thacher was not born a plebeian, but was determined to die one.[16]

Because of his reputation for virtue the Tories regarded Thacher as the most important member of the Whig party.[17] Adams con-

---

[12] Adams, II, 472 *n.*

[13] Mass. Hist. Soc., *Proceedings*, 1st Ser., XX, 46–47.

[14] Adams, X, 286–287.

[15] Thomas Hutchinson, *History of the Colony and Province of Massachusetts Bay* (Cambridge, 1936), III, 75.

[16] Adams, X, 285.

[17] *Ibid.*

sidered his influence in the revolutionary movement to be second only to that of Otis. When the town of Boston elected its four Representatives in 1763, Thacher won last place on the delegation. It was he who in that session of the House obtained the repeal of the vote to send Hutchinson to England to represent the case of Massachusetts against the Sugar Act.[18] He showed the same distrust by opposing the bill to place the militia at the call of the royal commander-in-chief.

At the next election, in March, 1764, Thacher stood first in the poll of Boston Representatives. In this session he served on the committee appointed to urge other colonies to oppose the Sugar Act. In October he presented to the House the draft of a petition to be sent to the Crown calling for the repeal of the new taxes. In it he argued that Massachusetts had poured out its treasure without stint in the French and Indian War, but now, crushed as it was already by the burden of its war debt, it was threatened by new taxes, levied without its consent, which would drain off its specie and drive it into bankruptcy. The stamp tax would be particularly burdensome because the great part of the people lived so scattered that they would have to travel forty or fifty miles to obtain a stamp to validate a document, and because it would fall hardest upon "the poorer sorts and those least able to bear that or any other tax." The new powers given to the Court of Admiralty would subject the colonists to the unregulated tyranny of the judges.[19] The House passed this bill but the Council rejected it and obtained the substitution of Hutchinson's milder petition. About the same time Thacher prepared a draft of a letter of instructions to the Provincial Agent in which to prove the financial exhaustion of the Province he listed its war expenditures since 1746. But nowhere did he even estimate the amount of the reimbursement received from the Crown.[20] These ideas he circulated by publishing a tract, *Sentiments of a British American*, in which he assumed that the colonists had all of the rights of Englishmen, and argued that England had benefited far more than had the colonies from the war, and therefore should pay for it. The argument was not of a rabble-rousing

[18] Hutchinson, III, 77.

[19] The text of his petition is printed in Mass. Hist. Soc., *Proceedings*, 1st Ser., XX, 49–52. The manuscript, with interesting revisions, is in the Thacher Mss. (Mass. Hist. Soc.).

[20] Mass. Hist. Soc., *Proceedings*, 1st Ser., XX, 52–56.

nature, but as sound as one could be which completely ignored the administrative problems created by the wars and the growth of the colonies.

Thacher's ideas did not satisfy Otis, who in the House "treated him in so overbearing and indecent a manner that he was obliged at times to call upon the speaker to interpose and protect him." [21] On the other hand, he was credited with holding Otis to the Whig party when he wavered. Perhaps it was the vote of the Otis faction which caused Thacher to drop to third place on the Boston slate in the election of 1765. This was his last. His wife had died of the smallpox at his father's home in Milton on July 4, 1764, and he himself never recovered from the physical shock of being inoculated in the Castle hospital. John Adams, visiting him to discuss taking over some of his business, asked him if he had seen the Virginia Resolves.

Oh yes — they are men! they are noble spirits! It kills me to think of the lethargy and stupidity that prevails here. I long to be out. I will go out. I will go out. I will go into court, and make a speech, which shall be read after my death, as my dying testimony against this infernal tyranny which they are bringing upon us.[22]

But instead he died on July 9, 1765, and was buried three days later with an elegant Masonic funeral.[23] A friend made this extempore eulogy:

> Once warm with Zeal in honest Virtue's Cause,
> That Tongue spoke free, and wielded Britain's Laws;
> With equal Eloquence, unwarp'd, display'd,
> For Wealth or Poverty, it's pow'rful aid;
> Alike to him, Worth could it's Charm impart,
> In King, or Beggar, touch'd his gen'rous Heart.
> From humble Birth, to Path's of just Renown,
> He dawn'd, he brighten'd to the Hour of Noon!
> Learn'd, yet not vain, in useful Science read,
> Fair Freedom's Cause with manly Strength he plead;
> A Patriot's Flame, with pious Zeal sustain'd,
> His Country's Rights, with jealous Care maintain'd;
> With grateful eye beheld the Glory past,
> Drop'd a sad Tear, and fighting breath'd his last.[24]

[21] William Gordon, *History of the Rise . . . of the United States* (London, 1788), I, 205.

[22] Adams, X, 287.

[23] Robert Treat Paine, Diary (Mass. Hist. Soc.), July 12, 1765.

[24] *Boston News-Letter*, July 18, 1765, 3/2.

It is proof of Thacher's virtue that he died leaving an estate of only £311, including a remarkable library. Some land in Franconia, New Hampshire, and Wilmington, Vermont, had no appreciable value. The list of his creditors, who held his notes amounting to £412, sounds like a Social Register. They and other friends pitched in to help Grandfather Thacher raise and educate the seven surviving children.

Thacher's seat in the House passed to Sam Adams. His library and manuscripts were scattered; the Thacher Mss. at the Massachusetts Historical Society contain only a few fragments. The so-called Smibert portrait is a fraud, misattributed both as to artist and subject.[25] The elegant portraits of Oxenbridge and Sarah attributed to Feke did not, as the books imply,[26] pass by inheritance in the family direct to the present owners; they were "discovered" by the dealer who found the "Smibert."

LEMUEL BRIANT

LEMUEL BRIANT, minister of the First Congregational Society of Quincy, Massachusetts, was baptized on February 25, 1721/2, a son of Thomas and Mary (Ewell) Briant of Scituate. During his first quarter at Cambridge he boarded with President Wadsworth.[1] He subscribed for the *Chronological History* of Thomas Prince (A.B. 1707), but he neglected Hebrew scandalously and slept through the hour set for his public admonition. The Faculty was irked:

Agreed, That Bryant Who saith (in Excuse for Himself, as to his Absence when he should have been admonish'd with the Rest for neglect of their Hebrew instructions) that He heard not the Bell for prayers being then fast asleep, be punished ten Shillings because He took not particular Care That Some One Should awake him, whereby he might have been present; And also that He Yet attend with his Confession in the Hall there to receive the admonition for the Crime above mentioned.[2]

---

[25] Henry Wilder Foote, *John Smibert* (Cambridge, 1950), pp. 245–246.
[26] Henry Wilder Foote, *Robert Feke* (Cambridge, 1930), p. 192. The history of the portraits was misrepresented to Dr. Foote.
[1] Benjamin Wadsworth, Commonplace Book (Mass. Hist. Soc.), p. 99.
[2] Faculty Records (Harvard University Archives), I, 96.

Thereafter Lemuel mended his ways. After graduation he took over the Plymouth school, and on July 5, 1741, he entered into full communion with the Second Church of Scituate. In August he returned to college where during the next year and a half he occupied successively Massachusetts 11, 19, and 8, his roommates being Hubbard, Rand, Wainwright, and Waldron of the Class of 1742 and Charnock of the Class of 1743. When he took his M. A. at the Commencement of 1742 he was prepared to hold the negative of "An Alicui Belli Causas ignoranti Inimicis Devastationem imprecari licitum sit?"

In November, 1742, Briant left Cambridge to supply the pulpit at Pembroke Village. Later he preached at Worcester and probably at Middleborough, for his signature on a petition of the inhabitants of that town was protested. In April, 1745, he declined a call to the second precinct of Carver, probably because the £46 salary offered was impossibly small. On July 29 the North Church of Braintree invited him to preach. Two weeks' trial convinced the church that it had found the minister it had been seeking. It waived the privilege of the initial call to the precinct, which the next day voted unanimously to have him. The only likely explanation for this enthusiasm is that they were getting a parson incredibly cheap, for he accepted a settlement of £100 and an annual salary of half that sum.

Briant was ordained on December 11, 1745, by Shearjashub Bourne (A.B. 1720) and Nathaniel Eells (A.B. 1699) of Scituate, Samuel Niles (A.B. 1699) of the South Church of Braintree, and John Taylor (A.B. 1721) of Milton. He took his time about choosing a wife, for it was not until August 23, 1749, that he was married to Abigail, daughter of Joseph and Mary Barstow of Scituate. The parsonage property, which occupied the site on which Adams Academy later stood, was described thus by the minister:

A Handsom Country Seat . . . containing besides a very commodious well finished House, a good Barn, Out-Houses, fine Gardens, and the best of Orchards, with about 40 Acres of choice Land belonging to it, just by the Meeting, not half a Mile from the Church [of England], and but about 2 Miles from the flourishing Settlement at Germantown.[3]

---

[3] *Boston Evening-Post*, Mar. 19, 1753, 2/1.

Here were born two sons, Lemuel on July 16, 1749, and Joseph, on November 23, 1751.

Soon the ministers who had ordained Briant had reason to wonder whether they had made a mistake. For a century the New England clergy had been formally orthodox, either ignoring or dismissing as mysteries the unpalatable parts of the logical structure of orthodox Christianity; but of late the New-Light preachers had taken to emphasizing the horrors of orthodox dogma in order to shock their congregations. Briant was particularly irritated by what he called the impiety of the New-Lights who declared that in the eyes of God the righteousness of the unelect was but "filthy rags," and he publicly took issue with them:

The unthinking Multitude may be best pleased with what they understand least, and be carried away into any Scheme, that generously allows them the Practice of their Vices, tho' every Article be a downright Affront to common Sense; yea, by a few rabble charming Sounds be converted into such fiery Bigots, as to be ready to die in the Defence of Stupidity and Nonsense . . . notwithstanding all this, I say; There always was and always will be some in the World . . . that have Sense eno' and dare trust their own Faculties so far, as to judge in themselves what is right.[4]

This blast, which was aimed particularly at the orthodox doctrine of Salvation by Grace alone, was first fired by Briant from Nathaniel Eells' pulpit. After the service the old Parson went up to his young colleague and said, "Alas! Sir, you have undone today, all that I have been doing for forty years." To this Briant replied, "Sir, you do me too much honor in saying, that I could undo in one sermon, the labours of your long and useful life."[5] To undo the damage Eells preached a series of sermons the doctrine of which, so it seemed to his amused parishioners, was hardly distinguishable from that of Briant.

Such heresy was by no means uncommon in New England and would probably have passed unnoticed had not Briant on June 18, 1749, delivered the same sermon in Jonathan Mayhew's pulpit in the great West Meetinghouse of Boston. The ministers of the important churches remained silent, but John Porter (A.B. 1736) of Brockton replied from the pulpit of the Third Church of Brain-

[4] Lemuel Briant, *The Absurdity and Blasphemy of Depretiating Moral Virtue* (Boston, 1749), p. 23.
[5] Samuel Deane, *History of Scituate* (Boston, 1831), p. 199.

tree with a sermon in which he correctly restated the orthodox doctrines of Predestination and Salvation by Faith, demonstrating, unintentionally, Briant's contention that they are incompatible with the goodness of God. According to New-Light custom he larded his sermon with references to "Heathenish morality" and the "filthy Rags" of "personal Rightiousness."[6] This reply was printed with an attestation by five country parsons, Nathaniel Leonard (A.B. 1719), Jonathan Parker (A.B. 1726), John Cotton (A.B. 1730), Solomon Prentice (A.B. 1727), and Elisha Eaton (A.B. 1729). Briant roared back with the charge that Porter had invaded Braintree in an effort to unseat him, and insinuated that Porter did not know enough Latin to understand Calvin correctly. He found particularly offensive the text, "All our Righteousnesses are as filthy Rags."[7] Porter kept his temper and in a moderate tone replied that Briant was denying him the divine right of private judgment on which, he said, all were agreed. He demonstrated that Briant's views were not orthodox Protestantism and reproved him for seeming to support James Foster (1697–1753), who denied the divinity of Christ.[8] Cotton replied that he was ready to respect Briant as a courageous, but mistaken, Arminian; but he resented his effort to prove himself more of a Calvinist than his critics.[9] Thomas Foxcroft (A.B. 1714), one of the important ministers of Boston, in a long footnote reproved Briant and "His Friend Dr. Mayhew" for charging the orthodox with ideas which they did not hold. The Braintree minister's second reply was a sad contrast to the moderate and scholarly arguments of the New-Lights whom he had called "fiery Bigots." In it his chief arguments were that Porter was a plagiarist and Cotton had a lecherous grandfather.[10] Of his replies his neighbor Niles said:

When he can't fairly grapple with an Argument, he knows how to shuffle and evade: And knows how to satyrise and reflect, when he can

---

[6] John Porter, *The Absurdity and Blasphemy of Substituting the Personal Righteousness of Men in the Room of the Surety-Righteousness of Christ* (Boston, 1750.)

[7] Lemuel Briant, *Some Friendly Remarks* (Boston, 1750).

[8] John Porter, *A Vindication of a Sermon Preached at Braintree* (Boston 1751).

[9] John Cotton, "A Appendix, Relating to the Same Subject," *ibid.*

[10] Lemuel Briant, *Some More Friendly Remarks on Mr. Porter & Company* (Boston, 1751).

no longer reason in any plausible Manner. And tho' at first he sat out with some specious Pretension of being Calvinist; yet it seems, he has thought fit to drop it.[11]

Even worse, by laying aside the catechism he was inflicting a dreadful injury on "the Souls of the poor Children, so shamefully and barbarously neglected."

Early in the controversy Briant had boasted that he had the solid backing of his congregation, but in time that began to weaken. His friend Mayhew privately reported the situation to his father:

Mr. Briant has lately met with considerable Trouble, by reason of his Wife's eloping from him. Some say she is distracted; and others that he did not use her well. His People are generally well satisfied with Him. Some indeed give heed to the evil Reports which his wife has spread about concerning him.[12]

John Adams (A.B. 1755), who was certainly sympathetic to Briant, thought that he was too "jocular and liberal," and "too gay and light, if not immoral." [13] For these reasons, he said, the controversy in the north parish "broke out like the eruption of a volcano and blazed with portentious aspect for many years." [14] When the discontented members of the congregation first asked the Parson to join them in calling an ecclesiastical council, he put them off. They then called a council of their own which met on December 5, 1752, at the house of Deacon John Adams, with no less a man than Joseph Sewall (A.B. 1707) of Boston as moderator. To this body the minority reported that "the Pastor not only denied but treated with Contempt, the fundamental Doctrines of the Christian Faith . . . and apostatized into such gross Errors of Arminianism," that they could no longer have communion with him. The council asked him to call a meeting of the church to select delegates to sit with them, but he refused to do so on the ground that some of the brethren were out of town. This, the council ruled, justified the aggrieved in calling another body to take formal action on the charges; but they urged the North Church to participate.

[11] Samuel Niles, *A Vindication of Divers Important Gospel-Doctrines* (Boston, 1752), pp. 11–12, first pagination.
[12] Jonathan Mayhew to Experience Mayhew (Boston University Library), Aug. 21, 1752.
[13] John Adams, *Works* (Boston, 1850–56), I, 41; X, 254.
[14] Mass. Hist Soc., *Proceedings*, 2nd Ser., VIII, 92.

When the second council met, on January 9, 1753, Briant refused to participate and his Standing Committee refused to open the meetinghouse. So the body sat at the house of Deacon Adams with Niles as moderator, for Sewall did not attend. It considered the charges and held that Briant had given offence by his theology, his refusal to attend public fasts, his refusal to answer his wife's charge that he had been guilty of "several scandalous Sins," his substitution of Pierce's unorthodox catechism for that of the Westminster Assembly and by his recommendation of a work by John Taylor (1694–1761), the Arian. Honest, as all these councils were, it reported that the minister had been unreasonable in his dealings with his critics, and that they had been too high in their charges. Some of these, we learn from private sources, were that he was guilty of "intemperance, gaming, neglect of family duties," and ill "Carriage towards his wife." [15] In its plea to him the council said:

Rev. Sir, We seriously and affectionately advise you, to do what in you lies to remove the several Stumbling-Blocks you have laid in the Way of your Brethren; particularly, that you take heed to your Self, and your Doctrine. We can't but think it a Duty incumbent on you, to endeavour to clear up your moral Character, as to the several scandalous Sins your Wife has charged you with, and to give your dissatisfied Brethren all Christian Satisfaction on these Points; and to behave so holy and unblameably in your Family, and before your People, as to be an Example of every Thing that is vertuous and of good Report. [16]

Briant had apparently no hope of conciliation with his wife, for in March, 1753, he advertised his house for sale. About the same time the North Church appointed a committee, headed by John Quincy (A.B. 1708), to sift the charges against the minister. Its report, made on April 14, both cleared him and nailed to the mast the flag of theological liberalism:

He is now ready (as soon as his Health permits) to teach our Children such Parts of the Catechism as he apprehends agreeable to the Scripture. — Nor can we think that any Christian Society ought to be so attach'd to any human Composure, as to make it a Crime in their Pastor to prefer pure Scripture Into it. . . . We have no Evidence of Mr. Briant's having made any particular Profession of his Faith at his Ordination, or that any such Thing was requir'd of him by the Council then present; or

[15] Josiah Cotton, Memoirs (Mass. Hist. Soc.), pp. 416–417.
[16] *The Result of a Late Ecclesiastical Council. . . . In the North Part of Braintree* [Boston, 1753], p. 4.

if he had made any such Profession, it cou'd not destroy his Right of private Judgment, nor be Obligatory upon him, any further than it Continued to appear to him agreeable to Reason and Scripture.

The committee went on to approve his recommending the heretical work of John Taylor so that all might judge for themselves what was the truth. They reported that they differed with him in doctrine, but upheld his "undoubted Right to judge for himself." [17] The church adopted the report by an overwhelming vote.

The Parson's victory was a triumph for a dying man. On October 10, 1753, he asked to be dismissed because of ill health. When his request was granted, eleven days later, a committee, headed by Edmund Quincy (A.B. 1722), was appointed to thank him for his services. Evidently he considered himself to be released from the ministry, for thereafter he described himself as "of Boston, gent." Judge Josiah Cotton (A.B. 1698) sighed with relief: "a good riddance of him, for he was never cut out for a gospel minister." [18] He died at Boston on October 1, 1754, and was buried with his fathers in the Norwell cemetery. His will mentions Abigail as being still his wife. His inventory consisted of a little money, a red coat, a bob wig, and an unimpressive library.

Briant has had a certain amount of fame because when W. E. Channing (A.B. 1798) laid down the tenets of Unitarianism in 1819, John Adams remarked that he had been familiar with them for sixty-five years, for in his boyhood he had heard his Braintree minister preach them. This was claiming too much for him. Briant made an outspoken attack on tenets of orthodoxy which many of his contemporaries privately felt were offensive to reason and insulting to the Deity, but he did nothing to develop a substitute philosophy which would satisfy the religious craving of men. The Unitarianism of Channing was not merely an attack on orthodoxy; it was a statement of man's relation to God which some men found more satisfactory than Calvinism.

[17] *The Report of a Committee of the First Church in Braintree, Appointed March, 1753* (Boston, 1753).

[18] Cotton, Memoirs, p. 422.

JOHN TUCKER

New England Puritanism differed from other contemporary Protes-
tant communities in the greater stress which it laid on the right and
duty of the individual to formulate his personal theology. The
resulting chaos of interpretations finally took shape in a vague
consensus which was called the Unitarian movement chiefly because
it was not Trinitarian. Tucker's violent controversies brought into
the open irreconcilable differences which had been ignored for a
century.

JOHN TUCKER, sixth minister of the First Congregational Church
of Newbury, was born on September 20, 1719, a son of Benjamin
and Alice (Davis) Tucker of the West Parish of Amesbury, now
the town of Merrimac. At college he waited on table until he won
a Hollis Scholarship, and was an exception to the general rule
that future theological liberals were bad boys in their undergraduate days. He had a speaking part at both of his Commencements, for the M. A. holding the affimative of "An Sanctio poenalis
sit Legi essentialis?" He read theology with his minister, Paine
Wingate (A.B. 1723), and became a member of the West Church
of Amesbury.

At that time the neighboring parish of Newbury was in an uproar because the aged minister, Christopher Toppan (A.B. 1691),
had lost his long battle with the New-Lights, many of whom had
left the church in order to sit under the preaching of Joseph Adams
(A.B. 1742), a wild revivalist. The ministers of the region and the
leading inhabitants of Newbury asked Tucker to undertake the
difficult task of reuniting the shattered church. If their critics are
to be believed, the Old-Lights forced the candidate upon old Parson Toppan, who repeatedly refused to take him as a colleague
because of his heretical views, and only by illegal voting did they
achieve a majority of twelve in the parish and two in the church.[1]
It was quite unheard of for a candidate to accept a call made by
so narrow a margin, so Tucker pondered it from July until October

---

[1] Marcus Shanger, *A Letter to the Reverend John Tucker* ([Boston], 1775),
p. 3.

JOHN TUCKER

before finally giving in to the pressure of his elders. The ordaining council met on November 20, 1745, heard and summarily dismissed the objections raised against Tucker, and proceeded to install him in his rickety pulpit. On this occasion William Johnson (A.B. 1727) made the opening prayer, Paine Wingate preached, Caleb Cushing (A.B. 1692) gave the Charge, John Lowell (A.B. 1721) gave the Right Hand of Fellowship, and Thomas Barnard (A.B. 1732) made the closing prayer. It was plain that the established clergy of the region supported Tucker, and they were bound by personal ties when on August 27, 1747, he married Sarah, daughter of the Reverend John (A.B. 1709) and Sarah (Martyn) Barnard of Andover.

Before the ordaining council had broken up, twenty-two members of the First Church issued this ultimatum to Tucker:

For the peace of our consciences, our spiritual edification and the honor and interest of religion as we think, we do now withdraw communion from you and shall look upon ourselves no longer subjected to your watch and discipline, but shall, agreeable to the advice given us, speedily as we may, seek us a pastor, who is likely to feed us with knowledge and understanding and in whom we can with more reason confide.[2]

Within the week the band of seceders had grown to more than a hundred. The church took the offensive and ordered them to appear before it to confess and repent their misdeeds. The New-Lights ignored the summons and joined in the organization of the First Presbyterian Church of Newburyport and the settlement of Jonathan Parsons (Yale 1729) over it. The Bradford association of Congregational ministers admitted Tucker but frowned upon Parsons. The lines which were to divide Essex for a generation had formed.

In Tucker the ministers of the New England establishment had acquired a vigorous and able champion:

He was a graceful speaker, in the opinion of all proper judges, a good pulpit orator. His voice was not loud, but clear and audible, soft and pleasing, even to the most delicate ear. His delivery was not vehement, but grave, serious and affectionate — at times *very* affectionate. . . . His sermons were excellent, adapted to the lowest capacities, and yet instructive to the highest. . . . The stile easy and natural, clear and striking; not very elevated and flowery, nor low and flat. . . . His

---

[2] Joshua Coffin, *A Sketch of the History of Newbury* (Boston, 1845), p. 216.

judgment was solid and penetrating. This seems to have been his predominant power.[3]

His significance in New England intellectual history lies in the fact that he championed the theological simplicity of seventeenth-century Puritanism against the scholastic philosophy of Rome, against Calvin, and against Jonathan Edwards. From his first sermon to his last he hammered at one idea — that the whole truth is plainly written in the Bible, and that all philosophy, creeds, and confessions of faith are glosses; interesting, but not binding on the individual. Every man must search out the truth for himself from the Bible. The inevitable heresies instead of being reprehensible or dangerous are useful. This liberalism he extended to the study of the Bible itself. He did not regard the Scriptures as infallible, but held to their sense and meaning rather than to the text. He was aware of such current Biblical criticism as the reassessment of the authorship of the Psalms, and he was ready to reject any New Testament passage which conflicted with the whole as an "old churchman's tale" which had become imbedded in the original text. When he was not angry, he preached these doctrines gently, clearly, and wisely.

Tucker was no modern, even by the standards of his own generation. He was untouched by the fever of science which was running in many of his clerical contemporaries. In many of these men the Lisbon earthquake shook the orthodox teaching of the agency of God, but his interpretation to his congregation might have come right out of the Old Testament:

He so orders it in the righteous administration of his providence, as to send great and sore evils upon a sinful people; such as expensive and destructive wars, consuming famine, grievous and wasting sickness, awful and desolating Earthquakes, &c. By these things God makes himself known in the earth: — By these he shews himself the moral as well as natural governour of the world: vindicates his dominion and authority over the children of men; and declares in dreadful language his displeasure towards sinners.[4]

He agreed with the scientific liberals that the sufferers in Lisbon were not the worst sinners in the world, but suggested that God

[3] Jonathan Eames, *Walking with God* (Newburyport, 1792), pp. 31-32.
[4] John Tucker, *Four Sermons* (Boston, 1756), p. 9.

was afflicting them in order to warn his chosen people in New England!

On the other hand, Tucker attacked Parsons for preaching that God arbitrarily chose the elect without regard to their moral character; this was, he said, to give God "the Character of a despotick and merciless Tyrant." When Parsons charged that the Old-Light ministers were leading their unfortunate parishioners to hell, Tucker replied that he was being as arbitrary as his god. He made fun of the Presbyterian's description of the wickedness of unconverted man, and strongly defended those unconverted souls who were trying to live a good life.[5] On the other hand he was unimpressed by Charles Chauncy's work on Universal Salvation — a doctrine which shook New England orthodoxy almost as badly as the earthquake shook Lisbon. Of Chauncy's work he said, "It is plausible, it is a splendid piece of theoretic reasoning; but it has no foundation in the Scriptures."[6]

The First Church of Newbury was no hotbed of liberalism. It was not until 1761 that the Parson induced it to adopt Tate and Brady, and not until 1769 that he obtained permission to read from the Bible as a part of the service. The Congregational clergy held his views in more esteem, and frequently employed him as a trouble-shooter. Thus when they were ordaining Amos Moody (A.B. 1759) at Pelham, a town notorious for driving out ministers, they appointed him to preach the sermon, knowing that he would hammer the congregation for its bigotry in attacking the clergy for theological deviations.[7] On the other hand, he many times asserted that the ministers were guides, not oracles; that their freedom to interpret the Bible did not in any way limit the right and the duty of the laymen to search the Scriptures for the truth. To limit the layman's right of interpretation of the Bible would repeat the error of Rome, which by keeping the "key of knowledge" out of the hands of the common people kept them "in blindness and ignorance."[8]

This right of interpretation was the one thing in which the New-

[5] John Tucker, *Observations* (Boston, 1757).

[6] William B. Sprague, *Annals of the American Pulpit*, I (New York, 1857), p. 453.

[7] John Tucker, *A Sermon Preached at the Ordination of . . . Amos Moody* (Boston, 1766).

[8] John Tucker, *Ministers of the Gospel* (Boston, 1767), p. 9.

Light members of the First Church agreed with Tucker. On February 11, 1766, forty-five of them presented to the parish a petition which opened a decade of bitter strife:

As we cannot adhere to his principles manifest in his preaching, especially of late, we cannot [but] think it our duty to ask the favour to be freed from paying any further taxes towards his support, or any other parish charges. We therefore . . . pray that you would . . . grant us the relief we might rationally expect in a nation where liberty of conscience is indulged to every sect . . . and in a land where a love of . . . liberty is born with us. . . . We think that every rational person must be convinced after about twenty years' trial, that we cannot enjoy any lasting peace in the parish while we thus continue.[9]

Informally they demanded that the minister call a church meeting to consider their motion to call an ecclesiastical council. He refused to do so until formal written charges had been presented for the consideration of the church. The New-Lights then issued a call for a council.[10] The church then resolved that it knew no grounds for such a council, and that the aggrieved were violating the ecclesiastical constitution in calling one. In this, the church was right. Some of the churches summoned sent excuses, and the delegates who were sent by others were obviously uncertain of their position. On March 31, 1767, the New-Lights presented to the council charges that Tucker had denied "the doctrine of original sin, as explained agreeable to the holy scriptures, in the protestant confessions of faith," that he had neglected "to preach explain and enforce the peculiar doctrines of grace," and that he had "openly preached and printed against all creeds and human confessions of faith as standards or summaries of christian doctrine, by which any, either ministers or private christians, should have their principles examined." [11] The members of the council offered to come to the parsonage to discuss the charges with the Parson, but he refused to receive them. The next day he formally notified them that the First Church had voted that there was no need for a council, and that what was needed in Newbury was not more strife but "a disposition to bear with one another in those differences of sentiment, which are incident even to good men, and

---

[9] Coffin, pp. 232–233.
[10] *A Brief Account of an Ecclesiastical Council, so Called, Convened in the First Parish of Newbury* [2nd ed.] (Boston, [1767]), p. 2.
[11] *Ibid.*, pp. 6–7.

which seem to be the unavoidable consequence of that freedom of thought, and right of private judgement all lay claim to." [12] The council then adjourned. To the New-Lights, Tucker protested that if any dissatisfied members could, without the consent of the church, call in other churches to question the pastor on points which had not been previously submitted for discussion, the result would be ecclesiastical anarchy. Without his knowledge, a committee of the First Church addressed a letter to the churches which had sent delegates, pointing out that all agreed that Tucker's life was blameless, that his views on the right of individual interpretation were old New England doctrine, and that without this liberty to break the bonds of narrow orthodoxy, the Christian church would never have grown from its original little flock.[13] He got a good press, one of the newspaper writers supporting him by pointing out that creeds were generally disliked by Dissenters in both Englands, and that the Calvinism of the Protestant churches of Europe was not of the dogmatic kind to which he objected.[14]

The council met again on April 22, 1767, and notified him that it would hear the charges brought by his critics. He rebuked it for assuming "a lordly power" over him and the First Church, and informed it that if he did participate in a hearing it would be before a more prudent and impartial body. The council censured him but, recognizing that it was not a "mutual" body in a legal sense, said that it would join him in "appealing to a more equal tribunal." Its Result justified its attention to the Newbury situation by pointing out that Tucker's settlement by so narrow a majority had been without precedent, that the number of regular communicants was only about one-third of what it had been in Toppan's day, and that the departure from the traditional confessions and catechisms, and from the doctrinal articles of the Church of England, was deeply disturbing. It advised the dissenters to make another attempt to agree with Tucker, and, when that failed, to proceed legally to obtain a separation.[15]

At this point Parson Tucker lost his temper, and it was several years before he found it again. He made a violent reply to what

---

[12] *Ibid.*, pp. 8–9.

[13] This letter is printed at the end of the first edition of the *Brief Account*; it is not in the second edition.

[14] *Boston Evening-Post*, supplement, Apr. 27, 1767, 1/1.

[15] *Brief Account* (2nd ed.), pp. 23–26.

he called the sly insinuations of the council, jeered at the "Grammatical defects" of the Result, and insinuated that its authors were not "gentlemen of Letters." [16] His chief fire was reserved for Aaron Hutchinson (Yale 1747), who in preaching to the council had used the usual language of Calvinism. Angrily Tucker demanded biblical proof of the doctrine that infants were guilty and polluted, and then made a bitter personal attack on the preacher, accusing him of an "arrogant kind of vanity," "Bigotry and self-sufficiency," and "malice and ill nature . . . set off with such infamous decorations of baseness and scurrility" as one would expect only in "a romish Priest" who "preaches up the doctrines of purgatory, transubstantiation, indulgencies, and other absurdities and fooleries of the popish religion." [17]

Hutchinson replied with an able defence of moderate Calvinism, jokingly calling Tucker a Don Quixote who tilted with his own straw men. While remarking on the "many low and sordid scurrilities" in the attack, he kept his own temper.[18] Tucker's answer was typical of his technique of personal controversy.[19] He attacked his opponent's literary style and made snide remarks about his learning, and then went on to imply that he was a liar, a plagiarist, and a clown. When in possession of his temper he wrote with clarity, economy, and logic, but in these screeds he split each hair a dozen ways. Hutchinson signed off with this public notice:

Advertisement.

Whereas the Rev. John Tucker of Newbury has printed some Considerations of my Reply to his Remarks on my Sermon; which appear to me to contain no answer to the arguments in said Reply; but serve only to shew that the subject, on his side, is exhausted, and that he is willing to have an end put to the controversy: and as there is nothing remarkable in his said Considerations, except a few Queries, such as may be, and often have been, started by the wit of men, respecting some of the most important doctrines of the Gospel; and some misrepresentations of my meaning, which are mostly so futile, that a very little degree of good sense, may easily correct them . . . I may be very reasonably excused from taking any further notice of the said Considerations.[20]

[16] Ibid., pp. 27–42.
[17] John Tucker, Remarks on a Sermon (Boston, [1767]).
[18] Aaron Hutchinson, A Reply to the Remarks of the Rev. Mr. John Tucker (Boston, 1768).
[19] John Tucker, The Rev. Aaron Hutchinson's Reply . . . Considered (Boston, 1768).
[20] Affixed to his Iniquity Purged (Boston, 1769).

Tucker let this controversy drop because he was, as one of his liberal friends put it, then under attack by "a host of bitter fanaticks." [21] One of these "fanaticks" was his neighbor, the Reverend James Chandler (A.B. 1728) of Georgetown. In a note in one of his published sermons Chandler had said that there were ministers who were artfully undermining the old Protestant faith by arguing that creeds and confessions were contrary to the scripture. Tucker took this as a personal attack and replied by demanding whether the "pious and learned divines" who had drawn up the creeds had "any peculiar privilege and authority to see, and determine, and settle points of faith for the rest of the world?" "As every man must believe for himself, must not every man see and understand for himself; and consequently must not every man examine and try doctrines, in order to judge of their truth?" [22] He charged that the Calvinist ministers were demanding that candidates accept the creeds. The Georgetown minister replied in a moderate tone, protesting that he had been misrepresented, that neither he nor any other minister would demand that a candidate accept any creed or confession, but that they ought to find out what the candidates' beliefs were in order to eliminate the hopelessly unfit among whom were, he implied, those who publicly stated that Calvinism "stank." [23] Tucker replied with an interesting description of the perennial problem of the young candidates who compromise their "integrity and uprightness" in order to obtain the necessary approbation of their elders. The practice of examination corrupted the candidates, and more:

It tends to hinder the increase of knowlege, and any farther reformation in the christian church. — It puts, as it were, shackles upon the supposed liberty of free enquiry: — It is a curb to the spirit of industry and diligence: — It deadens in the mind the love of truth, and aims to tie us down and fix us where we now are; and to forbid, as a christian church, our becoming any more knowing, wise, or virtuous. [24]

Chandler's reply was a moderate and reasoned statement of the necessity of doctrinal soundness in candidates, which did not, he

---

[21] William Bentley, *Diary* (Salem, 1905–14), I, 357.

[22] John Tucker, *A Letter to the Rev. Mr. James Chandler* (Boston, 1767), p. 6 *et passim*.

[23] James Chandler, *An Answer to Mr. Tucker's Letter* (Boston, 1767).

[24] John Tucker, *A Reply to the Rev. Mr. Chandler's Answer* (Boston, 1768), p. 54, *et passim*.

admitted, imply subscription to any particular creed.[25] Tucker
went further and insisted that "When . . . a candidate, having
been favoured with a christian and liberal education, freely pro-
fesses his belief of the christian religion, as contained in the new
testament, and his life appears correspondent to such a profession,
we ought to believe him a true christian; — to doubt of this, and
require farther evidence of it, is both uncharitable and unreasonable." [26] At this point Chandler dropped the controversy, for he
was a gentle man who suffered under the barbed personalities with
which the Newbury minister accompanied his arguments.

Tucker carried on with Jonathan Parsons the same controversy
about the horrors of Calvinism and the use of creeds and confessions of faith to establish a spiritual tyranny, as he put it; but
their relationship was further complicated by the fact that for
twenty years many members of the congregation of the Presbyterian church of Newburyport had been taxed for the support of the
First Church of Newbury because they lived within its parish.
Parsons finally launched a vigorous attack on the Massachusetts
ecclesiastical system, calling it a spiritual and civil tyranny because
it granted financial self-determination to Anglicans and Anabap-
tists but not, in this particular instance, to Presbyterian Calvinists.
Tucker replied, using typically bitter and unfair personal asides,
arguing that the Massachusetts system was not a spiritual tyranny
because it did not compel anyone to subscribe to the beliefs of the
majority of the inhabitants of a town, or to attend the services
which the town supported; and that it was not a civil tyranny,
because any dissenter was free to move to another parish where
his taxes would go for the support of a minister of his choice.[27]
The theological bitterness between the two churches was such that
the Newbury minister would not consider the obvious solution of
a poll parish which was used elsewhere in Massachusetts.

Parsons made no answer, but a member of his congregation
replied with a bitterness equal to that of Tucker:

Had I lived by the money extorted from those who never have attended my ministry, because of their different opinions in the essential
doctrines of christianity . . . Had I known my honest neighbours im-

---

[25] James Chandler, *A Serious Address* (Boston, 1768).

[26] John Tucker, *Remarks on the Revd. Mr. James Chandler's Serious Address* (Boston, 1768), p. 13, *et passim*.

[27] John Tucker, *Remarks on a Discourse* (Boston, 1774).

prisoned for taxes laid upon them for my support, when they solemnly declared that they esteemed my doctrines as poison to mens souls, as rats-bane to the human body . . . I should think the sensible part of mankind would call me an impudent villain to write in that unmannerly, false and abusive manner against any man, as you have wrote against Mr. Parsons.[28]

So bitter were these controversies, that to the participants the Revolution seemed a mere side show.

When not in debate, Tucker was one of the most gentle and charming of men. According to his friends, "No man was less opinionated, or discovered more pleasantness, good humour and good manners in social intercourse." [29] "He possessed a most amiable disposition — his temper was remarkably placid and even"; [30] he was "one of the sweetest men." [31] "What a happy temper — how calm and unruffled, in the storms which sometimes beat upon him — how meek and gentle towards all men — how patient under all the troubles of life." [32] Without doubt his failings as a controversialist were overlooked because he spoke for the great majority of the Harvard-dominated community. In 1768, when his battle was at its height, he was given the great honor of delivering the annual address to the convention of the Congregational clergy of Massachusetts. Since he had, really, only one sermon, with which all were familiar, this appointment was the invitation to deliver the keynote address for the Harvard-trained Congregationalists. As everyone expected, he asserted the right of each individual to interpret the Scriptures for himself, and to be accepted as a member of the Christian communion regardless of his attitude toward particular laws, creeds, doctrines, or rites.[33]

Three years later he received the other great honor, that of appointment to deliver the Election Sermon. In it he gave a clear and brilliant statement of the compact theory of government. He joyfully hailed the appointment of Thomas Hutchinson (A.B. 1727) as governor, and urged him to preserve the just prerogatives of the Crown as well as the liberties of the people. While firmly upholding popular rights, he turned toward the Whig

[28] Shangar, pp. 3, 4.
[29] John Eliot, *Biographical Dictionary* (Salem and Boston, 1809), p. 460.
[30] *Columbian Centinel*, Apr. 4, 1792, p. 27, 2/2.
[31] Bentley, IV, 205.
[32] Samuel Webster, *The Blessedness* (Boston, 1793), pp. 18–19.
[33] John Tucker, *Ministers Considered* (Boston, 1768).

politicians who sat before him in the pews of the Cambridge meetinghouse and lashed at "political zealots and pretended patriots":

There are found among the people, persons of a querulous and factious disposition. — Ever restless and uneasy, and prepared to raise and promote popular tumults. From the meer love of wrangling, or from ambitious views, — to rise from obscurity, to public notice, and to an important figure, they find fault with Rulers, and point out defects in the administration. — Small mistakes are magnified. — Evil designs are suggested, which, perhaps never existed, but in their own heads. They cry up liberty, and make a mighty stir to save the sinking state, when in no danger, but from themselves, and others of a like cast.[34]

Hutchinson must have inwardly cried "Amen," and Sam Adams must have writhed in his seat. It is said that the sermon "received a high compliment from the Earl of Chatham." [35] Although Tucker clearly saw the part which the Whig politicians played in bringing on the war, his belief in the right of the individual to judge for himself in religious matters carried over into civil affairs to bring him to acquiesce in the fact of the Revolution. In these larger matters he was entirely free from the bitterness which marked his personal controversies. At the height of the war he was invited to preach the Dudleian Lecture on the validity of presbyterian ordination. At a time when some American ministers of the Church of England, like his classmate Winslow, regarded their churches as branches of the British government, one might have expected this lecture to be a tirade against the Anglicans. Tucker, however, hailed the Church of England as a true and great branch of the Church of Christ, and objected only to her claim to an authority based on an ecclesiastical institution which had not developed in New Testament times.[36]

The last years of Tucker's ministry were peaceful and satisfying. So far as ecclesiastical controversy was concerned, he was regarded as the grand old man of the Harvard-trained Congregationalists. Discerning young men like John Quincy Adams (A.B. 1787) greatly enjoyed his preaching. He served as trustee and secretary of Dummer Academy, and in 1787 was given an S. T. D. by Harvard. But he had never been a strong man, he suffered

---

[34] John Tucker, *A Sermon Preached in Cambridge* (Boston, 1771), p. 34.
[35] Sprague, I, 453.
[36] John Tucker, *The Validity of Presbyterian Ordination* (Boston, 1778).

from sickness for several years, and was entirely incapacitated for some months before his death on March 22, 1792. His body was followed to the grave "by the Trustees, Preceptor, and Pupils of Dummer Academy, and many of the neighboring Clergy and Gentry." [37] His widow lived until August 31, 1814. He was survived by ten of his children: (1) Sarah, b. Jan. 5, 1749/50; m. William Stickney of Newburyport, Feb. 19, 1777. (2) Alice, b. July 30, 1751; d. unm. Nov. 1808. (3) John, b. Aug. 11, 1753; A.B. 1774. (4) Elizabeth, b. Oct. 3, 1755; d. unm. Mar. 11, 1844. (5) Mary, b. Sept. 11, 1757; m. Benjamin Rolfe (A.B. 1777), Oct. 14, 1795. (6) Barnard, b. Apr. 2, 1760; A.B. 1779. (7) Charlotte, b. Jan. 2, 1762; d. unm. Nov. 17, 1816. (8) Clarissa, b. Feb. 6, 1764; m. Nathaniel Thurston, May 27, 1802. (9) Catharine, b. Mar. 24, 1766; d. unm. Sept. 8, 1849. (10) Benjamin, b. Nov. 13, 1768; d. Jan. 31, 1832. The Parson's portrait is now in the possession of Gordon Hutchins (A.B. 1902), but his interleaved almanacs, formerly in the possession of the Reverend Charles L. Hutchins of Concord, are not now to be found.

[37] *Salem Gazette*, Apr. 10, 1792.

SAMUEL AUCHMUTY

SAMUEL AUCHMUTY, Rector of Trinity Church in New York City, was a son of Judge Robert Auchmuty of the Admiralty Court. The Judge had come from Scotland, and had become a vestryman of King's Chapel in Boston, and, with the elder Samuel Adams, a director of the Land Bank. When Sam was born, on January 16, 1721/2, the family was living in a handsome house in Essex Street, but in 1738 they sold that property and moved to a beautiful estate on the present Washington and Dudley streets in Roxbury. When he entered college he was placed fourth in his class on the strength of his father's position on the Bench. He roomed in Stoughton 3 with Chandler '43 and in Stoughton 12 with Brown '43, but in his Junior year left college without taking the trouble to resign. In June, 1741, the Corporation noticed that he should have long before been expelled for unexcused absence, and voted that "his Relation to the College" was "dissolved."

Auchmuty was in Boston at this time, for his signature appears on some Land Bank mortgages, but in 1741 he accompanied his father on a Province mission to England. He came back bringing an appointment as Register of Admiralty, replacing Andrew Belcher (A.B. 1724); but he did not hold the office long, for his father was detected in plotting against Governor Shirley, and the merchants protested that the son's appointment would give the father too much power. So the Admiralty Board cancelled his commission.

With a civil career thus prematurely cut off, Auchmuty gave thought to the urgings of his uncle, James Auchmuty, Dean of Armagh, that he should enter holy orders,[1] and to that end began to read under the direction of the Reverend Alexander Malcolm of St. Michael's Church in Marblehead. In 1745 the Corporation, at the request of the Judge, gave him his B. A., and the next year, after consulting with the Reverend Benjamin Bradstreet (A.B. 1725) of Gloucester as to his character and scholastic attainments, voted that he should have his M. A. with his classmates. Thus equipped, he sailed for England, where in 1747 he was admitted to holy orders by the Bishop of London.

[1] Mass. Hist. Soc., *Collections*, 6th Ser., X, 472.

On July 1, 1747, Auchmuty petitioned the S. P. G. for the vacant post of catechist in the Negro school in New York and parochial assistant to Dr. Henry Barclay of Trinity Church. With the hearty recommendation of Governor George Clinton, he was given the place. A New England friend, writing from London on October 28, reported that he was about to sail and described him as a "good pretty Behav'd Gentleman." [2] In January, 1748, he arrived in New York and took over his duties with energy and enthusiasm. In his first official report he described his work:

This is ... to inform you that I now constantly every Friday read a Lecture, after which I Catechise the Children; the Slaves not being able to attend on any Day but Sunday. It's with the greatest pleasure, that I can now acquaint you, that several of my black Catechumens make no small proficiency in the Christian Religion, and that the Number of them increases. I have baptized since my arrival here Five full grown Blacks, and at least Thirty Infants, and have now several Adults preparing, themselves for Baptism. [3]

He succeeded in interesting the masters in the Christianizing of their slaves, and soon had a regular class of sixty or seventy whites learning the catechism with the Negroes. As a result of the success of the school, the Bray Associates took it over in 1760, and provided him with the means of teaching sewing, knitting, and reading and writing.

To no small degree the public acceptance of the Negro school was due to the social position which Auchmuty established for himself by his marriage on December 13, 1749, to Mary, daughter of Richard Nicholls, widow of Captain Thomas Tucker, and sister of Mrs. Alexander Colden. This one stroke put him in the position of being a familiar friend of the successive governors, even of John Murray, Earl of Dunmore. This was very dear to the Parson, who had a sense of social position utterly unknown among the Congregational and Presbyterian clergy. Between himself and his parishioners, particularly those of St. George's Chapel, which was erected on Beekman Street in 1752 and placed under his charge, he saw a social gulf which could not exist in the churches of other denominations. Writing to his friend Samuel Johnson (Yale 1714) he said:

[2] Jeffries Mss. (Mass. Hist. Soc.), XI, 122.
[3] *Historical Magazine of the Protestant Episcopal Church*, VIII, 332.

SAMUEL AUCHMUTY

You can't think how very good and pious our boys and low life people are grown. We can match any of your new light in New England. I make myself very easy and tell them my mind very freely; and as they are of no consequence I neither care for, nor concern myself much with them. They towards me behave with great complaisance and at a proper distance I am determined to keep them. . . . N. B. The people I mean belong to the Chapel only.[4]

He was particularly harsh toward the "Knaves and Fools" in his congregation who participated in George Whitefield's popular revival. The Dutch he called "Loggerheads," and the Presbyterians he so abhorred that it made him sick to his stomach even to write about them.[5]

During the last illness of gentle old Dr. Barclay, Auchmuty brooded over the changes which he would effect when he had the authority:

No vestry called as yet, he not being able to attend even in his own house. . . . For want of one I am a most complete slave, which I would not submit to, was I not in hopes that one day or other I shall have it in my power to let some folks know that the rector of this parish is no such insignificant body as church wardens have hitherto thought him. . . . I am pretty sure there will be no danger of my not succeeding should it please God to deprive us of the good Doctor.[6]

He was right. Four days later, on August 20, 1764, the Doctor died, and the Vestry immediately and unanimously presented Auchmuty to Lieutenant Governor Cadwallader Colden for induction. He went down to Fort George, took the oath, subscribed to the declaration required by law, and was formally inducted on September 1 as rector.

In more ways than one the promotion was a relief to the Auchmutys. The Parson's inheritance from his father had been a mere £50, apparently far short of enough to settle his personal indebtedness in Boston, where his creditors rejoiced to hear that he had "succeeded to the best living of any upon the continent." [7] The next year he went up to Boston, settled his accounts, went fishing in "Monotomy Pond" with John Rowe, and "caught above sixteen dozen of pond and sea perch." [8]

---

   [4] Samuel Johnson, *Works* (New York, 1929), I, 348.

   [5] Samuel Auchmuty to Samuel Johnson (New-York Historical Society), Oct. 26, 1764, *et passim*.

   [6] Johnson, *Works*, I, 339, 340.

   [7] Mass. Archives, XXVI, 116.

   [8] Mass. Hist. Soc., *Proceedings*, 2nd Ser., X, 49.

Back in New York, the Parson moved into the old house of the charity school, which the parish had remodeled into "a large and elegant mansion" in which he entertained the aristocracy and even the successive governors. Now that he no longer had to pay rent, his salary was reduced to a still comfortable £250, New York currency.

Auchmuty had previously used his influence to obtain a salary increase and like favors for Samuel Johnson, and now the Doctor returned the favor by urging Thomas Secker, Archbishop of Canterbury, to obtain a doctorate for the Rector who was, he said "of nigh twenty year's standing of our Cambridge." Apparently Auchmuty did not suspect the source, for after he had received the degree he wrote to Dr. Johnson that he and Thomas Bradbury Chandler (Yale 1745) to please President Myles Cooper had taken their "Drs Degrees ad eundem from Oxford." [9] The next year, 1767, King's College gave the Rector its doctorate.

In 1766 Dr. Auchmuty presided over what was perhaps the most spectacular affair New York had ever seen, the consecration of St. Paul's Chapel. Newspapers throughout the colonies described it as "one of the most elegant edifices on the Continent," and carried the details of the service. [10] The use of "Violins, Bass Viols, French Horns, Flutes, Hautboys, etc." in the chapel attracted much attention. [11]

The Rector had now achieved for his office the distinction which he felt to have been so sadly lacking in the days of his predecessor. His congregation called him "the Bishop," and newspapers from Portsmouth to Philadelphia carried at length the story of his having intervened to save from immediate execution various criminals who were going to the gallows without proper preparation for their eternal state. They were executed after he had been allowed a week to give them "such spiritual Assistance as their unhappy Case required." In one instance a man condemned to death for stealing books from St. Paul's was allowed, at the request of the Rector, to plead benefit of clergy. [12]

It is not unlikely that the Doctor thought the poor thief more

[9] Samuel Auchmuty to Samuel Johnson (New-York Historical Society), May 25, 1767.
[10] New-York Gazette and Post Boy, Nov. 6, 1766, 3/2.
[11] Ezra Stiles, Itineraries (New Haven, 1916), p. 225.
[12] New York Journal, Aug. 6, 1767; New York Mercury, Aug. 17, 1767.

entitled to benefit of clergy than some of the "wretches" (as he called them) in holy orders whom he sent south, hoping that their characters would never be discovered by the people among whom they were to settle.[13]

The Rector's contacts with King's College were more pleasant. He served the college for five years as governor, resigning when he assumed responsibility for Trinity. He continued to serve the administrative board as clerk, recruiting teachers at Harvard when he found that he could not, to his great regret, obtain ordained Anglican clergymen from Oxford.[14] When President Cooper was out of town, the Rector was "daily plagued with the parents of the lads" who wanted to enter. Three of his own boys graduated at King's, and graduates intending to enter the Anglican ministry read under his direction. Using his personal friendship for Sir William Johnson as a lever, he tried to obtain a town in the New Hampshire Grants for the college.

The Doctor served also as an intermediary between the S. P. G. and Johnson in the efforts to place Anglican missionaries among the Indians. He thought that the religious principles of the savages were being "debauched by the stupid Bigots" who were being turned loose among them by the New England schools. As for "Wheelocks Cubs . . . surely such Wretches ought not to be suffered to go among the Indians." [15] Bitterly he complained to the S. P. G. of the success of his rivals:

I greatly fear Whelock and his Associates will engross all the Indian Country and lay the seeds of Schism so deep that it will hereafter be impossible to eradicate them. Is it not amazing an Indian Canter of a discarded Independent preacher should gull the people of great Britain out of £11,000.[16]

The Yankees would, he cried, corrupt even the Dutch of the upper settlements:

The Dissenters are worming themselves in among them, and unless something to the purpose is speedily done, they will in those extensive

---

[13] Samuel Auchmuty to Samuel Johnson (New-York Historical Society), Aug. 27, 1766, *et passim*.

[14] *Ibid.*, Oct. 26, 1764.

[15] *Historical Magazine of the Protestant Episcopal Church*, VIII, 5–36, *passim*.

[16] Samuel Auchmuty to Philip Burton (Dartmouth College Archives), July 9, 1768.

Countries have the Majority, which will render them dangerous to Church and State.[17]

A single Dissenter had, by accident, got on to the governor's council, and this error, Auchmuty warned Burton of the S. P. G., should not be permitted to happen again:

If you have a convenient opportunity I wish you would inform his Grace of Canterbury that there is one of his Majesty's Council for this province in a dangerous way, and its tho't near his end. We therefore intreat his Grace and the Friends of the Church that they will have a watchful Eye least we have another presbyterian run upon us. The whole Council (except one) belong to the Church. That one came in, in a Clandestine Manner; another may do the same if there is no care taken to prevent us. It is vile policy to trust avowed Republicans with posts under the British Government.[18]

Constantly he hammered his correspondents both in England and at home with his thesis:

His Majesty has very few subjects whose Loyalty he can depend upon on the Continent, *besides the members of the Church of England.* The rest are right down republicans, and If so, ah! very hard it is, that the most deserving shall be debarred of their just Rights, in order to please a rebellious set of people. Alas! Alas! [19]

When the New York Presbyterians attempted to obtain a charter of incorporation for their church, he fought the movement tooth and nail, both in the colony and in England, where, he complained, the bishops and the S. P. G. were entirely too tolerant of Dissenters. He rejoiced when the proposed charter was "knocked on the head" by the Board of Trade.[20]

The Dissenters of the city of New York were taxed for his support, and it was his fondest dream to see the system extended through the colony by an Act of Parliament "to erect every County at least, thro' -out the Government, into a parish, and make the Inhabitants pay Taxes, toward the support of a minister of the Established Church." [21] His foresight in keeping Presbyterians off

[17] Samuel Auchmuty to Samuel Johnson (New-York Historical Society), Mar. 21, 1769.

[18] Samuel Auchmuty to Philip Burton (Dartmouth College Archives), July 9, 1768.

[19] Samuel Auchmuty to Samuel Johnson (New-York Historical Society), Oct. 7, 1765.

[20] *Ibid.*, June 18, 1768.

[21] *Ibid.*, Oct. 26, 1764.

the Council was justified when that body refused assent to a Bill which had passed the Assembly to relieve Dissenters from taxation to support the Church of England. To muster his forces he led the organization of the clergy of the Church of England in Connecticut, New York, and New Jersey into a convention which, meeting for the first time at his house in New York on May 21, 1766, formally thanked him for his part in defeating the efforts of the Presbyterians.

Another of the Doctor's projects which met with success was the formation of a corporation for the relief of the widows and orphans of the clergy of the Church of England. An extract of the sermon which he preached on the occasion of the first meeting will give an example of his pulpit style:

And now methinks, when I lift up my Eyes to the glorious Expanse above which loudly proclaims the Greatness, the Majesty, and the Power of that august Being, who reared the mighty Fabrick, I behold the whole Hierarchy of Heaven looking down upon us with smiling Joy, to see us thus piously employed in Acts of Love, Pity and Compassion.[22]

His sermons after military defeats sounded like a recruiting sergeant, and those after victories made it plain that Divine Providence and the King's troops went hand in hand. He did not mention the contributions and the problems of the colonists.

Dr. Auchmuty's greatest dream was the appointment of a bishop for America, and to this end he gave his friends in England no peace. The opposition of the southern clergy to his plan he laid to the fact that it was "too notorious that no Bishop, unless a very abandoned one, could put up with the Lives they in general lead." [23] The Dissenters of New York he found to be more amenable:

I have the pleasure to find that upon talking with the better sort of them, they begin to be convinced of the necessity and Utility of having a Bishop. Nay the most Moderate of the Dissenters say, that if he is confined only to Ecclesiastical Matters relating to the Church of England, they are Contented. Pretty Dogs to be contented. Their Impudence is beyond all bounds.[24]

[22] Samuel Auchmuty, *A Sermon Preached before the Corporation* (New York, 1771), pp. 30–31.
[23] Samuel Auchmuty to Samuel Johnson (New-York Historical Society), Oct. 7, 1765.
[24] *Ibid.*, Apr. 30, 1766.

He was encouraged to obtain from Sir William Johnson a tract of land to endow an episcopate, and was bitterly disillusioned to have the government appoint a Roman Catholic Bishop of Quebec while ignoring his pleas for an Anglican bishop of the English colonies. He told a Province Agent who was about to leave for England that on this subject he could "trust the Archbishop; but no one Else." [25] There is nothing in his correspondence to support the family tradition that he was making arrangements to go to England to be consecrated Bishop of New York when the outbreak of political strife forbade. Bitterly he told the S. P. G. that there never would have been an American rebellion had his advice in regard to a bishop been taken twenty years before.[26]

To the Rector, the supremacy of King and Parliament was a matter of such simple religious faith that he appears never to have thought about the political issues. When speaking before the clerical convention of 1766 he "congratulated them upon the Repeal of the Stamp-Act," [27] which was apparently why he was roundly snubbed by the governor, Sir Henry Moore, the next day.[28] However, he recognized the non-importation agreement for the menace which it was:

Doubtless you have long heard of the New solemn League and Covenant, that the Old leaven here have entered into, and Methinks (unlawful I will be bold to say) they have pursued to draw in all the Dissenters . . . . This unlawful and cruel Combination has been kept a profound secret till last week. . . . I have not neglected acquainting our Superiors at Home of this unlawful Combination. I got some Copies struck off last Saturday Night, and put them in the Mail.[29]

When his Macedonian cries went unheeded, he bitterly denounced the selfishness of the Lords spiritual and temporal, and their indifference "to the distresses and injustice that the members of the best church in the world labor under, in America." [30]

All of Dr. Auchmuty's family ties bound him to the Tory party. His brother Robert was chosen by the Crown to defend the soldiers involved in the Boston Massacre. In 1774 his daughter

[25] *Ibid.*, Jan. 3, 1767.
[26] Samuel Auchmuty to ———, S. P. G. Mss. (Library of Congress transcripts), Sept. 12, 1774.
[27] *New-York Gazette* (Weyman), May 26, 1766, 3/1.
[28] *Historical Magazine of the Protestant Episcopal Church*, X, 140.
[29] Samuel Auchmuty to Samuel Johnson (New-York Historical Society), 1769.
[30] *Documentary History of New York* (New York, 1850–51), IV, 266.

Isabella married General William Burton, nephew of Bartholomew Burton, late Governor of the Bank of England.[31] And so it went. His friends were of the same sort. Governor Tryon presented him with "a complete Set of rich and elegant Hangings of crimson Damask for the Pulpit, Reading Desk, and Communion Table" of St. George's Chapel;[32] to the Dissenters this elegance was no less distasteful than the source. This all became more obvious when a letter from the Reverend Samuel Peters (Yale 1757) to the Rector was captured by the Whigs and during the month of October, 1774, published in the newspapers of the colonies. In it Peters called upon his friends to close ranks against the spiritual iniquity, bigotry, and villainy of the Dissenting clergy and magistrates, and told how the Sons of Liberty had near killed one Episcopalian, tarred and feathered two, abused others, and torn up his clerical gown "Crying out down with the Church, the Rags of Popery &c. Their Rebellion is obvious, and Treason is common, and Robbery is the daily Devotion."[33] The Connecticut Church of England clergy published a letter repudiating Peters and, by implication, Auchmuty.[34] Bitterly the Rector protested his friend's rashness:

I am now to inform you that your last Letter to me . . . with another to your Mother have been intercepted by the Connecticut Crew and published. . . . I am amazed that you would trust such Letters . . . to any person that was to go through the Land of oppression. You must imagine it is very disagreeable to me to have my Name bandied about by a parcel of Rascals. . . . Our Enemies before were many, but your Letters will increase the Number.[35]

Peters replied that he would rather have died than have his foolish rashness bring this trouble upon his friends, and he then accused Governor Trumbull (A.B. 1727) of inciting the mob and the "villainy of the saints" against the Episcopalians. This letter, like the first, fell into the hands of the Whigs and was published.[36]

Then, on April 19, 1775, the Doctor sat down and wrote this

---

[31] *Boston News-Letter*, Mar. 17, 1774, 2/3.

[32] *Ibid.*, Jan. 21, 1773, 1/2.

[33] *Boston Evening Post*, Oct. 24, 1774, 2/2–3. For a verse parody of the letter see the *Essex Journal*, Mar. 22, 1775, 4/1.

[34] *Boston Evening Post*, Nov. 14, 1774, 1/2.

[35] Samuel Auchmuty to Samuel Peters (New-York Historical Society), Oct. 27, 1774.

[36] *Connecticut Courant*, May 31, 1775, 4/2–3.

letter to Captain John Montresor, husband of his stepdaughter, Frances Tucker:

I must own I was born among the saints and rebels, but it was my misfortune. Where are your Congresses now? What say Hancock, Adams, and all their rebellious followers? Are they still bold? I trow not. We have lately been plagued with a rascally Whig Mob here, but they have effected nothing, only Sears the King, was rescued at the jail door. . . . Our magistrates have not the spirit of a louse, however, I prognosticate it will not be long before he is handled by authority.[37]

After the Nineteenth of April mail was not being delivered to Captain Montresor, who was shut up with the Regulars in Boston. The Whigs published the letter with commentaries expressing their opinion of the Rector:

From such servile wretches as the author of the above letter, do the British Administration receive informations relative to the state of America. . . . What must be the heart of the man who can jest, as doth this execrable Clergyman, with the miseries of his Country, and can exult at the thought of its being drenched in blood: the prospect of the arrival of Troops to answer this purpose affords him matter of triumph. This is a Tory Clergyman; to such men as these, our countrymen, (if we prevent them not by our own valorous exertions,) must we pay tithes of all that we possess; to such men as these must we become hewers of wood and drawers of water! . . . It hath been the misfortune of this Province to produce many such vermin as the author of this letter, many who have acted the part of parricides to their Country, and who would sacrifice that, together with their consciences, to their ambition and avarice.[38]

At this late date the Doctor announced that he had ceased writing letters. The Whigs noticed with amusement that after the publication of his last indiscretion he "had the satisfaction of preaching to almost naked walls." [39] The Connecticut wits let him off with gentle ribbing. Fortunately for him, the New York Whigs had no counterpart of the Boston committee for lying and vilifying Tories, but a group, with which on behalf of the church he had had a dispute over property, accused him of commissioning "a number of riotors" who broke down division fences "kicked a poor woman in the eye, and wounded her husband, who attempted to defend

[37] *Massachusetts Spy*, May 24, 1775, 2/2.
[38] *New England Chronicle*, May 25, 1775, 2/1.
[39] N. Y. Hist. Soc., *Collections*, 1886, p. 46.

her." [40] The Whig mobs ignored him although they made President Cooper run for his life. Probably they dismissed the Rector, as did an agent of the Crown, as "a warm, impetuous old Man, of no great Abilities, yet of some Weight from his Place and Standing." [41] In part he owed his immunity to the fact that the people of New York had not yet made up their minds which way to jump. In May, 1775, finding himself under pressure from the Whigs to observe a Fast Day, he appealed to Lieutenant Governor Colden to get him off the hook by proclaiming a legitimate one. He served on occasion as chaplain of the New York Provincial Congress, which was by no means as revolutionary a body as its Massachusetts counterpart. On one day on which he opened the session with prayer, the Congress heartily welcomed both General Washington and Governor Tryon.

As the year passed the cleavage became sharper. One day the Right Honourable William Earl of Stirling, whose chaplain Auchmuty had been in happier days, sent the Rector word by one of his sons that if he read the prayer for the King the following Sunday, he would be pulled out of the pulpit by a band of soldiers. Instead of delivering the message to his father, the lad gathered some of his classmates who went to church with concealed weapons determined to protect the Parson. When he commenced reading the prayers, Lord Stirling marched into the church with a body of soldiers and a band playing Yankee Doodle. The Doctor went through the prayers without faltering. The soldiers marched up one aisle and down another, and went out without offering violence. [42]

After this experience Dr. Auchmuty naturally felt "much indisposed." His wife and younger children had taken refuge with the Burtons at their manor, Buccleuch, in Brunswick, New Jersey, and on April 4, 1776, the Committee of Safety gave him permission to go through the lines to visit them. That summer his brother Robert was reported captured by the rebels, and his sons Robert and Samuel joined the British army. When the Regulars relinquished Brunswick to the Continentals, the Rector applied to the rebels

[40] Morgan Dix, *A History of the Parish of Trinity Church in the City of New York* (New York, 1898–1950), I, 367.

[41] B. F. Stevens' Facsimiles, 2045, p. 2.

[42] Wilkins Updike, *A History of the Episcopal Church in Narragansett* (Boston, 1907), I, 168.

for permission to return to New York. Refused, he took to the woods and after what must have been severe hardships for an "old" and sedentary man, he made his way through the lines and into the city early in November. He found his mansion and church burned to the ground. Weeping floods of bitter tears he dug through the ruins for the remains of the silver plate. St. Paul's Chapel still stood, and to the remnants of his flock gathered there he said:

When I reflect upon the dreadful scene of misery and Destruction this city and many of its inhabitants have undergone — when I reflect upon the banishment and cruel usage that many of his majesty's loyal subjects have suffered for some time past, my heart is filled with grief, the friendly Tear comes to my assistance and my steadfast trust and confidence in my God is my only comfort. Both my duty and my inclination prompted me to return to you, tho' at the risque of my life, and to participate in your joy for your deliverance from the cruel hands of your enimes. . . . We have already seen the salvation of the Lord in our deliverance from a worse than Egyptian bondage. But in order to insure further success . . . we must live and act agreeably to our profession . . . for are we not incompassed all around by our enemies . . . for neither the Tears of the tender sex, nor the lamentations of weeping children can soften the savage breasts of our enemies into tenderness and compassion.[43]

These last words had a poignant ring, for he believed that his wife and daughters were being held prisoners by the rebels. They were rescued, however, by an advance of the King's troops into Jersey; perhaps they had not dared use the pass which Washington had given to Mrs. Auchmuty.[44]

When the Rector and his family were reunited "His Spirits seemed to revive, his Health to mend, and he and his Friends indulged themselves in the pleasing Expectation of Peace and Happiness." [45] It was not for long. On February 24, 1777, he officiated at a wedding, but the next day he took to his bed with a cold and fever which carried him off on March 4. He was buried in St. Paul's chancel. There were long newspaper obituaries and memorial sermons which add nothing to our knowledge of the Rector, but in view of the character of the quotations which we have made from his letters, it is only fair to add this quotation from Charles Inglis, his assistant:

[43] Dix, I, 398, 399.
[44] Annette Townsend, *The Auchmuty Family* (New York, 1932), p. 63.
[45] *New York Gazette and Weekly Mercury*, Mar. 10, 1777, 3/2.

My Intimacy and Connection with him for nearly twelve Years, enabled me to know him well; and I can truly say, I scarcely ever knew a Man possessed of a more humane, compassionate or benevolent Heart. Often have I seen him melt into Tears at the Sight of Distress in others; and the distressed never sought his Aid in Vain.[46]

His loyalty was so strongly stressed in his obituaries that his critics were amused to see his house immediately taken over by female camp followers of the Regulars for a place of business. His widow went to England where she lived with the Montresors until her death on July 16, 1797. Her portrait, a miniature, was in 1927 owned by Mrs. Richard Tylden Auchmuty of New York City. The Rector's portrait is here reproduced from the original at Trinity Church. Their three sons graduated from King's and served in the British army. Richard Harrison died in an American prison after Yorktown, Robert Nichols settled in Newport, and Samuel remained in the army, became a general, and was knighted for the capture of Java in 1811.[47]

EDWARD BROMFIELD

EDWARD BROMFIELD was born in Boston on January 30, 1723/4, the eldest son of Edward and Abigail (Coney) Bromfield who lived in a handsome mansion which was the first dwelling erected on Beacon Street. The family was famous for its gentle, warm, and pious kindness, and the elder Edward's favorite public office was that of Overseer of the Poor. To these qualities the younger Edward added rare intellectual promise:

From his Childhood He was tho'tful, calm, easy, modest, of tender Affections, dutiful to his Superiours, and kind to all about him. As he grew up, these agreable Qualities ripened in him; and he appeared very ingenious, observant, curious, penetrating; especially in the Works of Nature, in mechanical Contrivances and manual Operations, which increased upon his studying the Mathematical Sciences, and also in searching into the Truths of Divine Revelation and into the Nature of genuine experimental Piety. His Genius first appeared, in the accurate Use of his

46 Dix, I, 406.
47 *Auchmuty Family*, pp. 75 ff.

EDWARD BROMFIELD

Pen; drawing natural Landscapes and Images of Men and other Animals, &c, making himself a Master of the famous Weston's short Hand in such Perfection, as He was able to take down every Word of the Professor's Lectures in the College Hall.[1]

At college he roomed in Stoughton 4 with his classmate Toppan and Downe '45, and remained in residence until May, 1743. When he returned for his M. A. he was prepared to hold the affirmative of "An omni praecepto divino, licèt Ratio ejus non appareat, obediendum sit?"

While an undergraduate, Bromfield joined the Old South and impressed Eleazar Wheelock as being "a really converted person."[2] His reputation in this direction induced John Walley (A.B. 1734) to seek him out and take him, in the fall of 1743, on a tour of Rhode Island and Connecticut. On the way they met "a young man educated at Yale College" and prayed with him. It was a pious journey, with New-Lightish criticism of the ministers whom they met. They hoped to have the last two days alone together so that they might discourse on holy things, but to their disappointment they were joined by Parson Cheney (A.B. 1711) of Brookfield.[3]

Apparently Professor Winthrop's neglect of the Hollis microscope turned Bromfield's attention to a field in which he soon far outdistanced his famous master.[4] According to his minister, Thomas Prince (A.B. 1707):

His clear Knowledge of the Properties of Light, his vast Improvement in making Microscopes, most accurately grinding the finest Glasses; and thereby attaining to such wonderous Views of the inside Frames and Works of Nature, as I am apt to think that some of them at least have never appeared to mortal Eye before. He carried his Art and Instruments to such a Degree; as to make a great Number of surprizing Discoveries of the various Shapes and Clusters contained in a Variety of exceedingly minute Particles . . . He seem'd to be making haste to the Sight of the Minima Naturalia, or the very minutest and original Atoms of material Substances.[5]

In an age avid for science, he soon became famous. He excited his

---

[1] *American Magazine*, III, 548 (Dec., 1746).
[2] *New Hampshire Repository*, I (1846), 15.
[3] John Walley, Diary (Mass. Hist. Soc.), final page.
[4] F. T. Lewis, "Advent of Microscopes in America," *Scientific Monthly*, LVII, 258.
[5] *American Magazine*, III, 549.

visitors by projecting on the wall an image magnified, Prince calculated, 388,800,000 times! Once he charmed a gathering of the clergy by showing them the internal workings of a "common" louse; one wonders where, in such company and in one of the finest houses in Boston, he found the material for the demonstration.

One young lady who went home from one of these demonstrations with a singing mind put her thoughts into verse:

> Bromfield, I join your raised Act of Praise:
> To your own Pitch my grov'ling Soul you raise.
> A Thousand untho't Glories you display
> In every Mote, by your inchanting Ray.
>   Such as your Glass the Resurrection Eye
> May be to Saints!" To you and me tho't I:
> The Transport of the Tho't could scarce contain,
> And scarce my Tongue the Tribute due restrain:
> Silent, in Extacy, my Soul ador'd
> The Wonders of my God, your Art explor'd.
>   I'll ever keep this well-spent Hour in View,
> The pleasing Subject will be ever new,
> And Midnight Dreams shall the gay Scene pursue.
> Late in the Night I'll loose it on my Tongue,
> And early waking raise my Morning Song:
> Then Charge my Eye and Heart no more to bend
> To Sight and Sense; — to Things unseen ascend.[6]

As long as the Bromfield mansion stood on Beacon Hill, awed generations pointed out the hole in the window through which the famous microscope had drawn its light.[7]

The heat gathered by the solar miscoscope convinced Bromfield that it would be a simple matter to heat a house with the sun's rays, but his attention was drawn off by less mundane matters. Tradition credited him with building the first organ constructed in America. He was not the first, but his work in that direction was remarkable:

He was well skill'd in Musick; — He for Exercise and Recreation, with his own Hands . . . made a most accurate Organ, with two Rows of Keys and many Hundred Pipes: his Intention being Twelve Hundred . . . The Workmanship of the Keys and Pipes surprizingly nice and curious, exceeding any Thing of the Kind that ever came Here from

---

[6] *Boston News-Letter*, Aug. 21, 1746, 1/2.
[7] Justin Winsor, *The Memorial History of Boston* (Boston, 1881), II, 521–522.

England. . . . And what is surprizing was, that He had but a few Times look'd into the inside Work of 2 or 3 Organs which came from England.[8]

In the summer of 1745 Bromfield carried a message to General Pepperrell at Louisbourg, but public affairs did not interest him; he was happily working in his study when stricken by the illness which carried him off, after but a few days, on August 18, 1746. In the obituaries and the funeral, the town showed its appreciation of its loss. Parson Prince, who had baptized Edward, went sadly back to the front room which was now emptied of its genius:

I see He has left in his Study, (1) Maps of the Earth in its various Projections drawn with his Pen in a most accurate Manner, finer than I have ever seen the like from Plates of Copper. (2) A Number of curious Dials made with his own Hands: One of which is a Triangular Octodecimal; having about its Center eighteen Triangular Planes, with their Hour Lines and Styles, standing on a Pedestal, though unfinished. (3) A Number of Optical and other Mechanical Instruments of his own inventing and making; the Designs and Uses of which are not yet known.[9]

The picture of Bromfield with his microscope, here reproduced, is sometimes attributed to Greenwood, and is owned by Harvard College.

[8] *American Magazine*, III, 549.
[9] *Ibid.*, 548–549.

WILLIAM RAND

WILLIAM RAND, a physician of Londonderry, New Hampshire, was born in Boston on July 6, 1723, the eighth child of Robert and Elizabeth (Welch) Rand. Robert was a member of the Second Church and a sailmaker by trade. Most of William's college bills were paid by an uncle, Dr. William Rand (1689–1758) of Boston. The lad helped a little by waiting on table, but he offset those earnings by incurring a fine of 60s for taking an unscheduled vacation of three months' duration.[1] When in Cambridge he roomed in Massachusetts 19 with Briant '39 and Tuttle '43. For his M. A. he prepared the negative of "An portenta, magorum Ægyptianorum, vi Deitatis immediatâ peracta fuerint?"

Apparently Rand served an apprenticeship with Uncle William, whose store in Essex Street offered, along with drugs, such commodities as varnish, oil, and turpentine. In 1750 and 1754 he was elected constable but excused from service, no doubt because of his status as a physician. However, for reasons which may be surmised, instead of succeeding to his uncle's lucrative business he moved to Londonderry where the quality of his medical practice was held in low repute.[2] Unable to earn a living in that way, he began in 1762 to keep school in the neighboring town of Chester. Failing here, he returned to Londonderry and turned his hand to other crafts, including counterfeiting, which in 1780 brought him to the Exeter jail. Here he was thrown into a cell with two other experts one of whom was Henry, son of Thomas Tufts (A. B. 1732), who practiced as an "Indian doctor" to cover his real trade of burglary.

Tufts rejoiced to see his associate and exclaimed that "Never, perhaps, did a more illustrious trio meet together within the same walls." In his *Narrative* he tells at great length how they made their escape and worked their way to the town of Palmer:

Here Rand has paid his devoirs to a young woman; but being at this time, wretchedly clad, and wishing to appear to better advantage, as he said, in the presence of his mistress, he was urgent that I should accommodate him with a suit of spare clothes, which I had hitherto

---

[1] Faculty Records (Harvard University Archives), I, 119.
[2] New Hampshire Hist. Soc., *Collections*, VII (1863), 383.

preserved through all difficulties. I hesitated, but on his promising, faithfully to restore them the next morning; and in the interem to provide me some place of abidance in security, I delivered him the whole suit. To do him a more particular kindness, I lent him linen, shoes and stockings, to which I added six crowns in money; that sum being every penny I could call my own. Rand was now accommodated to his wish; but as he was perfectly known in these parts, he durst not appear openly; he, therefore, prevailed on me to go to the habitation of his mistress, and to intreat her (in his name) to favor him with a visit in his present retirement. The girl honored the invitation, taking with her a pot of hot coffee, beefsteaks, and other ingredients for our morning repast. After Rand and I had made a plentiful meal, which, in our present exigent state, was indeed, epicurian; he expressed a wish to withdraw further into the bushes, under pretext of enjoying with his mistress a more private conference. They were absent nearly an hour, when the girl returned, and to my inquiries after Rand, made answer, that he had gone whither I should see him no more; she therefore advised me to shift for myself. . . . He had decamped with my best clothes, as well as all my money. I had, in truth, entertained but a slender opinion of this man's probity from our first acquaintance; little dreaming, however, that he would show me so scurvy a trick . . . I thought, at least, that the old adage, "honor among thieves," might have operated upon his feelings.[3]

Having thus bested Tufts, Dr. Rand might well have claimed to be the champion scoundrel of northern New England, but failing health and mind soon caused him to relinquish the title. The various towns in which he appeared marched him back to Londonderry, where he died, a town pauper, in 1787.

[3] Henry Tufts, *A Narrative of the Life, Adventures, Travels and Sufferings of.* . . . (Dover, 1807), pp. 205–206.

SAMUEL COOPER

Dr. Samuel Cooper of the Brattle Street Church in Boston was the second son of the Reverend William Cooper (A.B. 1712), the second minister of that church, and a grandson of Thomas Cooper, one of its founders. His mother was Judith, daughter of Chief Justice Samuel Sewall (A.B. 1671). He was born on March 28, 1725, and sent to Boston Latin where he suffered so much under the brutal tyranny of Master Lovell (A.B. 1728) that to his dying day his nightmare was the schoolmaster's ferule rather than British bayonets.[1] This mistreatment was in spite of the fact that even in grammar school he made his mark as a fine classical scholar. In college he became famous for his poetry, his prose composition, and his orations which glittered along on the smooth stream of a remarkable memory,[2] undisturbed, it was later remarked, by any depths of reason. He roomed in Massachusetts 24 and 28 with Waldo '43, Bulfinch '46, and was active in the undergraduate revival which was set off by Whitefield's visit. Once when he disturbed prayers in the Hall by breaking into laughter, the New-Lights attributed it to the instigation of the devil, but Sam frankly said that he was moved to it by Brinley's conversation.[3]

The death of Parson Cooper late in 1743 badly shook his son, but his church, charmed as almost everyone was by the lad's personality, on September 4, 1744, invited him to preach, if he found himself inclined to begin the career for which he seemed destined since childhood. The aged senior pastor, Benjamin Colman (A.B. 1692), was anxious to obtain Sam as his colleague, and the church, after hearing other candidates, on December 21 called him by a vote of 116 out of 138. In accepting he called attention to the fact that he was only nineteen years old:

Considering my Years, and great Want of Study and Experience I must ask of you, as you know I have a dear and engaging Example for it, that my solemn Separation to the Work of the Ministry may be deferred for a Year, and that in that time I may not be expected to preach but once in a fortnight, if it should please God to strengthen

[1] *The Common School Journal*, XII (1850), 311.
[2] John Eliot, *Biographical Dictionary* (Salem and Boston, 1809), p. 129.
[3] Henry Flynt, Diary (Harvard University Archives), 1740–41.

our Aged Pastor as he has hitherto done, and you think it convenient to grant me this Liberty.[4]

The "engaging Example" was a similar leave of absence granted to his father under like circumstances thirty years before. The church agreed, and Sam returned to the college, where he lived until the summer of 1745 when an epidemic which swept through his congregation called him back to devote his full time to the saddest part of his ministerial duty.

The church in Brattle Street was known throughout the English-speaking world as one of the most liberal in the Dissenting persuasion:

They neither require the making of publick Confessions, nor the owning a particular Church-Convenant, in order to admitting Persons to their Communion, as all the other Churches do; as likewise . . . they read the Scriptures, and recite the Lord's Prayer in their publick Worship.[5]

In his day Dr. Colman had been regarded as a dangerous liberal, and now the word went round the Brattle Street congregation that young Mr. Cooper was theologically unsound. On request he "gave in a sermon a confession of his faith to the general satisfaction." On the same day, April 6, 1746, he was admitted to membership in the church.

At the insistence of the senior pastor, the ordination was set for May 21, 1746. "Mr. Webb [A.B. 1708] began with Prayer, Dr. Colman preached, Dr. Sewal [A.B. 1707] gave the Charge — Mr. Prince [A.B. 1707], the Right Hand of Fellowship." [6] When Cooper returned to take his M. A. at the next Commencement, he was the envy of his classmates, for at the age of twenty-one he was installed in one of the famous pulpits of the Empire. On the program for the afternoon exercises he was listed as holding the thesis that the Patriarch Joseph had introduced the rite of circumcision among the Egyptians, but his part in the exercises was the Valedictory Oration. Admiral Warren and Sir William Pepperrell were in the audience.[7]

Cooper had chosen as his guardian Dr. Thomas Bulfinch, and

---

[4] *Records of the Church in Brattle Square* (Boston, 1902), p. 34.

[5] Daniel Neal, *History of New England* (London, 1747), II, 227.

[6] Andrew Eliot, Diary (Mass. Hist. Soc.), May 21, 1746. Colman's sermon was printed as *One Chosen of God* (Boston, 1746).

[7] R. T. Paine, Diary (Mass. Hist. Soc.), July 2, 1746.

SAMUEL COOPER

for his roommate during his last year at college, the Doctor's son, Thomas. On September 11, 1746, he married the Doctor's daughter, Judith. Her mother was Judith Colman. After the senior pastor's death in 1747 the young couple moved into his parsonage from which for years the Colman heirs tried to dislodge them by law. At one time or another they lived in several houses around Brattle Square. The congregation allowed them £80 a year for rent, and paid the Parson a salary of £10 a week. This was an ample sum which permitted him to subscribe for all of the magazines, to spend long vacations in the country, and to travel for pleasure. Cooper was far enough from his pioneer ancestors to describe a rough road as "a woody Romantic Way." [8] One of his more extended pleasure jaunts seems to have been a visit of the Harvard Club of Boston, or its equivalent, to the Yale Commencement of 1750. Hitherto Yale had awarded degrees to only six Harvard men, but on that day she made Masters of Arts William Ellery '22, Stephen Greenleaf '23, Samuel Cooper '43, Nathaniel Coffin '44, Thomas Cushing '44, James Bowdoin '45, and Ebenezer Storer '47. From New Haven they moved on into New Jersey where they were welcomed by Governor Jonathan Belcher (A.B. 1699) who wrote to a friend:

I duly received yours . . . per the hands of the Revd. Mr. Cooper who with the rest of the Gentlemen that traveld with him, I had the pleasure of Meeting at Amboy; where I got Mr. Cooper, to preach, both parts of the Sabbath, to a pretty numerous auditory, for that place, on which Occasion, I enjoyd more Satisfaction, than I can easily express. He is a Sweet, and most exi[   ] preacher, a great Honour to his late Fathers Memory and to himself.[9]

This being the opinion of a seventeenth-century Puritan, it is remarkable that it coincides with the views of liberals who lived into the nineteenth century:

In Brattle street men were charmed into the ways of wisdom by the eloquent, the graceful Doctor Samuel Cooper. With a voice melodious in the tones of a delicate flute, with an elegant address, in Attic diction, he allured his hearers to virtue, with soothing tenderness he poured the oil and the wine into the wounded bosom, and in persuasive language he recalled the vicious from the paths which lead to Death.[10]

---

[8] *American Historical Review*, VI, 310.
[9] Jonathan Belcher, Letter-Book (Mass. Hist. Soc.), Sept. 12, 1750.
[10] Ephraim Eliot, Commonplace-Book (in the possession of S. E. Morison).

One of the most useful things about the writings of John Adams (A.B. 1755) is that he was always ready to round out with illuminating criticism characters whom everyone else praised:

The Doctor's air and action are not graceful; they are not natural and easy. His motions with his head, body, and hands, are a little stiff and affected; his style is not simple enough for the pulpit; it is too flowery, too figurative; his periods too much or rather too apparently rounded and labored. This however, *sub rosâ*, because the Doctor passes for a master of composition and is an excellent man.[11]

One thing on which all of the Parson's friends agreed was that his knowledge, although very extensive, was very shallow; but so brilliant was his conversation that much more learned men listened to him with pleasure. Except where politics were concerned, he was always ready, smiling, to yield a point which others questioned. The only theological point to which he held firmly was the right of free inquiry, and when orthodox deacons tried to block the ordination of young liberals "Dr. Acquinas Cooper with a very mellifluous tone" would come to the rescue with light and reason as he saw them.[12] He read the writings of Jonathan Edwards, but they had no effect at all upon him.

The church in Brattle Street gladly followed their pastor down this primrose path, but due to the fact that he did not make a single entry in the minutes for twenty years after recording his victory in obtaining the introduction of Tate and Brady in 1753, the details are missing. The stark old meetinghouse was replaced by a magnificent structure, and John Hancock (A.B. 1754) contributed £1000 toward the beautification of the interior. The congregation, no longer warmed by threats of hell, asked the Parson to advise with his friend Franklin about obtaining "warming machines" for the meetinghouse. The philosopher replied that stoves were to be had in England "in the Form of Temples, cast in Iron, with Columns, Cornishes, and every Member of elegant Architecture," but suggested that they be cast in New England in the form of "a large Vase, or an antique Altar"; but he warned that they would not be satisfactory in so large a room.[13] So the

---

[11] John Adams, *Works* (Boston, 1850–56), II, 305. This criticism of the Parson's literary style is substantiated by a long rhymed imitation of Pope's Messiah in the Cooper Mss. (Henry E. Huntington Library).
[12] Mass. Hist. Soc., *Collections*, 6th Ser., IV, 225.
[13] Benjamin Franklin, *Works* (Smyth ed.), VI, 89–93.

Parson gave up the idea of stoves, but to keep warm or, as some suggested, to distinguish himself from the lesser clergy, he adopted what was described, with amusement, as a strange habit in the form of the Episcopal cassock, so full that one sleeve "would make a full trimm'd negligee." [14]

Cooper had, however, no leanings toward episcopacy. One of his most famous sermons, a Dudleian Lecture at Harvard, was a devastating attack on that "Man of Sin," the Church of Rome. Historians have been hard put to explain how such teachings could issue from the mouth of such a theological latitudinarian as he, but they were the inevitable outcome of the workings of a mind like his. Abnormally sensitive to tyranny, his delicate nostrils detected in every east wind the stench of the Inquisition smoldering at Lisbon. Widely read, he came to conclusions, similar to those of Max Weber, to the effect that liberty of mind, person, or estate cannot flourish where the Church of Rome is in control. In his Dudleian Lecture he warned Harvard College, which had nourished the transplanted seeds of liberty and enlightenment in America, to beware the "Romish Bishop" of Quebec.

If Popery, deceitfully assuming a milder form, seems to be less dreaded and abhored than it once was; let us be upon our guard, and remembering it is Popery still, be prepared to oppose it in every form. At best it is the extremest despotism. It decides all things at once, and by mere authority, and allows no examination of it's own mandates and decrees. It is a direct, an everlasting enemy to freedom of inquiry, and consequently to knowledge, and good literature. . . . Popery is incompatible with the safety of a free government. [15]

The students listened to Parson Cooper with respect, for as an Overseer since 1746 and a member of the Corporation since 1767 he had been very active in college affairs. In 1774 he was elected president but, to the great joy of his church, immediately declined. Instead he took on new duties as a trustee of the Hopkins Charity. Despite his prominence, his real contribution to Harvard was slight, for he regarded it as a tool to be used in the supremely important work of politics. [16]

[14] Anna Green Winslow, *Diary* (Boston, 1894), p. 14.

[15] Samuel Cooper, *A Discourse on the Man of Sin* (Boston, 1774), pp. 65–66.

[16] Thomas Hutchinson, Notes on Overseers' Meeting (Harvard University Archives), May 1, 1770.

Cooper's interest in politics went back at least to his appointment as chaplain of the House of Representatives in 1753. That same year he went to the Kennebec as chaplain of a commission to renew peace with the Indians, but, despite his subsequent membership in the Society for the Promotion of Christian Knowledge among the Indians, his real interest was not administration, but politics. As perennial chaplain of the House or the Council, in his prayers he mingled religion and politics so skillfully that he became the moral validation of the policies of the Whigs. His thesis was simple:

There is a close connection between civil liberty and true religion; Tyrants are commonly equal enemies, to the religious and civil rights of mankind; and having enslaved the bodies of their subjects, they affect also to enslave their consciences.[17]

With "Silver-Tongued Sam" speaking for the Deity in the General Court, and his brother William controlling Boston town meeting, it was said that the patronymic of the holy family was Cooper.

The Parson's first active participation in politics came in June, 1754, when he attacked a proposed province excise on wine and rum in an anonymous pamphlet entitled, *The Crisis*. In it he called those who favored the Bill "Bastards and not Sons" of the British constitution, and charged that the reports on consumption which it would require would destroy "the exclusive Right that every Man has to the innocent Secrets of his Family" and would justify the establishment of the Spanish Inquisition here. If excise taxes on rum and wine were permitted, taxes on small beer and soap would follow, and so liberty would be destroyed.

The Pulse of Liberty at this critical Conjuncture beats high in all Ranks: Those, whose Circumstances never allow'd, those, whose Inclination never prompted them to drink either Wine or Rum have for the pure Love of Liberty, joined the general Voice, and with one Consent, endeavour'd to Chase the deformed M——st-r back to the Den of Arbitrary Power, the Place of his Nativity.[18]

In the dispute between Charles Chauncy (A.B. 1721) and Thomas B. Chandler (Yale 1745) over the growing influence of the Church of England, Cooper was more rational, arguing serious issues, such as the "insolence of office" shown by an Anglican postmaster in

---

[17] Samuel Cooper, *A Sermon Preached in the Audience of His Honour Spencer Phips* (Boston, 1756), p. 9.
[18] *The Crisis* ([Boston], 1754), p. 11.

hindering the distribution of newspapers which printed Chauncy's writings.[19]

During the French and Indian War, the Brattle Street pastor used his silver tongue to spur on the war effort. The victories of the King of Prussia were, he said, the victories of God. However, he explained, the Deity did not interfere with the natural course of events, but only offered encouragement to his soldiers. The credit for the victory in America should go to Parliament:

These Colonies were a principal Object of those wise and vigorous Measures, that have given to the World so respectable and lasting an Idea of the British Policy and Power. How chearfully has our Mother Country employed her Riches and Strength for the Preservation of her tender and exposed Offspring! What Fleets and Armies have been sent for our Rescue! An Obligation which ought ever to be remembered with filial Respect, and with the warmest Gratitude.[20]

No one was warmer in his praise of the various members of the House of Hanover. On the accession of George III he told his congregation:

Who can forbear reposing as much Confidence in such a Monarch, and indulging as great and pleasing Expectations from his Government, as Humanity will allow? What Scenes of future Happiness do we now figure to our selves? Who does not hope to see the patriotic Plans, which employed the Cares of his royal Ancestors, happily perfected under his auspicious Reign?[21]

The political skirmishing which in Massachusetts preceded the Revolution was carried on in two spheres which hardly touched. Andrew Eliot (A.B. 1737) and Jonathan Mayhew (A.B. 1744) carried on the public debate, trying earnestly to discover the truth and to find a solution of the conflict of interest between the Empire and the colonies. Sam Cooper and Sam Adams, on the other hand, were the nucleus of an underground which waged unceasing war against the Crown's representatives in America. They had no constructive ideas, no comprehension of the issues involved, only a determination to turn back the clock to the old colonial system in

[19] Samuel Cooper to William Livingston, Livingston Mss. (Mass. Hist. Soc.), Apr. 18, 1768.

[20] Samuel Cooper, *A Sermon Preached before His Excellency Thomas Pownall* (Boston, [1759]), p. 31.

[21] Samuel Cooper, *A Sermon upon Occasion of the Death of . . . George the Second* (Boston, 1761), p. 39.

which Massachusetts had been nearly independent. The Brattle Street parson was credited by his contemporaries with many of the political essays which appeared in the *Boston Gazette* and, later, in the *Independent Ledger*, but these were wisely left unsigned because they were often contradictory. Any stick which came to hand would do to beat the Governor. Hutchinson was well aware that "Otis and the two Adams, Cooper and Church" went regularly every Saturday in the afternoon to set the Press, and that while publicly professing great friendship for him,[22] they privately denounced his "Malignity" with "Detestation and Horror." [23] There was an even closer inner cell, composed of Cooper, Otis, Hancock, and Sam Adams [24] which secretly met in a "smoke filled room" in the Brattle Street parsonage to cook up politics. At one of these meetings Silver-Tongued Sam cynically remarked that "an ounce of mother wit is worth a pound of clergy," [25] but his cloth made him an "excellent hand to spread a rumor." [26]

The same completely secular tone pervades Cooper's voluminous political correspondence with Benjamin Franklin and Thomas Pownall. Indeed the Cooper-Pownall correspondence is unique. Both men were democratic visionaries, seeing in America a haven from the social tyranny which prevailed in Europe. On specific social questions, such as child labor,[27] the Parson's views were identical with those of the most conservative Loyalists, but he had a sense of the dignity of man which they lacked. While unrealistically idealizing the rights of Englishmen for which he was prepared to fight, he was describing to the English Whigs a Massachusetts which never existed, an almost unanimous democratic Utopia of idealists. His distortions of fact were fantastic. He describes, for example, a "peaceful delegation of merchants" which was in fact a howling mob, hundreds strong, used as a tool to compel comformity or to silence criticism. On the basis of these misrepresentations, Pownall made political speeches in England which were sent back to Boston and read in the House of Representatives as illustrating the point of view of the English Whigs.[28]

[22] Mass. Hist. Soc., *Proceedings*, 2nd Ser., XX, 536.
[23] *American Historical Review*, VIII, 306.
[24] *Boston Gazette*, Mar. 25, 1782, 1/2.
[25] John Adams, *Works*, II, 262.
[26] *Ibid.*, X, 238.
[27] Samuel Cooper, *A Sermon Preached in Boston* (Boston, 1753), pp. 32–33.
[28] The original copies of these forty letters are in the British Museum; the

While such American Whigs as Andrew Eliot and even James Otis were being alienated by Sam Adams' doctrine and practice of violence, Cooper glibly explained away these political tools. Writing to Pownall about the effort to raise armed resistance to the Landing of the Troops he said:

I have nothing to say, as to the Propriety of the Vote respecting Arms — It had an ill Appearance upon which Account I dislik'd it; but that was all. It was strictly legal — For it was not, as has been maliciously represented, a Resolution to take up Arms, but only to comply with a Law that obliges the Inhabitants to be provided with them.[29]

On taxation he was as unreasoning and as violent in 1770 as he had been in 1754. He would have nothing to do with plans for intercolonial union; he frankly said that he was opposed to all innovations. His description of the Massacre was typical of the distortions which he fed to the English Whigs:

A party of Soldiers with Capt. Preston at their head fired upon the inhabitants of King Street, without a civil magistrate, without the least reason to justify so desperate a Step, and without any warning given to the people, who could have no apprehension of danger.[30]

He warned Pownall that a different version would be sent by others: "If it Should be represented that there was a great mob in King Street, and the Customhouse attacked, you may depend upon it nothing can be farther from the truth." He asked Pownall to obtain for him a list of those persons whose depositions in regard to the Massacre were contrary to his version. If they had told the truth, he said, they had nothing to fear: "I am an enemy to all disorders, and wish they could be prevented — But circumstances are candidly to be considered — and a country distinguish'd from a few obscure persons in it."[31] He attacked the Customs Com-

New York Public Library has transcripts. The substance of Pownall's letters is printed in Frederick Griffin, *Junius Discovered* (Boston, 1854). Cooper's letters are printed in the *American Historical Review*, VIII, 302–330. Cooper's copies, left in Boston in 1775, were carried to England by John Jeffries (A.B. 1763), from whom they passed to George III, who was so amused by their contents that he permitted copies to be handed around in Parliament. — John A. Schutz, *Thomas Pownall* (Glendale, 1951), p. 254.

[29] *American Historical Reivew*, VIII, 304.

[30] Samuel Cooper to Thomas Pownall (New York Public Library transcripts), Mar. 26, 1770.

[31] *Ibid.*, July 2, 1770.

missioners for contributing to the disorders by their "pretended fears" of violence and by their "retiring to the Castle tho' no insult or injury was Ever offered to their persons or anything belonging to them." [32] In periods of political quiet he labored with Sam Adams to stir up trouble by quibbling about trifles and ominously asserting that we must always be watchful of our liberties.

It was Cooper to whom Franklin sent the stolen Hutchinson letters, and in his reply the Parson frankly defended the duplicity involved in the publication of them:

The inconveniences, that may accidentally arise from such generous interpositions, are abundantly compensated by the reflection, that they tend to the security and the happiness of millions. I trust, however, that nothing of this kind will occur to disturb the agreeable feelings of those, who, in this instance, have done such extensive good. . . . I cannot . . . but admire your honest Openness in this Affair, and noble Negligence of any Inconveniencies that might arise to yourself in this essential Service to our injur'd Country. [33]

Cooper must have shown quite another face to those of his Whig friends who were distressed by such shoddy politics, for they believed, as one of them said, that "the Doctor was not to blame, and was much grieved at the consequence of the publication." [34] Such was his charm, that otherwise critical people found it hard to believe that his ethics were less strong than theirs. It is amusing to find him, in turn, so charmed by Commodore Gambier, who commanded the British fleet on the American station, that he became irritated at the super patriots who refused to join in a public tribute to him.

Very naturally the Doctor blamed the Tea Party on Hutchinson who, he said, "Seemed to choose that the Tea Should be destroyed and the exasperation of the countries heightened." [35] There is a tradition, which has little to commend it, to the effect that he put on paint and feathers to join what he called "the Assembly of the People" engaged in throwing the tea into the harbor.

The Hancock family attended Brattle Street and was strongly attached to the Parson. Thomas Hancock willed him £200 and a suit of clothes. There were several occasions on which John

[32] *Ibid.*, Nov. 5, 1770.
[33] Franklin, *Writings*, VI, 59, 272–273.
[34] Eliot, *Biographical Dictionary*, p. 131.
[35] Samuel Cooper to Benjamin Franklin (N.Y.P.L. trans.), Dec. 17, 1773.

Hancock, whom the Loyalists called the milch cow of the Whigs, might well have been lost to the party for good had it not been for the Parson's influence. It was he who solved this problem in March, 1774, by writing for Hancock's delivery a Massacre Oration which burned his bridges behind him.[36] He was reputed to have written Hancock's other public papers, and such was his reputation that in other colonies he was credited with the authorship of the Leonidas tract and other essays.

Dr. Cooper was slower than was Sam Adams to see that only by becoming independent could Massachusetts regain the freedom which she had enjoyed under the old colonial system. In 1770 he was startled to receive from Franklin a letter saying that the colonies were "originally constituted distinct states" and that Parliament had usurped authority over them;[37] but three years later he wrote of the revolution which he saw "in the Sentiments and hearts of the People." [38] The English Whigs fed his resolution, and one of them urged that the Americans take up arms, promising a rising in England as soon as blood was shed in America.[39] His reaction to the Port Bill was as violent as his reaction to the Excise Bill of nearly twenty years before. Because of his contacts with Europe his every word was weighed by his hearers. In the Thursday Lecture of November 18, 1774, he remarked that he was informed that the minds of the people in England were then more favorably disposed towards the Americans, and word immediately went round that he had just received news from Franklin.[40] Indeed one newspaper writer credited him with having an even more reliable source of information than Franklin:

> So Cooper speaks, enlight'ned from above,
> His teachings sure are right — for fill'd with love,
> He draws the Soul to Heav'n, as her abode
> Her friends, the Angels, and her center, God.[41]

He appears, in a less exalted role, in several other ballads of the period. The most popular of them said:

[36] Mass. Hist. Soc., *Proceedings*, XLIII, 155–156.

[37] Thomas Hutchinson, *History of Massachusetts* (Cambridge, 1936), III, 227–228.

[38] Samuel Cooper to Thomas Pownall (N.Y.P.L. trans.), Mar. 28, 1773.

[39] Mass. Hist. Soc., *Collections*, 4th Ser., IV, 371–372.

[40] *Ibid.*, 6th Ser., IV, 62.

[41] *Boston News-Letter*, Jan. 14, 1773, 2/1.

> In politics he all the tricks
> Doth wonderfully ken,[42]

which nobody could deny.

Indeed it was a standing joke for two decades that Dr. Cooper neglected his pulpit for politics. It was said that if "the Prime Minister" had his head in a political caucus at sermon time (and Sunday does seem to have been a favorite day for his practice of politics), he would thrust into his pulpit any cleric on whom he could get a hand. Charles Chauncy remarked that the Doctor seemed to think that a black coat qualified any man to preach, and Chauncy's slave once refused the gift of a black coat because, he said, he feared that Dr. Cooper would make him preach in the Brattle Street pulpit.[43]

The S. T. D. which William Robinson, the historian, in 1767 obtained for Cooper from Edinburgh at the behest of Franklin was typical of the political degrees with which the Whigs anointed their champions, and a London paper made a satirical attack upon it which was copied in the press throughout the colonies.[44] To the Loyalists within reach of the mobs Cooper's politics were not funny. Chief Justice Peter Oliver (A.B. 1730) described them in terms which, although obviously exaggerated, are interesting in showing the bitterness which he felt:

Dr. Cooper was . . . very polite in his Manners — of a general Knowledge — not deep in his profession, but very deep in the black Arts — his Behavior in Company was very insinuating especially among the fair Sex; and many of them, of his Acquaintance, had their Adams: — no Man could, with a better Grace, utter the Word of God from his Mouth, and at the same Time keep a two edged Dagger concealed in his Hand — his Tongue was Butter and Oil, but under it was the Poison of Asps — never was a Scholar of St. Omers, who was a more thorough Proficient in Jesuitism — he could not only prevaricate with Man, but with God also. . . . The Fluency of his Tongue and the Ease of his Manners atoned with some for all his Dissimulation; for his Manners were such, that he was always agreeable to the politest Company, who were unacquainted with his real Character; and he could descend from them to mix privately with the Rabble, in their nightly seditious Associations.[45]

[42] Ballad of the Boston Ministers, Am. Antiq. Soc. copy.

[43] Ephraim Eliot, Commonplace Book.

[44] *New Hampshire Gazette*, Feb. 10, 1769, 1/1; *Virginia Gazette* (Purdie & Dixon), Feb. 9, 1769, 1/3.

[45] Peter Oliver, The Origin and Progress of the American Rebellion (M. H. S. transcript), pp. 61–62.

With the Chief Justice feeling this way about Silver-Tongued Sam, it is not surprising that some persecution-maddened Loyalist urged the Regulars to put him to the sword the instant rebellion broke out.[46]

General Gage was keeping an eye on the Doctor when, on April 8, 1775, two swift vessels arrived at Marblehead bringing secret news which caused a great fluttering among the rebels. Cooper, who had preached that morning, thrust some other black coat into the pulpit for the afternoon and went into a huddle with the other politicians. By night they had all left town. The General wondered why for a week, but then he received his official dispatches ordering him to arrest the leaders. Obviously the English Whigs had leaked official information.[47]

The Doctor and his wife retired to the summer home of S. P. Savage at Weston where they remained until three in the morning of the Nineteenth of April when, being informed that the Regulars were at Lexington, they fled to Framingham. A few days later he went to meet the Committee of Safety at Cambridge while Mrs. Cooper went through the lines, marched up to the guards on Boston Neck, and inquired after her daughter Nabby, who had been left in town. The Regulars politely told her that Nabby was well, carried a letter to her, and brought her out to join her mother the next day.[48]

According to an article said to have been culled from a number of the Tory *Boston News-Letter* not now to be found, but then reprinted in the *Virginia Gazette*, the congregation of Brattle Street, indignant at Cooper's flight, called and ordained in his place "the Reverend Doctor Morrison," whose installation sermon "tended to show the fatal consequences of sowing sedition and conspiracy among parishoners." [49] Actually, "Doctor" Morrison was one William Morrison of Roxbury, a private who had deserted from the American army and was bent on a spoof; he seems to have at least confused Gage for a while.[50] A similar rag was a formal

---

[46] Mass. Hist. Soc., *Collections*, 2nd Ser., XVI, 92.

[47] Mass. Hist. Soc., *Collections*, 4th Ser., IV, 372. But see the contradictory account in the extract from Cooper's diary in Samuel Kirkland Lathrop, *A History of the Church in Brattle Street* (Boston, 1851), p. 102.

[48] *American Historical Review*, VI, 305.

[49] Frank Moore, *Diary of the American Revolution* (New York, 1860), I, 136.

[50] Mass. Hist. Soc., *Proceedings*, LIX, 133.

invitation sent out under flag of truce inviting Dr. Cooper and General Washington to attend a theatrical entertainment given by the British officers in Boston.

On May 3, 1775, the Provincial Congress invited Cooper to be its chaplain, but he was too busy with politics, propaganda, and preaching at Waltham to serve a body sitting daily at Watertown. Washington had brought him letters from the Adamses and Thomas Cushing (A.B. 1744). John Adams begged him to use his literary skill to draw a picture of the "Distresses of Boston," and the Provincial Congress appointed him to a committee to draw up an account of the battle of Bunker Hill to be sent to Great Britain, but the resulting document appears to have been the work of Peter Thacher (A.B. 1769), who had been an eyewitness of the struggle. The Doctor did spread certain misinformation about British atrocities.[51] He discussed military plans with Washington and gave Gates military advice, but because of his preoccupation with "the abominable Practice of prophane Swearing in our Army" it is not likely that he contributed much to the success of the campaign. He was personally responsible for the LL.D. awarded to Gates.

The Whig politicians were afraid that the Doctor would starve, and were trying to find an office for him when the Evacuation permitted him to return to his congregation. He found the town in a sad state:

A melancholy Scene. Many Houses pull'd down by the British Soldiery. The Shops all shut. Marks of Rapine and Plunder evr'y where. . . . Visited p.m. my House. Found all my Beds Bedsteds, Sheets Blankets Quilts and Coverlids, all my China Glass and Crockery Ware etc etc, plunder'd, 2 Lookin Glasses gone 2 broke, 1 Dressing Glass gone etc. Mrs. C and I supt and slept at Dr. Bulfinch's.[52]

The next day he began removing furniture from the houses of refugee Loyalists. Captain John Erving, the father of the Tory John Erving (A.B. 1747), who had fled to Halifax, helped the Doctor by giving him goods left behind by the refugees. Later the State legalized the transaction by declaring the goods confiscated and selling them to the Doctor.

The first year after the Evacuation was hard. Cooper had to

[51] Mass. Hist. Soc., *Collections*, LXXII, 73.
[52] *American Historical Review*, VI, 338.

petition the Council for enough cloth for a suit of clothes. The Harvard Corporation helped by paying him $31 for having attended its meetings for the three years past. The Council appointed him its chaplain, and the church gradually resumed paying his salary. In November, 1778, Lydia Hancock left to the church an elegant house on Court Street, and when the Parson moved into it he became the envy of the rest of the clergy. It was beautifully furnished with what Hutchinson estimated to have been half of his confiscated household goods. Another unexpected outcome of the war was the marriage of Nabby Cooper to Joseph Sayer Hixon, a Montserrat merchant who had been brought in on a Continental prize. Among the Parson's manuscripts is a long inventory of Hixon's slaves which would seem from the context to have contributed to his hearty approval of the match.

Now more than ever the Doctor functioned as the Prime Minister of Boston. Recognizing the influence which King's Chapel had exercised under royal patronage, he determined to sever it from the Church of England, and to that end he blocked the settlement of Nathaniel Fisher (A.B. 1763) and obtained the ordination of James Freeman (A.B. 1777), a man so liberal in theology that his church was bound to be an outcast from the Anglican establishment.[53] With "great solemnity and satisfaction" Dr. Cooper cut the last tie with the Empire by making the formal public reading of the Declaration of Independence. In 1777 he preached the first official Fourth of July sermon.

The Doctor's next goal was to bring France into the war, and to achieve his purpose he wrote to Franklin letters which, when placed in the proper hands in Paris, were credited with having accomplished that end.[54] His joy at hearing from Franklin the news of the treaty of alliance was certainly sincere, whatever his critics may have said about his Machiavellianism:

Your letter of the 27th February was handed to me on Sunday just as I was going to Divine service. Before getting into my carriage, I only had time to read the first paragraph, which assured me of the approaching signature of the treaties. I publicly gave thanks, and implored the blessings of Heaven for the King of France and his realm. This was something fresh in many respects, which excited in all the assembly a very

[53] William Bentley, *Diary* (Salem, 1905–14), II, 418.
[54] Mass. Hist. Soc., *Proceedings*, LXIII, 398; Franklin, *Writings*, VII, 407, VIII, 257.

agreeable surprise, and they joined very cordially with me in this act of devotion. . . . You cannot believe what joy the treaties with France have spread amongst true Americans, and the vexation which they cause to the small number of interested partisans and slaves of Great Britain who are amongst us. They had great hopes that France would not ally herself with us, and that they might be able to divide us, and bring us to a shameful reconciliation. . . . The English West Indies are lost. A thousand men could now take Jamaica. Never has France had such a good opportunity of crushing her rival.[55]

Throwing consistency to the winds he set himself to the task of selling the French alliance to the American people. Those who had heard "Silver-tongued Sam" denounce the King of France as a tyrant in the old days were startled to hear him tell the General Court:

The personal and royal accomplishments of Louis the Sixteenth are known and admired far beyond his own extended dominions, and afford the brightest prospect to his subjects and allies. The reign of this monarch diffuses new spirit through his kingdom, and gives freshness to the glory of France.[56]

Members of his congregation who had with patience suffered his neglect of his pastoral duties when engaged in local politics, became indignant when he devoted his time to campaigning on behalf of the French and devoted the sermon hour, it was said, entirely to praising France. Satirical advertisements began to appear in the papers offering rewards for copies of his sermons of the French and Indian War period when he had made it plain that the worst which could happen to the Americans was to share Gallic tyranny and misery with the oppressed people of France.[57] To this criticism he replied with redoubled efforts. He sent his grandson, Samuel Cooper Johonnot (A.B. 1783) to France, got Albert Gallatin a job teaching French at Harvard, and lavishly entertained the French naval officers and political delegations who visited Boston. Even his foes remarked that he was so neat in his dress, so handsome and amiable that no one who ever saw him could forget him, and this is substantiated by the fact that nearly every Frenchman who recorded his memories of Boston during the

[55] B. F. Stevens, Facsimiles of Manuscripts in European Archives (London, 1898), Nos. 826, 828.

[56] Samuel Cooper, *A Sermon Preached before His Excellency John Hancock* [Boston, 1780], p. 44.

[57] *Continental Journal*, July 18, 1782, 2/2.

Revolution gave some space to the Doctor. Some, like Chastellux, found him the ideal of republican openness and simplicity.[58] Others thought him pompous.[59] The Prince de Broglie recorded some of the most careful observations:

The Rev. Dr. Cooper, famous for his . . . discourses purely political, although delivered in the pulpit and in the church, his supple, insinuating and crafty spirit . . . is one of the men whose character and deportment struck me the most forcibly at Boston. His conversation is interesting, and although he expresses himself with difficulty in French, he understands it perfectly well, knows all our best authors, and has sometimes cited, even in the pulpit, passages from Voltaire and Jean Jacques Rousseau. . . . He writes sprightly verses, and carries certainly much cleverness under the immense wig of a clergyman, which he wears bigger and more heavily powdered than any of his brethren. He has his enemies among the clergy as well as the laity, and he is generally accused of a ductility quite macchiavellian.[60]

Lafayette, whom Cooper had introduced to Franklin, reported to Vergennes that they must place the Doctor at the head of the list of the friends of France in America.[61]

Many of Cooper's former friends were highly suspicious of his intimacy with the French. William Gordon (A.M. 1772) thought that French influence in Congress was threatening American sovereignty, and warned John Adams to be very discreet in his correspondence with him.[62] There were grounds for these suspicions. Three years before, in January, 1779, the French Ambassador, Gerard, in order to increase his influence in Congress, had offered to Tom Paine a pension of $1000 a year, and to Sam Cooper a fee of £200 sterling per annum. Paine indignantly rejected the offer as a bribe, but the Doctor's manuscripts indicate clearly that he accepted at least two payments and in return was regarded as a part of the French intelligence service. He reported frequently to "L. L." (La Luzerne) on such matters as ship movements and the contents of letters from Franklin and John Adams. A typical report shows him in action:

[58] François Jean de Chastellux, *Travels in North-America* (London, 1787), II, 281–283.
[59] Claude Blanchard, *Journal* (Albany, 1876), p. 182.
[60] *Magazine of American History*, I (1877), 378–379.
[61] Stevens Facsimiles, No. 1622.
[62] Mass. Hist. Soc., *Proceedings*, LXIII, 471.

Knowing Sir the Importance of the Affairs of the ensuing Campaign I have given them a particular attention, and continued to prepare the Minds of my Friends in Government as well as among the people for the most vigorous Exertions, and I am happy to find a prevailing Disposition to make such Efforts as our present circumstances will allow. Having had . . . a Letter from the M. de lay Fayette upon the Subject . . . I communicated it in Confidence to my worthy Friend and Parishoner Mr. Lowell [A.B. 1760] . . . a Member of the House of Representatives for this Town, and of very particular Influence in the Government. At his Request I gave Leave that part of the Letter should be read in the House which he did with an animating Spark . . . which had a good Effect.[63]

With this report the Doctor sent copies of newspapers to which he had contributed anonymous articles. In the French correspondence there is a suggestion that the Parson was being given a fund in order that he might hire a secretary, but there is no evidence that it was so employed, and plenty of evidence that both parties regarded the payment as an under-the-table affair.

Considering how indignant Silver-Tongued Sam had been when the Crown contributed to the salaries of the Judges of the Superior Court of the Province, his flexibility in this case is remarkable. There is some evidence that John Temple got wind of the transaction, and here Cooper was fortunate. Temple, trying to defend himself against the charge of having stolen the Hutchinson letters, publicly called upon the Parson to vindicate him, but, he complained, "the Rev. Doctor so far from making any kind of reply, either public or private," remained "totally silent upon the matter in question!" [64] Bitter although he was, Temple published nothing more damning than sarcastic remarks on the Doctor's reliability as "a Preacher of the Gospel of the TRUTH." [65]

In these years Cooper was refurbishing his reputation by his industry in helping to establish the new order. When Samuel Stillman (A.M. 1761), the Baptist minister, attacked the proposed constitution for Massachusetts because of the privileges which it accorded to the Congregational churches, "the silver-tongued Doctor . . . in his elegant and smooth way cut him up, and brought

---

[63] Cooper Mss., June 15, 1780. See also Henri Doniol, *Histoire de la Participation de la France à L'Établissement des États-Unis* (Paris, 1886–1892), IV, 34; Bernard Faÿ, *The Revolutionary Spirit in France and America* (New York, 1927), p. 133; and *Despatches and Instruction of Conrad Alexandre Gerard 1778–1780* (Baltimore, 1939), pp. 480–482.

[64] *Boston Evening Post*, Nov. 30, 1782, 2/2.

[65] *Continental Journal*, Oct. 3, 1782, 3/1–2.

his comments down to nothing . . . . then displayed a new lustre to his character by a compleat answer to every thing which he alledged." [66] The sermon which he preached at the inauguration of the new government won wide acclaim throughout America and was translated and reprinted by the Dutch as a significant state paper. Thomas Pownall reopened their correspondence by writing him a glowing letter of congratulation on the establishment of the independence of Massachusetts, and offering to endow a professorship at Harvard.[67] Unfortunately the lands which were to constitute the endowment had been confiscated.

The Doctor also delivered the opening address at the inauguration of President Willard, who like most of the able young men thought highly of him. However, "the French Dr." was regarded as Hancock's representative in the battle of the Overseers to obtain an accounting by that delinquent treasurer. Hancock, smarting with resentment at the treatment which he received in that affair, when offering to build a respectable fence around the Yard made the proviso that Dr. Cooper have the supervision of the work.

When the organization of the American Academy of Arts and Science was first proposed, Dr. Cooper protested that "it would injure Harvard College, by setting up a rival to it that might draw the attention and affections of the public in some degree from it." Won over by the argument that "the president and principal professors would no doubt be always members of it; and the meetings might be ordered, wholly or in part, at the college," he guided the project through the first legislature under the new constitution,[68] but not without political trouble. In the first bill the name of Samuel Adams preceded that of John Hancock, who indignantly took steps to withdraw from the Brattle Street Church. The Deacons went to the Parson and told him that this must not be, so the first bill was replaced by another in which the names appeared in strict alphabetical order.[69] The Doctor became the first vice president of the Academy. Never, it was remarked, did his skill in navigating between Scylla and Charybdis appear more brilliant than in these years. Although publicly a Hancock man, he generally sided in politics with Samuel Adams, and retained the

[66] Mass. Hist. Soc., *Collections*, 6th Ser., IV, 188.
[67] College Papers (Harvard College Archives), II, 91.
[68] John Adams, *Works*, IV, 260–261.
[69] Mass. Hist. Soc., *Proceedings*, LXIII, 445–446.

friendship of Washington. Openly a member of the Franklin camp, he avoided any personal difficulties with John Adams.

For years Dr. Cooper lived in terror of the mental breakdown which occurred in 1783:

He is supposed to have sacrificed his life to the inordinate use of Scotch snuff. His brain was first seriously affected, and his mind was much impaired before his physical powers failed. He told a friend who visited him a short time before the close of his life, "when you come again, bring with you a cord; fasten the ends of it in each corner of the room; let the cords cross in my head to keep it steady." [70]

His will, drawn on August 23, 1783, lacked the usual religious verbiage, and provided for the education of his grandson, Samuel Cooper Johonnot.[71] On September 10 he was "given over," but he lingered until December 29. At his funeral a broadside anthem by William Billings was distributed to the mourners.[72] The eloquent funeral sermon by John Clarke (A.B. 1774) went through two editions, and a verse elegy by Phillis (Wheatley) Peters was printed in pamphlet form. The long newspaper obituaries were without exception laudatory. Even the rising young literary lights, always prone to criticize their elders, felt that they had sustained a heavy loss. The contrast with the qualified mourning for Sam Adams was remarkable. When Lafayette returned to Boston in 1825 he refused to attend a Sunday service at the Catholic church, saying that he wanted once more to sit in the pew in the Brattle Street Church in which he had so often listened to his good friend, Dr. Cooper.[73]

The contemporaries of the Doctor regarded him as a great man, and certainly his contribution to the Revolution was far greater than that of Paul Revere, perhaps as great as that of Sam Adams. He has been forgotten chiefly because John Adams, who organized the pantheon of patriot gods in his old age, distrusted him, and assigned to him a less exalted place than others would have done.

Shortly after the death of Dr. Cooper, the church appointed a committee to publish a volume of his sermons, but it reported that his manuscripts were not in a usable condition. At its second

[70] William Sullivan, *Public Men of the Revolution* (Philadelphia, 1847), p. 66.
[71] For his descendants see the *New England Hist. Gen. Reg.*, XLIV, 57. Mrs. Cooper died in November, 1795.
[72] There is a copy of this broadside in the Harvard College Library.
[73] Josiah Quincy, *Figures of the Past* (Boston, 1883), p. 110.

meeting the Massachusetts Historical Society moved to obtain his collection of pamphlets, but showed no interest in his personal papers. The New York Public Library acquired thirty of his manuscript sermons, and the Henry E. Huntington Library purchased the rather disappointing remains of his correspondence and the surviving parts of his diary, the latter consisting of interleaved almanacs with line-a-day entries. The years 1753–54 have been printed in the *New England Historical and Genealogical Register*, XLI, 388–391, for 1764–65 and 1769, *ibid.*, LV, 145–149, and for 1775–76 in the *American Historical Review*, VI, 301–341. The portrait here reproduced is the Harvard Copley. For other copies see the Historical Records Survey, *American Portraits, 1620–1825, found in Massachusetts* (Boston, 1939), pp. 92–93. In 1784 libelous engravings were made from the Copley by John Norman in Boston and Valentine Green in London.

JAMES WARREN

GENERAL JAMES WARREN was born on September 28, 1726, the eldest son of Colonel James and Penelope (Winslow) Warren of Plymouth. At college he changed room and roommate annually but had a good record, except for one fine of 5s for drinking prohibited liquors. When he returned for his M.A. he held the affirmative of "An pro Hoste habendus est, Qui Hostibus Emolumenta Subministrat?"

Warren settled in Plymouth and entered into the family business of sending out in coastal and foreign trade small vessels some of which they owned outright. On November 14, 1754, he married Mercy, daughter of Colonel James and Mary (Allyne) Otis of Barnstable, and sister of James Otis (A.B. 1743). She was a woman whose strong character and never-quiet pen made her more famous than her husband. Untroubled by logic, reason, or perspective, furious in her prejudices, she poured upon the leading men of the times a confident and assertive correspondence which caused many a pitying glance to be cast toward her husband. He needed no sympathy, for in domestic matters she was as compliant as in public matters she was dogmatic. Far from forcing her views upon him, Mercy was reluctant even to express an opinion which differed from his on matters of family policy which concerned her vitally. In their later years they were generally regarded as a couple of troublesome porcupines, but there were no pricks in the warm and gentle love for each other which dominated their long lives.

James Warren was as moral a man as could be found in New England, but he never joined the church of which his father had been a pillar, and his correspondence was entirely free from the piety and religiosity which had been almost universal in the preceding generation. In place of the tags of scripture with which other men larded their letters, he used slang, two of his favorite expressions being "to take a lurch," meaning to have a desire, and "to have a breeze," meaning to have a quarrel.

During the first years of their married life, James and Mercy lived in the old Warren homestead on Eel River, but when failing health compelled the Colonel to begin to relinquish his civil offices, they acquired the Winslow mansion on the corner of North and

Main streets in Plymouth village. In 1756 James became Sheriff of Plymouth County, an office which his father and grandfather had held. At the same time he was appointed a Justice of the Peace, and the next year he bid in the farming of the liquor excise for the county. If he profited unreasonably from the opportunity afforded by the conjunction of his three offices, there was no loud complaint. Like his ancestors he became a colonel of militia. He entered politics in 1764 when he drafted the indictment of the Stamp Act which was voted by the town of Plymouth. In the next year, and those following, the town sent him to the General Court. There he voted with the Whigs against the liquor excise and against the recall of the Circular Letter resolution.

The Warrens were now regarded as being in the forefront of the resistance movement. James Otis was their beloved brother, and Mercy was the author of the Liberty Song which was on Whig lips. When Speaker Thomas Cushing (A.B. 1744) came down with the gout in 1769, Warren was elected in his place. The next House elected John Hancock (A.B. 1754) Speaker, and when Thomas Hutchinson (A.B. 1727) disallowed that choice, it substituted Warren, who was less objectionable to the administration. This called for some explaining by the Whig press, particularly as Warren had been detected in importing cloth in violation of the nonimportation agreement:[1]

It is not difficult to conjecture the Reason of his Honor's Disapprobation of Mr. Hancock; but why Col. Warren should be distinguished, is not so easy to account for. They have been equally firm in their manly Opposition to Ministerial Measures; and neither of them have in the least Degree swerved from Principles in the Judgment of their Constituents, of true Patriotism.[2]

As a matter of fact, Colonel Warren was, politically speaking, down at the mouth. Hutchinson was popular, and the public was weary with the spiteful war waged against him by Otis and Sam Adams (A.B. 1740). The Whig party was coming unglued, and Adams regarded Warren as almost his only ally outside of Boston. So Adams implored the Colonel to keep up the opposition, and nursed his dislike of "Julius Caesar," as they called Hutchinson. This was the right point at which to touch Warren, who had no

[1] *Boston Chronicle*, Aug. 24, 1769, 1/1.
[2] *Boston Gazette*, Apr. 23, 1770, 2/3.

JAMES WARREN

firm constitutional views. Describing the Plymouth situation to Otis, the Colonel said:

There seems no great happiness here but what the Tories have in Possession. They Enjoy themselves sweetly who while they sing Hallelujahs to their God [Hutchinson] (who like the Gods of Egypt was Born and Bred among them) have the double pleasure of praising him and Insulting all who wont worship at his Shrine.[3]

Typically he referred to Governor Hutchinson as "a Poor Sachem who had spent all his Wampum and must Appear in state equal to the Noble Blood running in his veins."[4] Apparently his dislike of the Governor had its origin in the Otis family feud, not in any personal feeling of social inferiority.

It was generally believed among the Whigs of that generation that the intercolonial system of committees of correspondence had its origin in a suggestion made by Warren to Adams when the latter visited Plymouth to mend his political fences in June, 1772.[5] The contrary view was most strongly advanced by a later generation engaged in glorifying Adams. At any rate, on November 4, 1772, Adams suggested to the Colonel that the town of Plymouth appoint a committee of correspondence, and when this was done, three weeks later, Warren was its chairman. In this capacity he regarded himself as the Old Colony agent of the Boston agitator, and begged for advice. With it came exhortations to shake off the despair with which he was encumbered. In many ways he served as Adams' voice, as on the committee to plan the action of Plymouth on the question of the salaries of the Judges of the Superior Court.

In May, 1773, Warren was appointed to the main Committee of Correspondence for intercolonial action. With Adams he drafted the circular letter which that committee sent out, and together they denounced the Boston Port Bill as an atrocity for which "the archives of Constantinople might be in vain searched for a parallel." At the end of the June, 1774, session of the General Court it was they who managed the appointment of delegates to the Continental Congress.[6] One of these, John Adams (A.B. 1755), was a close per-

---

[3] James Warren to James Otis, Apr. 13, 1771, in Otis Mss. (Mass. Hist. Soc.), 81.L.22.

[4] James Warren to Samuel Adams, Dec. 8, 1772, in Samuel Adams Mss. (New York Public Library).

[5] William Gordon, *History of the Rise . . . of the United States* (London, 1788), I, 312–313.

[6] *Ibid.*, p. 365.

sonal friend of Warren who now in their correspondence was compelled to organize his thinking on public affairs:

If I was enquired of, what I thought should be done with the Claim of Exemption from Parliamentary Legislation, as well as Taxation . . . I should answer that it was proper, practicable, expedient, wise, just, good, and necessary, that they should be held up in their full extent in the Congress at Philadelphia, and that means should be devised to support them. To determine on an annual Congress I think very important, both for the purpose of depressing the Scheems of our Enemies and raising the Spirits and promoting the Interest of our Friends.[7]

At home the Colonel continued his activity, in September, 1774, serving as chairman of the Plymouth county convention, and in October going to Salem to represent his town in the First Provincial Congress. When Mrs. Warren went up to Boston and he proposed to go into the town to see her, his friends warned him that the army would probably arrest him as a traitor. He went, and returned safely, but Harrison Gray demanded that he settle their accounts, saying that his treason made him a bad risk. At the moment he could not raise the money which he owed Gray, and he was too busy with public affairs to make the effort. He served on the watch-dog committee of seven which sat during the recess of the Provincial Congress, and by the beginning of 1775 decided that the Regulars were about to strike a blow. Writing to John Adams he anticipated the beginning of hostilities and urged stronger action:

I admire the Votes and Resolves of the Maryland Convention. They breath a Spirit of Liberty and Union which does Honour to them, and indeed the whole Continent. I am greatly puzzled to determine what Consequences the united force of all these things will produce in Britain. They must be infatuated to a degree I can hardly conceive of, if these things make no Impression, and yet in general I think, or rather fear, they will not. . . . Is it consistent with prudence that we should hold our Sessions at Cambridge? I am not more subject to fear than others; but if we mean to do anything important, I think it is too near the whole strength of our Enemies. If not, I shall repent leaving my own fire side at this severe Season.[8]

He attended that session of the Second Provincial Congress at Cambridge and was impatient with its long debates, its delays, and its failure to take decisive action.

[7] Mass. Hist. Soc., *Collections*, LXXII, 27–28.
[8] *Ibid.*, pp. 35, 36.

The session of the Congress which met at Concord early in April, 1775, was more to his liking, and his letters to his wife became happy:

We are no longer at a loss what is Intended us by our dear Mother. We have Ask'd for Bread and she gives us a Stone, and a serpent for a Fish. However my Spirits are by no means depressed. . . . All things wear a warlike appearance here. This Town is full of Cannon, ammunition, stores, etc., and the Army long for them and they want nothing but strength to Induce an attempt on them. The people are ready and determine to defend this Country Inch by Inch. . . . I hope one thing will follow another till America shall appear Grand to all the world.[9]

Just before the blow fell, he was sent to Rhode Island to urge that colony to raise troops for the common defence. After the Battle of Lexington and Concord he was chosen President of the Provincial Congress, but he asked to be excused and recommended Joseph Warren (A.B. 1759), who was elected. In the Third Congress he served on many important committees, always driving for quicker and more decisive action. Disillusioned by the unreliability of the Minute Men, he longed to propose that John Adams' friends Washington and Charles Lee be placed in command, but dared not make the suggestion. Looking around at the Congress, he was pleased to see that it contained a good proportion of "the Sense and property of the province," but was discouraged by what he considered its abominable "Timidity and slowness."[10]

Warren's differences with his colleagues arose from his simplicity and lack of perspective, qualities essential in a sincere revolutionary. He was the kind of man who bought lottery tickets as an investment and was mildly surprised when he failed to make a profit. His faith in paper money was equally simple. As a reader of newspapers he knew that French soldiers not only burned towns but sometimes put to the sword every living creature in them, but when the British set fire to Charlestown to drive out the snipers he denounced it as "a Savage Barbarity never practiced among Civilized Nations." Later in the war he was somewhat surprised to have the British confiscate vessels belonging to him and Hancock. On the other hand, he saw much more clearly what was immediately in front of him than did many of his colleagues with better perspective. He was

9 *Ibid.*, pp. 44, 45.
10 *Ibid.*, pp. 49–50.

bitter at the mismanagement of the Battle of Bunker Hill by Artemas Ward (A.B. 1748), and was sure that Washington or Charles
Lee could have made a glorious victory of it.[11]

After the death of Joseph Warren at Bunker Hill, the Colonel
was chosen President of the Provincial Congress in his place. Immediately he plunged into diplomatic correspondence with the
other colonies, urging their more active participation. To the delegates in the Congress at Philadelphia he wrote reproving that body
for not having "embraced so good an opportunity to form ourselves
a constitution worthy of freemen."[12] He had never dared propose
that Massachusetts call Washington to command its army, but had
urged the delegates in Congress to create a Continental army and
send him as its head. He headed the welcoming committee, and
was delighted at the amiability and the obvious wisdom and determination of the Virginian. On the other hand, he found Charles
Lee a great disappointment, being led to distrust him because of his
bad manners. As for the riflemen from the Virginia frontier, his
first impression that he had never seen "finer fellows" was soon
replaced by a conviction that they were "the most disorderly part
of the army."

With the creation of the Continental Army, President Warren
was relieved of the greater part of his military and diplomatic
burden, and was able to turn his attention to matters of patronage
and military vanity which were creating discord. In a letter to John
Adams he reported on his efforts:

I went yesterday for the first time this session to wait on the General.
. . . to see if I could serve the persons you recommended. . . . I find
the Colony, as you predicted, will suffer by referring the appointments
you mention to him. They will, I think, go to the southward. I am
amazed that the impropriety of his appointing was not sufficient to determine every one of your body. . . . When I was coming off, I took
the freedom to mention the sufferings and abilities of a number of gentlemen, and to ask the liberty to mention them, if he had any occasion for
them even in places of no great importance. He said there were many
gentlemen that had come some hundred miles, and as we had so large a
share of the places, they must be provided for.[13]

It was typical of Warren's grasp of what was immediately before

[11] *Ibid.*, p. 63.
[12] Mass. Hist. Soc., *Proceedings*, 1st Ser., XIV, 81.
[13] Mass. Hist. Soc., *Collections*, LXXII, 97.

him that he saw and said that the pressures which he and others were putting on Washington were something that so "great and good" a man should never have to bear. He was convinced that the Virginian was the best man who ever lived for the critical position which he held, and was determined to do everything in his power to make his burdens lighter. Historians have overlooked the importance of the fact that it was Warren who happened to be President of Massachusetts at this critical moment.

The President was always entirely without personal political ambitions, but John Adams knew that he was much abler than many of the men who were receiving important appointments, and was determined to place him in a high position in the Continental establishment. To this end he enlisted the aid of Sam Adams and the somewhat reluctant support of Cushing and Hancock. Warren, hearing of these efforts on the part of the Massachusetts delegation in Congress, wrote to disuade John Adams:

I am much obliged to you and my friend Adams for thinking of me. I am content to move in a small sphere. I expect no distinction but that of an honest man who has exerted every nerve. You and I must be content without a slice from the great pudding now on the table. . . . High Establishments will not be relished here, and I think bad policy in every view, and will lead us fast into the sins, folly and sufferings of our old impolitic and unnatural mother.[14]

Adams replied that the vast importance of the office of Paymaster General demanded the appointment of a man of "Family, Fortune, Education, Abilities and Integrity," and a record of long and faithful service in the American cause. On July 27, 1775, Warren was appointed Paymaster General of the Continental Army. If he feared jealousy, he must have been reassured by the laudatory verse which appeared in the press.[15]

Two weeks before, in order to emphasize the continuity of the new government of Massachusetts with the old, Warren had been elected Speaker of the House of Representatives and the office of President had passed to the senior member of the Council. The Speaker was regarded as a political leader as the President was not, so Warren increased his activities. He importuned Congress to create a Continental Navy of small sloops, and John Adams re-

[14] *Ibid.*, pp. 78, 79.
[15] *Boston Gazette*, Sept. 25, 1775, 3/2.

ported that his letters on this and other subjects were of great use in shaping congressional opinion. Repeatedly he urged the Massachusetts delegation to stiffen the resistance of Congress. There were two things which he dreaded, he said, smallpox in the army and, more dangerous, conciliatory proposals from England which would weaken the determination of the colonies.[16] During the first months of the war he continued to regard the Americans as the defenders of the system of English liberties and the Regulars as "the Rebels," but the news of the burning of Falmouth in October, 1775, drove him over the edge:

We have just heard that the pirates . . . have orders to destroy every sea port from Boston to Pemmaquid. This is savage and barbarous in the highest stage. What can we wait for now? What more can we want to justifie any step to take, kill and destroy, to refuse them any refreshments, to apprehend our enemies, to confiscate their goods and estates, to open our ports to foreigners, and if practicable to form alliances, etc., etc.[17]

Taking cognizance of the argument that foreign alliances would imply independence, he reproved the congressional delegation for continuing "to acknowledge a dependency on Britain or Britains King." Writing to Sam Adams in February, 1776, he said that he heartily agreed with the author of *Common Sense*, and urged that Congress declare independence before the House of Lords could act upon the American petitions for redress.[18] In every letter he needled Congress to act: "We looked for a declaration of *independence*, and behold, an indulgence to drink *tea*." When the Declaration was passed, he dismissed it as long overdue.

As Speaker of the House, Warren locked horns with the Council, which he accused to dragging its heels because of an exaggerated sense of its importance. To John Adams he complained:

The Board contend for the exclusive right, plead the Charter, and assert the prerogative with as much zeal, pride and hauteur of dominion as if the powers monarchy were vested in them and their heirs, by a divine, indefeasible right. This is indeed curious, to see a Council of this Province contending for the dirty part of the Constitution, the prerogative of the Governor. . . . I hate the monarchical part of our government, and certainly you would more than ever, if you knew our present mon-

[16] Mass. Hist. Soc., *Collections*, LXXII, 84.
[17] *Ibid.*, p. 154.
[18] Mass. Hist. Soc., *Proceedings*, 1st Ser., XIV, 280.

archs. . . . They have got a whirl in their brains, imagine themselves kings, and have assumed every air and pomp of royalty but the crown and scepter.[19]

This struggle with the Council contributed to break his health, and he was sick in Plymouth when the Regulars evacuated Boston. Riding back, he watched the departing British fleet from the heights of Braintree.

Warren had been an excellent paymaster, but he was anxious to be rid of the office now that the Boston crisis was over. Although he had urged Congress to take over the army in Massachusetts, when it did so he regarded it as foreign. In writing to Congress he regularly referred to "your army" and "your treasury," reproved it for not keeping them full, and justified the departure of the Massachusetts soldiers on the ground that they had "presumed so much on the public spirit of our countrymen."[20] However, Washington silenced him by authorizing him to pay Provincial regiments out of Continental funds when the General Court failed to provide. Immediately after the Evacuation, Warren tendered his resignation. The General said that he hoped that Warren would accompany the army to New York, but the paymaster replied that it would be "very disagreeable and scarcely honourable" for him to leave Massachusetts. After his resignation had been accepted and had gone into effect, he continued to pay out Continental funds to the mutinous troops until his hands were empty.[21] His courageous disregard of the letter of the law, and of his personal liability, was unique at a time marked by financial quibbling. A long and unfriendly audit of his accounts failed to turn up anything which could be used against him.

Immediately after Warren had resigned his paymastership, he was appointed to the Superior Court. John Adams, who usually insisted that only men trained in the law be appointed to the Bench, urged him to accept in order to give honor to the court, but he firmly refused.[22] When, two years later, the new Supreme Court was set up, he was appointed to it, and again he refused. His letter to John Adams explaining his reasons was typical of him:

[19] Mass. Hist. Soc., *Collections*, LXXII, 178, 183.
[20] Mass. Hist. Soc., *Proceedings*, 1st Ser., XIV, 277.
[21] Peter Force, *American Archives*, 4th Ser., VI, 830.
[22] Warren's statement of his reasons for refusal are entered at length in the Executive Records of the Province Council, XIX, 25.

So barren is our poor country that they have been obliged to appoint the most unsuitable man in the world. He had no suspicion of it before hand. He reasonably supposed that many blockheads might be hit on before it came to his turn; he had therefore no opportunity to prevent it. He is therefore embarrassed beyond measure. He fears your displeasure; he is puzzled with the solicitations of friends, or those who would get clear of this matter; but his conscience tells him he will by accepting injure his country and expose himself. He must therefore decline.[23]

In a similar way he refused to allow his friends to make him a candidate for the governorship. He was perfectly sincere in wanting to get back to Plymouth "to be a Farmer again."

Warren refused these offices because he doubted his ability, not because he shrank from the dirty work of revolution. Of his numerous Tory relatives he said, "Had ever any man so many rascally cousins as I have"; and he urged that they be punished. He was indignant at the defeat of legislation which would have confiscated the estates of Loyalists, exiled them, and hanged them if they returned. He denounced the Council and the town meetings for not exiling all who refused to take the oath of allegiance. His only remark on the narrow escape of Benjamin Church (A.B. 1754) from lynching was that it would have been derogatory to the authority of Congress. Of the Boston rioting of September 16, 1777, he said:

This Town was in a tumult all day yesterday carting out Rascals and Villains — small ones. This seems to be irregular and affords a subject for Moderate Folks and Tories to descant largely and wisely against mobs, but the patience of the people has been wonderful, and if they had taken more of them, and some of more importance their vengeance, or rather resentment, would have been well directed. It therefore seemed wrong to wish to stop them.[24]

He was furious that Burgoyne's officers — "Murtherers" he called them — were treated as gentlemen and guests and were allowed to corrupt the countryside. When, in the second year after the Declaration of Independence, British cruisers were so faithless as to seize American ships, he urged that Burgoyne and his army be held as hostages until every prize was released.[25]

Warren was appointed colonel of the First Plymouth Regiment in August, 1775, and the next year was promoted to be one of the

[23] Mass. Hist. Soc., *Collections*, LXXII, 240.
[24] *Ibid.*, pp. 368–369.
[25] Mass. Hist. Soc., *Collections*, 7th Ser., IV, 223.

Province's three major generals. In September, 1776, he was designated to take command in Rhode Island, but he pled that he could
not "support the fatigue." In his private correspondence he was
more frank:

The Militia is so despised, and I suppose is designed with all its officers
to be directed by the Continental Generals, that I intend to embrace the
first opportunity to quit it, that shall offer without any imputation. . . .
I congratulate you on the success of our Arms at the Northward and
Westward, very pretty affairs indeed, and to be done by the poor despised Militia too will give singular pleasure to some people.[26]

The prospect of major military operations in Rhode Island forced
him to a decision, and in August, 1777, he resigned his commission.
He thought it necessary to explain at some length to John Adams:

No body on that occasion was more embarassed than I was. I don't feel
afraid to fight, and I believe you are sensible nobody has more zeal for
the Cause then I have; but I have too much pride to submit to circumstances humiliating and degrading. Our Council ordered me to repair
there, and take the command of them and receive from General Spencer,
or such other officer as should be appointed to command there from time
to time, such direction as they should give me. . . . If we have no right
to appoint Major Generals we should not have done it. If we have they
ought to have their rank . . . when they came within the splendid orb
of a Continental Officer. . . . I forsee the Militia are to be considered
in the same light of inferiority with regard to the Continental Troops
that I have been used with indignation to see them with regard to the
British.[27]

At the same time General Warren was in a great heat because of the
failure of Congress to station a Continental army in New England
to protect it from a British reinvasion which he was sure was impending. Patiently Washington explained to him that Howe would
not attack New England because that was no way to defeat the
united colonies, and if he did, New England would be better able
to defend itself than would any other section.

General Warren had for two years been needling Congress to
establish a Continental Navy, so it was only natural that he should
have been appointed, in May, 1777, to the newly-created Navy
Board, Eastern District. John Adams, in informing him of the

[26] Mass. Hist. Soc., *Collections*, LXXII, 323, 364.
[27] *Ibid.*, p. 349.

appointment, told him that he would have to resign as Speaker and devote his whole time to the work of the Board, which would include "the building and fitting of all ships, the appointment of officers, the establishment of arsenals and magazines."[28] The General was unhappy because he had just made up his mind to retire to farming at Plymouth. Before he had received official notice of his appointment, there arrived in Boston two brigs loaded with naval stores consigned to him, so, guessing that they probably belonged to the United States, he went to work.

As General Warren began his Navy Board duties, a series of marine disasters shattered his hopes for the Continental Navy and proved him to be a false prophet, but he courageously gathered up the burned and shattered pieces and did his best. Throughout his term he was unhappy with the Marine Committee of Congress which, he felt, did not allow the Boston Board sufficient discretion to allow it to function well. Among his complaints was that Congress was sending officers to command ships although Boston officers and their families were idle and "starving on their bare pay." When the French fleet abandoned the Rhode Island campaign in 1778, he shared the common rancour at what was regarded as its treason, but as a matter of common sense he kept his opinions to himself and placed the Navy Board and its material at the disposal of the allies. Their visit served to widen the gulf between Warren and Hancock. After the latter's great dinner for the French naval officers, the General charged him with having short-changed Congress in the number of salutes fired after toasts. For a New Englander who had lived through the French and Indian War, Warren was amazingly free from prejudice against the French and the Church of Rome, and he charged Hancock with whipping up forgotten popular fears of popery with one hand while he entertained the allies with the other. Worse than this, he told Sam Adams, Hancock was corrupting the people by his extravagance:

All manner of Extravagance prevails here in dress, furniture, Equipage and Liveing, amidst the distress of the public and Multitudes of Individuals. How long the Manners of this People will be Uncorrupted and fit to Enjoy that Liberty you have so long Contended for I know not. . . . Folly and Wickedness stalk abroad with the same shameless rapidity and Confidence they ever have done and find Numbers to keep them

[28] John Adams, *Works* (Boston, 1850–56), IX, 465.

in Countenance. Assemblies, Gameing, and the fashionable Amusements Engage the Genteel People.[29]

While watching this Hancock-led extravagance, Warren was trying frugally to live on his Navy Board salary. When, in October, 1778, he decided that this was impossible, he sent his resignation to Sam Adams. The latter, however, refused to submit it to Congress unless the General would come to Philadelphia to take his place. Warren wanted nothing to do with Congress which, he charged, was making the Marine Board contemptible by going over its head: "I own I am Mortifyed and will not long submit to it."[30] To make matters worse, Hancock, in his effort to get authority over the Navy Board, charged the General with corruption in the conduct of its business. Warren did not think it worth while to make a reply, but on May 19, 1782, he submitted his resignation and asked that his accounts be audited.

In May, 1777, the month in which the General was appointed to the Naval Board, he announced that he would not accept reelection as Speaker, but he finally agreed to serve "for a few days." At the next annual town meeting, Hancock's supporters, making the most of Warren's unwillingness to serve in the army, combined with his Loyalist cousins to deprive him of his seat in the House of Representatives. Mercy indignantly reported to him:

Strange as it may appear one of the most subtile emissaries of Britain [Edward Winslow (A. B. 1736)] and the most malignant of your foes was suffered yesterday in full meeting of the town to stand up and cast the most illiberal reflections on a man whose primary object has been to rescue these People from the thraldom of a foreign yoke. . . . I have long felt so much indignation and disgust for the ingratitude and baseness of your constituents (a very few excepted) that I could scarce bear the reflection that you were sacrificing the best comforts of life, your domestic felicity to support the interests of those who stood ready to repay your indefatigable labour with undeserved execration.[31]

Under these circumstances the next House would ordinarily have elected such a distinguished lame duck as he to the Council, but that of 1778 did not chose him. He laid his double defeat to Hancock and to "the moderate Class," to which, he pointed out, he had never belonged.

[29] Mass. Hist. Soc., *Collections*, LXXIII, 59–60, 82.
[30] *Ibid.*, p. 93.
[31] *Ibid.*, pp. 16, 17.

Although Warren had no use for the "moderates," he was himself no leveller. In the early years of the revolutionary government he had attacked certain legislation as calculated to "drive every man of interest and ability out of office," and had denounced it as "the consequence of the leveling spirit." [32] He had sidetracked legislation to free the slaves for fear that it would offend the other colonies. To him the proposed constitution of 1778 was unsatisfactory because it did not contain a sufficiently high property qualification for the franchise. He had no regard for the wisdom of the people, and when the economic revolution put into chariots men who formerly, he said, might have brushed his shoes, the world seemed to him to be "turned topsy turvy." [33] In all this religion had no part, except that he remarked that the Virginia Act of Toleration demonstrated the fact that "Episcopacy and Liberty will not flourish in the same soil." [34]

In 1779 Warren was reelected to the House, and chosen to the Council, but he declined the latter honor. Two weeks later he was elected to Congress, but he again declined although he was worried by the strength of the "Aristocratic Party" in that body. He accepted election to the House of Representatives under the new constitution, having been begged by the people of Plymouth to serve them.[35] There being no popular choice of lieutenant governor, the General Court tendered the office to him, but he declined it, probably because he detested Governor Hancock, with whom he would have been obliged to serve. Of the new administration, he reported:

The Influence here is as uniform, and extensive as in England, and the Criterion to determine the qualifications for office much the same as in the most Arbitrary Governments, or in the most servile Nations. . . . Whether Pisistratus will be able to establish himself Perpetual Archon, or whether he will be able to convey that Honor and rank to his Family by hereditary right Time must determine. He has no Guards, yet established, but he has unbounded Adulation.[36]

In his letters to his family, the General called Hancock a despot, described his splendid balls as like those of the King of Prussia,

[32] Mass. Hist. Soc., *Collections*, LXXII, 219.
[33] *Ibid.*, LXXIII, 105.
[34] *Ibid.*, LXXII, 296–297.
[35] The petition is in the Mercy Warren Mss. (Mass. Hist. Soc.), Oct. 23, 1780.
[36] Mass. Hist. Soc., *Collections*, LXXIII, 150.

and denounced his administration for its lack of compassion for the distress of the country, and particularly for the sufferings of the soldiers. By way of contrast, he held up General Gates as "a true Republican of sterling virtue." Discouraged by the Massachusetts politicians, Mercy Warren reopened her correspondence with Martha Washington, and her husband revived his with the master of Mount Vernon. On both sides it was carried on in terms of the warmest respect.

Warren also opened a correspondence with Thomas Jefferson, but this had to do largely with his efforts to get a government job for his son Winslow. Not a man to seek favors for himself, he sought them for his sons in a most importunate way. An angry letter to Chief Justice William Cushing (A.B. 1751) after the Supreme Court of Massachusetts had failed to appoint one of the Warren boys to a clerkship lays down his views on patronage:

I claim the Merit of being at all Times Uniformly Steady in the Cause of Liberty and Virtue, both with regard to the Public and Individuals. I have had no Notice taken of me by the present Administration nor have I been mean enough to submit to those measures I knew would Insure success if I wished for it. You will Excuse my saying so much of myself. It is done only to Justify a Maxim I always practiced upon my self and wish to see Universal. That is to prefer . . . those Men and those Families who in our Extraordinary Times by their public Conduct and Exertions may have deserved some of the offices of Government. If the Judges of the Supreme Court are of Opinion that my Claims arise from Vanity, or that Others have a better Claim on the same principles, or that there is any Objection to the Character, Education or abilities of my Son, I shall give them my applause for deciding against him. If not my Opinion of their Justice and Patriotism will Encourage me to Expect a decision in his favour. I will Trouble you no further, on this subject, nor Can I be mortified with reflection of having made my first solicitations to men whose Characters, I did not revere.[37]

One personal favor he did ask for himself and have granted, that was the loan from the Council of a German prisoner to be his servant.

In some ways General Warren more resembled Washington and Jefferson than he did most of the Massachusetts Whigs. He was a founder of the American Academy of Arts and Sciences, and at one of its early meetings he presented a communication in which

---

[37] William Cushing Mss. (Mass. Hist. Soc.), Mar. 14, 1783.

he maintained that agriculture was the foundation of the ideal state, that it strengthened the mind, the morals, religion, and the economy as did no other profession.[38] Banking, internal improvements, and the opening of trade with the East had none of the interest for him which they excited in the contemporaries with whom he differed politically. His purchase of the Hutchinson house in Milton on January 28, 1781, was in part at least an expression of this ideal of bucolic simplicity. He contracted to pay £3000 for the house and because of the financial troubles of the times had difficulty in keeping up the payments, but he loved it as Thomas Hutchinson had done, and asked his son Winslow to send him handsome wallpapers from Europe.

When the new government was set up under the Articles of Confederation, General Warren urged that it be done with "all the plainness and simplicity Consistent with decency" and suitable to the "Circumstances and profession" of America.[39] Every move made to strengthen that government he associated with the "pompous parade" and "Magnificence of Monarchy" which he detested. The resolutions adopted by the Hartford convention of 1780 frightened him:

A Convention . . . solemnly Resolved to . . . Vest the Military with Civil Powers of an Extraordinary kind and . . . no less than a Compulsive power over the deficient States to oblige them by the point of the Bayonet to furnish money and supplies for their own pay and support. . . . General Washington is a Good and a Great Man. I love and Reverence him. But he is only a Man and therefore should not be vested with such powers.[40]

Similarly he denounced the feeble efforts to strengthen Congress, which he described as composed of "men vested with Imperial powers and . . . furnished with sources of Corruption equal to a King of Britain." [41] Shortly after this last outburst, on November 30, 1782, he was elected to Congress. He accepted, but three months later was still in Milton vainly trying to get his affairs in a shape to leave; he never succeeded. To his sons he wrote that it was just as well that he did not go to Philadelphia where he would be in the minority, without influence or favor in a genera-

---

[38] This communication is printed in the *American Museum*, II (1787), 344.
[39] Mass. Hist. Soc., *Collections*, LXXIII, 41–42.
[40] *Ibid.*, p. 152.
[41] *Ibid.*, p. 182.

tion which had forgotten the patriotism and virtue of the founding fathers and was dominated by a love of money. Publicly he denounced the efforts of Congress to obtain the power to lay federal taxes. What need was there now that the war was over? The public debts must be paid, but this could be done by a tax of a dollar an acre on land, or by assigning to each State its proper proportion and permitting it to chose its own way of raising the money. Interference with commerce by the States worried him, but he had no solution for that.

As General Warren hoed his potatoes on Milton Hill he brooded over the "French and Frankleian" influence on Congress. He feared the power placed in the hands of the Sage whom he regarded as "grown feeble by age" and "worse excesses." [42] "Morris is a King," he said, "and more than a King. He has the Keys of the Treasury at his Command, Appropriates Money as he pleases, and every Body must look up to him for Justice and for Favour." [43] He called Morris the "Premier, the King, or Grand Monarch of America," and accused him of cutting salaries in order to get rid of members of the government who would not support his "foreign Measures." To grant Congress the right to levy taxes would inevitably, he thought, lead to a monarchy. He regarded the Society of the Cincinnati as another monarchical threat, and encouraged the towns which protested against being taxed to pay pensions to Continental officers. To John Adams' attempts to calm his fears he replied:

You say there are as yet no Appearances of Artificial Inequalities of Condition, etc., That may be true, because the Barefaced and Arrogant System of the Cincinnati Association is not fully matured, but it is rapidly progressing. The People, who have no Stability, who equally forget the Benefits and Injuries, have almost forgot this Insolent Attempt at distinction and are Introducing the Members into the Legislature, and the first Civil and Military Offices. [44]

When Hancock retired from the governorship in 1785 he expected to make Cushing his successor, but the men in positions of political influence wanted no more of this "administration of Imbicility and weakness." According to Warren, he could have had the governorship had he been willing to be humble to Hancock,

[42] Mass. Hist. Soc., *Proceedings*, LIX, 86.
[43] Mass. Hist. Soc., *Collections*, LXXIII, 230.
[44] *Ibid.*, pp. 291–292.

but he preferred to "be honest and continue to despise his Caprise." Then, he heard, the Man of Beacon Hill said that he did not care who succeeded him so long as it was not the Man on Milton Hill.[45] More than Warren's usual reluctance to take high office was behind his refusal to stand for the governorship, for far more than most of his class he was aware that the economic chaos in the State was "verging to confusion and anarchy." His description of the conditions in Massachusetts was so like the complaints of the Shaysites that he was accused of being one of them. This was unfair, as the report of the Rebellion which he made to John Adams on October 22, 1786, shows:

We are now in a State of Anarchy and Confusion bordering on a Civil War. The General Court at their last Session could not, or would not, see the general Uneasiness that threatened this Event. However, they did not provide for the public Tranquility during their recess, but dosed themselves into an unusual Adjournment for six or seven months. . . . The Three upper Counties . . . have refused submission to the Government established by the Constitution and Obedience to the Laws made under it; that is, they have violated their Compact and are in a State of Rebellion. . . . I am mortified at the Triumph of our Enemies. . . . I wish everything may be so Conducted as to restore Order and submission to Government; but I fear it will be some time first.[46]

He charged that the policy of the Bowdoin administration was peevish rather than conciliatory, and that it violated the constitution for fear that Shays would destroy it, but that "After all the Apparatus of the Suspension of the Habeas Corpus, prosecution of some miserable Scribblers, Declarations of Rebellion, Acts of disqualifications, etc.," the situation was quite as unsettled as ever.[47] Disappointed in the administration of Bowdoin, and dismayed by the "arbitrary and despotic" demands of Sam Adams who would have hanged the insurgents, Warren found himself, to his embarrassment, somewhat relieved at the reelection of Hancock to the governorship. When the people of Milton asked him to represent them in the General Court, he accepted chiefly because he was stung by the old charges that he had always refused difficult posts in times of public need and crisis. In the House he was at once elected Speaker because he was known to sympathise with the dis-

[45] *Ibid.*, pp. 262–263.
[46] *Ibid.*, pp. 278–279.
[47] *Ibid.*, p. 292.

tressed masses, but accepted unhappily because he could see no way out of the troubles.

That the creation of a strong federal government combining the thirteen states into a single economic unit was the best solution for the troubles of the times never occurred to Warren. He was opposed to the Federal Constitution because of his pathological fear of Congress. In November, 1787, there appeared on the streets of Boston a handbill attacking the Constitution because it lacked a bill of rights, and asserting that if it went into effect, the trade and wealth would be drawn off to Philadelphia, leaving Boston to starve. It was at once suspected that Warren and James Winthrop (A.B. 1769) were the authors of this handbill. Apparently the General was the author of at least two temperate and searching letters in the newspapers. The reaction of the Federalists was anything but temperate. They called him an abettor of anarchy, "a finished monster of depravity," "a focus of public contempt," a person "spreading the poison of immorality" and exhibiting "the extremity of cowardice," "afraid to be a rogue and not wishing to be an honest man." [48] John Quincy Adams (A.B. 1787) was typical of his more moderate critics:

The General's political character has undergone of late a great alteration. Among all those who were formerly his friends he is extremely unpopular; while the insurgent and antifederal party (for it is but one) consider him in a manner as their head, and have given him at this election many votes for lieutenant governor. [49]

He was formerly a very popular man, but of late years he has thought himself neglected by the people. His mind has been soured, and he became discontented and querulous. He has been charged with using his influence in favour of tender acts and paper money; and it has been very confidently asserted that he secretly favoured the insurrections. . . . He has certainly given some reason for suspicion by his imprudence; and when in a time of rebellion a man openly censures the conduct in general, and almost every individual act, of an administration, an impartial public will always judge that such a man cannot be greatly opposed to a party who are attacking the same measures. [50]

Indeed so closely was Warren connected with Shays in the public mind that the antifederalist leaders were just as pleased that he

[48] See the *Columbian Centinel*, Nov. 24, 1787, and the references in Mass. Hist. Soc., *Proceedings*, LXIV, 149, 156.

[49] Mass. Hist. Soc., *Proceedings*, 2nd Ser., XVI, 413.

[50] John Quincy Adams, *Life in a New England Town* (Boston, 1903), p. 150.

did not publicly campaign against the Constitution. In the ratifying convention it was noticed with amusement that the speeches of some of the less literate opponents contained flourishes which came unmistakably from the pen of Mercy Warren. After the ratification of the Constitution Warren wrote to General Washington saying that it was "the duty of every good citizen to rejoice in every measure calculated to carry it into operation agreeably to the principle on which it was adopted." From Mount Vernon came the answer: "It gives me no small pleasure to find that former friendships have not been destroyed by a difference of opinion on this great political point." [51]

In 1788 Warren was again elected Speaker, but when in subsequent years he allowed his friends to run him as a candidate for the lieutenant governorship and for Congress he was defeated by what Mercy called the machinations of Hancock whom she blamed for having "destroyed his public influence." He held one minor State appointment, that of Land Officer for the Western Territory. When he visited the seat of the Federal government in 1790 it may have been in connection with Massachusetts claims to western lands, but he hoped for favors for himself or for his sons. Greatly disappointed in his expectations, he turned bitterly against the administration. Some time later John Adams tried to placate Mercy Warren by explaining why he had not been able to do anything for his old friend:

Had I been President and at liberty to act my own judgment, I should have nominated General Warren to the office of Collector for the port of Plymouth; for at that time all the obloquy I had heard, and all the extreme unpopularity into which he had fallen, had not shaken my opinion of his integrity. . . . But the conduct of General Warren at the time of Shays's Rebellion, whether truly or falsely represented, and his supposed decided and inveterate hostility to the Federal Constitution, had produced so determined a spirit against him, that if Washington himself had nominated him to any office he would surely have been negatived by the Senate. [52]

This was the wrong tone. Mercy Warren was furious that John Adams should have thought of insulting her husband by the offer of such a petty appointment, and she informed him that it was the General's influence and sage advice, once so humbly requested,

[51] George Washington, *Writings* (Fitzpatrick ed.), XXX, 331.
[52] Mass. Hist. Soc., *Collections*, 5th Ser., IV, 476.

that had started *him* on *his* path to glory and power. With this venom on her pen she wrote her *History of the Revolution* which ended the friendship of the Warrens and the Adamses.

The Warrens had real as well as imagined troubles. In 1791 their beloved son Winslow was jailed for debt in Massachusetts. Released, he joined St. Clair, and died in his defeat. Unable for financial reasons to keep the Milton house they had so much enjoyed, they sold it and returned to Plymouth. The General's only comfort was the love of Mercy, whom he often called his "saint." It flourished as brightly in their age as in their youth. Public sympathy put Warren on the Governor's Council for three years, beginning in 1792, but he took little interest in public affairs. Instead, he took to writing bitter letters contrasting the idealism of 1775 with the selfishness of later years. In his interpretation, the glories of the Revolution had become dust and ashes. "I have begun to think this world a farce," he wrote. The French Revolution seemed to be the only hope of mankind, and he looked eagerly "to see the downfall of Kings, and Congresses," only to be bitterly disappointed when Napoleon went whoring after the fleshpots of Egypt instead of striking prostrate the "Austrian, Russian and Turkish Tyranny."[53] He never shared Sam Adams' Gallophilism, but criticized all Americans who too enthusiastically supported either European camp.

In the Virginia Resolutions of 1799 General Warren saw a glimmer of hope for the future of the United States. "I think," he wrote, "the Virginia address a most capital performance worthy of the character of Americans in the brightest stage of her Existance when struggling for Liberty."[54] Among the leading figures of Massachusetts this view was unique, and served to increase his isolation. He regarded the defeat of John Adams and the Federalists in 1800 as a personal vindication and the liberation of the United States. On the inauguration of Jefferson he wrote happily to the new president:

I . . . sincerely congratulate you, on your Elevation to the first Magistry of the Union, and the Triumph of Virtue over the most malignant, virulant and slanderous Party that perhaps ever existed in any Country. Driven myself from active sceans of political Life, into neglect and obscurity, by the Malice of persecution, I have sat like a Man under the

[53] James Warren to Henry Warren, June 9, 1799, in Mercy Warren Mss.
[54] *Ibid.*, Feb. 14, 1799.

shade of a Tree unnoticed. . . . I have seen principles sacrificed to ambition, and consistency of sentiment to the Interest of the moment.[55]

During the next few years other Massachusetts Republicans came out from under the trees and joined him, so that in 1804 he had the pleasure of serving as a presidential elector and casting his vote for Jefferson.

This was General Warren's last public act, for failing health had compelled him to relinquish even his duties as a Justice of the Peace. He was happier, however, than for many years, and his letters breathe contentment:

I do not expect ever to recover more health. The season of the year is against it; my age is against it. I have had a long life, and have enjoyed a thousand blessings. I have uniformly endeavored to do my duty; I think I have generally done it, and wherein I have erred, I shall be forgiven. If death should make its approach this day, I should not be alarmed.[56]

Death did come in a matter of days, on November 28, 1808. So long had the General been out of the public eye that the newspapers noticed only the passing of "a distinguished Revolutionary character."

Abigail Adams longed to extend a sympathetic hand to Mercy Warren but could not forgive her attacks on John Adams in her *History*. If she wanted revenge, she had it, for the reason why James Warren has been almost unknown to later generations of Americans is that President Adams did not include him in the pantheon of saints of the Revolution who were canonized in his memoirs. Had the General died before Shays' Rebellion, his name would have been as familiar today as that of his brother Otis. Mercy's *History*, which told the Warren side of the story, was soon forgotten.

Mercy Warren died on October 19, 1814. Of the children, James was graduated at Harvard in 1776 and Charles in 1782.[57] Fortunately the family treasured and preserved the mass of family manuscripts. The letters with the greatest historical interest have been published in the *Proceedings* and *Collections* of the Massachu-

---

[55] James Warren to Thomas Jefferson, Mar. 4, 1801, *ibid.*

[56] Alice Brown, *Mercy Warren* (New York, 1896), p. 307.

[57] For the children see Mrs. Washington A. Roebling, *Richard Warren of the Mayflower* (Boston, 1901), pp. 27–28.

setts Historical Society, particularly in volumes LXXII and LXXIII of the latter series. There are many of the General's letters of similar content in the Samuel Adams Manuscripts and the Elbridge Gerry Manuscripts in the New York Public Library. The several collections of Warren Manuscripts in the Massachusetts Historical Society contain much unpublished family correspondence. The Copley portrait of the General here reproduced belongs now to the Boston Museum of Fine Arts. The so-called James Warren portrait in the Adams Mansion is believed by some experts to represent Dr. Joseph Warren.

EDWARD AUGUSTUS HOLYOKE

DR. EDWARD AUGUSTUS HOLYOKE was born in Marblehead on August 1, 1728, the first son of the Reverend Edward (A.B. 1705) and Margaret (Appleton) Holyoke. According to family tradition he was given his middle name because of the month of his birth; his elders and contemporaries always called him "Neddie." When he was seven his father became president of Harvard College, and his family moved into Wadsworth House. He was admitted to the college in July, 1742, the written part of his entrance examination being a Latin essay on the topic "Labor Improbus omnia vincit," on which he labored for a week. The next three weeks he spent in making the required copy of the college laws.

Neddie was below average height, but his small body was strong and agile, and fired by a vivacious disposition. The fact that he lived at home helped to keep him out of trouble except for one occasion after which he was fined for drinking prohibited liquors. To occupy his active hands and mind he took lessons in painting and French, and although that language was not a part of the curriculum he mastered it well enough to make a declamation in the college chapel.[1] He was fascinated by those problems of science which might yield to the skilled manipulation of apparatus, and for his *Quaestio* he held the negative of "An Concursus Axium opticorum per se sufficiat, at Solvendam simplicem Visionem?" Thanks in part to a Hollis Scholarship the total expense of his education was only £39 10s, but much of that remained on the college books unpaid for twenty years.

[1] Robert Treat Paine, Diary (Mass. Hist. Soc.), Nov. 29, 1745.

EDWARD AUGUSTUS HOLYOKE

During his undergraduate vacations Neddie kept the Lexington school, and after taking his first degree he moved on to a better post at Roxbury. In July, 1747, he began to read medicine with Dr. Thomas Berry (A.B. 1712) of Ipswich, with whom he remained for two years. In June, 1749, he hung out his shingle in Salem, where he boarded with Madam Turner on the corner of Essex and Union streets. It was, he soon found out, highly questionable whether that town could support a physician. For one thing, the traffic would bear no higher fees than 5s in paper or 8d in coin for a visit, and he soon became noted for his unwillingness to ask the poor to pay even that. After two years he was so discouraged that he would have abandoned the effort had it not been for fear of disappointing his father.

It was by incredible perseverence that Dr. Holyoke established his practice. He trudged five or six miles every day on the round of his patients and kept this up for more than fifty years, for when he could afford a horse he found that he could never learn to keep the animal from slipping its bridle. So assiduously did he apply himself to his work that during all of his hundred years he never traveled more than fifty miles from the place of his birth, and the only occasion on which he spent more than a night or two away from home was that on which he went down to Boston to take the smallpox inoculation. No doubt his financial circumstances were improved by his marriage on June 5, 1755, to Judith, daughter of Benjamin Pickman. She died in childbirth on November 19, 1756, and on November 22, 1759, he married Mary, daughter of Nathaniel and Mary (Simpson) Vial of Boston. In 1763 he bought the Bowditch house in Essex Street, where he lived for the rest of his life.[2]

Dr. Holyoke made a study of his practice which, so far as we know, has no rival in the archives of medicine. Although he sometimes made a hundred calls in a day, he made a record of each, filling eventually a hundred and twenty volumes of notes from which he compiled statistics. In 1755 he performed a delivery for the first time, and soon he drove the midwives of the town out of business. Troubled by the memory of the death of his first wife, he kept abreast of the latest European developments in practice,

---

[2] For a picture of the house and portraits of the members of the family see *The Holyoke Diaries* (Salem, 1911).

but because of sympathy for the mothers he differed sharply with the young physicians who used their instruments to effect "successful" deliveries regardless of results.[3] Among his prescriptions those for cold shower baths were unusual in his day. He felt, however, that water should be kept in its proper place, and argued that people heated by exercise or the weather should never drink "cold water or other weak small Liquids" without first having fortified themselves by taking "a dram of Rum or some spirit or a Glass or two of Wine." [4]

Few physicians disagreed with Holyoke's prescriptions, but John Warren (A.B. 1771), who tried to set up as a rival in Salem, complained that his practices were hurting the profession:

The People here are accustomed to being dealt so very easily with by their Physician, Doctor Holyoke having reduced the Fees to a very low Rate and never having troubled them with Accounts except when they troubled him for them. A Physician who should charge any thing nearly sufficient barely to support the Dignity of the Profession or should attempt to make any Innovations upon the ancient Usage of the Town would at once throw himself out of Practice.[5]

Holyoke, on his part, dismissed as "medical bucks" those physicians who regarded themselves as members of a superior caste. In this difference was fuel for fire, for his temper was hot and quick; but he was aware of this failing, and he never for a moment relaxed his control of his tongue. In consultations with younger colleagues he always expressed himself with diffidence and caution, and was slow to push his own ideas when he thought others wrong.

Dr. Holyoke, always social, became even more so as he became busier and older. Reluctantly he left off drawing and painting because they took too much time. He worked out the family genealogy, but his records, which he had lent to Thomas Hutchinson (A.B. 1727), were destroyed when the latter's mansion was sacked by the mob.[6] For a while longer he wrote verse, all of which had a medical flavor:

> When from serener skies and purer Air
> The gen'rous Zephyr drives the chilling blast,
> And poisonous Foggs and Vapors all disperse,

[3] E. A. Holyoke to James Jackson in James Jackson Mss. (Mass. Hist. Soc.), Jan. 14, 1800.
[4] E. A. Holyoke Mss. (Essex Institute), II, 52.
[5] J. C. Warren Mss. (Mass. Hist. Soc.), II, Jan. 29, 1775.
[6] *Memoir of Edward A. Holyoke* (Boston, 1829), p. 7 *n.*

The vital Fluid by our Lungs inhal'd
Revives the sluggish Blood with active Spring,
And swifter drives the purple Current round,
Replete with Life, with vigorous Health endow'd.[7]

He enjoyed skating and dancing, at which he was skillful, until he decided that his advancing years made these amusements unbecoming. Then he built up his library, had a personal bookplate engraved,[8] and joined in founding the Salem library. He was likewise a founder of the second and more dignified Salem club, and one of the most popular members of such social organizations. As a contemporary in that circle put it, he was always cheerful but never jovial or given to light conversation. His minister greatly admired his social manner:

He joined with facility in current conversation, and brought his ample stores of reading and reflection to illustrate the subject under remark; but never so much as thought of taking what is called the lead in any discussion. He regarded the ordinary intercourse of friends as a means of relaxation and of mutual improvement, and avoided, therefore, as much from principle as from native modesty, every thing resembling a dictatorial air, all premeditated dissertations, and parading exhibitions of his own resources. This was equally true of his intercourse with those who were younger and confessedly less informed than himself.[9]

To the tight Harvard community he contributed more than these social graces, for after the fire of 1764 he gave books and apparatus to repair the loss, and five years later he presented the college with a twenty-eight-foot telescope.[10]

Holyoke served the public in civil affairs as well, accepting appointment as a Justice of the Peace in 1761, and, four years later, to the town committee to instruct the Representatives to work for the repeal of the Stamp Act. He foresaw that the colonies would eventually become independent, but he thought that his own generation was incapable of the self-discipline necessary for self-government. So great was his distress at the violence of the agitators and the mobs that he fell into a fit of depression so profound that he really feared that he would die of it. He was without personal fear, however, for on April 26, 1774, he dined with

[7] Ibid., p. 80.
[8] Probably by N. Hurd; there is a copy at the American Antiquarian Society.
[9] John Brazer in E. A. Holyoke, An Ethical Essay ([Salem], 1830), p. xviii.
[10] Josiah Quincy, History of Harvard University (Cambridge, 1840), II, 490.

Governor Hutchinson and let himself be seen riding in his host's coach. A month later he signed the farewell address to the Governor, and in June he defied the mobs by signing the letter of welcome to General Gage.[11] Fervently he prayed that bloodshed might be avoided, and when it began he sent his family to Nantucket and tried to make up his mind to retire to the quiet inland town of Boxford, away from the violence of the Marblehead fishermen whose conduct he detested. He could not, however, bring himself to leave his patients, and through all the violence and excitement he never broke his daily round of visits and study. Such was his place in the community that despite his views no rioter dared touch his property, particularly after his public recantation of his error in signing the address to Hutchinson.[12]

A letter written to his wife on the day after Bunker Hill was typical of his attitude toward the war:

Well, my dear, I am heartily glad you are not here just at this time; you would, I know, be most terribly alarmed. We had an appearance yesterday of a most prodigious smoke, which I found was exactly in the direction of Charlestown and as we knew our men were entrenching on Bunker Hill there, we supposed the Town was on fire, and so in fact it proved, for in the evening . . . we were told the Regulars had landed at Charlestown under cover of the smoke from the buildings they had set fire to. . . . The commotion here was considerable, though none of our men went to the Battle . . . we had but one meetinghouse open.[13]

The greatest deprivation which the Doctor suffered during the war was the company of the men, mostly Loyalists now scattered, who had met in the Monday Night Club to discuss science and religion. Each Monday evening during the years of their absence he gathered his family and made a ritual of talking about their scattered friends.

Holyoke's correspondence with the refugees was circumspect, but it shows his resentment at what he regarded as wicked interference with freedom of communication:

Ever since I heard of your Arrival in England I have been about writing to You, but the Difficulty of communicating our Sentiments is so great, that I have been deterred from Troubling You with what must

[11] *Essex Gazette*, June 14, 1774, p. 180.
[12] *New England Chronicle*, June 1, 1775, 2/3.
[13] *Holyoke Diaries*, p. 90.

have been a mere Matter of Fact Letter. . . . I hope therefore You will not Attribute my not writing to any want of Friendship . . . but to the Unhappy Circumstances we are placed in, which almost necessarily render, every Thing which it is safe to write, very little worth Reading.[14]

If he could not safely express his thoughts, he could serve; and his work as head of the smallpox hospital was as great a service as any Salem man rendered to the community during the war years. In 1780 he was reappointed a Justice of the Peace, and some years later he was added to the Quorum.

The Doctor's dislike of the revolutionary movement arose from his detestation of the half truths, and the occasional deliberate lies, of the politicians. His passion for truth and accuracy was as rare as his open-mindedness. In his unceasing search for facts he performed autopsies on the bodies of his own children. He asked his exiled friends to send him the latest medical works and to have more accurate scientific instruments made for him in London. One good which the ill wind of the war brought to him was the opportunity to join in the purchase of the captured library of the Irish scholar, Dr. Richard Kirwan. Although his own loss in the smallpox hospital was a miraculously low one per cent, he was shaken by even this, and he became one of the first New Englanders to use vaccine, importing his stock from London. While other physicians jeered at Dr. Perkins' Points, he vigorously argued that, although apparently absurd, they should be judged only after thorough trial.[15] His own deductions were often shrewd, as when he connected the disappearance of the once-common complaint of the "dry belly ach" with the disuse of pewter with its high lead content.

Holyoke's passionate search for truth led him into pure science in which his work with the phenomena of color and vision brought no fruitful results,[16] but he did discover the power of evaporation to produce cold before that fact had been published in America. He tried his best to introduce an "American Thermometer" on which the freezing point of water was 100 and that of mercury was zero.

---

[14] E. A. Holyoke to Samuel Curwen, July 20, 1780, in Curwen Mss. (Am. Antiq. Soc.), I, 99.

[15] E. A. Holyoke Mss., I, 25.

[16] *Ibid.*, II, 53.

His demonstrations attracted the keen interest of men like Manasseh Cutler:

Mr. [President] Willard and I appointed this day to wait on Dr. Holyoke to see some experiments performed upon a new glass machine constructed for impregnating water with fixed air. The air passed through
capillary tubes, alkaline and vitriolic acids, in a state of effervescence to
the water, and gave it the taste of the acid, resembling beer or bottled
cider.[17]

The good Doctor would have been most distressed had he known
that future millions of hot Americans would consume oceans of
these "weak small Liquids" without a thought of first fortifying
themselves with rum.

In his youth Holyoke had been greatly impressed by displays of
the Aurora Borealis which much exceeded anything which earlier
generations of colonists had seen, and he never ceased to search for
an explanation of the phenomenon. His studies led him into
astronomy, and he made accurate observations of the transit of
Venus in 1769 and of the transit of Mercury in 1782, communicating his results to other scientists.

In October, 1781, the General Court designated Dr. Holyoke to
call a meeting to organize the Massachusetts Medical Society. He
presided at the meeting, served as first president, and to the
Transactions contributed papers drawn chiefly from his statistical
records of disease in Salem. He also contributed papers to the
Medical Repository, an excellent journal published in New York.

Dr. Holyoke was one of the charter members of the American
Academy of Arts and Sciences, and from 1814 to 1820 was its
president. One of the articles which he contributed to its *Memoirs*
was reprinted in a German magazine, and this and other contributions brought him into correspondence with men like Price,
Priestley, and Sir Charles Blagden. The Imperial and Royal
Agrarian Academy of Florence elected him a corresponding member.[18] However, the Doctor was too busy to cultivate these foreign
contacts; indeed the difficulty of getting him to Boston to attend
meetings of the Academy and the Medical society was a constant
source of trouble to them. This was why, probably, he was not

[17] Manasseh Cutler, *Life, Journals, and Correspondence* (Cincinnati, 1888),
I, 74–75.
[18] Holyoke, *Ethical Essay,* p. xv.

elected to the Massachusetts Historical Society although he gave many pamphlets to its library. On the other hand, he served for years as president of the Essex Medical Society, the Essex Historical Society, the Salem Athenaeum, the Institution for Savings, and the Salem Dispensary. Another of his local interests was the Danvers Iron Works, of which he was a proprietor. It was, however, not his interest in local industry but his Federalism which caused him to contribute $800 toward the building of the *Essex*. His participation in a Federalist caucus in 1803 caused his friend Bentley to protest that when he "lent his presence and name to the opposition," he "robbed himself in his old age of the confidence which had been placed through life in his discretion." [19] However, when the political heat had cooled Bentley said that the Doctor was "the most interesting character . . . in Salem, from his professional reputation and unspotted character and the warm affections of all." [20] This statement was made shortly after the Doctor had put himself at the head of a committee of thirty-six volunteers to enforce the by-law against the "Use of Segars in the Streets and public Buildings" because of the fire danger. [21] To this campaign the old gentleman contributed a poem condemning the "segar" as a

> Vile substitute for that white, slender tube
> Our fathers erst enjoy'd, in Winter's Eve,
> When the facetious jest, or funny pun,
> Or tales of olden time, or Salem Witch,
> Or quaint conundrum round the genial fire
> The social hour beguil'd. [22]

Neddie Holyoke had listened unmoved to the preaching of George Whitefield under the "Washington elm," and in his adult years he avoided revivalism and sectarianism like the plague. He was, however, very pious, and he read the Greek New Testament through each year. In 1770 he transferred his membership from the First to the North Church which he served as a ruling elder, and to which he gave a silver tankard which it still has. Regarding all confessions of faith as unpardonable deprivations of the sacred right of private judgment, and disliking mere negative criticism, he

---

[19] William Bentley, *Diary* (Salem, 1905–14), p. 17.
[20] *Ibid.*, IV, 30.
[21] *Ibid.*, p. 29.
[22] *Memoir*, p. 80.

set himself to the task of revising the traditional creeds and finally came up with one of his own, the essence of which was that "every Christian must judge for himself what these Truths and Precepts are." [23] He supported the S.P.G.N.A. because its missionaries were teaching Christianity and not sectarianism.

Israel Atherton '62 was the first of many medical students who learned their trade as his apprentices. To judge by his correspondence, those who went on to further training in London hospitals learned nothing there which lessened their affection and respect for their teacher. When the Massachusetts Medical Society was considering the establishment of minimum educational requirements for physicians it asked and received his opinion:

As to Your inquiry whether a collegiate Education be necessary? I answer No, but then, I am fully persuaded that at least a moderate Acquaintance with the Latin, and some even slight knowledge of the greek Languages, is necessary; and still more that enitiation into the newtonian Philosophy, and Chemistry, and in general into that Circle of Science which is taught by the Professor of Natural philosophy in the Apparratus Chamber. [24]

In 1783 he received the first M.D. granted by Harvard, and when this had become a common professional degree, the university further honored him, in 1815, with an LL.D. On the death of Peter Frye '44 in 1820 he became the oldest living graduate, and eight years later he became the first Harvard man to reach a hundred. This attracted considerable attention because English statisticians were denying that there was a single authenticated example of anyone living to that age. [25] Picking up a copy of the Triennial one day he checked off the names of thirteen graduates of the seventeenth century whom he had heard preach, and said that the oldest graduate whom he could remember as a prominent man was Thomas Dudley of the Class of 1685. [26] For many years he was a fixture at Commencement, and when he walked in the Yard the guests of Willard's Hotel came out to look at him.

The description of the Doctor in the fictitious journal of Marshal Soult probably reflects a visit which he paid to the author in jail:

[23] Holyoke, *Ethical Essay*, pp. 159–161.
[24] E. A. Holyoke Mss., II, 80.
[25] Mass. Hist. Soc., *Proceedings*, 1st. Ser., VIII, 438.
[26] J. L. Sibley, Private Journal (Harvard University Archives), I, 471.

He called on me . . . without a top coat, though it is a cold autumnal day. This sage has seen . . . a prodigious advancement in his profession. . . . He . . . was himself among the first on this side the Atlantic, to hail the progress of reason. His coadjutors were enlightened by his communications, and his numerous pupils were taught the best method of acquiring what was known, and trying what was not satisfactory; and the good man now reaps the reward . . . in the affection and reverence of the community. The monument of his fame is reared before his death; formed by the number, talents, and usefulness, of those he has instructed.[27]

When in his nineties Dr. Holyoke still performed delicate operations, and on the morning of his hundredth birthday he shaved and dressed himself and walked briskly to the Essex House to attend a great dinner in his honor.

He appeared in perfect health, and his firm and elastic step, his cheerful and benevolent looks, his easy and graceful manners, the model of the old school of gentlemen, his nicely powdered wig, his dress arranged with studied neatness, and just enough of antiquated fashion to remind one that he belonged to the generation gone by, but not outraging the proprieties of the present mode, his accustomed nosegay slipped through his button-hole, and his affectionate and grateful greeting of those who had assembled to do him honor, will never be forgotten. . . . He partook of the hilarity of the occasion with an evident zest.[28]

The Doctor walked home steadily and sat down to write out at great length the secret of his longevity. This was essentially moderation, with plenty of fruit. His drink was restricted to a half pint at dinner and another with his pipe afterward. It consisted of two parts of rum, three of cider, and ten of water. His nicotine was restricted to two pipes a day and "a small piece of pigtail tobacco" held in his mouth from breakfast to dinner, and another from dinner to tea.[29]

A friend who remarked on the Doctor's "energetic and rapid motion" that day, added:

There are some seams of care on his benevolent countenance, — a few traces of grief, dried up, — a little tremulousness of the voice, as he speaks of the past; but his whole visage betokens an unshaken confidence in himself.[30]

[27] [Samuel Lorenzo Knapp], *Extracts from the Journal of Marshal Soult* (Newburyport, 1817), pp. 78–79.

[28] *Memoir*, p. 33.

[29] Clipping in the Harvard University Archives.

[30] John B. Derby, *Reminiscences of Salem* (Boston, 1847), pp. 4–5.

There were, of course, good reasons for his grief. His wife Mary had died on April 15, 1802, and from that day he was cared for by his daughter Margaret, his dear companion who shared his interests. Her death on January 25, 1825, was a crushing blow. With great interest he kept a careful record of the progress of his dissolution, arguing with his fellow physicians about each symptom, and trying to draw conclusions about the general nature of senility. When death came, on March 31, 1829, all of the church bells of the town were tolled, an honor hitherto shown only to presidents of the United States. He must have been delighted when the autopsy showed, as he had maintained and his colleagues had doubted, that a fluid had formed in his brain. Following his instructions they made a thorough dissection.

Dr. Holyoke's children did not share his distinction.[31] His diaries for 1742–44 and 1746–47 are printed in *The Holyoke Diaries* with similar family records. That of his sister Priscilla for 1766, now in the American Antiquarian Society, is not included in this collection. His medical diaries and correspondence occupy thirty-two folio volumes in the Essex Institute, but unfortunately his more personal papers have not survived. The Essex Institute has oil portraits by James Frothingham and Charles Osgood, and a pastel by Benjamin Blyth is owned by Edward H. Osgood (A.B. 1938) and here reproduced by his kindness and by that of the Essex Institute.

[31] For the children see *The Holyoke Diaries*, pp. xv–xvi.

SAMUEL MOODY

SAMUEL MOODY, the first Master of Dummer Academy, was born on April 18, 1726, a son of the Reverend Joseph (A.B. 1718) and Lucy (White) Moody of York. The Parson, who was known to his contemporaries as "Handkerchief Moody" for his eccentricities, was the prototype of Hawthorne's "Minister of the Black Veil." His mental derangement compelled him to cease preaching in 1738, so Samuel had to struggle for an education. The college helped by giving him a waitership and Browne and Hollis scholarships. He was too busy to get into trouble, so the only considerable fine on his record is one for three weeks' unexcused absence. At a time when most scholars skimped Hebrew, he mastered it well enough to deliver a declamation in that language. His commonplace book shows a lively interest in science.[1] His name does not appear on the *Quaestio* sheet; perhaps he felt that he could not afford a Commencement. However, he received his M. A. with his class.

It was naturally expected that Moody would follow the family trade of preaching, but he had a low opinion of his abilities, and was discouraged by the fact that whenever he became excited he began to shake. He was not surprised when the people of the South Parish of Portsmouth, after hearing him, called another candidate. He attended the Falmouth Indian conference of 1749, and the next year spent a good part of the time preaching at Brentwood, New Hampshire. Early in 1751 he received a disappointment at Gloucester which was described by the successful candidate, Samuel Chandler (A.B. 1735):

Before I was sent for Mr. Samuel Moody had preached there Several months first as an assistant to his Grandfather White [A.B. 1698] who was infirm and unable to carry on the work of the Ministry, afterwards as a probationer for Settlement with Mr. Barnard [A.B. 1732], Mr. Bird [Class of 1744] and Mr. Dorby [A.B. 1747] and when they came to vote only Mr. Moody and Mr. Dorby were set up, one vote for Mr. Dorby, the Rest for Mr. Moody except the most of the principall men who were neuter and drew up a protest and sent to Mr. Moody by the committee that presented him with a copy of the

---

[1] The commonplace book is in the Harvard University Archives.

votes. The committee . . . came to York to treat with him. . . . Before they went out of Town Mr. Moody gave a negative answer.[2]

The next year he was preaching at Dunstable, and then on and off for several years at Gloucester.

Apparently no one took exception to Moody's character or theology, although it may well have been that there was not enough dogma and hell fire in his sermons to please the laymen of the region. He regarded himself as a Calvinist, but he held that it was both unprofitable and unbecoming to dwell upon the incomprehensible mysteries of Heaven. So firm were his views upon this point that in 1759 he boldly played David to the Goliath of the Calvinists by publishing *An Attempt to Point out the Fatal and Pernicious Consequences of the Rev. Mr. Joseph Bellamy's Doctrines.*[3] Bellamy's classic *True Religion Delineated* had attempted to show that only Calvinism provides a logical explanation of sin in a world created by a good God. Moody argued that "Mr. Bellamy had no Right to be so violently confident. . . . Great Dishonour is done to God, and infinite Prejudice to Religion, by being over curious and positive in Doctrines and Dispensations abstruse and mysterious. . . . God is holy; — hates Sin — cannot be the Author of it." With clarity and precision he demonstrated that the God of the Calvinists was a nauseating creature; far better to assume that man had created sin by exercise of the freedom of the will. Bellamy in *The Wisdom of God* replied that Moody was denying the absolute power of God who had created sin for His glory.

By this time Moody had given up all idea of following the ministry. In 1754 he had been appointed a Justice of the Peace for York, and he busied himself with such civil activities as witnessing the Penobscot treaty of that year. Much time he spent in trying to obtain from the General Court compensation for the sufferings of his grandfather, Samuel Moody (A.B. 1697), at Louisbourg and Port Royal. In 1760 the First Parish of York voted "to Mr. Samuel Moody with the concurrence of the Revd. Mr. Lyman [Yale 1747] Liberty and Priviledge of Erecting a House for the Instruction of Youth in the Lerned Languages" on parsonage land,

---

[2] Samuel Chandler, An Account of my Settlement (Essex Institute).

[3] The work is anonymous, but the views are those of Moody, and Harvard has what appears to be the printer's copy in his hand.

near the pound.[4] Here he soon made a name for himself, not so much for his skill as a teacher as for his will to teach. Typically, he met a poor boy on the street and said, "Willard, you must go to college." [5] This dictum he enforced by preparing the boy at York and sending him to Harvard College, of which he eventually became one of the better presidents.

Lieutenant Governor William Dummer had left an endowment for the establishment of a "Free Grammar School" which was established at Byfield and opened with twenty-eight boys on hand on February 27, 1763. Samuel Moody was unanimously chosen master by the trustees, who then retired from the picture, thinking that they had no further authority unless the master should lead "a profligate and wicked life," in which case they were to bring the matter to the attention of the Overseers of Harvard College. The Master's brother Joseph and his wife ran the academy farm and boarded the boys.

Moody at this time was a "stout, stalwart man" with strong features and a number of eccentricities, the most noticeable of which was his usual costume, a long green flannel gown and a tasselled smoking cap. He was a superb teacher of Latin and Greek, he had no use for mathematics, and he taught no science because in those days that was strictly a college subject. On the other hand he hired a Frenchman to teach dancing. Tradition has it that when the dancing master arrived, he and the preceptor stood bowing and scraping, each insisting that the other precede until Moody lost his temper, seized the little man by the collar and the slack of his pants, and threw him into the room.[6] The introduction of dancing into the curriculum created a great public uproar, and the opponents even took to verse:

> Ye sons of Byfield now draw near,
>   Leave worship for the dance,
> Nor farther walk in wisdom's ways,
>   But in the ways of France.[7]

Other innovations were an insistence on cleanliness and the en-

---

[4] Charles Edward Banks, *History of York, Maine* (Boston, 1935), II, 271.
[5] Sidney Willard, *Memories of Youth and Manhood* (Cambridge, 1855), I, 18–19.
[6] *Old-Time New England*, XLIV (1954), 48.
[7] Sarah Anna Emery, *Reminiscences of a Nonagenarian* (Newburyport, 1879), p. 161.

couragement of exercise. The Master in warm weather kept watch of the state of the tide and when that was favorable suspended classes to go swimming with the boys. During study hours he permitted the boys to read their lessons out loud and to walk around the room talking to each other; one rap on his desk was all that was needed to restore order. He never spoke evil of anyone, and when he erred it was in the direction of excessive praise. His account of a clash with one of his boys is typical:

[Edward] Preble was standing by the fire-place or stove of the school-room, in violation of a rule, and was ordered peremptorily to take his seat; but not promptly obeying, the master, provoked by the delay, approached him angrily, seized the shovel, and by his attitude with the weapon threatened to break the boy's head. "But the boy," said Moody, "neither flinched nor winked; he disarmed me; I looked him full in the face and exclaimed, — Preble, you are a hero." [8]

He used to boast that in thirty years of teaching he had never had to use the rod.[9] He had a genius for inspiring boys to do their best, telling them, "Crede quod Possis, et potes." His boys at Byfield worshiped him, and at college found that they were better prepared than their rivals. Out in the world "the habits of promptness, independence of thought, exactness, and thoroughness which he formed in them" persisted and made them successful. Consequently after a few academic generations Master Moody "was the subject of greater veneration and applause than the President of any College in America," in some circles at least.

Dummer Academy during the period of Moody's control contributed 25½ percent of the students entering Harvard, and a lesser but considerable proportion in other colleges.[10] In 1771 he and Governor John Wentworth (A.B. 1755) attended the first Dartmouth Commencement and rejoiced "to see the solitary gloom of the wilderness give place to the light of science, social order and religion." [11] He led the movement to have the friends of Dartmouth College buy a handsome piece of plate for President Wheelock. Appropriately, Dartmouth gave him an M. A. in

[8] Willard, I, 21 *n*.
[9] Charles C. P. Moody, *Biographical Sketches of the Moody Family* (Boston, 1847), pp. 121–122.
[10] James Duncan Phillips, "Harvard College and Governor Dummer's School," in Mass. Hist. Soc., *Proceedings*, LXIX, 199 ff.
[11] David McClure, *Diary* (New York, 1899), pp. 23–24.

1779. He was interested in the settlement of the town of Dummer, New Hampshire, which he named.[12]

In 1779 Master Moody made claims against Harvard on behalf of Dummer, but nothing serious developed.[13] He loved to attend Commencement at Cambridge and to be greeted by his former pupils. When in 1787 the Corporation voted to have a private Commencement, he protested bitterly to President Willard:

A Public Commencement I conceive by far the most glorious Show in America, and when the Exhibitions are fine and well-conducted, reflect the greatest Lustre on the Officers of the College, raise to the highest pitch the Ambition and Efforts of the Youth, carry the Applause of a large Assembly of brillant, learned and respectable Spectators, spread the Fame of the University far and wide, and make large Accessions to the Numbers.[14]

According to the Willards, Master Moody was not always happy at the Harvard Commencements:

At one of these annual dinners he was sorely disturbed and scandalized at the poor, mean, and cold provisions on the tables, and left them, accompanied by Professor Pearson, early enough to repair to the President's house in season for the dessert. But his complacency was not immediately restored by the happy faces and inviting fruits before him. He began upon his grievances forthwith, and after exhausting his vocabulary for terms of denunciation of the dinner in the hall, as if in despair of reaching the climax, he called upon Mr. Pearson, as professor of languages, for epithets of execration strong enough to supply his deficiencies.[15]

Moody took a substantial interest in public affairs, particularly after his appointment as Justice of the Peace in 1768. Two years later he was campaigning for a subscription to endow a writing school under the wing of the academy, the money to lapse to the college if the plan failed.[16] Not succeeding in this, he personally gave £100 to the town grammar school. At the time of the Boston Port Bill he gave 5 Guineas to the poor of Boston and collected another £7 from his boys.[17] In August, 1775, the town of New-

[12] For Moody's correspondence in regard to these lands see Benjamin Cutler, *History of the Cutler Family* (Boston, 1871), pp. 316 ff.

[13] Corporation Records (Harvard University Archives), III, 46.

[14] John Louis Ewell, *The Story of Byfield* (Boston, 1904), pp. 311–312.

[15] Willard, I, 21 *n*.

[16] There is a subscription form in the Harvard University Archives, U. A. I.5.120 (1770).

[17] *Boston Post-Boy*, Nov. 21, 1774, 3/2.

bury elected him to the House of Representatives, and the new government appointed him a Justice of the Peace, an office which he held until his mind failed. He was a charter member of the American Academy of Arts and Sciences, and a Federalist in politics.

By 1782 the family weakness of mind was showing up in Master Moody. For example, he once begged a loaf of bread, saying that his boys were starving, and taking it to the academy he beckoned them to come out and share it with him. In 1778 he had boasted that Dummer was filled to its capacity of sixty boys, and that he was each year turning away enough to fill Phillips Andover,[18] but a dozen years later it was reported that Dummer was empty and Andover flourishing. The Reverend Charles Chauncy (A.B. 1721) led the movement to have the charter changed in order to give the trustees more authority, but it was not until 1790 that they accomplished the delicate and painful task of obtaining the Master's resignation.

Moody returned to his old home at York and had some idea of accepting the preceptorship of Berwick Academy, but he obtained too much pleasure from traveling about to visit his former students to settle long anywhere. Sidney Willard left a description of these always welcome visits:

I remember him as a frequent visitor at President Willard's. . . . He generally arrived late in the afternoon, on horseback, rode into the yard, called for the male servant, gave him directions for the care of his horse, brought his portmanteau into the house, and entered the parlor, as he well knew, a welcome guest. After tea, if he was not disposed to sally forth on a visit to others of his former pupils, he would call for bootjack and slippers, and robe himself in his study-gown and belt; as much as to say, "I'll now have a good, cosey time." If an enterprising fit seized him, he will call upon the Professors, and, if so inclined, summon them to supper at the Presidents.[19]

More and more, however, these trips became the irrational wanderings of a man with a disordered mind. On December 24, 1795, returning from an extended visit to Boston, he arrived at the Exeter home of William Parker (A.B. 1751) in good spirits but exhausted by his struggle against a severe storm.

[18] Benjamin Guild, Diary (Mass. Hist. Soc.), Nov., 1778.
[19] Willard, I, 20 *n*.

He drank a Cup of Tea and complained of being sick at his Stomach. Proposed a walk . . . appeared to be weak and totering, but still in good Spirits . . . he got up and after walking several times across the room, fell suddenly and expired in about five minutes.[20]

His body was carried to the meetinghouse at York, and thence to the old graveyard of the parish. He never married, but he founded a remarkable pedagogical line.[21] Without him, Dummer Academy almost withered away.

PEASLEE COLLINS

PEASLEE COLLINS was born on February 12, 1728/9, a son of Elijah and Abigail (Peasley) Collins of Boston. They were Quakers, and Peaslee was the first member of that denomination to be admitted to college since George Hussey of the Class of 1715, and it was probably for that reason that he was favored in the placing of the class; but the Faculty soon found that like his predecessor he was troublesome. At various times he was fined for "Cursing," for making "tumultuous Noises," and "for defiling the uninhabited parts of the College." When ordered to stand up in the college Hall to receive public admonition for these sins, he "Smiled and fleered in a most contemptuous Manner." For this he was degraded to next to last place in his class, only the charity student being below him, and in that position he was graduated. A year later the Faculty soundly rejected a petition that he be restored:

Whereas Sir Collins stands degraded . . . . there was this Day Sent in to Us a Paper from him (who is at Rhode Island) which he designed should be taken by us as a Confession of his Fault, but as there is nothing of that Nature in it, it was rejected; so that the Said Collins abides in the Place to which he was degraded.[1]

In June, 1750, a more humble petition resulted in his restoration in time for Commencement, when for his M. A. he offered the

[20] Jonathan Sayward, Diary (Am. Antiq. Soc.), Dec. 1795; Salem Gazette, Dec. 29, 1795.
[21] "Pedagogical Genealogy" in Harriet Webster Marr, The Old New England Academies (New York, 1959), p. 74.
[1] Faculty Records (Harvard University Archives), I, 270.

affirmative of "An Innovationes in Religione ex Affectionibus hypochondriacis, non rarò Ortum ducant?"

This topic is significant in connection with Collins' later life, for he went to Europe to study medicine and, according to President John Adams, "came back a disbeliever of every thing; fully satisfied that religion was a cheat, a cunning invention of priests and politicians; that there would be no future state, any more than there is at present any moral government."[2] Such views were no bar to social advancement in colonial Boston, however, for on November 12, 1754, he registered his intention of marrying Mary, daughter of John (A.B. 1731) and Mary (Deming) Avery. This marriage was of short duration, for the bridegroom died on April 30, 1755. His obituary said:

This Morning died here of a Consumption, Mr. Peaslee Collins, a Physician of this town: He was a young Gentleman (26 years of Age) liberally educated, and esteem'd by the Gentlemen of his Generation [?], for his uncommon natural Powers, as well [as?] Learning, and the Knowledge of Physick; he was universally beloved and esteemed by all those who were well acquainted with him, for his Humanity and Benevolence, as also for his great Integrity, as well as Gentleness of Manners, and indeed for all the Virtues that are the ornament of social Life.[3]

His widow married John Collins of Newport on May 23, 1757.

[2] John Adams, *Works* (Boston, 1850–56), II, 13.
[3] *Boston Weekly News-Letter*, May 1, 1755, 2/1.

ISAAC GARDNER

MAJOR ISAAC GARDNER, the first Harvard man to fall in the Revolution, was born in Brookline on May 9, 1726, a son of Deacon Isaac and Susannah (Heath) Gardner. Early in his Freshman year he was fined for "drinking prohibited Liquor" while serving as a waiter at a meeting of the Senior class, and later in his career he was punished for drinking, gambling, "disorderly behaviour, etc." For his M. A. he chose to argue that a marriage dominated by love of money is unworthy. On April 26, 1753, he was married to Mary, daughter of Thomas and Mary (Oliver) Sparhawk.

Gardner settled as a farmer in his native town. During his first years of married life he apparently lived in the southern part, for he joined the Roxbury Church. In 1767 he inherited the family homestead nearer the Brookline meetinghouse. Here it was long remembered that he "was a delightful singer and lead the Choir." From 1758 he served the town as well as the church, holding perennially the offices of clerk, treasurer, selectman, moderator, and custodian of the Devotion School Fund. In 1761 he was appointed a Justice of the Peace. In the militia he attained the rank of major. So essential had "Squire Gardner" become that when the homestead burned in 1768 leaving its eighteen inhabitants without a roof, the town voted him £100 to assist in rebuilding, and the ladies held spinning bees for the family.[1]

In the political activities which preceded the Revolution, the Major took an active part. He was a member of the Committee of Correspondence and of the Suffolk Convention of September, 1774, and served on the delegation of the latter which went to General Gage to protest the fortification of Boston Neck. He had information of the impending march on Concord, and one of his daughters in later years made a part of local tradition the story of the night spent in making cartridges, of the last meal, at which neither parent could eat, and of the parting made agonizing by premonition.[2]

[1] *Boston News-Letter*, Sept. 8, 1768, 2/3; *Boston Evening-Post*, May 8, 1769, 2/2.
[2] Brookline Historical Society, *Publications*, I, 19–20.

After the column of Regulars had passed within sight of the Gardner home, the militia gathered at the meetinghouse, where the Major took command of one of the two irregular companies of minutemen. They followed the British, passing the Charles on the trestles of the bridge, and moved to Watson's Corner, about a mile north of the college. There they stationed themselves behind some empty casks in order to ambush the returning Regulars. The Major had left his company to get a drink at a well when he was surprised by a company of light infantry. He fell with a dozen balls in his body, but as the troops paused to bayonet his riddled body all of the other Brookline men escaped. Dr. William Aspenwall (A.B. 1764), who was to marry the Major's daughter Susanna six weeks later, carried the riddled body back to his home where he kept it hidden for two days lest its mangled condition arouse the town to a frenzy. Only the eldest son, Isaac, who later became a general, saw the remains before they were buried.

In the colony, broadside verse lamented the Major's death as that of a patriot in arms, but the English Whigs were fed a different story:

The groundless and inhuman reflection cast upon I. Gardner Esqr; one of his Majesty's Justices of the Peace who is said to have been killed, fighting against his Sovereign, and is held up as a specimen of New England Magistrates, ought, in justice to the deceased, as well as to truth, to be set right. This unfortunate Gentleman was not in arms, but returning to his family from a long journey, and lodged at Lexington the night preceeding the action; early in the morning of which fatal day he set out for home, and on the road, being unarmed, he was barbarously shot in cold blood by a Scotch grenadier of the King's own regiment, though he begged for mercy, and declared solemnly he had taken no part in that days disturbance. He has left a widow and a large family of young children, who, it is hoped his most gracious Majesty will provide for.[3]

An angry refutation of this tale in the British press contained a still more colorful story:

Isaac Gardner, one of His Majesty's Justices of the Peace, was not killed as he was *peaceably* riding along, but was killed in the *very act of attacking* the King's troops. The rebels in their own accounts, confess this, and confute Mr. Potatoe Head's falsehoods. Their account,

___

[3] [John Almon], *The Remembrancer . . . for the Year MDCCLXXV* (London, n. d.), p. 83.

dated the 24th of April, says that Isaac Gardner took 9 prisioners, that 12 soldiers deserted to him, and that his ambush proved fatal to Lord Percy and another general officer, who were killed the first fire.[4]

All of this was small comfort to his widow, who died on December 26, 1778. Of their eight surviving children, Sibil married Cornelius Waters (A.M. 1788).[5] Of the Major's manuscripts, the only significant remnant is his interleaved almanac for 1754, now in the library of the New England Historic Genealogical Society.

EBENEZER STORER

EBENEZER STORER, Treasurer of Harvard College, was born on January 27, 1729/30, the fifth child of Captain Ebenezer and Mary (Edwards) Storer of Boston. The family had for generations fought Indians in Maine, but the Captain, weary of this strife, had moved to Boston where he became a merchant, a Justice of the Peace, and a Deacon of the Church in Brattle Square. His kindness, charity, and devotion to good works were famous. His son Eben from his college days showed the same earnest piety, an introspective religious concern which belonged to the previous century rather than his own.[1] When taking his M. A. in 1750 he naturally chose to hold the negative side of the *Quaestio*, "An Ex Appetitu sensitivo, major sit Voluptas quam ex rationali?" That same year he went down to the New Haven Commencement with a group of young Harvard graduates who were graciously given *ad eundem* degrees by the Yale Corporation.

Storer settled in Boston where on July 17, 1751, he was married to Elizabeth, daughter of Joseph and Anna (Pierce) Green. He entered his father's warehouse as an apprentice, but by 1754 the concern was known as Ebenezer Storer & Son. In 1759 the Cap-

---

[4] *Gazetteer and New Daily Advertiser*, July 4, 1775, quoted in John Gould Curtis, *History of the Town of Brookline* (Boston and New York, 1933), p. 148.

[5] For the children see the *New England Magazine*, XII (Mar., 1895), 108–109.

[1] Storer's *Admitatur* is in the Harvard University Archives. His "diary" (at the New England Historic Genealogical Society), covering the years 1750–1764, consists of nothing but religious meditations.

EBENEZER STORER

tain, feeling death approaching, deeded to Eben the great three-story wooden mansion which he had built years before at the corner of Sudbury and Portland streets, with pleasant gardens running to Hanover Street. This contained the office in which the Storers negotiated the sale of the goods in the warehouse, sold tickets for the Newbury Bridge Lottery, and the like. Among similar business houses their holdings of Boston real estate were relatively large, and their investment in Maine and New Hampshire wild lands, relatively small.

In 1751 Eben joined the Church in Brattle Square, in which he took over his father's prominent place, serving, for example, on the committee to rebuild the meetinghouse. Although he resigned as Deacon on August 1, 1773, he was known by that title for the rest of his life, and in subsequent years he represented the Church in Brattle Square at no less than thirteen ordinations. When in the next century a new street was run through his property it was named Deacon Street.

Storer, like his father, was an equally faithful servant of the town. In 1752 he paid a fine rather than serve as constable, but in later years he served as warden, Selectman, and particularly as an Overseer of the Poor, his father's favorite office. He served on the regular committees, such as those to visit the poor and to inspect the schools, and on special committees such as that to consider what steps should be taken "to give a check to the progress of Vice and Immoralities in this Town, and to promote a Reformation of Manners." Perhaps connected with the last was his service on the committee to introduce street lights. On the one occasion when he was on the wrong side of the law, he was called before the Selectmen for having for nine years sheltered one Priscilla Hayden of Braintree instead of warning her out of town.

The Deacon was famous in his generation for his orderly mind and his meticulous industry, so he was called upon to serve on many organizations like the Trustees of the Hopkins Foundation and the Massachusetts Society for Propagation of Christian Knowledge among the Indians of North America. The latter organization was killed by the jealous Episcopalians, but the New England Company, a British missionary organization, appointed Storer one of its American commissioners. Probably he most enjoyed the very small and select Company for Promoting of Good Order (Re-

ligious) which was headed by Dr. Joseph Sewall (A.B. 1707).[2]
He served also on the Society for Encouraging Trade and Commerce, and joined in discouraging it by signing the non-importation agreements. His signature appears on the political Appeal to the World issued by the town of Boston in 1769, and he served on the committee "to counteract the designs of those inveterate Enemies among us, who there is reason to think are still continuing their misrepresentations, and using their Endeavours to increase the present unhappy misunderstanding between Great Britain and the Colonies." As a member of the town committee on the Boston Massacre he visited Captain Thomas Preston in jail to protest his testimony, and as a member of the committee dispatched to bring Salem, Marblehead, and Newbury back into the non-importation fold, he was chased back into Boston with threats of tar and feathers. In July, 1774, he was with his Loyalist brother-in-law, Francis Green (A.B. 1760), when the latter was mobbed in Windham, Connecticut.[3] Storer was no rubber-stamp Whig, for when Nathaniel Ray Thomas (A.B. 1751) was accused of treasonable utterances, he flew to his defence with a vigor which caused John Hancock and Dr. Warren to drop the matter hastily.[4]

There is one strange story told of Deacon Storer which would be dismissed as impossible had it not been carefully reported by Ezra Stiles after he had dined one day in 1773 with Chief Justice Peter Oliver (A.B. 1730) and others who would have known the facts:

Mr. Storer of Boston suffered in the Stamp Act 1765 and went home for Redress, The Ministry put him off, till he should obtain Governor Hutchinson's Recommendation, and indeed it was finally referred to the Governor to provide for him some provincial office. It has not been done. Mr. Storer to have a Rod over &c procured 18 letters of Lieutenant Governor Oliver and half a dozen of Governor Hutchinson to one of the Secretaries of some of the Ministerial Boards in London, as a specimen of their Correspondence for 15 years past urging and recommending the present arbitrary Government over the Colonies. The Governors Hutchinson and Oliver were last year given to understand that Mr. Storer had them in his power by means of a Collection of these Letters, and that the only Condition of not exposing them was his

[2] For a list of members see the Smith-Carter Mss. (Mass. Hist. Soc.), Nov. 1761.
[3] *American Archives*, 4th Ser., I, 629–634.
[4] Papers laid in Eliza S. Quincy, Ms. Memoir (Mass. Hist. Soc.), I, 110a.

being provided for. The matter was neglected. Judge Oliver now here took occasion to ask the Governor whether there was any Danger &c. when the Governor said he was under no Apprehensions.[5]

The only evidence supporting this story is the public notice taken of Storer's sudden resignation of his deaconship at the time of the publication of the Hutchinson-Oliver letters.

These were bad years for the Deacon. On December 8, 1774, his wife died, "very greatly and justly lamented." She "excell'd in all the Characters of Wife, Parent, Friend, Mistress, and Benefactor. Those who knew her the most intimately, loved her the most sincerely. If she had a Fault, it was a too tender Heart, and too great a fondness for her Friends."[6] Driven from Boston by the outbreak of war, the Deacon shocked Ebenezer Parkman (A.B. 1721) by traveling on the Sabbath while searching for shelter.[7] He spent the first winter of the war in Needham, and in February, 1776, went back into Boston under a flag of truce. He was immediately elected Overseer of the Poor and Selectmen, and reluctantly agreed to serve one year in order to get the town back onto its feet.

Harvard College called upon Deacon Storer to perform an even more difficult task, for John Hancock, as Treasurer, had grossly neglected his duties, and now, with his usual arrogance, refused even to return the records in his possession. The Deacon was known to the Corporation as a generous donor of books and scientific apparatus, but particularly for his financial skill. When he took office as Treasurer on July 14, 1777, the currency was swiftly becoming worthless and accounts had to be kept in a complicated manner because of the different media in circulation. This is not the place to record the remarkable story of his success,[8] which was due in no small measure to his "tolerant and pacific" disposition and his "naturally mild and social temper." He got on well with the Corporation, and with the very touchy and retiring President Langdon. His friendship with President Willard was par-

---

[5] Ezra Stiles, *Literary Diary* (New York, 1901), I, 380–381.

[6] *Boston News-Letter*, Dec. 16, 1774, 3/3. For an elegiac poem see the *Boston Evening-Post*, Dec. 12, 1774, 3/1.

[7] Ebenezer Parkman, Diary (Am. Antiq. Soc.), Oct. 15, 1775.

[8] See particularly Andrew McFarland Davis, "The Investments of Harvard College: an Episode in the Finances of the Revolution," in *The Quarterly Journal of Economics*, XX (May, 1906), 399–418.

ticularly warm, and the two wings which he added to Wadsworth House survive as a reminder of it.

A host of new friendships resulted as a result of the Deacon's marriage on November 13, 1777, to Hannah, daughter of Colonel Josiah (A.B. 1728) and Hannah (Sturgis) Quincy, and widow of Dr. Bela Lincoln (A.B. 1754). From the ceremony, which was performed at the Colonel's mansion in Braintree, they returned to Boston and the problem of keeping soul and body together in a difficult period. The Deacon took out a license to sell wine and spirits, and assumed such odd jobs as serving as paymaster and clothier of the Second Massachusetts Regiment. In 1788 he sold off great quantities of Boston real estate. The fact that he appears to have sold and then shortly bought back his mansion house suggests that he may not have been acting because of financial stringency but in order to take advantage of currency fluctuations. During these years he had no time for town offices, but in 1779 he did serve as a member of the constitutional convention. Service to institutions took more of his time than ever. He was a founder of the American Academy of Arts and Sciences, and its first Treasurer. He was a charter member of the new Society for Propagating the Gospel in North America, and for some seventeen years its Treasurer. Unfortunately he was not elected to the Massachusetts Historical Society, but he did present an old broadside to its library. Of the new business corporations, he was active only in the Boston Tontine Association.

The Deacon's good works included social entertaining, and the great old house in Sudbury Street became popular for its hospitality, particularly on Sunday evenings, and particularly for the clans united by the Quincy tie.[9] One member of the group left a fond picture of the old house:

From the gate, a broad walk of red sandstone separated it from a grass-plot which formed the court-yard, and passed the front-door to the office of Mr. Storer. . . . The vestibule of the house, from which a staircase ascended, opened on either side into the dining and drawing rooms. Both had windows toward the court-yard, and also opened by glazed doors into a garden behind the house. They were long, low

[9] For an account of such a Sunday evening see Edmund Quincy, *Life of Josiah Quincy* (Boston, 1867), p. 45.

apartments; the walls wainscoted and panelled, the furniture of carved mahogany. The ceilings were traversed through the length of the rooms by a large beam, cased and finished like the walls; and from the centre of each depended a glass globe, which reflected, as in a convex mirror, all surrounding objects. There was a rich Persian carpet in the drawing-room, the colors crimson and green. The curtains and the cushions of the window-seats and chairs were of green damask; and oval mirrors and girandoles, and a teaset of rich china on a carved table, completed the furniture of that apartment. The wide chimney-place in the dining-room was lined and surrounded by Dutch tiles, and on each side stood capacious armchairs . . . for the master and mistress of the family.[10]

The Storer business had never recovered from the war, and the Deacon was too busy with his unpaid treasurerships to have paid much attention to it anyway, so the Quincy clan assumed the responsibility of finding a regular income for him. In November, 1797, Abigail Adams announced that he had been appointed inspector of Massachusetts district No. 3 of the Excise Office of the United States:

There has been no opening until the *removal* of Mr. [Leonard] Jarvis. The publick will be *no looser* I presume by the exchange. His Successor's honour, integrity, and probity are there Shure pledge and Security. Without these qualities the president would not knowingly appoint to office any man, however nearly connected, or otherwise bound to him.[11]

Abigail would probably not have been so complaisant about this appointment had she known of the power which Hannah (Quincy, Lincoln) Storer still possessed to stir the Presidential heart.[12]

Deacon Storer held this office until replaced by a Jeffersonian, and then he served for two years as Treasurer and Collector of Taxes for the Town of Boston. It was a contemporary wonder that such an old gentleman could keep the complex business of his treasurerships from confusion, and the secret of his success seems to have been that he did everything the moment that it should have been done.

The last day of his life was marked by the same noiseless regularity and attention to business, which had distinguished his whole career. In his

---

[10] *Memoir of the Life of Eliza S. M. Quincy* (Boston, 1861), pp. 89–90.
[11] Mass. Hist. Soc., *Proceedings*, LXII, 24.
[12] Josiah Quincy, *Figures of the Past* (Boston, 1883), pp. 64–65.

usual health, he retired to rest on the night of the 5th inst. [January 5, 1807] and sunk quietly into a sleep, from which he never awoke.[13]

The Deacon's extensive library and collection of scientific instruments was sold at auction, and many of the latter were bought by Josiah Quincy, who when president gave them to the college. In 1843 the Deacon's "comet-catcher" turned out to be the best instrument in Cambridge to catch the comet of that year.[14] His widow lived until 1826. He is said to have been survived by four children by his first marriage and three by his second, but only one of the latter has been identified: (1) Ebenezer, b. Aug. 10, 1752. (2) Ebenezer, b. Feb. 17, 1754; m. Eunice Brewster, 1780. (3) Elizabeth, b. Feb. 23, 1756; m. John Atkinson, Dec. 2, 1773. (4) Mary, b. Nov. 11, 1758; m. Seth Johnson of New York, Apr. 21, 1796. (5) Charles, b. Mar. 4, 1761; A.B. 1779. (6) George, b. Aug. 19, 1764; m. Anna Bulfinch, May 26, 1795. (7) Hannah, b. May 10, 1779. There are two Copley portraits of the Deacon, one owned by the Metropolitan Museum of Art, New York, and the other owned by John P. Sedgwick (M.B.A. 1922) of Wilton, New Hampshire, and deposited at the Worcester Art Museum.

[13] *Columbian Centinel*, Jan. 10, 1807, 2/5.
[14] *Memoir of Eliza Quincy*, p. 90.

SAMUEL FRENCH

SAMUEL FRENCH was born on July 2, 1729, a son of Samuel and Bathsheba (Beal) French who lived on Weymouth Road in Hingham. At college he roomed with his classmate Lesslie, waited on table, and held Fitch and Hollis scholarships. After graduating he kept school at Littleton for a time and then moved to Hingham where he taught while reading for the ministry with his pastor, Daniel Shute (A.B. 1743). Either diffidence or poor health kept him from returning to Cambridge to take his M. A. with his class. He died on May 21, 1752, and was buried in Fort Hill Cemetery where his stone says, "Here lyes the body of young Samuel worn out with Study into dust it fell who did in knowledge and in virtue shine a learned schoolar and a good divine." A century later his memory was still warm in Hingham:

An old lady who was not born till nearly twenty years after his death . . . remembers his room, which, with affectionate care, was long kept in the condition in which he left it, and was called by the children of the household, "Uncle Sam's Study". . . . Another old person relates this anecdote of him. Too modest to preach his first sermon in his native town, he accepted an invitation to do so in an obscure parish in Scituate. On returning in the evening, he met Dr. Shute . . . who enquired how he had succeeded in his duties of the day? His reply was, "I did not wait to see." [1]

[1] *Our Old Burial Grounds* (Hingham, 1842), p. 12.

JONATHAN SEWALL

JUDGE JONATHAN SEWALL, a son of Jonathan and Mary (Payne) Sewall of Boston, was born on August 24, 1729, and baptized at the Brattle Street Church a week later. His father, an unsuccessful merchant, died young, leaving his family destitute. Little Jonathan, however, "early discovered a pregnant genius," and was educated on funds which his pastor, William Cooper (A.B. 1712), wrung from his wealthy parishioners. His uncle, Chief Justice Stephen Sewall (A.B. 1721), met part of the expense of sending him to college, where he received Mills and Hollis scholarships. He did well enough until Senior year, when he was degraded for burning fences on Guy Fawkes' eve. Upon a "humble Confession" made with "proper Decency" he was restored to his place before graduation.

Sewall went to Watertown to keep the town school, but on the night of May 11, 1749, was back in Cambridge when there were "several loud Hurras made in the College Yard" and a large stone was thrown through the window of Tutor Mayhew with the evident intent of hitting that gentleman as he lay in bed. Apprehended, Sewall vigorously denied any participation in the disorder, and was allowed to return to Watertown. Mayhew, however, calculated the parabola of the missile and demonstrated that it must have been thrown from the point at which witnesses placed Sewall. The Faculty then recalled him from Watertown and threatened him with civil action, upon which he confessed and was expelled.[1] A sadder and wiser man, he went back to work, and during the next two years he kept the Woburn and Salem schools. When the time for his second degree approached he presented a petition for restoration which was unanimously rejected by the Faculty as unsatisfactory. A second and "much more humble Confession" met the same fate, but when he returned with a third document and stood in the college chapel in "a very becoming Manner" while it was read, the Faculty and Corporation voted that he should be restored. For his *Quaestio* he prepared the affirmative of "An in Amicitiâ inæquali, sit plus tribuendum et retribuendum majori, et minus minori?"

[1] Faculty Records (Harvard University Archives), I, 285–292.

During the next five years Sewall kept the Salem school and frequented the law office of John Higginson where he performed small services in return for instruction and access to the records. He never forgot the kindness of the Higginsons and their generosity with "Wine, Beer, Tea, etc., etc," at a time when his small income would not permit him to enjoy such luxuries otherwise.[2] Through Higginson he met Judge Chambers Russell (A.B. 1731) of Charlestown, who took him into his home, furnished him with books, instructed him, and prepared him for the practice of the law. Even with this backing, such were his doubts of his ability to present a case without collapsing in a panic, that he wanted to begin his practice in a modest manner in the lower courts, but he was induced reluctantly to the "courageous undertaking of a cause of great importance," which he won.[3] Thus encouraged, he opened a law office in Charlestown where he took over the practice of Judge Russell, who had moved to Concord.

Sewall soon lost his diffidence about the practice of the law. He never shared the earnestness of the scholarly lawyers, but was wont to describe the law with the light gayety which was his normal attitude toward everything:

It is fathomable only by those intellects whose naturally squabthick powers have been, as it were, wire-drawn to a sufficient longitude by diurnal and nocturnal porings, poundings, dreamings, sleepings, cursings and swearings over codes, pandects, digests, institutes, abridgments . . . in infinitum. This is but a faint glimpse of the immense sea of legal — what shall I call it? — chaotical confusion of primary principals of abstract ratiocination, thro' which for the tedious space of 3 lives, or 21 years, we toil, tug, labour, plunge, paddle, scramble, wallow, and from which . . . we at length emerge with souls fitly enlarged and enlightened for assuming the guardian ship of the lives, liberties, and properties of our ignorant fellow men.[4]

His correspondence is larded with tags from the better poets and gaily bad verse of his own. With his Salem friend Thomas Robie he hoaxed the Charlestown minister, Simon Bradstreet (A.B. 1728)[5]; it is not surprising that he found that sober town unfriendly.

[2] Robie-Sewall Mss. (Mass. Hist. Soc.), May 10, 1757.
[3] Edmund Quincy to Samuel Sewall, Quincy Mss. (Mass. Hist. Soc.), Oct. 23, 1778.
[4] Mass. Hist. Soc., *Proceedings*, 2nd Ser., X, 409.
[5] Robie-Sewall Mss., 1757.

Attracted by the personality shown in some letters written by a young law student named John Adams (A.B. 1755), Sewall opened a correspondence with him which developed into a warm friendship. They followed the Supreme Court together, frequently sharing a bed and the crumbs of practice which fell from the table of James Otis (A.B. 1743). Adams found his friend charming:

He possessed a lively wit, a pleasing humor, a brilliant imagination, great subtlety of reasoning, and an insinuating eloquence. . . . I know not that I have ever delighted more in the friendship of any man, or more deeply regretted an irreconcilable difference in judgment in public opinions. He had virtues to be esteemed, qualities to be loved, and talents to be admired.[6]

He particularly admired the older man because he kept his social graces in proper subjection; he could dance well, but not well enough to make puritanical minds suspicious of him.

In their early years together these two men saw eye to eye on public questions. Meeting Adams one day in the Boston town house, Sewall took him aside and said:

These Englishmen are going to play the devil with us. They will over-turn every thing. We must resist them, and that by force. I wish you would write in the newspapers, and urge a general attention to the militia, to their exercises and discipline, for we must resist in arms.[7]

Adams replied that Sewall, who was much the better writer, should undertake this task, but the older man said that if he did so, Attorney General Trowbridge (A.B. 1728) would break him. Sewall foresaw the troubled times to come, and realized that a man might rise to fame in them. Writing to Adams in 1760 he pointed out that a man's fame rested on the greatness of his country, his field of activity, and said that his friend might be remembered in future ages after New England had "risen to its intended grandeur."[8] For his part, perhaps because he remembered the poverty of his youth, he would play safe and not antagonize the administration.

After the death of Chief Justice Sewall in 1760, Jonathan was drawn into the Tory camp. He had dearly loved his uncle, and he thought to save his memory from the stain of bankruptcy by getting the General Court to assume the debts of the estate. He

---

[6] John Adams, Works (*Boston*, 1850–56), II, 78.
[7] *Ibid.*, IV, 6.
[8] *Ibid.*, I, 51.

asked the Otises to present his petition, and when it was rejected he accepted the suggestion of Trowbridge and Thomas Hutchinson (A.B. 1727) that he had been doublecrossed by his agents. The administration, quick to encourage loyalism, appointed him a Justice of the Peace for Middlesex in January, 1762, and after he had donned barrister's habits in August of that year,[9] saw to it that he had profitable business. It would be unfair, however, to conclude that he undertook the public defence of Bernard because of these bribes, for he never could stomach politics on the level of the untruthful and vilifying attacks of the Otis faction on the Governor. John Adams replied to him on a level on which he was glad to argue, and their controversy in the pages of the *Boston Gazette* and the *Boston Evening-Post* attracted wide attention. Over the signature "Philanthrop" Sewall contributed an argument which filled the entire first page of the *Evening-Post* of January 26, 1767, and to the satisfaction of the administration disposed of the "calumniators" of the Governor. Naturally he was accused of being the author of the savage "Jemmibullero" attack on the younger Otis in the *Evening-Post* for May 13, 1765, but as he pointed out he did not fight in that way. In his private correspondence he would denounce an opponent in a newspaper controversy as "a mean Cowardly Scoundrel" who had made a "low, pitiful, Grubstreet, Billingsgate, rascally, dastardly, Indian-like, c——d, d——'d impudent Attack upon our right noble selves,"[10] but he always had his tongue in his cheek when he indulged in such violent language, and in his published writings he maintained that decorum which the native Tories thought that public conduct demanded.

During these years of public controversy, Sewall and Adams remained close personal friends. The older man turned over to the younger his sheep-stealing clients, and tried to get Governor Bernard to appoint him a Justice of the Peace in spite of his political position.[11] He frequently stayed with the Adamses in Braintree from Saturday to Monday, and he employed the time to court Esther Quincy, daughter of Edmund (A.B. 1722) and Elizabeth (Wendell) Quincy, and sister of Dorothy who married John Hancock (A.B. 1754). According to John Adams, Esther was "pert,

---

[9] Josiah Quincy, *Reports of Cases . . . in the Superior Court . . . of Massachusetts* (Boston, 1865), p. 35.

[10] Paine Mss. (Mass. Hist. Soc.), June 8, 1764.

[11] Mass. Hist. Soc., *Proceedings*, XLVI, 406.

sprightly, gay," and "celebrated for her beauty, her vivacity, and spirit," but not much given to thinking or reading. Jonathan and Esther were published at Boston on January 21, 1764. They settled in Charlestown near the Woburn line, on some property which had belonged to Chief Justice Samuel Sewall (A.B. 1671). Their first child was born and buried in that town, but their son Jonathan was baptized at Christ Church, in Cambridge, in 1766. Apparently they had by then rented the Lechmere house at the corner of Brattle and Sparks streets which they bought in 1771. Sewall served as a member of the Cambridge school committee and a warden of Christ Church. He was now in a position to re-pay to others the kindness which rich men had shown him when he was an impecunious student, and he befriended Ward Chipman (A.B. 1770) and several other boys as Judge Russell had befriended him.

Some of the poor boys who achieved such success and a place in one of these aristocratic Church of England circles were awed by their new environment, but Sewall retained his perspective and his sense of gentle amusement at the world. Writing to his old friend Tom Robie, he said:

O! by the bye, why didn't you come to Commencement — you cant conceive what exquisite Enjoyment a right true-blue speculative Mind would have found in sitting in a Crowd, where the Heat was seven times more intense than that of Nebuchadnezar's Furnace, and hear a learned forensick Disputation upon the Cause, or manner of Engendring, of Frost, which was one of the Subjects of yesterday's Exercises. To hear Syllogisms like a Storm of Snow and Hail, ratling just over your head while you yourself were scorching or rather melting, not per Spicula nitrosa, but from the incontestable Warmth of a surrounding atmos-phere, compounded, of every kind of Effluvia that you can suppose to be emitted from any and every part of the Bodys of 1500 Male and Female, black, white, and orange Yahoos, all sweating profusely. . . . [12]

His secure place in the Cambridge aristocracy did not change one whit his sense of justice to others, be they white, black, or orange. His neighbor, Richard Lechmere, had a Negro slave, James, who claimed his freedom. Sewall took the Negro's case in James vs. Lechmere, argued that under the terms of the Massachusetts Charter all persons born or residing in the province were free, re-

---

[12] Jonathan Sewall, Letters (Mass. Hist. Soc.), p. 27.

gardless of color, and won his client's freedom.[13] This was in 1769, two years before the Somerset Case, argued along parallel lines, established the same point in England. Had not the Revolution cut short his legal career in this country, his influence on American law would have been considerable.

On March 25, 1767, Sewall was appointed "Special Attorney General," a new office to which Jeremy Gridley (A.B. 1725) objected, arguing that it was a contradiction in terms.[14] In June his title was changed to Solicitor General. The Whigs cried out that he was being bribed, but the Attorney General needed an assistant, which is what he was. In November he became Attorney General, having previously been appointed Advocate General of the Court of Vice Admiralty. Even John Adams had to admit that these appointments, from a purely legal point of view, were good ones:

Mr. Sewall had a soft, smooth, insinuating eloquence, which, gliding imperceptibly into the minds of a jury, gave him as much power over that tribunal as any lawyer ought ever to possess. He was also capable of discussing before the court any intricate question of law, which gave him at least as much influence there as was consistent with an impartial administration of justice. He was a gentleman and a scholar, had a fund of wit, humor, and satire, which he used with great discretion at the bar.[15]

Adams should also have pointed out the intellectual integrity with which Sewall carried out his duties. Early in 1768 the Customs Commissioners requested his support in compelling coasting vessels sailing from one colony to another to post bonds. This was a doubtful interpretation of the law and would have been such an outrageous imposition that he gave his opinion against the bonds.[16] In April the Commissioners directed him to institute proceedings against John Hancock for not permitting the search of the *Lydia* for smuggled goods. He refused to act, saying that to do so would be to invade the province of the grand jury, and that hitherto such "information" was used as a basis for proceedings only when the case was clear and the Judges had given their approval. As little

[13] Mass. Hist. Soc., *Collections*, 4th Ser., IV, 334–335.
[14] Quincy, *Reports*, pp. 241–242.
[15] John Adams, *Works*, IV, 7.
[16] Oliver M. Dickerson, *The Navigation Acts and the American Revolution* (Philadelphia, 1951), pp. 214–215.

as the Attorney General liked John Hancock, he thought that, like James, he ought to have justice:

In the present Case it is most certainly a doubtful point whether *any* offence against Law has in fact been committed or not, and tho Mr. Hancock may not have conducted so prudently or courteously as might be wished, yet from what appears it is probable his Intention was to keep within the boundaries of the Law. No personal Insult or Injury appears to have been offered to either of the Tidesmen by Mr. Hancock or any of his Company.[17]

The angry Commissioners appealed to the Treasury Board in London and tried to have the Attorney General removed for neglect of duty.[18] Although he was convinced that they had lied to him about the evidence which they had against Hancock, he was so sure that they could have him removed from his extremely lucrative job that he told Governor Bernard that he would resign. However, Bernard, Hutchinson, and Admiralty Judge Robert Auchmuty (Class of 1746) rallied to his support, and the worst that the Customs Commissioners could do was to fire their deputy, Samuel Venner, for having leaked the news of their plan to get rid of the Attorney General.[19]

Hancock now became careless, or the Commissioners more careful, and on October 29, 1768, Sewall, no doubt with satisfaction and apprehension, libeled him before Judge Auchmuty on the basis of the evidence discovered in the sloop *Liberty*.[20] Before the resulting uproar had died down, Sewall in January, 1769, received a commission as Judge Commissary, Deputy and Surrogate of the Vice Admiralty Court for Quebec, Newfoundland, and Nova Scotia, an office which carried the handsome salary of £600, payable from fines and forfeitures. The headquarters of the Judge Commissary was Halifax, and when in June, 1769, he sailed for Nova Scotia to appoint a deputy to represent him there, the newspapers flayed him for having "listed under the Banners of Corruption":

[17] *William and Mary Quarterly*, 3rd Ser., IV, 504.
[18] *Mississippi Valley Historical Review*, XXXII (1946), 532–534.
[19] Bowdoin-Temple Mss. (Mass. Hist. Soc.), VII, 231–244, 270–293.
[20] Dickerson's charge (*Navigation Acts*, p. 241 and Index) that Sewall acted because of the fee is as forced as the rest of his statements about him. The documents are in Quincy, *Reports*, pp. 457 ff.

What Benefit a Province can reap from a Non-Resident's Salary of £600 per Annum, when all his Deputies can do the Business for about the sixth Part of that Sum divided among them, we leave to our œconomical Ministry to point out. . . . This foreign Judge has so long after his Appointment acted among us in the several Characters of Attorney-General, Advocate-General, &c. and discerning People cannot but highly applaud the Wisdom of our Superiors in multiplying Posts and Pensions in America, and making the Expence of Government, in the new Settlements and Colonies, bear a goodly Proportion to the civil Establishment of the Mother Country.[21]

In the "Journal of the Times" written for circulation through the colonies, the Whigs made various charges against him, one being that he falsified court records to protect the Regulars who were spilling American blood in the streets of Boston.[22]

John Adams admitted that Sewall never favored the British policies which brought on the Revolution:

He always lamented the conduct of Great Britain towards America. No man more constantly congratulated me . . . upon any news, true or false, favorable to a repeal of the obnoxious statutes and a redress of our grievances; but the society in which he lived had convinced him that all resistance was not only useless but ruinous.[23]

Opposed to all violence, he wanted to prosecute the case against John Robinson for the assault on Otis.[24] He drew the indictment of the soldiers involved in the Boston Massacre, but then disappeared, leaving the prosecution to a less deeply dyed Tory, Samuel Quincy (A.B. 1754) and a Whig, Robert Treat Paine (A.B. 1749). John Adams, who conducted the defence, was sharply critical of him for withdrawing,[25] but there were obvious and sound reasons for his action. When Samuel Adams (A.B. 1740), disappointed at the bloodless outcome of the trials, published over the pseudonym "Vindex" articles giving a distorted version, Sewall replied over his old pseudonym "Philanthrop."[26] John Adams, now bitter against his old friend, accused "Philanthrop" of "propagating as many lies and slanders against his country as

[21] *Essex Gazette*, Aug. 15, 1769, 1/3.
[22] *Ibid.*, June 20, 1769, 1/2.
[23] John Adams, *Works*, IV, 9.
[24] Mass. Hist. Soc., *Proceedings*, XLVII, 212.
[25] John Adams, *Works*, X, 201.
[26] Thomas Hutchinson to Israel Willliams, Israel Williams Mss. (Mass. Hist. Soc.), II, 166.

ever fell from the pen of a sycophant, rewarded with the places of Solicitor-General, Attorney-General, Advocate-General, and Judge of Admiralty, with six thousands a year." [27] Adams was as wrong as to the truthfulness of the Judge's articles as he was as to his salary. Sewall's less angry father-in-law said that Jonathan was "caught in the snare of £600 a year." [28] Adams, however, was implacable:

I thought [him] the best friend I had in the world. I loved him accordingly, and corresponded with him many years without reserve. But the scene is changed. At this moment I look upon him as the most bitter, malicious, determined, and implacable enemy I have. God forgive him the part he has acted, both in public and private life! [29]

Sewall refused to be rebuffed and did his best to save his friend from what he was sure would be the result of his folly. When they were attending the Casco session of the Superior Court shortly after Adams had been elected a delegate to the Continental Congress, he invited him to take an early morning walk and took the opportunity to plead with him not to oppose the irresistible power of Great Britain. Adams replied sadly, "I see we must part, and with a bleeding heart I say, I fear forever; but you may depend upon it, this adieu is the sharpest thorn on which I ever set my foot." [30]

With the great majority of Harvard-educated lawyers in Boston, Sewall signed the farewell address to Hutchinson, and later the address of welcome to Governor Gage. Most of the signers who lived beyond the protection of the Regulars were promptly visited by mobs, and the friends of the Judge feared for his safety in Cambridge. One of the last things Hutchinson had done was to tell the Council that Hancock was for raising the mob "to take off that brother-in-law of his." [31] Gage, fearing to lose a good man, ordered Sewall into Boston, so on September 1, 1774, he rode in, visited briefly with his father-in-law, and then disappeared into Headquarters. That night the mob appeared at the Sewall mansion. According to the Whig versions, they did no damage when they found that he was not at home. According to Loyalist ver-

[27] John Adams, *Works*, II, 251.
[28] Edmund Quincy, Commonplace Book (Mass. Hist. Soc.), p. 34.
[29] John Adams, *Works*, II, 302–303.
[30] *Ibid.*, IV, 8–9.
[31] *Ibid.*, II, 325.

sions, the mob smashed windows and did other damage but was driven off by the brave defense put up by the young gentlemen of the family.

Sewall was immediately put to work as Gage's private secretary and adviser,[32] so the Whigs were probably right when they accused him of hardening the unhappy Governor's stand and writing his proclamations. Such was his reputation as the voice of the Loyalists that in the "Massachusettensis" controversy Adams thought that his opponent was his old friend, and that idea persisted for a generation although some contemporaries knew that the Tory champion was Daniel Leonard (A.B. 1760). Sewall was accused by the Whigs of converting both Leonard and General William Brattle (A.B. 1722). In *The Group* he was the character "False Philanthrop," [33] and in *M'Fingal* he was described at some length as "that wit of watergruel." [34]

Discouraged by the course of events, the Judge on December 11, 1774, reported to Hutchinson that the well-meaning but deluded majority would never again be loyal subjects until convinced by vigorous methods that forcible opposition could not succeed.[35] He denounced the taking of the munitions from the fort at Portsmouth as treason, and after the Nineteenth of April he urged General Gage to take care lest another such prize fall into the hands of the rebels. [36] At the request of General Frederick Haldimand he drew up a long report on the American situation in which he held that "the hidden spring of this wonderful movement is the ancient republican spirit brought by the first emigrants" and kept alive by the form of government which the New England colonies enjoyed.[37] His ideas are most succinctly formulated in a tract which he published at this time under the title *The Americans Roused, in a Cure for the Spleen*. In the form of a barbershop dialogue he makes the points that the king never had the authority to make a compact with the colonists placing them beyond the authority of Parliament; that Parliament had for more than a century legislated for the colonies, and they had accepted it without ques-

[32] Hugh Edward Egerton, *The Royal Commission on the Losses and Services of American Loyalists* (Oxford, 1915), p. 233.
[33] Sewall is identified in the Boston Athenaeum copy of Mercy Warren's play.
[34] John Trumbull, *M'Fingal* (New York, 1864), pp. 40, 42.
[35] *Dartmouth Papers*, II (Hist. Mss. Comm., 14, Pt. 10), 238.
[36] Jonathan Sewall to Thomas Gage (W. L. Clements Library), May 7, 1775.
[37] *Dartmouth Papers*, II, 305.

tion; that the tyranny from which the Americans were supposed to be suffering was imaginary; that the colonists had been grossly deceived by lies about proposed Parliamentary legislation; that the rebels were madly leading the people into a suicidal war against the most powerful nation in the world.[38]

Of his own situation in Boston, Sewall wrote on June 7, 1775, to Robie at Halifax:

If you have your flocks and Hurds, as well as your little Ones, with you, you are well off. I wish I had mine — but I am not discouraged — I don't fear starving . . . cheer up Robie, I think I see Daylight, tho' it has been a long, dark, stormy night — I begin to hope the storm has almost spent itself. . . . I am scituated in Tom Boylston's House in School Street, formerly Col. Wendell's. . . . I have a very convenient house and garden and my Family are in good health and spirits. . . . I spend my Time in scolding, mourning, laughing, cursing, swearing and praying; so that what the Body wants in variety is amply made up to the Mind.

He went on to describe "the Worshipers of Liberty," some "plunging themselves, their Wives and children, in certain poverty and destruction, quitting or wasting their Substance, strolling about like pilgrims not knowing whither they are bound, — others plundering and destroying all around them, killing horses, stealing cattle, sheep, money, and goods, burning houses, barns and hay — the whole troop rushing into the arms of Slavery; and all, in honour of the aforesaid Goddess Liberty." [39] If Sewall heard that he, along with Gage and a few others, was excepted from the full and free pardon offered by the Provincial Congress, he was not concerned. Inasmuch as he was not a member of the Mandamus Council, he must have owed this special attention to his association with Gage in Headquarters.

After the Battle of Bunker Hill, Sewall began to doubt that the dawn was about to break:

Musketry, bombs, great guns, redoubts, lines, batterys, enfilades, battles, seiges, murder, plague, pestilence, famine, rebellion, and the Devil have at length brought me to a determination to quit a scene with which I am thoroughly cloy'd, and to retreat to the cold climate of Halifax, — a spot which I flatter myself will afford peace at least, because it is not

[38] This tract is attributed to Sewall by John Trumbull in the Hartford, 1792, edition of *M'Fingal*, p. 19.
[39] Jonathan Sewall, Letters, p. 31.

worth quarrelling about. It is not despair which drives me away, but because I am heartily tired. I have faith like a mountain of mustard seeds that rebellion will shrink back to its native Hell, and that Great Britain will rise superior to all the gasconade of the little, wicked American politicians.[40]

By the middle of August, Sewall had changed his mind and determined to sail for England where, he told Robie, he would be found by anyone who inquired in Westminster "for Squire Wronghead and Family lately from the Country of Independents." "I am really going Bag and Baggage — I am discouraged at the prospect before me — I see Infatuation, Destraction and hairbrained Blindness on one Side, and determined resolution and resistless power on the other, and for my Wife and Children I dread the shock. . . . I have seen, heard and felt enough of the rabies — I wish to be out of the noise." [41] Accordingly he sailed for London on the twenty-first, and arrived a month later. Mrs. Sewall and the three children were immediately inoculated for smallpox.

Governor Hutchinson met the Judge and apparently told him that Lord North, appreciating his services, had intended to make a place for him at Halifax, but that his having left Boston without the permission of the Crown had cooled his friends in the government. Sewall pled that his declining health had compelled him to leave America.[42] Unlike Hutchinson, who was depressed and made homesick by the luxury of London, the Judge was delighted. Writing to Edward Winslow (A.B. 1765) on January 10, 1776, he said:

You can have no idea what a noble Country this is for a Gentleman — every Thing is upon an immense Scale — whatever I have seen in my own Country, is all Miniature, yankee-puppetshow. I was at Court the Day before yesterday . . . and I believe in my Conscience, the prime Cost of the Dresses I saw there, was sufficient to have purchased our whole Continent — the Wealth of this Country is truly astonishing, but unless a Gentleman can get his Share of it, he has no Business here — £600 per Annum is but a Drop in the Ocean . . . I wish to stay here for the Sake of giving my Boys a Chance for the grand prizes which every profession presents to view — however, I fear, the cold, inhospitable, Lilliputian Region of Halifax will finally bring me up; for

[40] Mass. Hist. Soc., *Proceedings*, 2nd Ser., X, 413.
[41] Jonathan Sewall, Letters, p. 39.
[42] Hutchinson-Hardwicke Letters, Gay trans. (Mass. Hist. Soc.), Sept. 22, 1775.

as to Massachusetts Bay, I wish it well, but I wish never to see it again till I return at the Millenium. . . . I hate the Climate where Rebellion and Fanaticism are engendered — and I would shun it as I would a country infested with the plague.[43]

Sewall joined in founding the New England Club of refugees at the Adelphi, and perhaps the accents there did make him a trifle homesick, for at times he wrote that of a Saturday night his mouth watered for salt fish, and for "Newtown pippins, Shagbarks and Cranberrys." He had a sad jolt when his servant was taken up for burglary and, in spite of all that he could do for him, was hanged at Tyburn.[44] He spent the winter of 1776–1777 in the Tory circle at Brompton, but it was an unhappy time, for he lost his only daughter, Betsey, and in his bitterness he imagined that those "unfeeling brutes . . . the American pirates" with their usual "barbarous, inhuman, unchristian, diabolical sentiments" were intercepting his letters and enjoying his suffering.[45]

Some of his intercepted letters addressed to old friends still in the colonies were embarrassing to them. He bitterly reproved David Sewall (A.B. 1755) for accepting a seat on the Massachusetts Council:

When the Seeds of Sedition were first began to be sown in the province of Massachusetts-Bay, I was nearer the hands which scattered them, than you were; and could better discover the principles of Ambition and Envy which prompted them and the delusive Artifices by which the Passions of the Multitude were inflamed, and their Judgments obscured, in order to seduce them to Disorder, Anarchy and Ruin!

He could understand that the common people in the freest and happiest nation in the world might be misled, but educated men like David could only have been corrupted by greed of power.[46] Usually these letters ended with threats of the wrath to come, as this to John Lowell (A.B. 1760):

I saw with less surprize than grief your name to several acts of your usurped Council as Deputy Secretary pro Tem. . . . with equal emotions, I heard of your having assumed the office of Attorney General, and of your having taken possession of Mr. Paxton's house and furni-

---

[43] *Winslow Papers* (St. John, 1901), pp. 13–14
[44] Jonathan Sewall, Letters, p. 40.
[45] Mass. Hist. Soc., *Proceedings*, 2nd Ser., X, 418.
[46] Jonathan Sewall to David Sewall, Misc. Mss. (Mass. Hist. Soc.), Apr. 14, 1777.

ture as if it had been your own property. Thus far you have been triumphant — but beware how you proceed one step further; depend on it a dreadfull storm is gathered, and is approaching towards you; a terrible summer is before you; and a day of strict reckoning hastening on. . . . The Government from which you have inconsiderately Revolted is too powerfull to oppose; but it is characteristically marked with clemency and mildness — it never meant to abridge you of your just Rights and Liberties — it wishes not to punish to utter destruction; but will force your antient constitutional submission.[47]

Disappointed at the outcome of the military events of 1777, he urged his American friends to take up the British offer of reconciliation, which he described as wise and sure to be attended with the happiest consequences. If the Americans still insisted on independence, he would be satisfied to live and die in England. His great purpose in life was the proper launching of his two boys, to whom he was abnormally devoted. He wrote constantly of them, but almost never mentioned his wife, whose letters to her father suggest that she was anxious to forget the whole business and go back to live in Massachusetts.[48]

The Sewalls spent much of their time in Bristol, where living was less expensive than in London, in order to keep young Jonathan in Hackney School. When the Judge in May, 1778, heard that Massachusetts had confiscated the estate which he had hoped to leave to his sons, he was furious:

Notwithstanding all I have known and all I have believed of the persecuting tyrannical Spirit of our American patriots, I confess, the Act of Assembly which you mention is a fresh proof of such a total want of principle, such an open disregard to common Justice, and those natural and social rights of Mankind, of which they have set themselves up as the Defenders, that I am really astonished; to rob and plunder, and then to banish from their native Country, upon pain of Death, those whose only Crime has been the Exercise of the right of private Judgment! to deny the privilege of breathing the common Air to those who have only withdrawn themselves from the horrors and Calamities of Civil War! [49]

The Judge was formally proscribed six months later by the Act of October 16, 1778. His brother-in-law, Thaddeus Mason (A.B.

---

[47] Mass. Hist. Soc., *Proceedings*, XIV, 182–183.

[48] Edmund Quincy Mss. (Mass. Hist. Soc.), Aug. 22, 1777.

[49] Jonathan Sewall to Isaac Smith, Smith-Carter Mss. (Mass. Hist. Soc.), May 9, 1778.

1728), battled without success to save the Lechmere-Sewall property in Cambridge.

The Judge spent the summer of 1778 in Sidmouth in a state of depression. After his experience, he thought that George III had "too much of the milk of human kindness," that if we had had "an unfeeling, politic king of Prussia or Empress of Russia" willing to lay waste the cities and villages of America, this policy would have "opened the eyes of the deluded and changed the hearts of the deluding, and brought back the surviving remnant to duty and happiness." [50] By the end of the year he was feeling better and sure that the suppression of the rebellion was nearer than it had ever been. The entrance of France and Spain into the war did not shake his confidence:

We beat them both in the last War, and we will do it again — Some gloomy Croakers indeed cry out, with long faces, aye, but then we had America on our side — but I deny that this makes any alteration to the disadvantage of Great Britain — it cost her more during the last War to protect America against France and Spain, than it will now cost her to keep the american Cur at a distance, while she is whipping the french and spanish Dogs. . . . The force now in America is less than what this Kingdom maintained there the last War, while she also was conquering french America, at immense Expense, in Germany. [51]

As for the Bostonians, they were "Sanctified Hypocrites" who "Having familiarized themselves with Treason, Robbery, Murder, Sacriledge, perjury, and all other lesser Crimes," had now "absolutely arrived to such a pitch as to ride through the State of Holy Boston, at Noonday, with kept-Mistresses!" Moreover, "Complaisance to their great and good Ally" had "led them to sacrifice their wives and Daughters to french Gallantry." [52]

During the rest of the war Sewall hailed each favorable turn of events as the beginning of the end of the rebellion. "The Rebel Arnold's return to his Allegiance" he took "as a proof that the fabric of Congress' social Tyranny" was "mouldering and tottering to its downfall." [53] When England gave up the struggle, he watched expectantly for the beginning of the civil wars between the States: "Great Britain will have no more hand in it — no, shee

[50] *Winslow Papers*, p. 36.

[51] Jonathan Sewall to Isaac Smith, Smith-Carter Mss., June 22, 1779.

[52] Jonathan Sewall, Letters, pp. 44–59.

[53] Jonathan Sewall to Elisha Hutchinson, Hutchinson-Watson Mss. (Mass. Hist. Soc.), Nov. 25, 1780.

will now sit still and see them cut each other's Throats, and bless God, as I and every good Christian will most devoutly and sincerely, for every throat that is cut — and the sooner they begin the glorious work the better, say I." [54]

The Judge had never been interested in political theory, but like many other educated Europeans he was fascinated by the American experiment in a federal republican government. Writing to Mrs. John Higginson in September, 1783, he said:

I want your opinion upon the present appearance respecting the future government of the united States considered as one confederated people. Must there not be a supreme and controuling power lodged somewhere? Must not this supreme head be vested with a discretionary power to make war in the name of the 13 States? to provide in time of peace against such an event? to discharge the public debt contracted in the late war? and to these ends must not he or they, be it Protector, Stadholder, or Congress, have power to levy taxes and to enforce the payment of them? Will a people who so lately took up arms to avoid a 3 penny duty on an acknowledged luxury . . . patiently submit to a tax on all the luxurys and many of the necessaries of life levy'd by creatures of their own creation but yesterday? [55]

Soon after this Judge Sewall drew, apparently at the request of some official, a belated plan of colonial union under the Crown. Pointing out that it was distance between England and America which had caused the breakdown of the machinery of government, he proposed to lodge as much of the authority of the Crown and Parliament as possible in a President and intercolonial Council which was to be hand-picked from the American aristocracy to bind it by interest to England as was the ruling class of Ireland. The legislatures should be clipped of their authority and elected triennially, but the property qualification for membership was to be kept low by English standards. Town meeting powers should be sharply curtailed, and all lawyers and college officers compelled to take an oath of loyalty to the States. College students should be compelled to take the same oath before receiving their degrees:

In all colledges and other public seminaries of learning Caution ought to be used to prevent the principal trusts being lodged in the hands of gentlemen whose religious tenets point them decidedly to republicanism. The heads of all societies have a great influence over the youth belong-

---

[54] Robie-Sewall Mss., Aug. 25, 1783.
[55] Mass. Hist. Soc., *Proceedings*, 2nd Ser., X, 425.

ing to the same. A slight retrospection of the late conduct in America will not only substantiate this observation, but evince the danger thereby intended to be avoided. However I would not be understood as aiming at any illiberal distinction between different opinions among protestants.[56]

In all this Sewall had no idea of himself returning to Massachusetts, and he marked the severance by changing the traditional New England spelling of his name. The Herald's Office convinced him that this should be "Sewell," which seemed reasonable to him because, he said, the fact that his ancestors had emigrated to America proved that they didn't know enough to spell their own name.[57] His finances were, however, pressing him toward Halifax. So long as he remained in England a quarter of his salary had to go to support his substitute in Nova Scotia, and what was left kept him on such a short tether despite his typically New England financial prudence that only the interest of his boys kept him from going out. He enquired of Ward Chipman as to the likelihood of his obtaining an appointment to the Nova Scotia Court of Common Pleas, saying that he venerated juries and would rather work with them than in the Admiralty Court without them.[58] To his old friends in Massachusetts he now wrote kindly letters saying that he had never held any enmity toward them, and that he regretted the war-time letters which had hurt friends like Lowell. To Samuel Curwen (A.B. 1735), a friend in exile now returned to Salem, he wrote in June, 1784:

I don't absolutely despair of seeing you again in this strange world, for upon my soul, though I was born and bred I am a stranger in it; but my design is to go out to Nova Scotia this autumn or early in the spring; there, if you wish, you may see me, but while the unjust, illiberal, lying Act of 1779 remains unrepealed, never will I set foot on the Territories of the Thirteen United Independent States. I feel no resentment against them. I wish them more happiness in their unnatural independence than my judgment allows me to hope for them. . . . I wish my judgment may still be erroneous.[59]

[56] Julian P. Boyd, *Anglo-American Union* (Philadelphia, 1941), p. 170; *Canadian Historical Review*, XXXII (1951), 34–42.
[57] Robie-Sewall Mss., Feb. 14, 1783.
[58] Joseph Wilson Lawrence, *The Judges of New Brunswick* ([St. John], 1905), p. 81.
[59] *Journal and Letters of Samuel Curwen* (Boston, 1864), pp. 449–450.

In October he wrote to another friend, "The Devil drove me to England, and the Devil has kept me here hitherto — but if I don't escape next Year, the Devil of Devils must be in the luck." [60] The news of a plan to found a college at Halifax made him all the more anxious, for that would solve the problem of educating his boys. A friend who saw him in London at that time wrote:

Mr. Sewall was in Town . . . he is much altered, he not only looks older but his Face is full of Carbuncles - whether arising from Intemperance I know not, but it has that Appearance - he intends to go out to Halifax in the Spring and to take his two Boys with him - what is to become of her [Mrs. Sewell] I cannot tell.[61]

Spring came and went, and two others followed it, but Sewell remained in Bristol, complaining bitterly that once the war was over, Lord North and his successors had washed their hands of the Loyalists. However, in the spring of 1787 he went up to London to embark for St. John. John Adams, who was then American minister, on hearing of the arrival of his old friend drove at once to visit him:

I ordered my servant to announce John Adams, was instantly admitted, and both of us, forgetting that we had ever been enemies, embraced each other as cordially as ever. I had two hours conversation with him, in a most delightful freedom, upon a multitude of subjects. He told me he had lived for the sake of his two children; he had spared no pains nor expense in their education, and he was going to Halifax in hope of making some provision for them.[62]

Mrs. Adams pressed Mrs. Sewell to have dinner with them, but although time would not permit, the Judge was happy in the renewal of their friendship: "if he could but play backgammon, I declare I would chuse him, in preference to all the men in the world, my fidus Archartes, in my projected assylum." [63]

Sewell was in St. John in July, 1787, where he heard a rumor that Pitt was going to abolish the Vice Admiralty Court, which left him wondering whether he had come to hang pirates or to turn pirate for a living. The blow soon fell, leaving him with a pension of £200 to be paid "out of the monies arising from the sale of old

[60] Jonathan Sewall, Letters, p. 66.
[61] Thomas Aston Coffin in Coffin Mss. (Mass. Hist. Soc.), Dec. 3, 1784.
[62] John Adams, Works, IV, 9.
[63] Mass. Hist. Soc., Proceedings, LXXI, 514. Sewell's account of the reunion is printed in John Adams, Works, I, 57.

naval stores." Insulted, furious, unable to raise £10 to pay a bill, he shut himself in his house for eighteen months and sat and fumed. He was unable, however, to keep from looking out of the windows to watch the wonderful growth of St. John, where the only cloud was, he thought, the immigration of freed Negroes. Of these he wrote to his old friend, Judge Joseph Lee (A.B. 1729) of Cambridge:

I remember an opinion you once sported — that Negroes seemed to be intended for Slaves, from their rank in the Scale of being — I combatted the Opinion then, but I adopt it now, I believe the Maker of all never intended Indians, Negroes or Monkeys, for Civilization.[64]

When the Loyalist claims were settled in London, Sewell received £1600 for property losses which he maintained amounted to £5793. This was insulting, but it was cash, and he emerged to practice law in St. John. In July, 1789, he was thinking of joining his son Jonathan in Quebec. There, according to Canadian sources, he was "retained by the city members" of the House and achieved a success at the bar unparalleled in America; one wonders if this is not in part confusion with his son. After a few years he returned to St. John where, after a long decline, he died on September 26, 1796, and was buried in the vault of James Putnam (A.B. 1746). There is a memorial brass in Trinity Church, St. John. Esther Sewell was now free to return to Cambridge, where in 1797 she was living in her old home. She returned to Canada, however, where her son Jonathan had become Chief Justice, and Stephen, Solicitor General.[65] She died at Montreal on January 21, 1810.

The only sizable lots of the Judge's manuscripts remaining are in the collection of his letters, chiefly to Robie, and the Robie-Sewall collection in the Massachusetts Historical Society. A small lot formerly in the possession of Mrs. A. R. F. Hubbard of Quebec could not be located.

[64] Adams Mss. (Mass. Hist. Soc.), Sept. 21, 1787.
[65] For the genealogy see Edward Elbridge Salisbury, *Family Memorials*, I (1885), 181 ff.

SAMUEL HAVEN

SAMUEL HAVEN, the fourth minister of the Second, or South, Congregational Church of Portsmouth, was born on August 4, 1727, the son of Joseph and Mehetabel (Haven) Haven of Framingham. He was the oldest son and moderately introspective and pious, so his father, whose means were substantial, sent him to college, where he roomed with Robert Treat Paine. In his Freshman year he was fined for drinking prohibited liquors and for fetching them into the college for the use of his friends, but he steadied down and turned his attention to versifying, which was always one of his hobbies. Typical is this verse written in his Sophomore year on "The Happy Student":

> Not he Who thoughtless of his Time
> forgits those things that are Sublime
> in Sports Consumes Each day
> his careless heart will ne're obtain
> A blessing by high Jack and Game
> till God Absends his sway.
>
> Nor he whose brutish mind demands
> a flowing bowl and flowing Cans
> to quench his Sordid Thirst
> Such drunkards Ne're Can Soar above
> Nor live With Men in peace and love
> Whose Earthly Souls are Curst.[1]

In this spirit he volunteered to inform on any of his fellows who used profanity. He joined the revived "Phinphilenici" society, a literary and speaking club. The class elected him thesis collector, and the college gave him both degrees in course.

After the class had scattered, its members began a brisk correspondence in which Haven's letters were incredibly sentimental and stilted. Once when he had a bad cold he decided to take as his motto "Memento Mortis" because the frequent consideration of death would rob it of its terrors for him. On another occasion he came to what he seemed to think was the novel conclusion that he should live each day as if he were to die that night.[2] His read-

[1] R. T. Paine, Diary (Mass. Hist. Soc.), Dec. 24, 1747.
[2] R. T. Paine Mss., *passim.*

SAMUEL HAVEN

ing, which ran to books like Jonathan Edwards' biography of
Brainerd, did nothing to change his attitude.

In 1749 Haven kept a school of seventy or eighty scholars in
Groton and boarded with the mother of Oliver Prescott (A.B.
1750). The next year he took the school in the west parish, where
he boarded with the minister, Joseph Emerson (A.B. 1743). He
was drawn to the ministry by the fact that when about twelve he
had heard the revivalist Whitefield preach a sermon which so impressed him that to the day of his death he could repeat it almost
verbatim.[3] Writing to Paine from Groton, he reported:

I am Just dipping in the study of Divinity among my other Studies and
am ready to guess it is a profound abyss but a delectable scene of Christ-
exalting wonders which may fill the soul with sweet surprise.[4]

This was his approach to religion, for he never experienced a definite conversion.

For his instruction in divinity, Sam chose the Reverend Ebenezer
Parkman (A.B. 1721) of Westborough, and his father readily
agreed to pay his board there. As a matter of course he was expected to help the Parson with his haying and other farm work.
Sam was delighted with his situation and his tutor:

I enjoy not only the Advantage of his most Learned as well as Christian Conversation; but also the pleasure and profit of perusing a Collection of Books not inferior to the best in the hands of Any Country
Minister nor do I imagine my priviledges less noble than those to be
enjoyed at College.[5]

After only three months Parkman decided that his pupil was ready
to preach and took him to a Worcester meeting of the Association,
which on July 18, 1750, approbated him. An aged minister who
was there that day wrote in his diary, "A promising young man
may God bless him and make him a blessing to his people." [6]

During the next two years Haven preached widely in eastern
Massachusetts and New Hampshire, and received several calls.
The best one came from Brookline, but he declined it because
three dissenting votes were cast against him, and chose instead to

³ Timothy Alden, *A Collection of American Epitaphs* (New York, 1814),
II, 193.
⁴ R. T. Paine Mss., Nov. 4, 1749.
⁵ Samuel Haven to R. T. Paine, *ibid.*, June 21, 1750.
⁶ Israel Loring, Diary, XVI (Mass. Hist. Soc.), 13.

go to the South Church of Portsmouth, which was unanimous. There he was ordained on May 6, 1752, and immediately appointed to serve as one of the two chaplains of the House of Representatives. On January 11, 1753, he married Mehitable, daughter of the Reverend Nathaniel (A.B. 1712) and Margaret (Gibbs) Appleton of Cambridge. Her environment had given her an unusual education, piety, charm, and social and physical graces which made her popular in Portsmouth. They built a handsome house on Pleasant Street [7] and filled it with eleven children. This was not an entirely urban environment then, for the town gave the Parson the use of the nearby training field for a cow pasture.

Portsmouth had acquired a man of suavity of manner, sweetness of countenance, and the kind of fervent oratorical powers popular in that generation. He was a man of unusual dignity, but he broke through the barrier of reserve and formality which usually separated a minister from his flock, and became intimate and familiar with even the poorest members of his congregation. He studied medicine and gave his services to the poor with such discretion that the physicians of the town were not alienated. As his contemporaries put it, he was a man of such "great sensibility" that his love extended to the wronged Africans, and he early became a foe of slavery regardless of the frowns of the wealthy.[8] Fortunately Portsmouth had no part in the slave trade.

In religion he showed the same "sensibility" and defended the ordination of the blind preacher Joseph Prince against a majority of the Association of Ministers of New Hampshire. Toward members of other denominations he showed such a mild and pleasant countenance that in 1757 a dozen members of his congregation seceded, charging him with heresy. To the church they stated their case:

We have left you for our better edification, and that because we cannot profit under Mr. Haven's preaching For we judge Mr. Haven's preaching generally tends to incourage saints and sinners in a generall way to think that, if they exercise the natural strength and power they have, that God will be sure to have mercy on them for Christ's sake, which, we judge, tends to make persons think that God's decree in

[7] For pictures and an account of the house see *Old-Time New England*, XXXIV, 6–8.
[8] Alden, II, 197.

Election depends upon the conduct of the creature; which would at once overturn the doctrines of free Grace and Election.[9]

In reply Haven elaborated on his views, and a majority of the congregation voted that they were acceptable. He was a Calvinist, but he could not believe that God's love for mankind was less than his, so he found some curious explanations of the horrors of Geneva doctrine. He said, for example, that God had put evil into the world to prepare sinful man gently for hell. In that day of weak Trinitarianism, his failings in this direction passed unnoticed. He regarded Christ as the mirror in which to see God, and not as God Himself.

Haven's popularity was very wide, and it was said that no contemporary was more frequently called upon to preach on special occasions in other churches. In that generation many Massachusetts people went to New Hampshire to be married, and the pastor of the South Church was a favorite for this purpose. So frequently were his sermons printed in Boston, that New Hampshire people began to wonder whether this indicated that they lacked religious enthusiasm.[10] Letters in the Boston newspapers defended or criticized his choice of guest preachers for the Portsmouth pulpit. On controversial issues he tended to be on the popular side, as when he opposed theatrical performances in Portsmouth.

From his first settlement in New Hampshire, Haven had advocated the establishment of an academy or college in that province. He was greatly interested in Dartmouth, and he urged President Wheelock not to boggle at the appointment of the Bishop of London to the Board of Trustees.[11] By keeping his ear to the ground in Portsmouth, he was able to warn Wheelock when the rumblings against college food neared revolution pitch. Dartmouth gave him an S. T. D. in 1773, but it was not his first. Colonel Samuel Sherburne (A.B. 1765), in order to repay the Parson for the funeral sermon which he had preached for his father, Judge Henry Sherburne (A.B. 1728), enlisted the aid of Benjamin Franklin to obtain for him an Edinburgh S. T. D., which was granted in 1770.

[9] Alfred Gooding, *The Theological History of an Old Parish* (Boston, 1901), pp. 14–15.
[10] *New Hampshire Gazette*, June 14, 1771, 3/3.
[11] Dartmouth College Archives, Oct. 27, 1769.

Haven was for many years the clerk of the New Hampshire Association of Ministers, and in 1760 he preached its formal Convention Sermon. He served on the committee of the Association which drew up a happy address to George III on his accession. In a sermon on this event he said that God "took the kingdom of our British Israel from the unhappy family of the Stewards, and (having a favour for our nation) placed it in the family of Brunswick, and said, let *George be King*." Of George, the Parson said:

What joy swells every breast, where religion and love of our country have any place, while we hear him from the throne, bestowing his royal influence and authority for the encouragement of piety and virtue, and discountenancing vice and prophaness? [12]

In his philosophy, England and Protestantism were identical, and in reviewing history he pointed out how time and again God's hand had delivered them from the "great papal apostasy." The struggle of England and Prussia against Spain and France he described as the last scene in this long drama. Of the defeat of Braddock he said, "How black was the cloud which then hung over the Protestant church!" Finally the fall of Canada, "the American Carthage," demonstrated God's support of His church.[13]

The new colonial policy adopted by the ministers of George III soon made an ardent Son of Liberty of the Doctor, but as late as 1773 he came publicly to the defence of Governor John Wentworth (A.B. 1755), asserting that the province could not wish for a better, and insisting that His Excellency would not countenance any acts of injustice or oppression which he had the power to prevent.[14] The Parson's honesty was so obvious that no one seems to have pointed out that at the same time he was trying to obtain confirmation of large grants of wilderness land which the Governor had made to him. When the news of the Battle of Lexington reached Portsmouth, the Havens sat up most of the night casting bullets. Later the Parson turned to the manufacture of saltpeter for gunpowder, using earth which he dug from under the meetinghouse. He signed the Association Test and heartily supported the new government. At the time of the invasion alarm of 1776, he shouldered his fowling piece and marched with his parishioners,

---

[12] Samuel Haven, *The Supreme Influence* (Portsmouth, 1761), pp. 17, 24.
[13] Samuel Haven, *Joy and Salvation* (Portsmouth, 1763), *passim*.
[14] New Hampshire Hist. Soc., *Collections*, IX, 318.

his cartridge box slung on the wrong side, as his wife used to re-mind him.[15]

Mrs. Haven was spared the worst of the war, for she died on September 9, 1777, so widely mourned that even the Boston news-papers noticed her passing. Left with seven children to care for, the Doctor joined forces with a brave young widow who had two daughters of her own. She was Margaret, daughter of George Marshall, and widow of Captain William Marshall of Portsmouth; she became Mrs. Haven on June 2, 1778. She added six more children to the crowd in the parsonage, so the time came when thirteen children and forty grandchildren gathered for Thanks-giving. It was no wonder that the Doctor, whose salary was $300 a year, was famous for his poverty. He inherited a considerable estate, probably from his father, who died in 1776, but this was quickly distributed among the needy of Portsmouth. It was said that he kept weekly track of the level of the meal barrel in every home in his parish. The war struck Portsmouth cruelly; at one time there were forty widows in his congregation, and no public funds to care for them and the other unfortunates. Although the Parson's salary was far in arrears, he was the only support which many families had. His successor, many years later, heard the story from the parishioners:

I never visited them, without their speaking to me of his sympathy, gifts, and efficient services, in the stress of their need. His name was on their lips in the very agony of death. Some of them have told me, that, in the absence of the public and private charities now so liberally dispensed, his care and generosity were all that stood between them and utter des-pair.[16]

The Revolution broke many of the leading families of the South Church, revivalists drew off the more excitable members of the congregation, and the popular and orthodox young minister of the North Church, Joseph Buckminster (Yale 1770), attracted most of the new families in Portsmouth. It was typical of the difference between the two ministers that when the news of Corn-wallis' surrender arrived, the Doctor was all for rushing into the

[15] Charles Chauncy Haven to John Langdon Sibley, Harvard University Ar-chives, Feb. 16, 1867.
[16] Andrew P. Peabody, *Sermons Connected with the Re-Opening of the Church of the South Parish* (Portsmouth, 1859), pp. 69–70.

meetinghouse to sing a Puritan *Te Deum*, but the young man coldly said "that it would be early enough when we had authentic or official intelligence." [17] Against these influences and the obvious decline in the powers and popularity of the old Parson, his swarming family, now including some prominent people, rallied, and by their ardent support restored the South Church to its old position.

In the new Piscataqua Association the Doctor held a respected but not very influential position. In 1785 he proposed a plan to modify the Congregational ecclesiastical polity in a Presbyterian direction, but only four churches in the Association adopted it. Four years later he proposed that the Association sponsor a brewery to the end that the use of strong beer in place of ardent spirits might be encouraged, but his motion was unanimously tabled. He was more successful in getting the Association to adopt a manual compiled under his guidance and published as the *Prayer Book, for the Use of Families* (Portsmouth, 1799).

Although the legislature had moved to Concord, the Doctor was reappointed chaplain, and in 1786 he preached the Election Sermon. He was appointed to the town committee to welcome President Washington in 1789, and hearing a discussion as to the title to be used in addressing the visitor, he struck off this impromptu verse, which was greatly admired:

> Fame spread her wings and loud her trumpet blew;
> Great Washington is near! What praise his due!
> What title shall he have? She paus'd and said,
> "Not one; his name alone strikes every title dead." [18]

The President himself left a record of their meeting:

I was visited by a Clergyman of the name of Haven, who presented me with an Ear and part of the stalk of the dyeing Corn, and several pieces of Cloth which had been dyed with it, equal to any colours I had ever seen, of various colours. [19]

When the news of Washington's death reached Parson Haven, he was as stunned and overwhelmed as if he had suffered the most crushing and unexpected family tragedy. [20]

The colored cloth shown Washington was one of the Doctor's

[17] Mass. Hist. Soc., *Collections*, 6th Ser., IV, 218.
[18] Alden, II, 198.
[19] J. C. Fitzpatrick, ed., *Diaries of Washington* (Boston, [1925]), IV, 45.
[20] New Jersey Hist. Soc., *Proceedings*, 2nd Ser., I, 98.

few scientific interests; the mystery of the White Hills which hung upon his horizon was another.[21] His frequent visits with the Appletons and President Langdon at Cambridge did not keep him in touch with the new science or the new ways in divinity. Successive generations of college students described his preaching as an affected imitation of Whitefield, who had long been out of favor.[22] The conclusion of the Dudleian Lecture which he delivered in 1798 is a sample:

And now, adieu — Dear Alma Mater, adieu. May you long remain an ornament to our rising empire — our happy Republic — may you nurture and send forth many sons, trained in the paths of virtue, and the most important branches of literature, pillars to the state, and able ministers of the New Testament. Thus may you be to the end of time a nursing mother both to the church and state.[23]

In one way Dr. Haven was touched by the new trends of thought, and he showed it by his longing inclination toward Universalism. He named one son Charles Chauncy (A.B. 1804), but his tendency toward the heresy of Dr. Chauncy was less a matter of personal infection than of his own warm human sympathy. In a volume of verse which he published in 1798 he put the problem thus:

"The omniscient God entuitively knew
The man, he form'd — and what that man would do,"

so how could he institute a system under which

"Millions of helpless souls be doom'd to dwell,
Enchain'd with that arch fiend in endless hell." [24]

The difficulty was, he saw, the hopeless contradiction between the revelation of the goodness of God and the revealed system of theology. It was his intention "modestly to hint" the happy doctrine of universal salvation "to the public mind for their consideration," but for himself, he hastily assured the coldly orthodox Buckminster, "he never meant to risk his salvation upon that ground." [25]

In the last decade of the century, the Doctor became too feeble

---

[21] Haven's observations on the White Hills are in the Belknap Mss. (Mass. Hist. Soc.), 161.A.104.

[22] J. Q. Adams, Diary (Mass. Hist. Soc.), May 21, 1786.

[23] Samuel Haven, *The Validity of Presbyterian Ordination* (Boston, 1798), p. 24.

[24] [Samuel Haven], *Poetic Miscellany* (Portsmouth, 1798), p. 24.

[25] *Portsmouth Oracle*, Mar. 15, 1806, 3/2.

to climb into the high pulpit, but he regularly preached from below it until 1799. The next year Timothy Alden (A.B. 1794) was ordained as his colleague, but he left five years later. The Unitarians said that the old parson's parsimoniousness made the situation impossible for the young man, but actually the trouble seems to have been theological. In 1804 the Doctor began to fail rapidly, both in body and in mind, but he lingered until March 3, 1806. His widow, who was then seemingly in perfect health, closed his eyes, took to her bed and died within thirty-six hours. The Parson had prepared directly beneath the pulpit a comfortable tomb into which he had gathered the remains of his first wife and other members of the family, leaving a place for himself and his second wife. Here they were interred after the best-attended funeral the town had ever seen. When the meetinghouse was sold in 1879 their remains were removed to the cemetery on South Street.

The record of the Haven children is somewhat confused, and this list is tentative: (1) Samuel, b. Aug. 4, 1754; A.B. 1772; m. Abigail Marshall; d. Aug. 1825. (2) Joseph, b. Dec. 11, 1757; m. 1st Eliza Wentworth, 2nd Mrs. Sarah Appleton. (3) Margaret, b. Aug. 24, 1759; d. young. (4) Mehitabel, b. May 27, 1761; d. young. (5) Nathaniel Appleton, b. July 19, 1762; A.B. 1779; m. Mary Tufton Moffat. (6) Elizabeth, b. July 2, 1764; m. John Adams of Portsmouth; d. 1833. (7) John, b. Apr. 8, 1766; m. Ann Woodward. (8) Henry, b. June 30, 1768; m. Emma Cullum; d. Sept. 18, 1823. (9) William, b. July 30, 1770; m. Sophia Henderson. (10) Joshua, b. Apr. 2, 1779; m. 1st Olivia Hamilton, 2nd Mary Cunningham. (11) George, b. Mar. 1781; d. at Havre de Grace, Apr., 1796. (12) Thomas, b. Mar. 2, 1783; m. 1st Eliza Hall, 2nd Mehitabel Jane Livermore, 3rd Ann Furness. (13) Mehitabel, b. Apr. 24, 1785. (14) Charles Chauncy, b. Aug. 17, 1787; A.B. 1804; m. 1st Prudence Griswold, 2nd Catharine Jeffries. (15) Mary, b. Sept. 6, 1789; m. Jacob Sheafe of Portsmouth; d. Jan. 1, 1840. The Doctor's will provided that at the death of the last of his descendants his house should be taken down and the land given to the city for a public park; this was done in 1898.[26] The portrait of Haven which illustrates this sketch belongs to the Massachusetts Historical Society.

[26] Harold Hotchkiss Bennett, *Vignettes of Portsmouth* (Portsmouth, 1913), pp. 22–23.

BENJAMIN MARSTON

BENJAMIN MARSTON was born in Salem on September 22, 1730, the eldest son of Colonel Benjamin (A.B. 1715) by his second wife, Elizabeth (Winslow) Marston. When he was eight his parents moved to Marblehead, and he sometimes visited his grandfather, Colonel Isaac Winslow of Marshfield. A letter from his grandfather to his father gives a glimpse of him there:

This is to . . . acquaint you of the indisposition that dear little Bennee hath been under. He was taken the last Friday; was weak at night, with a strong fever, which continued upon him till Monday, when we sent for Dr. Otis . . . who . . . judged it to be the intermitting fever, and thought it would be best to bleed him in the arm, which he immediately did, which he bore like a hero, held out his little arm, let the doctor prick it, see the blood run from it, and did not so much as whimper in the least. The doctor came the next day and gave him a vomit, and stayed with him till it had done working.[1]

Bennee suffered from the fever while an undergraduate, and it may have had something to do with his taking an unscheduled vacation of seven weeks for which the Faculty fined him 52s, an enormous sum. Soon after he had taken his first degree he was involved with some other students in giving "several loud Huzzas" as a stone was thrown through Tutor Mayhew's window. At the hearing before the Faculty, his classmate Sever maintained that *he* had done no shouting, but Ben would have none of that:

Sir Marston . . . declared . . . concerning the Persons who huzza'd . . . as to Sir Sever . . . that He tho't he stood at his right Hand, when He stood by Williams's Window with his face to the East, at the Time of the Huzzaing and that the rest of the company stood to the East of him, except Parker who stood westward . . . And both Sir Marston and Parker declared that they had no Doubt but that the whole company . . . joined in Huzzaing and that they knew nothing of any ones leaving the company before this Action was done.[2]

With the other members of the party, Ben was degraded seven places in his class, but two years later the Faculty agreed that he should be restored if he behaved himself "with a proper Decency" when his confession was read in the college hall.

[1] Mass. Hist. Soc., *Proceedings*, 2nd Ser., II, 230–231.
[2] College Papers (Harvard University Archives), I, 87 (179).

Marston's failure to take his M.A. before 1753 is no doubt explained by the fact that he was sailing in his father's ships. When the Colonel died in 1754 leaving Ben a quarter of the substantial family estate, he retired from the sea, and on November 13, 1755, he married Sarah Swett, daughter of Joseph and Hannah Swett of Marblehead. He built what was later known as the Watson House at the head of Watson Street in Marblehead, had a store in King Street, and in partnership with Jeremiah Lee and Robert Hooper sent ships to England, Spain, and the West Indies. For his store he imported every thing from lead bars and fish lines to English periodicals.[3] The Marston-Watson clan was large, prosperous, and congenial, and showed a rift only when Ben tried to get the others to share the cost of having the Marston coat of arms registered and emblazoned at the Herald's Office. The family correspondence shows an unanimous agreement that he was an energetic and absolutely honest business man, a charming and generous gentleman, and a scholar who loved, read, and annotated the handsome library which he built up.[4]

Marston served the public without stint, and was perennial Assessor, Fireward, Selectman, Overseer of the Poor, and Moderator of town meeting. In 1761 he was appointed a Justice of the Peace. He was a pious and devoted member of the Church of England, and that fact gave rise to the family tradition that he was a Loyalist because of his religion, but there is no indication in his voluminous writings of any such sectarian feeling on his part. In 1769 he served on the town committee to protest the Greenwich Hospital tax, but thereafter he was left off important committees although regularly elected Moderator. In 1771, foreseeing troubled times ahead, he began to sell off his property. The next year he publicly protested against a set of resolutions published as representing the almost unanimous opinion of a town meeting whereas, he said, only about twenty had voted, and they had not been given the opportunity to know upon what they were voting. He protested the assumption of the resolutions that the entire British government was wicked and corrupt:

These resolves seem to be dictated, and the whole proceeding of the said meeting to be animated by the little, low, and narrow spirit of bigotry

---

[3] Benjamin Marston's account book is in the Marblehead Historical Society.
[4] This library was in 1864 in the possession of John Lee Watson (A.B. 1815); it does not appear to have been scattered by sale.

and party zeal; as one of the said resolves speaks to this effect, viz. That whoever differs in opinion from these resolves, is a deserter from his country and its honest cause.[5]

Still Marston continued to be elected Moderator until May, 1774, when he signed the valedictory letter to Governor Hutchinson (A.B. 1727). After that his house was visited by a "committee" which without legal authority broke open his desk, embezzled his money and notes, and carried off some of his books and accounts.[6] On April 29, 1775, he tried to make his peace by giving the Selectman a formal recantation:

Whereas I was a Signer to an Address to Governor Hutchinson, when he left this Province, and as said Address has given much Offence to People in general, and has been otherways attended with evil Consequences; I do now totally renounce said Address, and declare that I am very sorry I ever signed it. But as I did it with a good Intention, hoping it might have a Tendency to serve us, so I now hope and desire my Countrymen will forgive me all the Injuries which has been undesignedly done by it.

And as it has been reported of me, that I have made frequent Journies to Boston, for these Six or Eight Months past, to inform the Governor of what was doing here: I do now declare upon my Word and Honor, that I never did, either directly or indirectly, give any such Information to the Governor, or to any other Person whatever: But that all my Journies thither were entirely on my own private Affairs.[7]

The publication of this recantation by the Committee of Safety did nothing to make Marston feel safer. For some months he hid out in the country homes of his friends, leaving his wife to watch over his Marblehead property. She died during the summer of 1775, however, and he, seeing no other way to avoid being "forced into Rebellion," on the night of November 24, 1775, set out in an open boat with one companion from "Colonel Fowle's farm" and made his way safely into Boston.[8] There he collected £250 which was owed to him and tried to invest it in a voyage to the West Indies to obtain supplies for the garrison. Failing in this, he left for Nova Scotia with the army in March, 1776.

[5] *Boston Post-Boy*, Dec. 28, 1772, 2/1.
[6] *New England Hist. Gen. Reg.*, XXVII, 392.
[7] *The Recantations of Jacob Fowle, Benjamin Marston. . . .* (n. p., n. d.), p. [1].
[8] E. Alfred Jones, *The Loyalists of Massachusetts* (London, 1930), pp. 210–211.

From Windsor, Nova Scotia, Marston sent to his sister Lucia at Plymouth a miniature of himself and a bit of verse which is typical of a great quantity which he ground out:

> Speed, little picture, quickly hence, and go,
> A Brother's likeness to his Sister show;
> Full to her view disclose his features all,
> And tell her thus appears th'original.
> Health and content enlivening his face,
> Show that within his breast dwells balmy peace;
> And tho' now exiled from his native land,
> Driven from his home by Faction's cruel hand,
> He still looks down on fickle fortune's power,
> Nor lets her frowns his equal temper sour.[9]

Nothing but this cheerful equanimity would have carried Marston through the bitter years which followed. He had hoped to "go into the Military line," but failing there he went into trade. In June, 1776, he sailed from Halifax for Dominica in the schooner *Earl Percy*, of which he was part owner. On the return voyage they were taken by the privateer *Eagle* and carried into Plymouth, where he had a hearing before the Committee of Correspondence. He left an amusing description of that revolutionary body which he describes as consisting of a man who owed "his existance to the very people he is now insulting," "a pious whining body," a "gentleman with a ragged jacket and, I think, a leather apron," "a good sort of man, made a tool of to serve the purpose of the occasion," "a simpering how-do-you-do-sorry-for-your-loss kind of a body," and "one that has been handsomely and kindly entertained at my house. He can do dirty work." [10] Marston was slightly surprised to be found a dangerous person and ordered confined to the Plymouth jail, but perhaps the committee knew what it was doing, for on this or a later occasion he was employed to land a spy in the rebelling colonies.[11]

From this time on much of Marston's life was passed in jails, and his meditations behind the bars at Plymouth were typical of his attitude in these circumstances over the years:

This is the 3d Sunday I have spent in this place, and have not been once inside of a meeting house. So cruel are my Enemies they deprive me of

[9] *New England Hist. Gen. Reg.*, XXVII, 393.
[10] New Brunswick Hist. Soc., *Collections*, VII, 84.
[11] *Second Report of the Bureau of Archives for Ontario* (1904), p. 605.

the pleasure of hearing their pious Good ministers preach and pray. I believe they think I have no soul, or They don't care what becomes of it, Or they think that going to meeting will not do me any good. I think so too. But Still I Should be glad to go now and then, for a little variety's sake.[12]

His brother-in-law, John Watson (A.B. 1766), got him out of jail and obtained liberty for him to walk in his yard and garden week days and to go to meeting on Sunday. Soon after he had reported that he was getting fat in this comfortable situation, he was carried off to the Bristol County jail, whence he was transferred to Boston.

All that Marston could see from behind his prison bars convinced him that the rebelling colonies were wrong:

This miserably deceived people are made to believe they can support independency. . . . There is now an order for draughting every fourth man to relieve the army, whose term of service is within a few days of expiring. . . . Their army is now broken to pieces. Their general is not to be found. . . . They have likewise lost a very great part of their cannon, tents and baggage. And yet the managers of the game . . . still push the draughting of every fourth man to relieve the army, who are every day running home, sick, lowzy, ragged and full of all manner of nastiness. Nay, General Washington, who moves the puppets of this place, has the affrontry to give out that a French Fleet and Army will be over early in the Spring. A Fleet from France! There will be one from the Moon as soon.[13]

Even the bitter winter in unheated jails did not shake his belief that a man who had patience, resignation, and a clear conscience had no reason to complain of the hardness of his lot. To his friends he wrote cheerful nonsense verse:

> I, poor de'il, am here confin'd,
> (A state which no way suits my mind)
> For being, — you know all the story, —
> A sad, incorrigible Tory.
> And being now so left i' the lurch,
> I cannot even go to Church.
> However, even let it run, —
> 'Tis a long lane that has no turn.[14]

To the Council he petitioned for liberty to leave the State, and was finally given permission to depart "in the earliest Vessel now at

---

[12] Benjamin Marston, Diary (New Brunswick Museum transcript), Oct., 1776.
[13] New Brunswick Hist. Soc., *Collections*, VII, 86.
[14] Marston, Diary, Feb., 1777.

Marblehead he giving his Parole to use his Endeavor to return a person belonging to this, or some other American State of like Rank." [15] So on March 9, 1777, he sailed on a cartel for Halifax where he was overjoyed to find his "dear Miss C——." To his diary he confided, "The pleasure of again seeing that dear Girl has abundantly rewarded me for all the disagreable feelings of a 6 months imprisonment. Gracious Heaven, grant me to be but so fortunate as to be able to provide for that dearest Girl an easy Situation in Life, and ye cannot make me happier." [16] One assumes that this Miss C—— was "Eliza, dearest maid," to whom he wrote joking verse when leaving Halifax to "ply the Seaman's art."

In the spring of 1778, Marston made a voyage to the West Indies and returned by way of Philadelphia. A summer voyage was less fortunate, for in August he was taken in the schooner *Polly* by the privateer *General Gates* and carried into Boston, where he was lodged in the guard ship. The next day his old friend Samuel White obtained his release and carried him home where he lived pleasantly until some small-minded Marbleheader wrote to the Council that he was "such an inveterate enemy to the Country that it would be dangerous" to leave him at large. So he found himself again in the guardship where he took out and polished his prison philosophy:

I have Caulked the malice of my Enemies — in spite of their ill nature — I spent the time of my confinenent chearfully and profitably — For I have learned that a man may enjoy himself in prison — I would not change the reflections of my own mind on the matter for all the pleasure they may have received from the gratification of a mean Revenge — Poor! Miserably Poor Divels. [17]

In September, 1778, the State again released Marston, but hardly had he sailed for Halifax than it included his name in the formal act of proscription. His property was confiscated, but his nephew, Marston Watson, who had been about to enter Harvard in the Class of 1779 when the outbreak of war drew him into the army, did succeed in recovering the mansion in Marblehead. [18] The exile had no desire to return, however, for he was entirely out of sympathy

---

[15] Executive Records of the Council, XX, 306–307.

[16] Marston, Diary, Mar. 18, 1777.

[17] *Ibid.*, Sept. 8, 1778.

[18] The confiscated property is itemized in the *Independent Chronicle*, Mar. 2, 1780, 2/3, and the *Salem Gazette*, May 1, 1781.

with the new ways. Particularly he was discouraged to hear that a list of all the male inhabitants had been presented to a town meeting which voted on each name by show of hands to determine who were Tories. This was, he said, "being tryed by one's Peers with a Vengeance."[19]

What little capital Marston had managed to salvage was now gone, and he was in serious financial difficulties before he found a job, in December, 1778, as supercargo on the brig *Ajax*. At Santa Cruz in the Danish West Indies he was shocked by the sight of slavery in the raw:

I saw here a cargo of these poor creatures landed out of a King's ship . . . drove like so many cattle to a large yard, men and women, boys and girls all together, each as naked as God made them, saving a piece of coarse linen. . . . Each slave with a wooden tally tyed upon it . . . Great God! what must be the feelings of a sensible human being to be torn from all that is reckoned valuable and dear, and to be condemned to the most servile drudgery and infamous uses without the least hope of relief. . . . I fancy that there is some mistake in the very trite maxim "that all men are by nature equal."[20]

This and two similar voyages were successful, but in February, 1781, he was in the *Ranger* when it was taken by the privateer *Ariel* and carried into Philadelphia. There his jail fare was bread and water for days on end, but he felt himself fortunate by contrast with the British seamen who in winter weather were confined in rooms from which the window sash had been taken by order of Congress, he said, in retaliation for the carrying of American prisoners to England.[21] He would have suffered greatly had not the good people of Philadelphia daily sent in "fresh meat and vegetables, fruit, milk, eggs and goodies"; and for him in particular a friend named Collins daily sent a plate from his own table. The harsh jailer was replaced by a civil one, and Marston was beginning to feel comfortable again when he was released in March, 1781. Proceeding to New York he bought and refitted the brig *Britannia,* but in obtaining a cargo he ran afoul of General Benedict Arnold who, having lost some tobacco to the rebels in Virginia, was compensating himself by seizing all of that commodity which the owners could not prove was *not* his. Having protested this inversion of the law and

[19] Marston, Diary, Apr. 3, 1779.
[20] New Brunswick Hist. Soc., *Collections*, VII, 89–90.
[21] Marston, Diary, Feb., 1781.

recovered his tobacco, Marston dismissed the experience with the remark that "military men are generally bad Legislators." [22]

The *Britannia* was ill fated, and its northward voyage was miserably uncomfortable and unsuccessful. As he huddled in the cold cabin Marston wrote in his diary not of Eliza in Halifax, but of Sally in Marblehead:

> Ah — That was once my happy lot
> When I with house and home was blest
> I'd then a fair Companion got
> Of many a female Charm possest.
>
> Yes — *dearest Sally* — you was fair
> Not only fair but kind and good
> Sweetly together did we share
> The blessings heaven on us bestow'd
>   * * *
> 'Till base Rebellion did display
> Her banners foul with false pretence
> Then Kindly Heaven took you away
> From evils which have happened since. [23]

Worse was yet to come. In December the *Britannia* was caught in the ice near Canso and had to be abandoned. Marston, the crew, and the pet dog set out overland for Halifax, but the journey proved too much for the Captain's strength:

Today so lame, I could go no further, so parted with my people, who left me very unwillingly. Gave them my share of Tiger, whom we killed last night, poor faithful animal. I lay in the woods two days and two nights, with no other sustenance than some dried moose, when I was relieved by the mate and two Indians. [24]

It was typical of him that he observed that the Catholic Indians were "kept in great decorum by their present priests, and in point of morals . . . in general not worse men than their better instructed British neighbours." Particularly in their favor was the fact that they were "always inimical to the Rebel privateers" which the profit-hungry English were not. [25]

For a year after the loss of his ship, Marston was unable to find work, and his funds shrank to two and a half dollars and one guinea

[22] *Ibid.*, June 16, 1781.
[23] *Ibid.*, Sept., 1781.
[24] New Brunswick Hist. Soc., *Collections*, VII, 106.
[25] *Ibid.*, VIII, 270.

which no one would accept because it had been clipped. His diary for the year 1782 is a mixture of misery and hope:

Still laying by, no business offers, nor do I Know where to go to look for any. *Go.* I cant Stir from this place, I have not the means of transporting me a single days journey. Heaven knows what is to become of me. . . . However I have one thing to thank Heaven for, my hopes do not yet fail me.[26]

Undeterred by pride, he industriously sought out trifling odd jobs so that his friends would not have to feed him. He had a mild contempt for the Loyalist officers, "riotous Vagabounds," he called them, who expected the Crown to support them for their loyalty. Unfailing in his own attachment to king and country, he was sharply critical of the soldiers and civilians around him who thought of the war and the peace which followed it only in terms of their personal profit.[27]

Edward Winslow (A.B. 1765) and his group of Loyalists appreciated Marston, and in April, 1783, without his knowledge, they induced Governor Parr to appoint him "chief Magistrate, or a kind of Governor-General," of the new Loyalist colony at Port Roseway, now Shelburne, Nova Scotia.[28] This was a fine dream of a new New York, and it did for a short time become the fourth city in size in North America, exceeding the combined populations of Montreal, Quebec, and Three Rivers. Although there was a large proportion of college graduates among the Loyalist refugees from New England, there were very few in this last wave from the middle colonies, and not one educated man among the Roseway settlers. Hence, it was argued, the importance of the appointment of Marston, although in cold fact his commission was only that of the surveyor for the colony.

The actual management of the Roseway settlement rested in the hands of "16 illiterate men" whom Sir Guy Carleton had, Marston thought, appointed because there was no one person in the entire group capable of command. The settlers he described as "a collection of characters very unfit for the business they have undertaken. Barbers, Taylors, Shoemakers and all kinds of mechanics, bred and used to live in great towns, they are inured to habits very unfit for undertakings which require hardiness, resolution, industry, and

[26] Marston, Diary, Oct. 7, 1782.
[27] *Ibid.*, Feb. 24, 1783.
[28] W. O. Raymond, *Winslow Papers* (St. John, 1901), p. 98.

patience." [29] They were indolent, clamorous, and inclined to be mutinous, for they suspected, with good reason, Marston thought, that the captains were chiefly interested in feathering their own nests. Working as chief surveyor and director of the settlement but without authority, he had a miserable time. His diary is one long complaint like this:

The multitude object to the place which the Captains and Chief men have chosen . . . so they propose to mend the matter by choosing three men from every company to do the matter over again. That is to commit to a mere mob of sixty what a few judicious men found very difficult to transact with a lesser mob of twenty, so this day has been spent in much controversial nonsense. This cursed republican, town-meeting spirit has been the ruin of us already, and unless checked by some stricter form of government will overset the prospect which now presents itself of retrieving our affairs. Mankind are often slaves, and oftentimes they have too much liberty. [30]

Under this system he was unable to prevent the laying out of the settlement in a manner which would permit the members of the first wave of refugees to exploit the necessities of the later arrivals, although he did assume the authority to strike the names of obvious speculators from the list of those who were to draw for lots. After recording the greed of the captains he said, "Sir Guy's commissions have made many men here *gentlemen,* and of course their wives and daughters *ladies,* whom neither nature nor education intended for that rank." [31] He was the more irritated at what was going on around him when he received a letter addressed to him as "a gentleman in power" from the widow of Pelham Winslow (A.B. 1753) describing the beggary to which she and her children were reduced in Massachusetts. [32]

Although Marston built himself a house in Shelburne and attended balls and card parties, he made no friends among the settlers. He was a Mason, a member of St. John's Lodge in Boston, but he did not enjoy the company of the many members of the fraternity among the refugees:

Tuesday last was St. John's day a free mason festival which was cele-brated here by such members of that worshipfull Fraternity as are here

[29] New Brunswick Hist. Soc., *Collections,* VIII, 214.
[30] *Ibid.,* p. 211.
[31] *Ibid.,* pp. 219–220.
[32] Mass. Hist. Soc., *Proceedings,* 2nd Ser., II, 240.

with their usual enebriety. The D — l is among these people. Last night there were two boxing matches, in one of which a Captain was Concerned. . . . They are a miserable lot. They have no men of education among them, none to whom to look up for advice and direction. One of their agents . . . used to plume himself that they were going to effect a settlement without the assistance of the clergy, intending to have none of that order among them for the present. It would be better for them if they had one or two sensible discreet ministers, and that they would believe in them.[33]

Still, he was content with his job because it kept him from being a burden on his friends; and through the bitter winter weather he labored with his surveying until his head swam with "Triangles, Squares, Parallelograms, Trapezias, and Rhomboidses."

In the spring of 1784 Marston received the good news that he had been appointed Register of Probate, but before that could develop he was driven from Shelburne by disorders for which he was in no way responsible. On June 27 the disbanded soldiers rose against the free Negroes because of their economic competition, drove them out of town, and pulled down their houses. A mob of "villanous scoundrels" then decided to take advantage of the situation and lynch the surveyor, who took refuge in the barracks and then took ship for Halifax. Governor Parr went down to Shelburne to discover the cause of the trouble, and was quickly convinced that the surveyor had been taking bribes.[34] It is quite obvious that this was untrue, but the relations between Parr and Marston had never been good, in spite of Edward Winslow's admonitions to the latter on the necessity of keeping governors in good humor. The surveyor was, he indignantly reported, turned out of office without a hearing:

Presented a memorial to Governor Parr this day and date [September 7, 1784], requesting a publick inquiry to be made into my conduct while Chief Surveyor at Shelburne. He says only in general that every body accuses me of the most corrupt partial conduct while in my office of Chief Surveyor. He has ordered me to wait upon him tomorrow at 12 o'clock. He will then tell me if I shall be heard or not. I find he has sent my character home under all these infamous accusations. . . . Saturday, 11. Having waited on the Governor at the time appointed to recieve his answer to my memorial of the 7th, missed seeing him, he being gone out — waited upon him this morning: met the Secretary of

---

[33] Marston, Diary, June 26, 1784.
[34] Public Archives of Canada, Nova Scotia "A", CV, 240–247.

the Province at the door; he told me his Excellency had referred my
memorial to the Board for locating of lands at Shelburne, I asked him
if I must look upon that as the Governor's answer? He told me yes.
I told him I looked upon that as a denial of the petition: for referring the
matter to the people, who perhaps were some of them raisers of the
slanders against me, is altogether an *ex parte* business, which I shall not
submit to. I have prayed to have my accusers face to face.[35]

Marston never did get his hearing. The Winslows at Halifax
took him in and set about to find him another job. He was already
looking longingly toward New Brunswick, which was to be settled
by New Englanders, and educated ones at that. Sir John Wentworth (A.B. 1755), who was a stiff-necked Yankee himself, listened
willingly to the Winslows, and in December, 1784, appointed
Marston Deputy Surveyor of the King's Woods in charge of the
New Brunswick area. Before his commission was actually signed,
he took off for his new post, pausing to visit General Ruggles (A.B.
1732) on the way. Edward Winslow met him at St. John, introduced him to Governor Carleton, and found him a place to live
with Ward Chipman (A.B. 1770).

Here at last was company which Marston could enjoy:

Last Wednesday we exhibited at the Hall, under the Auspices of General
Chippy, a monstrous great Ball and fine supper to about 36 Gentlemen
and Ladies such as Govenours, Secretaries, Chief Justices, Chancellors
and such kind of people with their wives and daughters. We ate, drank,
danced, and played cards till about 4 o'clock in the morning. We had
everything for supper.[36]

His enthusiasm was not chilled by the backward spring of 1785
which he jovially laid to the "bad Neighbourhood" of the new
province, meaning New England. He urged that the Reverend
Mather Byles (A.B. 1751) move to New Brunswick where he
would be welcome.[37]

The one fly in the ointment at St. John was the lack of profitable
employment, so in June, 1785, Marston accepted appointment as
Deputy Surrogate, Deputy Surveyor of the Woods, and first Sheriff
of Northumberland County, and immediately set out for the still
wilder western country. Here, at Miramichi, he found disappointment:

[35] New Brunswick Hist. Soc., *Collections*, VIII, 266.
[36] *Winslow Papers*, p. 269.
[37] Robie-Sewall Mss. (Mass. Hist. Soc.), May 15, 1785.

The condition of this River respecting the numbers of Inhabitants has been greatly mis-represented — to me at least. There are not above 100 families, if so many, upon it at present. . . . My appointments here will be mere sound and not much more. The emoluments of them will never make it worth my while to remain here after I have done those particular kinds of service which I came hither to execute. These I shall finish at all events. This makes it more necessary for me to get into some other line of business for a lively-hood and not depend more upon Government for employment.[38]

For the time being there was work at Miramichi. Marston surveyed until he and his assistants collapsed in the August heat. Then came the dangers of the winter woods, not the least of which was the threat of a settler to murder the surveyor if he did not get the lot he wanted.[39]

Determined to get out of government employment as soon as possible, Marston built a sawmill, opened a store, and tried, hopefully but in vain, to obtain a contract to supply masts for the navy. In March, 1786, he resigned as Sheriff and went to Halifax where he bought irons for his mill and all kinds of odd lot remainders for his store. One of these was a lot of wine so old that the owner could not remember where he obtained it and so bad that "a Halifax pig would not have drank" it; but it served his purpose, he said, when on Sunday, July 24, 1786, he published the charter of Northumberland County, and the population of Miramichi got drunk at his expense.[40] Although the people of the settlement were a drunken and illiterate lot of salmon fishermen, he decided to spend the rest of his days among them.

First, however, Marston would go to Boston to obtain proof of the losses which he and Uncle Winslow (A.B. 1736) had suffered by the Revolution. He arrived there on March 27, 1787, and was collecting old debts when some of his creditors had him taken up and lodged in the too-familiar jail. Three days later Marston Watson arrived in town and obtained Uncle Ben's release by assuming the debts himself. Now came a blow, however, for the Boston agents of Lane & Co. of London informed Ben that they could not make the firm of Marston and Watson of Miramichi their outlet agents for New Brunswick There was now nothing left for Mars-

[38] *Winslow Papers*, p. 309.
[39] Marston, Diary, Feb. 18, 1786.
[40] New Brunswick Hist. Soc., *Collections*, IV, 97.

ton to do but to go to England to raise cash from his claims for war losses. So he visited Plymouth and then went on to New York where he reported that the city was in great economic distress because it had not yet adjusted from the contacts of the British occupation to those of the United States.

From New York, Marston sailed for London where the Commissioners on Loyalist Claims advanced him £45 while they studied his modest statement of losses.[41] Many of the Tories who had made no effort to support themselves in exile had had their losses reimbursed practically in full, but honest Marston was compensated for only ten per cent of his losses, and that only after such delays that his debts for living expenses since arriving in London consumed all the cash which he received. Meantime three promising business connections fell through. Most important of these was his invention of an artificial horizon for Hadley's quadrant which would permit the determination of the sun's altitude when the real horizon was invisible. His instrument was patented, manufactured, and tried out at sea, but he obtained no substantial benefit from it.[42] He was eager to return home, but he was unable to find the means to eat regularly, to say nothing of paying his passage to America. Of his situation, a friend wrote:

Sometimes he was whole days without bread, and weeks together his daily expenditure amounted only to three half-pence, — a penny-worth of bread, and a half-penny worth of figs. Too noble to beg, yet willing to work; but, unknown and friendless in England, no one would employ him. Thus did this good man struggle in poverty for ten years, in that country which he had fought for. . . . I never heard this good man rail at, or say hard things of that country by which he had been so ill treated; he bore all patiently.[43]

This account of his poverty was somewhat exaggerated, for from time to time he obtained temporary employment which enabled him to clear up his debts, but he never could get ahead. Finally in March, 1792, he obtained an appointment which he described with enthusiasm in his last letter to his sister Elizabeth:

I am engaged to go out with a large Company who are going to make a Settlement on the Iland Bulam on the coast of Africa, as their Land Surveyor General, on a pretty good lay. I have £60 stirling per annum

[41] *Ontario Archives*, II, 605–608.
[42] *Winslow Papers*, pp. 368, 375–376.
[43] Philip Beaver, *African Memoranda* (London, 1805), pp. 115–116.

and Subsistence . . . and 500 acres of Land. . . . My expectations are chiefly from events which This Settlement will give rise to, The great Object of which is to found a great commercial System with the Native Africans on reciprocal advantages, To cut up by the roots that most wicked traffic, The Slave trade, which all flesh in this country are strongly setting their faces against — West India planters, and Guinea Merchants excepted — and which will most certainly be eventually abolished. . . . No expedition could have hit my taste and humour more exactly than such an one as this promises to do. It is so much of the Robinson Crusoe kind, that I prefer it vastly to any employment of equal emolument and of a more regular kind that might have been offered to me in this country. In fact I am truly glad that I can leave England, of which I am heartily tired. It is in most respects inferior to every country I have ever seen, — excepting what the Art, Skill and Industry of Its Inhabitants have done for it, — which has not yet, — nor can — procure for it Bright Suns and Serene Skies. . . . As to gratifying your wish in making my native country the residence of the remainder of my days, it is not at present in my power to do, for want of means. But was that otherwise. . . . That Rambling humour which was born with me, — and which has never yet been fully gratifyed — being now unrestrained by any local connexions, will be yet prompting me to engage in adventures which will carry me to new scenes, especially while I have vigor of body and mind capable of fatigue and application. . . . In this I follow my natural bent, for there is not remaining the least resentment in my mind to the Country, because the party whose side I took in the late great Revolution, did not succed, for I am now fully convinced It is better for the world that they have not. For it is the foundation, — the first step, to what has since followed in France, — and of many others yet in Embryo in the other European Kingdoms . . . and it will proceed till all Usurpations, all Lording of one over many, both in Spirituals and Temporals, will be entirely wrot off and despumated; and Man be left master of himself. . . .

I don't mean by this to pay any compliment to the first instigators of our American Revolution. Although it has eventually been of such advantage to Mankind, I should as soon think of erecting monuments to Judas Iscariot, Pontius Pilate, and the Jewish Sanhedrin for betraying and crucifying the Lord of Life, because that event was so importantly and universally beneficial.[44]

To Ward Chipman he wrote that "God in His merciful providence" had at last opened for him "a door to escape out of England."

On April 14, 1792, Marston sailed for the island of Bulama off the coast of Sierra Leone on the *Calypso*, which carried 275

44 *New England Hist. Gen. Reg.*, XXVII, 399–400.

colonists and joyful hopes. They arrived at their destination in June and found, not noble savages and an Eden, but savage men and a worse climate. Marston died of disease on August 10. Captain Beaver entered into his journal an appreciative biographical note of his friend, ending:

Should this Journal, by any accident, ever reach Marblehead, it may be a consolation to some of his friends and family to know what became of him, and at the same time to know, that, if he did not die a rich, he died a good man.[45]

When the news reached Massachusetts over a year later, a Boston paper in reporting it called him "A man of most amiable and mild disposition, whose soul possessed the magnet of every manly virtue, which never failed to attract him a generous esteem from all who fell within the circle of his acquaintance." [46] Up in Nova Scotia, Ward Chipman sadly opened the chest which his old friend had left in his care, and from it he sent to Edward Winslow the Marston diary and letter book for the years 1776–1787 which is now at the University of New Brunswick; sections of the diary are printed in the New Brunswick Historical Society, *Collections,* IV, 95–108, VII, 79–112, VIII, 204–277. A mass of family manuscripts in the possession of John Lee Watson (A.B. 1815) in 1872 cannot now be found. The miniature was in the possession of Benjamin Marston Watson (A.B. 1839), but that, too, is lost. Robert Southey found Marston's story in the Beaver diary and published it in the *London Quarterly Review* of July, 1829.

[45] Beaver, p. 116.
[46] *Independent Chronicle,* Oct. 14, 1793.

NATHAN TISDALE

Nathan Tisdale, the famous schoolmaster of Lebanon, Connecticut, was born in that town on September 19, 1732, the eldest son of Ebenezer, a blacksmith, and Hope (Basset) Tisdale. At college, where he held a Stoughton scholarship, he preferred fishing in Fresh Pond to riots, champagne parties, and student clubs. The fact that he never took the trouble to qualify for his M. A. illustrates the same indifference to custom.

About 1743 Jonathan Trumbull (A.B. 1727), the leading citizen of Lebanon, had begun looking about for a means of educating his eldest son, Joseph (A.B. 1756). With the encouragement of the minister, Solomon Williams (A.B. 1719), a dozen inhabitants signed an agreement "for the education of our own children, and such others as we shall agree with. . . . each one to pay according to the number of children that he sends and the learning, they are improved about, whether the Learned tongues, Reading and writing, or Reading and English only." [1] Soon after his graduation, Tisdale was placed in charge of this school, which his genius as a teacher soon built up into a famous academy, drawing students, it was said, from every North American colony and from the West Indies. For nearly forty years the brick schoolhouse on Lebanon Green regularly held from seventy to eighty scholars who under the guidance of Master Tisdale alone were learning their A B C's, studying surveying and navigation, or preparing for admission to Harvard and Yale, which often gave them advanced standing. There is a strong tradition that Yale accepted some of his scholars without examination. If the experience of Jeremiah Mason (Yale 1788) was typical, the speed of college preparation at Lebanon was amazing:

I was very backward for my age in all school learning. I read but poorly and spelt worse; my handwriting was bad, and in arithmetic I knew little. . . . In the course of a few months I commenced the study of the Latin, and soon after that of the Greek language. In less than two years I was declared by Master Tisdale to be fitted for college. [2]

---

[1] Robert C. Armstrong, *Historic Lebanon* (Lebanon, 1950), p. 62.
[2] Jeremiah Mason, *Memoir and Correspondence* (Cambridge, 1873), p. 7.

No doubt one reason for this speed was the "excellent rule of the school to have no vacations, in the long idleness and dissipation of which the labors of preceding months might be half forgotten." [3]

Of the hundreds of boys who passed through the hands of Master Tisdale, one of the last, Dan Huntington (Yale 1794), left the best account of his teaching methods:

> I found myself pleasantly ensconced with the good old Master Tisdale, in the "brick school-house," where I fitted for college upon very good terms. After prayers, hearing the Bible class, and seating the older scholars, for writing and ciphering, the younger ones came under the more immediate notice of the master. After hearing them read and spell their lesson, he would occasionally indulge himself in a little chat with the children, in their A, B, C. When through reading, he took me, and in "great A, little a, -ron, Aaron," in my turn, betwixt his knees, saying, "Dan, what do you intend to be, a minister, or a plough-jogger?" Without hesitating at all, I replied, "A minister, sir." He burst out into a broad laugh. "Well," said he, rubbing my head with his hand, and patting my shoulder, "sit down, Dan; study your lesson; be a good boy, and we will see about your being a minister." [4]

In the autumn of 1773 Master Tisdale was crippled by paralysis, and had to place the school in the hands of John Trumbull (A.B. 1773) until spring. Obviously he needed personal care, and about three years later he was married to Zerviah (Wadsworth) Porter, widow of Captain Nathaniel Porter (Yale 1749), and mother of four children, one of whom, John (Yale 1776), had been fitted for college in the brick schoolhouse. Like the rest of his contemporaries whose income was based on contracts, Master Tisdale suffered severely from the inflation resulting from the Revolution. In billing the fathers of some of his pupils for tuition he suggested that if payment be made in Continental bills, the "Candor, Generosity and Justice" of the debtor would make allowance for the depreciation of the paper.[5] This faith in human nature was misplaced, however, and in 1786 he had to petition the proprietors of the school for some £27 which were due him:

> In this business, gentlemen, I have continued nearly the space of forty years, with almost uninterrupted application to the duties of my charge. . . . I have educated a large number of youth who have done an honor

---

[3] John Trumbull, *Autobiography* (New Haven, 1841), p. 4.

[4] [Dan Huntington], *Memoirs, Counsels, and Reflections* (Cambridge, 1857), p. 47.

[5] Nathan Tisdale to Jonathan Deming, Yale University Library, Feb. 27, 1779.

to this school, who have gone forth into the world and have become bright ornaments to society. I have now spent the prime of life, the flower of my days, in this service; but I have acquired no fortune — and perhaps I may say that I have been more profitable to the community than to myself.[6]

The proprietors voted to put this matter over to another meeting, but on February 5, 1787, two days after another stroke, Master Tisdale died. William Williams (A.B. 1751) reported sadly:

Died, about 7 o'clock in the morning, my valuable friend and intimate acquaintance . . . a gentleman of learning and virtue; for more than 30 years, master of the grammar school . . . a wise, faithful, learned and useful instructor, benevolent and kind . . . of a 2d or 3d paralytick shock.[7]

He lies in the Torrey Hill cemetery under a stone which eloquently describes his virtues.[8]

[6] Daniel Coit Gilman, *An Historical Discourse Delivered in Norwich* (Boston, 1859), p. 123.
[7] John Langdon Sibley's Letters Received (Harvard University Archives), VI, 6. For an obituary see the *Connecticut Journal*, Mar. 7, 1787, 2/2.
[8] For the text of the inscription see the *New England Hist. Gen. Reg.*, XII, 62.

GEORGE WASHINGTON

PRESIDENT GEORGE WASHINGTON's first contact with Harvard College probably occurred in February, 1756, when he visited Boston to consult with Governor Shirley on the aftermath of Braddock's Defeat. He lodged at Cromwell's Tavern in School Street, and could hardly have avoided being taken out to Cambridge "to see the colleges," but what he recorded was the amount of money which he lost in playing cards with the sons of the Puritans.[1] Perhaps there was a lingering memory of this visit in the mind of John Adams (A.B. 1755) when at Philadelphia in 1775 he urged that the Virginian be appointed commander-in-chief of the army before Boston. But for the New Englanders' influence, the choice might well have been some more dashing and glamorous man; Washington's solidity and sobriety were virtues which they regarded highly.

When the Provincial Congress was informed of this appointment, it ordered that "the president's house in Cambridge, excepting one room reserved by the president for his own use" "be cleared, and furnished for the reception of the Commander-in-Chief and General Lee." This was no particular hardship for President Langdon (A.B. 1740), who was keeping bachelor quarters in Wadsworth House. It is Cambridge tradition that when Washington rode into Cambridge early in the afternoon of July 2, 1775, he caught sight of a colored boy swinging on a gate, watching the bright uniforms. Perhaps the black face touched a chord of homesickness; the General called out, "Boy, would you like to work for me?" The little Yankee turned the matter over in his mind for a moment, and then asked, "For what wages?" The General snorted incredulously: "Pay wages to a black boy?" So one of the oldest families of Cambridge lost its opportunity to be in the spotlight of history.

The next morning Washington looked around and remarked that Cambridge was "in the midst of a very delightful country, and . . . a very beautiful place itself." The President's House was then a lovely place, with trees, flowering shrubs, and a garden in front, and a spacious yard behind with stable and coach house.

---

[1] John C. Fitzpatrick, *George Washington, Colonial Traveller* (Indianapolis, 1927), p. 94.

It would have been too much to expect, however, that President Langdon would get along with Charles Lee's dogs. On July 6 the Provincial Congress instructed the Committee of Safety to ask General Washington if there was any Cambridge house which he would prefer. He suggested the John Vassall (A.B. 1757) place, now known as the Longfellow-Craigie House, which then had a fine view of the river rather like those of so many Virginia plantations, and out of his own pocket he paid for having it cleaned of the debris of a Marblehead regiment which had occupied it.

The contemporary references to Washington in Cambridge are for the most part trivial, few even as interesting as this account by William Palfrey of the time when he was drafted to read prayers in Christ Church:

What think you of my turning parson? I yesterday, at the request of Mrs. Washington, performed divine service at the church at Cambridge. There was present the General and Lady . . . and they were pleased to compliment me on my performance. I made a form of prayer, instead of the prayer for the King, which was much approved.[2]

In the next century John Langdon Sibley (A.B. 1825) and others made a point of interviewing every inhabitant of Cambridge who could remember the Revolution, but came up with nothing more important concerning Washington than the fact that he enjoyed the society of the Tory ladies whose husbands or fathers were in exile, and went out of his way to assure them that he would see to it that they were not molested.

Of the New England soldiers and political figures who came in touch with Washington in 1775 and 1776, only one, Artemas Ward (A.B. 1748), had serious reservations. Curiously enough, these two saw eye to eye on political matters, but it was James Warren (A.B. 1745, q. v.), who was at the other end of the political spectrum, who said of the Virginian: "He is certainly the best man for the place he is in, important as it is, that ever lived."[3] When on April 3, 1776, the Corporation and Overseers voted him the degree of Doctor of Laws, it was only the second time that Harvard had conferred this honor. President Langdon hastened home and wrote a diploma in picturesque Latin which the newspapers translated as follows:

[2] Colonial Soc. Mass., *Publications*, XXV, 374–375.
[3] Mass. Hist. Soc., *Collections*, LXXII, 186.

The Corporation of Harvard College in Cambridge, in New-England, to all the faithful in Christ, to whom these Presents shall come, Greeting.

Whereas Academical Degrees were originally instituted for this Purpose, That Men, eminent for Knowledge, Wisdom and Virtue, who have highly merited of the Republick of Letters and the Common-Wealth, should be rewarded with the Honor of these Laurels; there is the greatest Propriety in conferring such Honor on that very illustrious Gentleman, GEORGE WASHINGTON, Esq; the accomplished General of the confederated Colonies in America; whose Knowledge and patriotic Ardor are manifest to all: Who, for his distinguished Virtue, both Civil and Military, in the first Place being elected by the Suffrages of the Virginians, one of their Delegates, exerted himself with Fidelity and singular Wisdom in the celebrated Congress of America, for the Defence of Liberty, when in the utmost Danger of being for ever lost, and for the Salvation of his Country; and then, at the earnest Request of that Grand Council of Patriots, without Hesitation, left all the Pleasures of his delightful Seat in Virginia, and the Affairs of his own Estate, that through all the Fatigues and Dangers of a Camp, without accepting any Reward, he might deliver New-England from the unjust and cruel Arms of Britain, and defend the other Colonies; and Who, by the most signal Smiles of Divine Providence on his Military Operations, drove the Fleet and Troops of the Enemy with disgraceful Precipitation from the Town of Boston, which for eleven Months had been shut up, fortified, and defended by a Garrison of above seven Thousand Regulars; so that the Inhabitants, who suffered a great Variety of Hardships and Cruelties while under the Power of their Oppressors, now rejoice in their Deliverance, the neighbouring Towns are freed from the Tumults of Arms, and our University has the agreeable Prospect of being restored to its antient Seat.

Know ye therefore, that We, the President and Fellows of Harvard-College in Cambridge, (with the Consent of the Honored and Reverend Overseers of our Academy) have constituted and created the aforesaid Gentleman, George Washington, who merits the highest Honor, Doctor of Laws, the Law of Nature and Nations, and the Civil Law; and have given and granted him at the same Time all Rights, Privileges, and Honors to the said Degree pertaining.[4]

One of these privileges was having his name enrolled at the foot of the Class of 1749. Samuel Cooper (A.B. 1743) took the diploma to headquarters the next morning, but found that Washington had already left for New York.

Through the war years and Washington's presidency, he had the hearty support of the vast majority of the Harvard community,

---

[4] *New England Chronicle*, Apr. 25, 1776, 1/1–2.

the most significant exception being John Hancock (A.B. 1754), who for other reasons was very unpopular with the college. It was Hancock who was chiefly responsible for the story that Sam Adams (A.B. 1740) had been plotting against the Commander-in-Chief. Governor James Bowdoin (A.B. 1745) and his wife became the closest of the New England friends of the Washingtons.

In 1781 Washington was elected to the American Academy of Arts and Sciences, and when he visited Boston in 1789, Washington Street was renamed in his honor. On this occasion the Harvard Corporation presented this address:

> Permit us, Sir, to congratulate you on the happy establishment of the government of the union, on the patriotism and wisdom which have marked its public transactions, and the very general approbation which the people have given of its measures. At the same time, Sir, being fully sensible that you are strongly impressed with the necessity of religion, virtue, and solid learning for supporting freedom and good government, and fixing the happiness of the people upon a firm and permanent basis, we beg leave to recommend to your favorable notice the University entrusted to our care, which was early founded for promoting these important ends.
>
> When you took command of the troops of your country you saw the University in a state of depression — it's members dispersed — it's literary treasures removed — and the Muses fled from the din of arms, then heard within it's walls. Happily restored, in the course of a few months, by your glorious successes to it's former privileges and to a state of tranquility, it received it's returning members, and our youth have since pursued, without interruption, their literary courses, and fitted themselves for usefulness in Church and State. The public rooms, which you formerly saw empty, are now replenished with the necessary means of improving the human mind. . . . While we exert ourselves, in our Corporate capacity, to promote the great objects of this institution, we rest assured of your protection and patronage.[5]

Four days later President Washington replied formally to the Corporation address:

> It gives me sincere satisfaction to learn the flourishing state of your literary Republic. Assured of its efficiency in the past events of our political system, and of its further influence on those means which make the best support of good government, I rejoice that the direction of its measures is lodged with men, whose approved knowledge, integrity, and patriotism give an unquestionable assurance of their success. That the Muses may long enjoy a tranquil residence within the walls of your

[5] Corporation Records (Harvard University Archives), III, 348–349.

University, and that you, Gentlemen, may be happy in contemplating the progress of improvement through the various branches of your important departments, are among the most pleasing of my wishes and expectations.[6]

On October 29, 1789, Washington set out from Boston on his tour to the eastward, and was accompanied to Cambridge by Vice President John Adams, James Bowdoin, and a great number of other gentlemen. President Willard welcomed the party and took it on a tour on which Washington was most impressed by "the Philosophical aparatus, and amongst others Pope's Orary (a curious piece of Mechanism for shewing the revolutions of the Sun, Earth, and many other of the Planets), the library, (containing 13,000 volumes,) and a Museum."[7] A few days later Willard wrote to the President asking permission to have his portrait painted for the college, and in time received a courteous reply:

Sir: Your letter . . . was handed me a few days since by Mr. Savage, who is now engaged in taking the Portrait which you, and the Governors of the Seminary over which you preside, have expressed a desire for, that it may be placed in the Philosophy Chamber of your University. I am induced, Sir, to comply with this request from a wish that I have to gratify, so far as with propriety may be done, every reasonable desire of the Patrons and promoters of Science. And at the same time, I feel myself flattered by the polite manner in which I am requested to give this proof of my sincere regard and good wishes for the prosperity of the University of Cambridge.[8]

Eighteen months later Edward Savage, the artist, made a present of the portrait to the college.[9]

During the years which remained to him, Washington had no direct contact with Harvard, but he used to recommend it to parents with sons of college age:

Inform me how far to the Eastward you would consent that your Sons should go; to come at the *best* schools. There are two or three private Academies in the State of Massachusetts that are spoken very favorably of; the College in that State is also in good repute; but neither in that, nor at Yale College in Connecticut, do they admit boys until they are qualified by a previous course of education. This, however, is not the

[6] Corporation Papers (*ibid.*), UA I.5.120.1789.

[7] John C. Fitzpatrick, *The Diaries of George Washington* (Boston and New York, 1925), IV, 39.

[8] George Washington, *Writings* (Fitzpatrick ed.), XXX, 483.

[9] College Book VIII (Harvard University Archives), 323.

case with the Seminary here [at Philadelphia], nor I believe that at Princeton.[10]

What is best to be done with him, I know not. My opinion always has been that the University in Massachusetts would have been the most eligable Seminary to have sent him to, 1st, because it is on a larger Scale than any other; and 2nd, because I believe that the habits of the youth there, whether from the discipline of the School, or from the greater attention of the People, generally, to morals and a more regular course of life, are less prone to dissipation and debauchery than they are at the Colleges South of it.[11]

It seemed to most Harvard men of the next generation that the course of events proved the political wisdom of Washington, and in 1832 the college formally celebrated the centennial of his birth.

THOMAS DUDLEY

THOMAS DUDLEY of Roxbury was born in that town on September 9, 1731, the eldest son of the Honorable William (A.B. 1704) and Elizabeth (Davenport) Dudley. The family was one of the most distinguished in New England, and the Class of 1750 was known as "Dudley's Class." [1] Tom, who was a boy of very ordinary abilities, bitterly resented having such distinctions thrust upon him, and developed a painful inferiority complex. Even his disorders were undistinguished, for the only fine against his name on the college records is a small one for "forcing open Jonses's door." As the head of the class he had to deliver the oration at the Commencement, and he did so poorly as to attract considerable attention.[2] His performance at the Commencement of 1753 at least attracted no attention, but his *Quaestio* (the affirmative of "An Agricultura magis profit Reipublicae quam Venditio Mercium"?) expressed his firm intention of being a plain farmer.

Tom had inherited the magnificent estate of his uncle, Judge Paul Dudley (A.B. 1690), which covered the area between the modern Washington and Roxbury streets, and his brother Joseph

[10] Washington, *Writings*, XXXIV, 44.
[11] *Ibid.*, XXXVI, 136.
[1] R. T. Paine, Diary (Mass. Hist. Soc.), Mar. 12, 1746/7.
[2] Ebenezer Parkman, Diary (Am. Antiq. Soc.), July 4, 1750.

(A.B. 1751) urged him to take it over and live in elegance like his ancestors. To demonstrate the fact that he did not have the necessary dignity, Tom had his oxen yoked to the fine Dudley coach and drove into Roxbury village, loudly "geed round," and drove home again. It was plain that the townspeople would never take Tom seriously after that, so his mortified brother traded him their father's "farm in the woods" of West Roxbury for the Paul Dudley estate and proceeded to get the entails broken in order to relieve him of the burden of his dignity. This defection from the New England aristocracy of public service was a blow which leaders like Thomas Hutchinson (A.B. 1727) resented bitterly. Tom was the worse, but both Dudley boys "seemed to prefer the manners of ordinary life, and very soon were mingled with the people who make up the common mass of human society." [3]

On April 26, 1753, Thomas Dudley married Hannah Whiting. Their life was hardly that of farmers in the woods of West Roxbury, for they owned a great deal of property and built a house in Dedham. Their residence remained in Roxbury, however, for that town sent Dudley to the House of Representatives for a single term in 1766. He also accepted appointment as a Justice of the Peace. He called himself a farmer, but that occupation would never have paid for the complicated legal suits arising from his inheritance which were finally carried to the King in Council.[4] To make matters worse, he died intestate on November 9, 1769. His widow married Colonel Joseph Williams of Roxbury on April 5, 1770. The Dudleys had seven children, but none of them had any Harvard connection.[5]

[3] John Eliot, *Biographical Dictionary* (Salem, 1809), pp. 162–163.
[4] Edmund Trowbridge to William Bollan, Dana Mss. (Mass. Hist. Soc.), Nov. 18, 1763.
[5] Dean Dudley, *History of the Dudley Family* (Wakefield, 1886), I, 286.

HENRY GARDNER

HENRY GARDNER, the first Treasurer of the State of Massachusetts, was a son of the Reverend John (A.B. 1715) and Mary (Baxter) Gardner of Stow, where he was born on November 14, 1731. As an undergraduate he waited on table, served as a Scholar of the House, and held Fitch and Hollis scholarships. He volunteered to serve on a committee to report students who used profanity, but he had his own failings, for he was fined and publicly admonished for playing cards. At the Commencement of 1750 he made a very favorable impression as a respondent,[1] and at the Commencement of 1753 he had a speaking part, the affirmative of "An Anima Natura sua sit immortalis?"

Between his two degrees Gardner kept the Worcester school, but he settled in Stow, which sent him to the General Court every year that it was represented from 1757 to the Revolution. In that body he voted for the liquor tax which the patriots opposed, but otherwise regularly supported the Whigs. He opposed the measure to make the militia subject to the call of the royal commander-in-chief, voted not to rescind the circular letter resolution, supported the extreme demands of the House for authority over the Executive, but in October, 1770, voted to abandon the legislative strike against Governor Hutchinson (A.B. 1727). On the other hand, he supported the efforts to impeach Chief Justice Oliver (A.B. 1730).

Gardner took an active part in the Middlesex County Convention of August, 1774. He had been appointed a Justice of the Peace in 1761, and in September, 1774, he was one of the Justices signing the agreement not to open the Inferior Court of Common Pleas.[2] In October he attended the First Provincial Congress and was very active on committees having to do with military supplies. The Congress appointed him Receiver General of provincial taxes and ordered the towns and districts to remit their taxes to him and not to Harrison Gray, the Treasurer of the Province. Governor Gage issued a formal counter-proclamation ordering all officials to pay their taxes to Gray rather than to Gardner. He was active

[1] Annotated Thesis sheet in the Harvard University Archives.
[2] *Boston Gazette*, Sept. 19, 1774, 2/3.

in the Second Provincial Congress and on the Nineteenth of April nearly lost the chest containing his public cash and archives when the Regulars searched the Jones Tavern but gave over too easily when a lady claimed it as her personal property.[3] After the battle he joined Dr. Warren (A.B. 1759) in composing the account of the day which was rushed to England.[4]

In May, 1775, Gardner tried twice to be excused from serving longer as Receiver General, but his petitions were tabled and his town returned him to the Third Provincial Congress. In July this body, acting as the General Court, unanimously elected him Treasurer and Receiver General, and annually reappointed him. The difficulties of the wartime treasurer were serious indeed. When he took over Gray's books he discovered that many of the towns had showed their patriotism by not paying their share of the provincial taxes since 1771; only the depreciation of the currency permitted the clearing of the accounts. The bookkeeping problems created by the different kinds of money in circulation proved to be insoluble, and eventually the General Court had to vote him absolution.

The new government renewed Gardner's appointment as a Justice of the Peace and Quorum, and in this capacity he was responsible to the towns for certain fines collected. After considerable negotiation the town of Boston begrudgingly accepted Mr. Justice Gardner's accounts but voted a resolution that "a regular account ought to be kept by the Justices in the future." [5]

In 1776 Gardner represented the State at a Micmac conference down east. This was the last year in which the town of Stow returned him to the House of Representatives. About that time he moved into rented quarters in Boston, and in 1778 he served on the committee to visit the schools of that town. On September 21 he married Hannah, daughter of Colonel Ebenezer and Elizabeth (Hall) Clap of Dorchester. The bride, who was just half his age, had Dorchester property which they used as a summer home, but in April, 1779, the General Court gave them permission to move their home and the treasury office into the

[3] Mass. Hist. Soc., *Proceedings*, LVI, 87.
[4] *Ibid.*, LIX, 295.
[5] *Boston Town Records, 1778 to 1783* (Record Commissioners, XXVI), 195.

Province House, which was then occupied by the committee on accounts. Their sons were born there.

Gardner was elected to the Council in 1777, and he served on that body until the adoption of the new constitution under which as Treasurer he was disqualified from legislative office. As a member of the Council he had been a member of the Board of Overseers, and after his retirement he was asked to attend Commencement with other Justices "for the prevention of Disorders." He attended the meeting called to organize the American Academy of Arts and Sciences at the college, and was chosen a Councilor of the new society. He was universally respected, and when he died on October 7, 1782, newspapers as far away as Philadelphia noticed the "loss of a courageous, uniform, industrious patriot, and a discreet, humane, upright Judge." Those nearer home published long obituaries and verse eulogies such as this:

> Blest Gardner, thou art from our Region fled,
> And left thy mortal Part amongst the Dead:
> Angelic Guards around thee do convey,
> Thy kindred Spirit to eternal Day;
> Where thou with Angels join'd, in Chorus sing,
> Eternal Praise to the almighty King.[6]

Hannah Gardner on December 28, 1784, became the third wife of the Reverend Moses Everett (A.B. 1771) who had married her to her first husband. The Treasurer had two sons, Henry, who was graduated in 1798, and Joseph, in 1802.

---

[6] *Continental Journal*, Oct. 17, 1782, 1/3.

INDEX

These sketches are fully indexed in the volumes of *Sibley's Harvard Graduates.*
Single dates in parentheses indicate Harvard classes. Page numbers are omitted
in the case of birthplaces and residences; these are easily found in the sketches,
which are arranged chronologically by class and alphabetically within each class.

WALLOOMSACK

BENNINGTON WILMINGTON

MOHAWK R.

FORT MASSACHUSETTS

ALBANY

GREENBUSH FORT PELHAM FORT SHIRLEY

MOUNT WACHUSETT

HUDSON R.

DEERFIELD

HARV

NEW LEBANON PONTOOSUC BARRE
FORT ANSON RUTLAND

WEST STOCKBRIDGE NORTHAMPTON NORTHBOROUC
WESTBORC
WARE WORCESTER
STOCKBRIDGE BROOKFIELD MARLE
HOPH
PALMER GRAFTO

CONNECTICUT R.

SALISBURY BEL

SIMSBURY

HARTFORD

WINDHAM

LEBANON

NORWICH

NEW MILFORD

STONINGTON

NEW HAVEN KILLINGWORTH
GUILFORD
FAIRFIELD STRATFORD

LONG ISLAND